Logic Without Borders

Ontos Mathematical Logic

Edited by
Wolfram Pohlers, Thomas Scanlon,
Ernest Schimmerling, Ralf Schindler,
Helmut Schwichtenberg

Volume 5

Logic Without Borders

Edited by
Åsa Hirvonen, Juha Kontinen, Roman Kossak
and Andrés Villaveces

DE GRUYTER

ISBN 978-1-61451-772-6
e-ISBN (PDF) 978-1-61451-687-3
e-ISBN (EPUB) 978-1-61451-932-4
ISSN 2198-2341

Library of Congress Cataloging-in-Publication Data
A CIP catalog record for this book has been applied for at the Library of Congress.

Bibliographic information published by the Deutsche Nationalbibliothek
The Deutsche Nationalbibliothek lists this publication in the Deutsche Nationalbibliografie; detailed bibliographic data are available on the Internet at http://dnb.dnb.de.

Printing: CPI books GmbH, Leck
♾ Printed on acid-free paper
Printed in Germany

www.degruyter.com

To our friend and colleague Jouko

From the editors

The list of invited speakers at the meeting in honor of Jouko Väänänen's Sixtieth birthday at the University of Helsinki in September of 2010 can be seen as a perfect illustration of the present state of affairs in mathematical logic. Twenty five speakers gave talks on the history and philosophy of mathematics, set theory, model theory, and theoretical computer science, and topics within each of these disciplines varied greatly. Despite all that diversity, there was a unifying theme, a ground common to all the talks given. They all were on various aspects of *Mathematical Logic*. Another important uniting factor was personal: all speakers found natural ways to connect their talks to the diverse body of work of Väänänen. Many speakers were his former students, and many were his co-authors. Overall, it was an impressive display of logic without borders.

This volume is dedicated to Väänänen, and presents a collection of articles, both stemming from lectures given during the conference, and at the same time brightly displaying the diversity and depth of contemporary mathematical logic; in the intention of the editors, it also invites readers to a discussion on the state of the discipline. The contributions are all by authors who are important members of the "avant-garde" of contemporary mathematical logic, and the subjects of their writings constitute an impressive "state of the art" of a good part of the discipline as a whole.

Åsa Hirvonen
Juha Kontinen
Roman Kossak
Andrés Villaveces

Preface – Unity and Diversity of Logic

This short essay serves as an informal introduction to the volume and an invitation to the discussion on the state of the discipline. Mathematical logic has a distinctive place between mathematics its foundations and its philosophy. In the last four decades, several branches of logic have evolved into highly specialized technical disciplines. There are strong arguments for autonomy and separation of some areas, but there are also strong arguments for unity of all those diversified developments. We comment briefly on some historical aspects and some recent developments.

Logic today: Is there unity? What is its role within mathematics?

What is mathematical logic today? How does it connect with its historical roots? How does it continue to serve as foundations of mathematics, and how does it impact mathematics in general? *Does it continue to serve as the foundations of mathematics at all?* What distinguishes advanced areas of mathematical logic from other branches of mathematics? What parts of mathematical logic should be considered philosophy, and what parts evolved into independent subdisciplines of algebra, analysis or computer science? The article by Juliette Kennedy in this volume addresses some of these issues directly, as does Jouko Väänänen's personal account of the development of his interests in mathematical logic. Other articles in the volume might be construed as providing partial responses to these questions, of course not necessarily in a direct way, but through the connections and links they explore, both internally within logic and externally between logic and other disciplines.

There is no question that mathematical logic is diverse. The current pressing subject of discussion is its unity. Now, the problem of unity of mathematical logic cannot be separated from the similar problem for mathematics in general: as mathematics grows and its subdisciplines become highly specialized, so do many areas of set theory, model theory, proof theory, and computability theory. The "unifying glue" in the past was the foundational character of the problems motivating research. This, of course, to some extent is still true today, but the philosophical motivation, the driving force behind the early developments, has to large extent been demoted from the center of concerns of the mathematical community at large.

The issue of unity of mathematical logic has become ever more pressing in recent years: even the question of the *existence* of unity of mathematical logic has been raised by mathematical logicians themselves across the wide spectrum of subareas: from some prominent model theorists who intend to regard their own subfield as a sort of extension of algebraic geometry and no longer a part of "Logic"[1]. While some model theorists have pulled towards generalized forms of algebraic geometry, some set theorists have taken a similar step toward topological dynamics, Ramsey theory applied to Banach space theory or even the classification of operator algebras, C^*-algebras, von Neumann algebras, etc. Others, notably some recursion (or computability) theorists have made strides towards both computer science and recursive model theory[2]. Also, proof theorists are doing proof-mining and have been formalizing proofs in ways that add seriously to this variety and diversity of perspectives. Add to all this pull in different directions the sharp splinter of unity brought about by the emergence during the past three or four decades of a *category theory*-grounded logic, far away from set heory, and now successful by some measures[3].

The odd case of model theory

The specific case of model theory and its relationship with Logic in general provides an interesting case study[4]. In earlier days, and especially before 1970, model theory's main problems were very set-theoretic in nature: several important questions of the time (two-cardinal theorems, end extension methods, generalized omitting types) were dependent on the specific model of set theory where the model theorist is working. Extremely rich interplay between the two subdisciplines created such offspring as Ehrenfeucht-Mostowski models, Jensen's covering lemma, Morley's categoricity theorem and its enormous off-spring reaching now the domains of mathematical physics through quasi-minimal excellent classes, Chang's and Vaught's conjectures, etc. However, this sort of logical interplay was not necessary under stability assumptions: Shelah's work of the 1970s provided model theorists with a "paradise" with rich dividing lines, rich

1 An "independence from the rest of logic" - and in particular from set heory - granted in the first order case by spectacular results in stability theory, itself deeply connected to set theory in wider contexts, creating a paradox of set theory independence in logic.

2 The name "recursion theory" is now falling out of favor among many of its practitioners - the more fashionable "computability theory", a name that is both a description of new emphasis and a declaration of intent, has now been the official name of the subarea for more than two decades.

3 Homotopy type theoretic foundations for parts of mathematics is by far the largest subfield in 2014 in terms of grants in the United States!

4 See the papers by Baldwin and Kennedy in this volume for more on this subject.

structural properties, and where many of the questions that in general would have needed set-theoretical assumptions did not need them. Stability theory for first order logic opened a large playground for applications of model theory to branches of mathematics different from logic, and for wonderful and deep internal connections between definability issues, combinatorial characterizations and independence notions. This playground has increased in size, covering areas outside stable theories, with enormous success accumulating over more than four decades. One side result of this story of success, for stability theory is that large groups of model theorists do not feel the need for more interaction with set theory. In an interesting shift of events, the ultimate expansion of stability theory to abstract elementary classes and to connections with second order logic, via new interplay with large cardinals and forcing and its connections with analytic structures, in the work started by Zilber around the turn of the century, is again pushing model theory to areas where understanding of its connections with set theory is central.

The dynamics of grounding

When Andrzej Mostowski outlined mathematical logic in "Thirty Years of Foundational Studies," an impressively comprehensive summary of the history of the subject from 1930 to 1964, he saw it as a continuation of efforts of the three major schools in philosophy of mathematics: "the intuitionism of Brouwer, the logicism of Frege and Russell and the formalism of Hilbert." In the introduction, Mostowski acknowledged failures of each program, but stressed that the " [...] activity of these schools brought about a great number of important new insights and discoveries that have deepened our knowledge of mathematics and its relation to logic. As it often happens, these by-products have turned out to be more important than the original aims of the founders of the three schools." Mostowski's statement is still applicable today: we are proud every time a development in an advanced area of mathematical logic finds an important application in mainstream mathematics or in computer science. Still, much work in the discipline itself concerns directly foundational issues, that may be not, and quite often are not, well understood by outsiders. What makes us unique in the mathematical community, is our understanding of the role of formal languages and their interpretations. It is the Hilbert-Russell-Gödel-Gentzen-Tarski legacy, that gives us a clear separation of syntax and semantics, and shows how investigation of one informs the other. At a risk of oversimplification, one could say that while modern mathematical logic branched into a vast landscape of diverse subfields, what keeps us together, what constitutes our unity, is exactly this understanding of the syntax/semantic separation line, and understanding of the importance of foundational questions that this separation motivates.

Grounding and Set Theory.

Another significant, and perhaps overlooked by some nowadays, aspect for the unity of mathematical logic is the **grounding** of mathematics in set theoretic formalism. Set theory grew out of the need to provide a formal system for mathematical arguments involving actual infinity. It started with early efforts of Bolzano, Cantor and Dedekind, and then followed a remarkable path of success. Let us just mention in passing the Zermelo and Fraenkel axiom system, Gödel's proof of relative consistency of the axiom of choice and the continuum hypothesis, Cohen's invention of forcing with all further advances in the understanding of the independence phenomena in set theory, the role of large cardinal axioms, applications of descriptive set theory to classification problems in classical mathematics, Shelah's classification program in model theory; the list could go on an on. All those developments necessarily result in specialization. Still, there is little doubt that all those successes depend strongly on the solid unity and strength of mathematical logic.

In their normal practice, most (almost all) mathematicians *who are not logicians* tend to completely ignore most of these issues. It is a very strong measure of the *success* of the foundational program (in spite of all the limitations it may - and does - have) that algebraists, analysts, algebraic topologists, algebraic geometers, mathematical physicists, number theorists may do their work without having to pause and ponder about foundational issues underlying their disciplines, the nature of real numbers being a prime example of this phenomenon. This phenomenon of *robust and healthy ignorance*, due to - and witness of - the very success of the program, produces the odd effect of many mathematicians outside of logic wondering what on earth logicians do. Setting aside this negative consequence of the success, we have a weird witness to the uncanny power of set theory as grounding of a large part of mathematics.

Growth and Evolution.

Of course, mathematics has evolved at a very fast pace, both inside and outside logic, in the past few decades. And the (perhaps) "healthy and robust" ignorance of logic could well turn upon its own head. Shifting the focus from foundational issues to other kinds of interaction between logic and the rest of mathematics, there are interesting developments. In recent decades, whole areas that initially seemed remote from logic have started interaction, at deep levels, including some important conjectures in number theory, or Galois theory of differential equations, or algebraic dynamics, to mention a few of these areas, with developments grown inside mathematical logic, model theory, of set theory. What is striking is that those more recent developments, originally grown "inside the garden" of logic, and independently from considerations from those other disciplines, include ideas and tools seriously sophisticated, among them geometric stability theory, ideas around one-basedness[5], combinatorial properties such as "dependence", etc.. It is at that advanced level that those applications have been appearing. The connections with mathematical physics drawn from Zilber's work, itself oddly partially grounded in Shelah's work in the model theory of non-elementary classes - or the success of motivic integration through model theory again, or the recent interactions with the combinatorics of approximate subgroups, through stable groups, all bear witness of this new wave of interaction. The list is longer and should probably include whole chunks of category theory in odd interactions with, at the same time, algebraic topology and large cardinals, with forcing and universally Baire sets or generic absoluteness or... this is *grounding* of a kind different from the original foundational program.

These interactions seemingly pulling logic in many different directions, and perhaps creating a new phase of "grounding", witnesses enormous vitality and apparent challenges to unity. However, mathematical logic has some *internal themes* that appear in many faces throughout the more sophisticated, advanced developments. One of them is **definability**. Another is **categoricity**[6]. Yet another is **stability**[7].

Jouko Väänänen has described **definability** as a wormhole that somehow travels through all of mathematical logic. This beautiful metaphor on the ubiquity of definability in logic (its appearances in disguised form sometimes, linking first order, second and higher order logic, linking model theory, proof theory, set theory and computability theory, appearing in theorems both classic and cutting

5 See the article by Chatzidakis in this volume for more on these applications.

6 Since Łoś and Morley categoricity has been central; categoricity transfer is at the heart of Shelah's development of stability theory - in some cases it interacts in interesting ways with set theory - yet more recently Zilber's new focus on categoricity in connection with pseudo-analytic structures and constructions coming from mathematical physics has provided new avenues of interaction.

7 Stability as a concept seems central to model theory, yet it has serious antecedents in the work of Grothendieck, Krivine and others in Banach space theory.

edge) is an important reminder of what unites logic, beyond and together with the shifting dynamics of its internal movements and external interactions.

Roman Kossak
Andrés Villaveces

Contents

Juliette Kennedy

On the "Logic without Borders" Point of View

Juliette Kennedy: Department of Mathematics and Statistics, University of Helsinki, Finland

> *...as the Bhagavad-Gita teaches, one achieves knowledge and indifference at the same time.*—Andre Weil.

> *Definability is like a wormhole from one field of logic to another.*—Jouko Väänänen.

1 Introduction

Finitism, intuitionism, constructivism, formalism, predicativism, structuralism, objectivism, platonism; foundationalism, anti-foundationalism, first orderism; constructive type theory, Cantorian set theory, proof theory; *top down* principles or *building up from below*—framework commitments, that is, ideology, permeates the logician's mathematical life. Such commitments set in early and are—usually—final: lines are drawn in the sand, and a working life is mapped out.

Of course, not all logicians are attracted to dogma. Some are fascinated by the space between theories, by points of data downplayed by this or that theoretical stance, or left out altogether. Their approach is pantheistic and ecumenical, and, with respect to foundations in particular, opportunistic and localized. Their attitude is critical, not toward any particular logical method, but toward the idea of *omniscience*. Neutrality is not a goal in itself; border-crossing logicians are willing to take ideology seriously where they find it effective—it is just that they rarely find it so.

* This paper is based on a series of conversations with Jouko Väänänen.

We might call such a perspective the "Logic without Borders" point of view. In the below we will recount some episodes in border-crossing, pieces of mathematics chosen almost arbitrarily from the work, career and conversation of the dedicatee of this volume, Jouko Väänänen.

We will see that a key concept turns out to be definability—and indeed, what is more important than the question of what we can say? As Väänänen puts it in his "Pursuing Logic without Borders":[1]

> We were persuaded by the idea that model theory, set theory and recursion theory are just different approaches to the same goal, understanding definability.

2 First Episode: Model Theory

Should model theory have borders? Countability, or more precisely, the border between the "genuinely" uncountable as opposed to the "only apparently" uncountable, separates pure model theory, i.e. that part of the subject which is relatively free of entanglement with set theory, from the rest. Logicians on one side ask the question: why resort to set theory for studying the apparently marginal, the outlying cases, when there is so much work to do in the cases where set-theoretic methods are (in general) not needed, i.e. in the area of (sufficiently) stable and also in the area of o-minimal models? While the border-crossing logician will take a different view of the term "marginal", at least insofar as it is used as a synonym for "entangled with set theory".

In Shelah's classification theory, for example, uncountable models of a sufficiently stable first order theory[2] can be analyzed in terms of dimension-like invariants, somewhat reminiscent of analyzing a vector space in terms of the size of its basis. Since countable first order theories with infinite models have models in all uncountable cardinalities, just as there are vector spaces and algebraically closed fields of any dimension, the (only apparently) uncountable cardinalities of such models must arise from their dimensions, in a kind of stretching from the countable case. This places constraints on how complicated such models can be. On the other hand, countable complete first order theories which fail to be sufficiently stable[3] have, in every sufficiently large cardinality, models which are non-isomorphic, but they are so close to each other that one cannot imagine analyzing

1 in this volume

2 i.e. superstable, NDOP, NOTOP

3 i.e. they are unstable, or stable but unsuperstable, or superstable with DOP or OTOP

the models in terms of dimension-like invariants. What "so close to each other" means can be expressed in exact terms in several ways. Originally Shelah showed that such models of size κ can be $L_{\infty\kappa}$-equivalent.[4] Using the method of transfinite Ehrenfeucht-Fraïssé games and their approximations by trees[5] the original result of Shelah has been greatly improved. In recent work by Kangas-Hyttinen-Väänänen such models are constructed in suitable cardinalities which are even $L^2_{\kappa\omega}$-equivalent.[6] This is the best possible result in the sense that for the classifiable case each model of cardinality κ (for suitable κ) can be characterized in $L^2_{\kappa\omega}$ up to isomorphism.

The phrase "problematic set-theoretical content" occurs in the literature in connection with debates about the foundational role of second order logic;[7] but it also seems to have been found useful in connection with the question, which structures should one study? Following standard mathematical practice, for model theorists avoiding pathological cases—however this may be defined in a particular context—has become the rule.

Important oppositions, such as those between tame and nontame, classifiable and nonclassifiable, o-minimal or not, admitting geometric invariants or not, decidable and undecidable, or sometimes simply countable and uncountable, emerge and become harmonized with the oppositions between "nonpathological" and "set-theoretical", or, finally, "interesting" and "too general". But this is precisely where ideological borders emerge. As to undecidability, the border-crossing logician sees undecidability as a richness, a welcome elaboration of the basic picture. As for "interesting" and "too general", for the border-crossing logician "too general" is never a term of criticism, if all that is meant by "too general" is that one's reply to the question, what structures should one study? is simply "all of them".

2.1 A Remark of Sacks

Sacks expressed the conundrum thus in 1972:

> B. Dreben...once asked..."Does model theory have anything to do with logic?" It is true that model theory bears a disheartening resemblence to set theory, a fascinating branch of mathematics with little to say about fundamental logical questions, and in particular to

4 See [14].
5 a method developed by the Helsinki Logic Group in cooperation with Saharon Shelah
6 See [4].
7 See below.

> the arithmetic of cardinals and ordinals. But the resemblance is more of manners than of ideas, because the central notions of model theory are absolute, and absoluteness, unlike cardinality, is a logical concept.[8]

Among other things Sacks is referring to the fact that when one analyzes countable models in terms of finite partial isomorphisms, one uses countable ordinals as invariants and the set-theoretic aspect is submerged, because the concept of an ordinal is absolute.[9] In detail, consider the Ehrenfeucht-Fraïssé game of length ω between non-isomorphic countable models $\mathfrak{A}$ and $\mathfrak{B}$. Clearly player I[10] has a winning strategy because he can enumerate the models in ω moves and then he must win because no isomorphism exists. Consider now the restriction to games of finite length, not to any fixed finite length, but modifying the game by adding the clause that player I has to count down an ordinal α while he plays. This ordinal is like a clock which ticks down from the ordinal α and stops when it hits zero. He can count down only finitely many steps so the game is finite, but it is potentially infinite in the sense that there is no bound on the length of the game.[11] In general, for "small" α player I will not have a winning strategy. So how big must α be in order that I wins this harder game? A simple argument shows that if player II[12] has a winning strategy for all countable α, then II has a winning strategy in the original Ehrenfeucht-Fraïssé game of length ω, and the models are isomorphic. As the models are not isomorphic, there must be a countable α such that player II does not have a winning strategy, and then by the Gale-Stewart theorem, which implies that these games are determined, player I does have a winning strategy. The smallest such α measures the distance of the models $\mathfrak{A}$ and $\mathfrak{B}$ from being isomorphic. The bigger α is the closer they are to being isomorphic. This gives a hierarchy in terms of countable ordinals. On each level α of the hierarchy there are the pairs of countable models where α is the "watershed", the boundary where the advantage in the Ehrenfeucht-Fraïssé game slides from player II to player I. With all clocks less than α player II, the "isomorphism player", is able to survive without losing, but once the clock is started from α (or bigger), player I is able to find the difference in the models and manifest the non-isomorphism by winning the game. All elements of this game are quite absolute.

---

8 [11]

9 See Scott, [12].

10 the anti-isomorphism player, sometimes called the "spoiler"

11 The idea of thinking of ordinals as measures of potential infinity and of trees as measures of potential countability, is presented first in [6]; see also [18].

12 the isomorphism player, sometimes called the "duplicator"

The point is that this fails in uncountable models. When one uses countable partial isomorphisms to investigate uncountable models, one needs trees (an analogue of ordinals) which have non-trivial set theoretic properties. So set theory becomes entangled here with model theory.

To see this, consider the Ehrenfeucht-Fraïssé game of length ω_1 between non-isomorphic models $\mathfrak{A}$ and $\mathfrak{B}$ of cardinality $\aleph_1$. Clearly player I has a winning strategy, as before, because he can list the models in ω_1 moves and then he must win because no isomorphism exists. In analogy to the countable case we again modify the game by adding the clause that player I has to go up a tree, which has no uncountable branches, while he plays. This tree is like a clock which ticks up the tree and stops when the branch ends. He can go up only countably many steps so the game is countable, but it is potentially uncountable in the sense that there is no bound on the countable length of the game. As before, in general, for small trees player I will not have a winning strategy. How big must the tree be in order that I wins this harder game? It can be shown that for any non-isomorphic $\mathfrak{A}$ and $\mathfrak{B}$ there are trees such that player I wins, and of course there are trees S such that II wins (because one can start with small trees). Clearly we are thinking of trees here as analogues of ordinals. How far this analogy reaches is an interesting set-theoretical question. The structure of the class of trees with no uncountable branches is much more complicated than the structure of the class of all ordinals but as before, there is a hierarchy in terms of such trees and one can use properties of such trees to chart the area where the advantage in the Ehrenfeucht-Fraïssé game (with a tree as a clock) moves from II to I. Interestingly, there is a gray area where neither player has a winning strategy because the Gale-Stewart theorem (or Borel Determinacy) does not give determinacy for these games.

Here is an example of a non-trivial and novel set-theoretical analysis which was and is needed to work out the properties of such trees, and this has immediate implications for the model theory of uncountable structures. For example, as the work of Hyttinen, Shelah, Tuuri and others has shown, the extent to which the non-isomorphism of uncountable elementarily equivalent models can measured by trees is closely related to the stability theoretic properties of the first order theory of the models.[13]

13 See [5].

3 Second Episode: The Symbiosis between Model Theory and Set Theory

Symbiosis is the relationship between model theory and set theory in which one on the one hand exploits set-theoretical results to prove theorems in model theory, and on the other hand, one uses model-theoretic considerations to force interesting concepts and problems in set theory out into the open. Symbiosis was developed by Väänänen in order to, as he puts it, "expose the nature of the logic"; to "uncover the set-theoretical commitments of the logic, its content, its strength, even its reference". Symbiosis signifies *co-dependence*—in the benign sense of the term—and is a form of *entanglement*. Recent debates about the foundational virtues of second order logic vs. set theory, for example, decry the entanglement of set theory with second order logic, insofar as it is admitted to exist at all.[14] In fact, as Väänänen shows, not only is there nothing pernicious here, second order logic (denoted *SOL*) is actually symbiotic with set theory, even in the technical sense of the term defined below—a predicament, possibly, for those who feel compelled, on foundational grounds, to make a choice between the two formalisms.[15]

The technical definition of symbiosis is as follows. First some notation. By a "predicate" we mean a formula of set theory, typically "x is a cardinal" or "x is the power-set of y". If a predicate P is added to the language of set theory as a (definable) new symbol, then a $\Sigma_1(P)$-predicate means a Σ_1-formula in the vocabulary $\{\in, P\}$. A $\Delta_1(P)$-predicate is a $\Sigma_1(P)$-predicate $Q(x_1, \dots, x_n)$ for which there is another $\Sigma_1(P)$-predicate $Q'(x_1, \dots, x_n)$ such that $\forall x_1 \dots x_n(Q(x_1, \dots, x_n) \leftrightarrow Q'(x_1, \dots, x_n))$ is true (in V). If ϕ is a sentence of a logic L^* and L' is a subset of the (many-sorted) vocabulary of ϕ, then the *projection* of ϕ to L' is the class of reducts of models of ϕ to L'. A model class is said to be *Δ-definable* in L^* if it is a projection of a sentence of L^* and also its complement is. Now the definition:

Definition 3.1. A logic L^* is symbiotic with a predicate P of set theory if the predicate "$\phi \in L^*$" and the predicate "$M \models_{L^*} \phi$" are $\Sigma_1(P)$ and $\Delta_1(P)$ respectively, and in addition, a model class K_P describing P (see below) is Δ-definable in L^*.

What symbiosis tells us about a logic L^* is that its truth predicate is "recursive" in the predicate P, in the generalized sense of being $\Delta_1(P)$. The class K_P is defined as follows:

---

14 See for example [13].

15 See below for the proof that *SOL* is symbiotic with set theory. See also [15].

Definition 3.2. Suppose P is n-ary. The model class K_P consists of models

$$(M, E, a_1, ..., a_n)$$

isomorphic to some $(M', \in, a_1', ..., a_n')$ such that M' is a transitive set and

$$P(a_1', ..., a_n')$$

holds.

Barwise’s concept of an *absolute logic* is related to symbiosis but is not the same.[16] An absolute logic as defined by Barwise requires the satisfaction predicate to be Δ_1, but without extra predicates. In the generalization of the concept introduced by Väänänen one adds the predicate P as a kind of “oracle”.

For example, there is a symbiosis between the Härtig-quantifier and the predicate $x = Cd(y)$ (“the cardinality of y is x”). First of all, as Lindström showed in [9], the class of well-ordered models is a relativized reduct (i.e. a reduct in the sense of many-sorted logic) of a model class definable by means of the Härtig-quantifier, for a linear order $(A, <)$ is a well-order if and only if there are sets X_a, $a \in A$, such that for all $a, b \in A$: $a < b \leftrightarrow |X_a| < |X_b|$. The satisfaction predicate “the assignment s satisfies the formula ϕ in the model M” for the extension of first order logic by the Härtig-quantifier can be defined in set theory by a $\Sigma_1(Cd)$-formula. On the other hand, the class of models (up to isomorphism) of the form $(M, \in, a, b)$, where M is transitive and $a = |b|$, is Δ-definable in the extension of first order logic by the Härtig-quantifier. We get hold of transitive models $(M, \in)$ because we can characterise well-foundedness. We can make sure that $a = |b|$ because we can use the Härtig-quantifier.

Let us take an example of an application of set theory in model theory via symbiosis. One of the cornerstones of set theory is the Levy Reflection Principle, which states that the hereditarily countable sets (denoted HC) form a Σ_1 elementary submodel of V. One can now observe that if a logic L^* is symbiotic with set theory per se (i.e. without any added predicate), then it has the Löwenheim-Skolem Theorem down to $\aleph_0$. To see this, note that if $\phi \in L^* \cap HC$ has a model, then the statement “ϕ has a model” is a true Σ_1-sentence of set theory, hence it is true in HC. Hence ϕ has a countable model. In general, if symbiosis exists with a predicate, then we have a Löwenheim-Skolem Theorem down to κ as soon as $H(\kappa)$[17] has enough reflection. For example, we noted that there is a symbiosis between the Härtig-quantifier and the predicate $x = Cd(y)$ (“the cardinality of y is x”). Thus as soon as $H(\kappa)$, which

---

16 See [2].

17 $H(\kappa)$ denotes the set of sets of hereditary cardinality less than κ. Thus HC is $H(\aleph_1)$.

is always a Σ_1 submodel of V, is also a $\Sigma_1(Cd)$-elementary submodel of V, we have a Downward Löwenheim-Skolem theorem for the Härtig quantifier: If $\phi \in L(H)$ has a model, then ϕ has a model of cardinality $< \kappa$.

Let us consider an example in the other direction, from model theory to set theory. The compactness of the infinitary language $L_{\kappa\kappa}$ is a well-known case. This logic is symbiotic with the predicate "x is the set of sequences of length $< \kappa$ of elements of y". There are two different versions of compactness for this logic. The first says that any theory T of size κ all subsets of smaller cardinality of which have a model, has itself a model. The other drops the cardinality assumption that T has size κ. The first concept leads to the concept of a weakly compact cardinal, the second leads to the concept of strongly compact cardinal. Both concepts have become important in set theory, independently of the relation to infinitary languages.

A kind of ultimate symbiosis is the symbiosis of second order logic with the power-set operation "x is the power set of y", as was mentioned, and as is proved below.

Once one begins to look for symbiosis, one finds it everywhere! For example, as Väänänen points out, symbiosis is not limited to infinite models, but happens in the area of finite models too. To see this, recall that a class of finite models is recursive iff it is Δ-definable in the above sense in first order logic (denoted FO) in the context where all models considered are finite.

So here a computational definability concept, namely "recursive", coincides with a model-theoretic definability concept. Moreover, due to a result of Fagin, a class of finite models is existential second order definable iff it is NP.[18] Finally, a class of finite ordered models is definable in Fixpoint logic iff it is PTIME.[19]

A beautiful case of symbiosis for a logic which is strictly between first order logic and second order logic is the extension of first order logic by the well-ordering quantifier

$$WxyA(x,y) \leftrightarrow \{(a,b) : A(a,b)\} \text{is a well-ordering}$$

denoted $FO(W)$. This logic is symbiotic with set theory per se, i.e. with the predicate $x = x$. This is because a model class is Δ_1 if and only if it is Δ-definable in $FO(W)$. The well-ordering quantifier is simple and yet it hits exactly the Δ_1 level in set theory.

Symbiosis often goes deeper than just the level of model classes. In many examples model-theoretic invariants, such as Löwenheim number, Hanf number, decision problems and so on have natural set theoretic characterizations. For ex-

---

18 See Fagin, [3].

19 See N. Immerman [7] and M. Vardi [20] and [8].

ample, the Löwenheim number of second order logic is the supremum of all Π_2-definable ordinals, and similar results hold for the Hanf number and the decision problem of second order logic. Conversely, in many cases set-theoretic invariants such as cardinals where one has reflection, have natural model theoretic applications. For example, by a result of Magidor, the first supercompact cardinal is the first cardinal where second order logic has a strong form of a Löwenheim-Skolem Theorem.[20] The symbiosis here is informal, but what is also interesting is that in each case one can look at the symbiosis more carefully and sharpen the results about Löwenheim-Skolem Theorems, Hanf numbers and so on to be able to prove symbiosis in the technical sense also.

4 Third Episode: The Entanglement of Second Order Logic with Set Theory

That second order logic is entangled with set theory, is, under this perspective, of no concern. Väänänen's precisification of the notion of entanglement is the above-mentioned concept *symbiosis*. The concept was isolated in his 1978 [16], in which the following is proved: Second order logic is symbiotic with the predicate "$y = P(x)$" (the powerset operation). This means that a model class (closed under isomorphism) is definable in set theory by a Δ_1 formula in the vocabulary which includes the binary predicate "$y = P(x)$" in addition to $\in$, if and only if the model class is Δ-definable in second order logic.

Why is second order logic symbiotic with the power-set operation? We reproduce here Väänänen's reply to this question,[21] together with some of his thoughts on the first order case:

> First of all, since second order logic is closed under negation, it is enough to show that the satisfaction predicate is Σ_1 with respect to the power-set operation. This is entirely standard. The power-set operation is needed for the semantics of the second order quantifiers. Conversely, we have to show that the class of transitive sets equipped with the power-set operation is definable in second order logic with extra predicates. The first observation is that well-foundedness, and hence transitivity, can be defined (up to isomorphism) in second order logic. After this is it easy to use a second order quantifier to say that that power-set operation on the transitive set is really the full power-set operation.

20 See Magidor, [10].
21 personal communication

> In the first order case the symbiosis of *FO* with set theory takes place on the level of Kripke-Platek set theory with urelements, denoted KPU^-, without the axiom of infinity.[22] So a model class is definable in first order logic (when we think the universe of each model consisting of urelements) iff it is Δ_1 in KPU^-. The moral of the story is that first order logic is so weak that its symbiosis with set theory takes place on the level of the weak set theory KPU^-.

Another way of presenting the entanglement of second order logic with set theory is laid out by Väänänen in his 2001 and 2012 papers [17], and [19]. Here set theory and second order logic are presented as *analogs*, or reflections of each other—*avatars*, to use Michael Harris's terminology[23]—kindred logical productions crystallized by the intention of the logician as he inclines toward this or that precisification of the mathematician's natural language discourse. As Väänänen puts it in that paper:

> We study two metatheories of mathematics: first order set theory and second order logic. It is often said (e.g. in [13]), that second order logic is better than first order set theory because it can in its full semantics axiomatize categorically N and R, while first order axiomatization of set theory admits non-standard, e.g. countable models. We show below that this difference is illusory. If second order logic is construed as our primitive logic, one cannot say whether it has full semantics or Henkin semantics, nor can we say whether it axiomatizes categorically N and R.[24] So there is no difference between the two logics: first order set theory is merely the result of extending second order logic to transfinitely high types.

The border-crossing logician does not pronounce, he proves. Which is to say, this is just the starting point of the analysis. For example, Väänänen asks, how do we

---

22 See J. Akkanen, [1].

23 and Grothedieck's, Deligne's and others. Apparently the word has become a term of art in the field. Harris quotes Grothedieck on avatars:

> Inspired by certain ideas of Serre, and also by the wish to find a certain common "principle" or "motif" for the various purely algebraic "avatars" that were known, or expected, for the classical Betti cohomology of a complex algebraic variety, I had introduced towards the beginning of the 60s the notion of "motif".

and Weil (translated from the French by Harris):

> ... those obscure analogies ... disturbing reflections of one theory in another.

Though as Harris notes, the reference to avatars is only implicit in the Weil quote. See Harris's forthcoming "Not Merely Good, True, and Beautiful," Princeton University Press.

24 Of course as Väänänen notes in [19], one can prove *internal* categoricity in these cases. See below.

recognize second order characterizable structures? Which structures are second order characterizable in the first place?[25] What does it really mean to say that a theory fails to be categorical?[26]

We briefly consider the last two questions. Recall that a theory is κ-categorical if all models of the theory of size κ are isomorphic. A key observation in [19] is this: the question whether arithmetic in its second order construal is categorical is, as it stands, *ill-defined*. The well-defined question is whether two Henkin models (of second order arithmetic) which have a common expansion to a model of the Comprehension Axiom are isomorphic. If such an extension exists, then the two models are easily proved to be isomorphic in that common extension, that is, without any set-theoretical metatheory in the background.[27] If no such expansion exists, that is if the language in question is so impoverished that it cannot describe both structures, e.g. by lacking certain predicates, then the question whether the two structures are isomorphic makes no sense. Put another way, if we cannot even say what the two structures are, then how can we ask whether the one is isomorphic to the other?

Which structures are second order characterizable? Answer: almost all of the structures the mathematician encounters in his or her working life are second order characterizable. It is consistent that some are not, e.g. it is consistent that the reals with a Hamel basis, when the reals are considered as a vector space over the rationals, is not second order characterizable.[28] Second order characterizable structures actually form a hierarchy. If a structure of cardinality κ is second order characterizable, then so is κ (as a structure of the empty vocabulary), and all second order characterizable structures of the same cardinality are Turing equivalent in the sense given in [19]. In fact the following are proved in that paper:

Proposition 4.1. *The first inaccessible (Mahlo, weakly compact, Ramsey) cardinal is second order characterizable. If κ is the first measurable cardinal, then 2^κ is second order characterizable. All second order characterizable cardinals are below the first strong cardinal.*[29]

25 A structure M is said to be second order characterizable if there is a second order sentence θ such that M is, up to isomorphism, the only model of θ, that is, θ categorically characterizes M.

26 p. 97, [19].

27 This is called "internal categoricity" in [19].

28 The result uses Cohen forcing. See [4] for details.

29 The result concerning inaccessible cardinals is due to Zermelo, and the result on measurable cardinals is due to D. Scott. The result concerning strong cardinals is due to Magidor. The large cardinal sequence begins above ω. See [10].

Proposition 4.2. *If κ is second order characterizable, then so are κ^+, 2^κ, $\aleph_\kappa$ and $\beth_\kappa$. More generally, if κ and λ are second order characterizable, then so is κ^λ.*

Thus, as Väänänen observes, the first non-second order characterizable cardinal is a singular strong limit cardinal.

The analysis goes deeper. One can ask, of the predicate "ϕ is a second order characterization of a structure", what is its complexity? It is shown in [19] that the predicate is the conjunction of a Σ_2-complete and a Π_2-complete property of ϕ. As such, it is not itself Σ_2 or Π_2, a fact which has very important foundational consequences:

> So recognizing whether a candidate second order sentence is a second order characterization of some structure is so complex a problem that it cannot (by the Proposition above) be reduced to truth [of the form] $A \models \phi^*$ in any particular second order characterizable structure A. It encodes a solution to propositions of the type $\not\models \phi^*$. So in complexity it is above all the particular truths $A \models \phi^*$ and on a par with, but not equivalent to $\models \phi^*$. The whole framework of the second order view takes the concept of a second order characterizable structure as its starting point. In the case of familiar classical structures we can easily write the second order characterizations. But if we write down an arbitrary attempt at a second order characterization, the problem of deciding whether we were successful is in principle harder than the problem of finding what is true in the structure, if the sentence indeed characterizes some structure.

This actually supports, because it emphasizes the individuality of particular structures, forms of structuralism which specialize to the second order view as laid out in [19]—a view which is built on the idea that second order validity is not reducible to truth in any one second order characterizable structure. At the same time a weakness of the view becomes visible: does the number theorist who is looking for integer solutions to Diophantine equations really work with a completely different structure than the analyst who works with complex numbers? Are not the integers rather a substructure of the complex numbers? It seems counterintuitive that, e.g. Dedekind's embedding of the natural numbers into the integers, and of these into the rationals, and of these into the reals, and finally of these into the complex numbers, is a wrong picture.

5 Conclusion

We began with Weil's recommendation of a passage from the Bhagavad-Gita. We want to say that here too one achieves knowledge and indifference at the same time—or perhaps it would be more correct to say that for the border-crossing logician, his indifference is actually the *source* of his knowledge. For this is what makes *his* logical life worth living: the fine-structural, set-theoretical focus both on logic, and on logics; the development of logical frameworks which are not so much *groundings* as systems of *avatars*; and finally the attempt to expose matters of reference and of content, while all the while remaining unmoved by ideological pressures.

Bibliography

[1] Jyrki Akkanen. Absolute logics. *Ann. Acad. Sci. Fenn. Ser. A I Math. Dissertationes*, (100):83, 1995.

[2] K. Jon Barwise. Absolute logics and $L_{\infty\omega}$. *Ann. Math. Logic*, 4:309–340, 1972.

[3] Ronald Fagin. Generalized first-order spectra and polynomial-time recognizable sets. In *Complexity of computation (Proc. SIAM-AMS Sympos. Appl. Math., New York, 1973)*, pages 43–73. SIAM–AMS Proc., Vol. VII. Amer. Math. Soc., Providence, R.I., 1974.

[4] Tapani Hyttinen, Kaisa Kangas, and Jouko Väänänen. On second-order characterizability. *Log. J. IGPL*, 21(5):767–787, 2013.

[5] Tapani Hyttinen and Heikki Tuuri. Constructing strongly equivalent nonisomorphic models for unstable theories. *Ann. Pure Appl. Logic*, 52(3):203–248, 1991.

[6] Tapani Hyttinen and Jouko Väänänen. On Scott and Karp trees of uncountable models. *J. Symbolic Logic*, 55(3):897–908, 1990.

[7] Neil Immerman. Upper and lower bounds for first order expressibility. *J. Comput. Syst. Sci.*, 25(1):76–98, 1982.

[8] Harry R. Lewis, Barbara B. Simons, Walter A. Burkhard, and Lawrence H. Landweber, editors. *Proceedings of the 14th Annual ACM Symposium on Theory of Computing, May 5-7, 1982, San Francisco, California, USA*. ACM, 1982.

[9] Per Lindström. First order predicate logic with generalized quantifiers. *Theoria*, 32:186–195, 1966.

[10] M. Magidor. On the role of supercompact and extendible cardinals in logic. *Israel J. Math.*, 10:147–157, 1971.

[11] Gerald E. Sacks. *Saturated model theory*. World Scientific Publishing Co. Pte. Ltd., Hackensack, NJ, second edition, 2010.

[12] Dana Scott. Logic with denumerably long formulas and finite strings of quantifiers. In *Theory of Models (Proc. 1963 Internat. Sympos. Berkeley)*, pages 329–341. North-Holland, Amsterdam, 1965.

[13] Stewart Shapiro. *Foundations without foundationalism*, volume 17 of *Oxford Logic Guides*. The Clarendon Press Oxford University Press, New York, 1991. A case for second-order logic, Oxford Science Publications.

[14] Saharon Shelah. Existence of many $L_{\infty,\lambda}$-equivalent, nonisomorphic models of T of power λ. *Ann. Pure Appl. Logic*, 34(3):291–310, 1987. Stability in model theory (Trento, 1984).

[15] Jouko Väänänen. Abstract logic and set theory. I. Definability. In *Logic Colloquium '78 (Mons, 1978)*, volume 97 of *Stud. Logic Foundations Math.*, pages 391–421. North-Holland, Amsterdam, 1979.

[16] Jouko Väänänen. Abstract logic and set theory. I. Definability. In *Logic Colloquium '78 (Mons, 1978)*, volume 97 of *Stud. Logic Foundations Math.*, pages 391–421. North-Holland,

Amsterdam, 1979.
[17] Jouko Väänänen. Second-order logic and foundations of mathematics. *Bull. Symbolic Logic*, 7(4):504–520, 2001.
[18] Jouko Väänänen. *Models and games*, volume 132 of *Cambridge Studies in Advanced Mathematics*. Cambridge University Press, Cambridge, 2011.
[19] Jouko Väänänen. Second-order logic or set theory? *Bull. Symbolic Logic*, 18(1):91–121, 2012.
[20] Moshe Y. Vardi. The complexity of relational query languages (extended abstract). [8], pages 137–146.

Samson Abramsky

Arrow's Theorem by Arrow Theory

Abstract: We give a categorical account of Arrow's theorem, a seminal result in social choice theory.

Keywords: mathematics of social choice, Arrow's theorem, category theory

MSC: 91-XX, 03, 18

Samson Abramsky: Department of Computer Science, University of Oxford

1 Introduction

Arrow's theorem [2, 3] is a seminal result from 1950 which founded the modern field of *social choice theory*. It is a no-go theorem for combining individual preferences: under two very plausible assumptions, the only possible 'social welfare function' is a dictatorship! Closely allied results, notably the Gibbard-Satterthwaite theorem [6, 10], relate to *voting systems*: when there are at least three alternatives, the only non-manipulable voting systems are dictatorships.

Our aim in this note is to cast this result at a more abstract, categorical level, to expose common structure with no-go theorems in other fields, such as the foundations of quantum mechanics. This is motivated by our recent work with Adam Brandeburger on a sheaf-theoretic approach to quantum non-locality and contextuality [1], and also by the work of Jouko Väänänen and others on Dependence/Independence Logic [13, 7].

We shall assume only the most basic notions of category theory: categories, functors, and natural transformations. Moreover, these will be used in a simple, concrete setting. Thus the categorical language should not be an impediment to readers interested in a structural account of Arrow's theorem.

1.1 What Arrow's Theorem Says

Let A be a set of *alternatives*, and $\mathfrak{I}$ a set of *individuals*.

$\mathbb{P}(A)$ is the set of *preference relations* on A. These are usually taken to be weak orders (transitive and connected relations), sometimes linear orders (transitive, connected and irreflexive).

$\mathbb{P}(A)^{\mathfrak{I}}$ is the set of *profiles* or *ballots*, which assign a preference relation on the alternatives to each individual – a "vote".

A *social welfare function* is a map

$$\sigma : \mathbb{P}(A)^{\mathfrak{I}} \longrightarrow \mathbb{P}(A).$$

Such a map produces a single ranking on alternatives – a social choice – from a profile.

Two conditions are standardly considered on such functions:

- **Independence of Irrelevant Alternatives** (**IIA**). The social decision on the relative preference between two alternatives a, b depends only on the individual preferences between these alternatives. It is *independent* of their rankings with respect to other alternatives.
- **The Pareto or Uniformity Principle** (**P**). If every individual prefers a to b, then so should the social welfare function.

We can now state Arrow's Theorem more formally as follows.

Theorem 1.1 (Arrow's Theorem). *If* $|A| > 2$ *and* $\mathfrak{I}$ *is finite, then any social welfare function satisfying* ***IIA*** *and* **P** *is a* dictatorship*: i.e. for some individual* $i \in \mathfrak{I}$*, for all profiles* $p \in \mathbb{P}(A)^{\mathfrak{I}}$ *and alternatives* $a, b \in A$*:*

$$a\,\sigma(p)\,b \iff a\,p_i\,b.$$

Thus the social choice function, under these very plausible assumptions, simply copies the choices of one fixed individual – the dictator.

An extraordinary number of different proofs, as well as innumerable variations, have appeared in the (huge) literature. For a small selection, see [3, 5, 9, 12].

A closely related result is the *Gibbard-Satterthwaite theorem* [6, 10] on voting systems:

Theorem 1.2. *If* $|A| > 2$ *and* $\mathfrak{I}$ *is finite, then any voting system*

$$v : \mathbb{P}(A)^{\mathfrak{I}} \longrightarrow A$$

which is non-manipulable *is a dictatorship.*

The following quotation from the recent text [12] nicely captures the significance of the result:

> For an area of study to become a recognized field, or even a recognized subfield, two things are required: It must be seen to have coherence, and it must be seen to have depth. The former often comes gradually, but the latter can arise in a single flash of brilliance. ... With social choice theory, there is little doubt as to the seminal result that made it a recognized field of study: Arrow's impossibility theorem.

The further contents of the paper are as follows. In Section 2, we shall present a fairly standard account of Arrow's theorem which will fix notation and serve as a reference point. In Section 3, we will reformulate Arrow's Theorem in categorical terms, and in Section 4, we shall give a development of the proof which uses the categorical formulation to emphasize the structural aspects. Section 5 concludes.

2 A 'standard' account of Arrow's theorem

The aim of this section is to give a clear, explicit presentation of a fairly standard account of Arrow's theorem and some related notions.

The arguments in Section 2.1 follow [3], with some clarifications and refinements due to [5]. In Section 2.2, we follow [9].

2.1 Preference Relations

We consider a number of properties of binary relations $R \subseteq A^2$ on a set A. These are all universally quantified over elements $a, b, c \in A$:
Reflexivity aRa
Irreflexivity $\neg aRa$
Symmetry $aRb \Rightarrow bRa$
Antisymmetry $aRb \wedge bRa \Rightarrow a = b$
Transitivity $aRb \wedge bRc \Rightarrow aRc$
Connectedness $aRb \vee bRa$.

A *weak preference relation* is a transitive connected relation. We write $\mathbb{P}(A)$ for the set of all weak preference relations on A. Given a weak preference relation R, we can define two other relations:
Strict Preference $aPb := aRb \wedge \neg bRa$.
Indifference $xIy := aRb \wedge bRa$.

Then P is a strict ordering (transitive and irreflexive), while I is an equivalence relation (reflexive, symmetric and transitive). These relations satisfy the following properties:
Trichotomy $aPb \vee bPa \vee aIb$.
Absorption $aIb \wedge bPc \wedge cId \Rightarrow aPd$.

A weak preference relation is *linear* if it additionally satisfies antisymmetry. We write $\mathbb{L}(A)$ for the set of linear preference relations on A. If R is linear, then the associated indifference relation I is just the identity relation, while P is a strict linear order.

Given $A \subseteq B$, we can define a restriction map $\mathbb{P}(B) \to \mathbb{P}(A) :: R \mapsto R|A$, where $R|A := R \cap A^2$. Note that the truth of any property of R expressed by a universal sentence is preserved under restriction, so this is well-defined; moreover, the same operation also defines a map $\mathbb{L}(B) \to \mathbb{L}(A)$.

2.2 Social choice situations

We shall define a class of structures which provide the setting for Arrow's theorem.

A *social choice situation* is a structure $(A, \mathcal{I}, \mathcal{D}, \sigma)$ where:

- A is a set of *alternatives.*
- $\mathcal{I}$ is a set of *individuals* or *agents.*
- $\mathcal{D} \subseteq \mathbb{P}(A)^{\mathcal{I}}$ is the set of allowed *ballots* or *profiles* of individual preferences.
- $\sigma : \mathcal{D} \to \mathbb{P}(A)$ is the *social choice function.*

We write p_i for the weak preference relation of the individual i in a profile p. We write $p_i^>$ for the strict preference relation associated with p_i. Similarly, we write $\sigma(p)^>$ for the strict preference relation associated with $\sigma(p)$. We extend restriction to profiles pointwise: $(p|A)_i := p_i|A$.

We shall now define a number of properties of social choice situations.

UD Unrestricted domain:

$$\forall a, b, c \in A. \forall p \in \mathbb{P}(\{a, b, c\})^{\mathcal{I}}. \exists q \in \mathcal{D}. q|\{a, b, c\} = p.$$

P Pareto:

$$\forall a, b \in A. \forall p \in D. (\forall i \in \mathcal{I}. a p_i^> b) \Rightarrow a\sigma(p)^> b.$$

WP Weak Pareto:

$$\forall a, b \in A. \forall p \in D. (\forall i \in \mathcal{I}. a p_i^> b) \Rightarrow a\sigma(p) b.$$

IIA Independence of irrelevant alternatives:

$$\forall a, b \in A. \forall p, q \in \mathcal{D}. p|\{a, b\} = q|\{a, b\} \Rightarrow \sigma(p)|\{a, b\} = \sigma(q)|\{a, b\}.$$

D Dictator:

$$\exists i \in \mathcal{I}. \forall a, b \in A. \forall p \in \mathcal{D}. a p_i^> b \Rightarrow a\sigma(p)^> b.$$

We can now state Arrow's theorem.

Theorem 2.1 (Arrow). *Let* $(A, \mathfrak{I}, \mathfrak{D}, \sigma)$ *be a social choice situation with* $|A| \geq 3$, *satisfying* **UD**, **IIA** *and* **P**. *Then if* $\mathfrak{I}$ *is finite it also satisfies* **D**, *i.e. there is a dictator.*

2.3 Proof of Arrow's Theorem

In this section we shall fix a social choice situation $(A, \mathfrak{I}, \mathfrak{D}, \sigma)$ satisfying the following conditions: $|A| \geq 3$, **UD**, and **IIA**.

2.3.1 Decisiveness, Neutrality and Monotonicity

In this subsection, we shall assume that our social choice situation satisfies the weak Pareto principle **WP**.

Given $U \subseteq \mathfrak{I}$ and $p \in \mathfrak{D}$, we introduce the notation $ap_U^{>}b := \forall i \in U.\, ap_i^{>}b$. Given a set $U \subseteq \mathfrak{I}$ and distinct elements $a, b \in A$, we define

$$U_{ab} \;:=\; \{p \in \mathfrak{D} \mid ap_U^{>}b \;\wedge\; bp_{U^c}^{>}a\}.$$

We define a relation D_U on A by

$$aD_Ub \;:=\; a \neq b \;\wedge\; \forall p \in U_{ab}.\, a\sigma(p)^{>}b.$$

We read aD_Ub as "U is decisive for a over b".

Proposition 2.2. *For all* $a, b, c \in A$:
1. *If* $c \neq a$, *then* $aD_Ub \;\Rightarrow\; aD_Uc$.
2. *If* $c \neq b$, *then* $aD_Ub \;\Rightarrow\; cD_Ub$.

Proof. For (1), if $b = c$ there is nothing to prove. If $b \neq c$, we consider a profile $p \in \mathfrak{D}$ such that for all $i \in U$, p_i restricts to the strict chain abc, and for all $i \in U^{\mathsf{C}}$, p_i restricts to the strict chain bca. Such a profile exists by **UD**. Note that $p \in U_{ab} \cap U_{ac} \cap \mathfrak{I}_{bc}$. Since aD_Ub, $a\sigma(p)^{>}b$, while by **WP**, $b\sigma(p)c$. By transitivity if $b\sigma(p)^{>}c$, or by absorption otherwise, $a\sigma(p)^{>}c$. Now consider any profile q such that $q \in U_{ac}$. Then $q|\{a, c\} = p|\{a, c\}$, and by **IIA**, $a\sigma(p)^{>}c \;\Rightarrow\; a\sigma(q)^{>}c$; thus aD_Uc. The argument for (2) is similar. □

As pointed out in [5], the following purely relational argument allows us to conclude Neutrality from the previous proposition.

Proposition 2.3. *Let* R *be an irreflexive relation on a set* X *with at least three elements, such that, for all* $a, b, x \in X$:
1. *If* $x \neq a$, *then* $aRb \;\Rightarrow\; aRx$.

2. *If $x \neq b$, then $aRb \Rightarrow xRb$.*
If x, y are any pair of distinct elements of X, then $aRb \Rightarrow xRy$.

Proof. If $y \neq a$, then $aRb \Rightarrow aRy \Rightarrow xRy$. If $x \neq b$, then $aRb \Rightarrow xRb \Rightarrow xRy$. Otherwise, $x = b$ and $y = a$, and we must prove $aRb \Rightarrow bRa$. In this case, since X has at least three elements, we can find $c \in X$ with $a \neq c \neq b$. Then:

$$aRb \Rightarrow aRc \Rightarrow bRc \Rightarrow bRa.$$

□

As an immediate consequence of Propositions 2.2 and 2.3, we obtain

Theorem 2.4 (Local Neutrality). *For all $a, b, x, y \in A$ with $x \neq y$:*

$$aD_U b \Rightarrow xD_U y.$$

We now define a relation E_U on A by:

$$aE_U b \;:=\; \forall p \in \mathcal{D}.\, ap_U^{>} b \Rightarrow a\sigma(p)^{>} b.$$

Thus we ask only that the individuals in U strictly prefer a to b; there is no constraint on those outside U. Clearly, $aE_U b \Rightarrow aD_U b$. The converse is an important property known as *monotonicity*.

Proposition 2.5 (Monotonicity). *For all $a, b \in A$, $aD_U b \iff aE_U b$.*

Proof. We shall prove $aD_U b \Rightarrow aE_U b$. Suppose we are given a profile p such that $ap_U^{>} b$. We can find an element $c \in A$ with $a \neq c \neq b$. We consider a profile $q \in \mathcal{D}$ such that for all $i \in U$, q_i restricts to the strict chain acb, and for all $i \in U^{c}$, c is strictly preferred to both a and b in q_i, while $q_i|\{a, b\} = p_i|\{a, b\}$. Such a profile exists by **UD**. Note that $q \in U_{ac} \cap \mathcal{I}_{cb}$, and $q|\{a, b\} = p|\{a, b\}$. Since $aD_U b$, by Proposition 2.2 $aD_U c$, and so $a\sigma(q)^{>} c$. By **WP**, $c\sigma(q)b$. By transitivity if $c\sigma(q)^{>} b$, or by absorption otherwise, $a\sigma(q)^{>} b$. Since $p|\{a, b\} = q|\{a, b\}$, by **IIA** we conclude that $a\sigma(p)^{>} b$, and hence $aE_U b$ as required. □

2.3.2 The Ultrafilter of Decisive Sets

In this subsection, we assume the strong Pareto principle **P**, which serves as a basic existence principle for decisive sets. Note indeed that **P** is equivalent to the statement that $\mathcal{I}$ is a decisive set.

We define

$$\mathcal{U} \;:=\; \{U \subseteq \mathcal{I} \mid \exists a, b \in A.\, aD_U b\}.$$

Theorem 2.6 (The Ultrafilter Theorem). *$\mathfrak{U}$ is an ultrafilter.*

Proof. (F1) As we have already noted, **P** implies that $\mathfrak{I} \in \mathfrak{U}$.

(F2) Now suppose that $U \in \mathfrak{U}$ and $U \subseteq V$. By Proposition 2.5, we can conclude that $V \in \mathfrak{U}$, since clearly $U \subseteq V$ implies that $E_U \subseteq E_V$.

(F3) Now suppose for a contradiction that U and V are both in $\mathfrak{U}$, where $U \cap V = \varnothing$. Consider a profile $p \in \mathfrak{D}$ such that $ap^{>}_{U}b$ and $bp^{>}_{V}a$. By Proposition 2.5, we have both $a\sigma(p)^{>}b$ and $b\sigma(p)^{>}a$, yielding a contradiction.

(F4) Finally suppose that $U \in \mathfrak{U}$ can be written as a disjoint union $U = V \sqcup W$. We shall show that either $V \in \mathfrak{U}$ or $W \in \mathfrak{U}$. Consider a profile $p \in \mathfrak{D}$ such that for each $i \in V$, p_i restricts to the strict chain bca, for each $i \in W$, p_i restricts to the strict chain cab, while for each $i \in U^c$, p_i restricts to the strict chain abc. Such a profile exists by **UD**. Note that $p \in U_{ca} \cap V_{ba} \cap W_{cb}$. We argue by cases:

- If $c\sigma(p)^{>}b$, then by **IIA**, for all $q \in W_{cb}$, $c\sigma(q)^{>}b$, and hence $cD_W b$, and $W \in \mathfrak{U}$.
- Otherwise, we must have $b\sigma(p)c$. Since $p \in U_{ca}$ and $U \in \mathfrak{U}$, using the Neutrality Theorem 2.4, we must have $c\sigma(p)^{>}a$. By absorption, $b\sigma(p)^{>}a$. By **IIA**, for all $q \in V_{ba}$, $b\sigma(q)^{>}a$, and hence $bD_V a$, and $V \in \mathfrak{U}$.

The conditions (F1)–(F4) are easily seen to be equivalent to the standard definition of an ultrafilter, given as (F1) and (F2) together with:

(F5) $\varnothing \notin \mathfrak{U}$.

(F6) $U, V \in \mathfrak{U} \;\Rightarrow\; U \cap V \in \mathfrak{U}$.

(F7) $\forall U \subseteq \mathfrak{I}. U \in \mathfrak{U} \;\vee\; U^c \in \mathfrak{U}$. □

We define the set of ballots which are linear on the alternatives a, b:

$$\mathcal{L}_{ab} \;:=\; \{p \in \mathfrak{D} \mid p|\{a,b\} \in \mathbb{L}(\{a,b\})\} \;=\; \{p \in \mathfrak{D} \mid \exists U \subseteq \mathfrak{I}.\, p \in U_{ab}\}.$$

We now show that $\mathfrak{U}$ completely determines σ on linear ballots.

Proposition 2.7. *For all $a, b \in A$, $p \in \mathcal{L}_{ab}$:*

$$a\sigma(p)^{>}b \iff \{i \in \mathfrak{I} \mid ap^{>}_{i}b\} \in \mathfrak{U}.$$

Proof. The right-to-left implication is immediate, since if $U = \{i \in \mathfrak{I} \mid ap^{>}_{i}b\}$, $p \in \mathcal{L}_{ab}$ implies that $p \in U_{ab}$. For the converse, we use property (F7) from Theorem 2.6. □

We also show that social choice functions map linear ballots to strict preferences.

Proposition 2.8. *For all distinct alternatives $a, b \in A$:*

$$\forall p \in \mathcal{L}_{ab}.\, a\sigma(p)^{>}b \;\vee\; b\sigma(p)^{>}a.$$

Proof. Immediate from the previous proposition and property (F7) from Theorem 2.6. □

2.4 Arrow's Theorem

Theorem 2.9 (Arrow). *Let $(A, \mathcal{I}, \mathcal{D}, \sigma)$ be a social choice situation with $|A| \geq 3$, satisfying **UD, IIA** and **P**. Then if $\mathcal{I}$ is finite it also satisfies **D**, i.e. there is a dictator.*

Proof. By the Ultrafilter Theorem 2.6, $\mathcal{U}$ is an ultrafilter. Since $\mathcal{I}$ is finite, $\mathcal{U}$ must be principal, consisting of all supersets of $\{i\}$ for some $i \in \mathcal{I}$. Then $D_{\{i\}}$ is decisive, or equivalently by Proposition 2.5, i is a dictator. □

3 Categorical Formulation of Arrow's Theorem

Given a universe $\mathbb{A}$ of possible alternatives, where the cardinality of $\mathbb{A}$ is ≥ 3, we consider the category $\mathcal{C}$ whose objects are subsets of $\mathbb{A}$, and whose morphism are injective maps; and its posetal sub-category $\mathcal{C}_{\mathsf{inc}}$ with morphisms the inclusions. We shall use $\mathcal{C}^{(k)}$ and $\mathcal{C}^{(k)}_{\mathsf{inc}}$ to denote the full sub-categories of $\mathcal{C}$ and $\mathcal{C}_{\mathsf{inc}}$ respectively determined by the sets A of cardinality $\leq k$, for $k \geq 0$. We write $\mathcal{C}^{\mathsf{op}}$ for the opposite category of $\mathcal{C}$.

For any notion of binary preference relation axiomatized by *universal* sentences (universal closures of quantifier-free formulas), we get a functor

$$\mathbf{P} : \mathcal{C}^{\mathsf{op}} \longrightarrow \mathbf{Set}.$$

$\mathbf{P}(X)$ is the set of preference relations on X, and if $f : X \rightarrowtail Y$ and $p \in \mathbf{P}(Y)$, then we define

$$x\,(\mathbf{P}(f)(p))\,x' \iff f(x)\,p\,f(x').$$

Note that injectivity ensures that $(X, \mathbf{P}(f)(p))$ is isomorphic to a sub-structure of (Y, p), and hence the truth of universal sentences is preserved [8].

Also note that $\mathbf{P}$ cuts down to a functor $\mathbf{P}_{\mathsf{inc}} : \mathcal{C}^{\mathsf{op}}_{\mathsf{inc}} \longrightarrow \mathbf{Set}$.

We shall use $\mathbb{P}$ to denote the functor induced by the notion of weak preference relation introduced in the previous section, and $\mathbb{L}$ for the subfunctor of linear preference relations.

3.1 Categorical formulation of UD

The functor $\mathbb{P}^{\mathcal{I}}_{\mathsf{inc}}$ is defined as the product of $\mathcal{I}$ copies of $\mathbb{P}_{\mathsf{inc}}$; thus for each A, $\mathbb{P}^{\mathcal{I}}_{\mathsf{inc}}(A) := \mathbb{P}_{\mathsf{inc}}(A)^{\mathcal{I}}$. This gives the set of all possible profiles over a set of alternatives A for the agents in $\mathcal{I}$. We shall assume we are given a subfunctor $\mathcal{D}$ of $\mathbb{P}^{\mathcal{I}}$. Thus $\mathcal{D} : \mathcal{C}^{\mathsf{op}} \longrightarrow \mathbf{Set}$ is a functor, with a natural transformation $\mathcal{D} \stackrel{\cdot}{\longrightarrow} \mathbb{P}^{\mathcal{I}}$ whose components are inclusion maps. $\mathcal{D}$ restricts to a functor $\mathcal{D}_{\mathsf{inc}} : \mathcal{C}^{\mathsf{op}}_{\mathsf{inc}} \longrightarrow \mathbf{Set}$.

The axiom **UD** can be stated in these terms as follows:

(CUD) (i) For $A \in \mathcal{C}^{(3)}$, $\mathcal{D}_{\mathrm{inc}}(A) = \mathbb{P}^{\mathcal{I}}_{\mathrm{inc}}(A)$.
(ii) $\mathcal{D}_{\mathrm{inc}}$ preserves epis.

The requirement that $\mathcal{D}_{\mathrm{inc}}$ preserves epis means that inclusions $\iota : A \hookrightarrow B$ are mapped to surjections $\mathcal{D}_{\mathrm{inc}}(\iota) : \mathcal{D}_{\mathrm{inc}}(B) \longrightarrow \mathcal{D}_{\mathrm{inc}}(A)$.

3.2 Categorical formulation of IIA

Next, we make the observation that **IIA** is equivalent to the following statement:

(CIIA) The social welfare function is a natural transformation $\sigma : \mathcal{D}_{\mathrm{inc}} \stackrel{\cdot}{\longrightarrow} \mathbb{P}_{\mathrm{inc}}$.

Explicitly, this says that for each set of alternatives A we have a map

$$\sigma_A : \mathcal{D}_{\mathrm{inc}}(A) \longrightarrow \mathbb{P}_{\mathrm{inc}}(A)$$

such that, for all inclusions $A \subseteq B$ and profiles $p \in \mathcal{D}_{\mathrm{inc}}(A)$:

$$\sigma_A(p|A) = \sigma_B(p)|A$$

More precisely, we have the following result.

Proposition 3.1. *We assume that $\mathcal{D}_{\mathrm{inc}}$ is a subfunctor of $\mathbb{P}^{\mathcal{I}}_{\mathrm{inc}}$ satisfying **CUD**.*

1. *If $\sigma_{\mathbb{A}} : \mathcal{D}_{\mathrm{inc}}(\mathbb{A}) \longrightarrow \mathbb{P}_{\mathrm{inc}}(\mathbb{A})$ is a function satisfying **IIA**, then it extends to a natural transformation $\sigma : \mathcal{D}_{\mathrm{inc}} \stackrel{\cdot}{\longrightarrow} \mathbb{P}_{\mathrm{inc}}$.*
2. *If $\sigma : \mathcal{D}_{\mathrm{inc}} \stackrel{\cdot}{\longrightarrow} \mathbb{P}_{\mathrm{inc}}$ is a natural transformation, then for every $A \in \mathcal{C}$, σ_A satisfies **IIA**.*

Proof. 1. Firstly, note that $(\mathbb{A}, \mathcal{I}, \mathcal{D}_{\mathrm{inc}}(\mathbb{A}), \sigma_{\mathbb{A}})$ is a social choice situation in the sense of the previous section. We are assuming that this structure satisfies **IIA**.

We note that **IIA** implies the following, more general statement: for all $A \subseteq \mathbb{A}$, and $p, q \in \mathcal{D}_{\mathrm{inc}}(\mathbb{A})$,

$$p|A = q|A \Rightarrow \sigma_{\mathbb{A}}(p)|A = \sigma_{\mathbb{A}}(q)|A.$$

This holds because any binary relation on a set X is determined by its restrictions to the subsets of X of cardinality ≤ 2.

Given $A \subseteq \mathbb{A}$ and $p \in \mathcal{D}_{\mathsf{inc}}(A)$, by **CUD** there is $q \in \mathcal{D}_{\mathsf{inc}}(\mathbb{A})$ such that $q|A = p$. We define $\sigma_A(p) := \sigma_{\mathbb{A}}(q)|A$. By our previous remark, this is independent of the choice of q. For naturality, if $\iota : A \hookrightarrow B$ and $p \in \mathcal{D}_{\mathsf{inc}}(B)$, then for any $q \in \mathcal{D}_{\mathsf{inc}}(\mathbb{A})$ such that $q|B = p$, $q|A = \mathcal{D}_{\mathsf{inc}}(\iota)(q|B)$, and hence

$$\sigma_A \circ \mathcal{D}_{\mathsf{inc}}(\iota)(p) = \sigma_{\mathbb{A}}(q)|A = (\sigma_{\mathbb{A}}(q)|B)|A = \mathbb{P}_{\mathsf{inc}}(\iota) \circ \sigma_B(p).$$

2. For the converse, if $\sigma : \mathcal{D}_{\mathsf{inc}} \xrightarrow{\cdot} \mathbb{P}_{\mathsf{inc}}$ is a natural transformation, $A \subseteq \mathbb{A}$, and $\iota : \{a, b\} \hookrightarrow A$, then $p, q \in \mathcal{D}_{\mathsf{inc}}(A)$ with $p|\{a, b\} = q|\{a, b\}$ means that $\mathcal{D}_{\mathsf{inc}}(\iota)(p) = \mathcal{D}_{\mathsf{inc}}(\iota)(q)$. Using naturality, we have

$$\begin{aligned}\sigma_A(p)|\{a, b\} = \mathbb{P}_{\mathsf{inc}}(\iota) \circ \sigma_A(p) = \sigma_{\{a,b\}} \circ \mathcal{D}_{\mathsf{inc}}(\iota)(p) &= \sigma_{\{a,b\}} \circ \mathcal{D}_{\mathsf{inc}}(\iota)(q) \\ &= \sigma_A(q)|\{a, b\}.\end{aligned}$$

□

3.3 Categorical formulation of P

Consider the standard diagonal map $\Delta_I : X \longrightarrow X^{\mathcal{J}}$. An arrow $f : X^{\mathcal{J}} \to X$ is *diagonal-preserving* if $f \circ \Delta_I = \mathsf{id}_X$. The Pareto condition is essentially a form of diagonal preservation.

Firstly, note that given a functor $F : \mathcal{C}_{\mathsf{inc}}^{\mathsf{op}} \longrightarrow \mathbf{Set}$, we can define the restriction $F^{(2)} : (\mathcal{C}_{\mathsf{inc}}^{(2)})^{\mathsf{op}} \longrightarrow \mathbf{Set}$. Now Δ_I induces a natural transformation $\mathbb{L}_{\mathsf{inc}}^{(2)} \xrightarrow{\cdot} (\mathbb{P}_{\mathsf{inc}}^{\mathcal{J}})^{(2)}$. Using part (i) of **CUD**, this factors through the inclusion $\mathcal{D}_{\mathsf{inc}}^{(2)} \hookrightarrow (\mathbb{P}_{\mathsf{inc}}^{\mathcal{J}})^{(2)}$. Thus we obtain a natural transformation $\Delta_I : \mathbb{L}_{\mathsf{inc}}^{(2)} \xrightarrow{\cdot} \mathcal{D}_{\mathsf{inc}}^{(2)}$. Also, $\mathbb{L}_{\mathsf{inc}}^{(2)}$ is a sub-functor of $\mathbb{P}_{\mathsf{inc}}^{(2)}$, with inclusion $e : \mathbb{L}_{\mathsf{inc}}^{(2)} \xrightarrow{\cdot} \mathbb{P}_{\mathsf{inc}}^{(2)}$. The categorical formulation of the Pareto condition is now as follows:

(CP) $\sigma \circ \Delta_I = e$.

Diagrammatically, this is

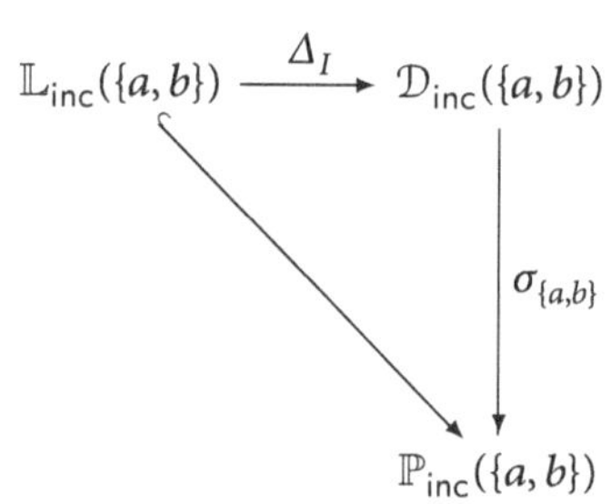

Proposition 3.2. *Let $\mathcal{D}_{\mathsf{inc}}$ be a subfunctor of $\mathbb{P}^{\mathcal{I}}_{\mathsf{inc}}$ satisfying **CUD**, and*

$$\sigma : \mathcal{D}_{\mathsf{inc}} \stackrel{\cdot}{\longrightarrow} \mathbb{P}_{\mathsf{inc}}$$

*a natural transformation. Then σ satisfies **CP** if and only if for every $A \subseteq \mathbb{A}$, $(A, \mathcal{I}, \mathcal{D}_{\mathsf{inc}}(A), \sigma_A)$ satisfies **P**.*

3.4 Categorical Formulation of Arrow's Theorem

A *categorical social choice situation* is given by a set $\mathbb{A}$ determining a category $\mathcal{C}$, a set $\mathcal{I}$ of individuals, a subfunctor $\mathcal{D}_{\mathsf{inc}}$ of $\mathbb{P}^{\mathcal{I}}_{\mathsf{inc}}$ satisfying **CUD**, and a natural transformation $\sigma : \mathcal{D}_{\mathsf{inc}} \stackrel{\cdot}{\longrightarrow} \mathbb{P}_{\mathsf{inc}}$.

Theorem 3.3 (Arrow's Theorem: Categorical Statement). *Let $(\mathbb{A}, \mathcal{I}, \mathcal{D}, \sigma)$ be a categorical social choice situation where $|\mathbb{A}| \geq 3$, $\mathcal{I}$ is finite, and σ satisfies **CP**. Then $\sigma = \pi_i$ for some fixed $i \in \mathcal{I}$, where $(\pi_i)_A : p \mapsto p_i$.*

More colloquially, this can be stated as:

The only diagonal-preserving natural transformations $\sigma : \mathcal{D}_{\mathsf{inc}} \stackrel{\cdot}{\longrightarrow} \mathbb{P}_{\mathsf{inc}}$ are the projections.

3.5 An Analogous Result in Type Theory

We remark that when Arrow's theorem is formulated in this way, it displays an evident kinship with a well-studied genre of results in functional programming and type theory [14, 4]. These results use (di)naturality constraints to show that the behaviour of polymorphic terms are essentially determined by their types.

We illustrate these ideas with an example.

3.5.0.1 Question

What natural transformations

$$t_X : X^2 \stackrel{\cdot}{\longrightarrow} X$$

can there be in the functor category [**Set**, **Set**]?

3.5.0.2 Answer

The only such natural transformations are the projections.

3.5.0.3 Sketch

Use naturality to show this first for two-element sets, then to lift it. E.g.

$$\begin{array}{ccc} \{a,b\}^2 & \xrightarrow{\pi_1} & \{a,b\} \\ \downarrow & & \downarrow \\ X^2 & \xrightarrow[t_X]{} & X \end{array} \qquad\qquad \begin{array}{ccc} (a,b) & \overset{\pi_1}{\longmapsto} & a \\ \downarrow & & \downarrow \\ (x,y) & \underset{t_X}{\longmapsto} & x \end{array}$$

3.5.0.4 Exercise

Show that the same result holds for natural transformations $X^{\mathcal{I}} \stackrel{\cdot}{\longrightarrow} X$.

4 Categorical perspective on the proof of Arrow's Theorem

We shall now revisit the proof of Arrow's theorem from the categorical perspective.

We shall assume throughout this section that we are given a categorical social choice situation $(\mathbb{A}, \mathcal{I}, \mathcal{D}, \sigma)$ satisfying **CP**. Note firstly that, for each $A \subseteq \mathbb{A}$, $(A, \mathcal{I}, \mathcal{D}(A), \sigma_A)$ is a standard social choice situation satisfying **UD**, **IIA** and **P**. We shall use results from Section 2 freely.

4.1 Neutrality as Naturality

The property of Neutrality, which in the concrete setting was stated in a 'local' form in Theorem 2.4, becomes a form of naturality. To state this properly, we need to consider the subfunctor $\mathcal{D}^L$ of $\mathcal{D}$, where $\mathcal{D}^L(A) = \mathcal{D}(A) \cap \mathbb{L}(A)^{\mathcal{I}}$. Thus $\mathcal{D}^L(A)$ is the set of admissible *linear* ballots. By Proposition 2.8, σ cuts down to a natural transformation $\sigma^L : \mathcal{D}^L_{\mathsf{inc}} \stackrel{\cdot}{\longrightarrow} \mathbb{L}_{\mathsf{inc}}$.

The key neutrality property becomes the following:

Proposition 4.1 (Neutrality: categorical version). *The social choice map σ^L extends to a natural transformation $\sigma^L : \mathcal{D}^L \stackrel{\cdot}{\longrightarrow} \mathbb{L}$.*

The assertion of this proposition is that a social welfare function is natural with respect, not just to inclusions, but to injective maps. Note that the family of maps $\{\sigma^L_A\}$ is the same: we are claiming that additional naturality squares commute.

Proof. Let $\alpha : A \rightarrowtail B$ be an injective map. We must show that $f = g$, where

$$f := \sigma^L_A \circ \mathcal{D}^L(\alpha), \qquad g := \mathbb{L}(\alpha) \circ \sigma^L_B.$$

Note that, for $r \in \mathbb{L}(B)$ and $a, a' \in A$,

$$a\,(\mathbb{L}(\alpha)(r))\,a' \iff \alpha(a)\,r\,\alpha(a'),$$

and for $p \in \mathcal{D}^L(B)$,

$$(\mathcal{D}^L(\alpha)(p))_i = \mathbb{L}(\alpha)(p_i).$$

For $p \in \mathcal{D}^L(B)$, and distinct $a, a' \in A$, by Proposition 2.7:

$$\begin{aligned} a\,f(p)\,a' &\iff \{i \in \mathcal{I} \mid a\,(\mathcal{D}^L(\alpha)(p))_i\,a'\} \in \mathcal{U} \\ &\iff \{i \in \mathcal{I} \mid \alpha(a)\;p_i\,\alpha(a')\} \in \mathcal{U} \\ &\iff \alpha(a)\,\sigma^L_B(p)\,\alpha(a') \\ &\iff a\,g(p)\,a'. \end{aligned}$$

□

4.2 The Factorization Theorem

We now recast the Ultrafilter Theorem into an arrow-theoretic form. Firstly, we recall some standard notions from boolean algebra [11].

Proposition 4.2. *Given a set $\mathcal{I}$, there is a bijection by characteristic functions between families $\mathcal{U}$ of subsets of $\mathcal{I}$, and functions $h : \mathbf{2}^{\mathcal{I}} \longrightarrow \mathbf{2}$, where $\mathbf{2} := \{0, 1\}$. A family $\mathcal{U}$ is a ultrafilter if and only if the corresponding function h is a boolean algebra homomorphism. If $\mathcal{I}$ is finite, the only such homomorphisms are the projections.*

We can define maps

$$\phi : \mathbb{L}(A) \longrightarrow \mathbf{2}^{A^2} \;::\; \phi(r)(a, a') = 1 \iff a\,r^{>}\,a'$$

and

$$\psi : \mathcal{D}^L(S) \longrightarrow (\mathbf{2}^{\mathcal{I}})^{A^2} \;::\; \psi(p)(a, a')_i = 1 \iff a\,p_i^{>}\,a'.$$

The following is a restatement in arrow-theoretic terms of the Ultrafilter Theorem.

Theorem 4.3 (Factorization Theorem). *For any social choice function*

$$\sigma : \mathcal{D}^L \overset{\cdot}{\longrightarrow} \mathbb{L},$$

there is a boolean algebra homomorphism

$$h : 2^{\mathcal{I}} \longrightarrow 2$$

such that the following diagram commutes:

$$\begin{array}{ccc} \mathcal{D}^L(A) & \xrightarrow{\sigma_A} & \mathbb{L}(A) \\ \Big\downarrow{\scriptstyle \psi} & & \Big\downarrow{\scriptstyle \phi} \\ (2^{\mathcal{J}})^{A^2} & \xrightarrow[h^{A^2}]{} & 2^{A^2} \end{array}$$

The content of this result is that all the information needed to compute the social welfare function σ is contained in the boolean algebra homomomorphism h.

The categorical form of Arrow's theorem, Theorem 3.3, follows immediately from the Factorization Theorem and the last remark in Proposition 4.2.

5 Discussion

One of our motivations in undertaking this study of Arrow's theorem was to see if common structure could be identified with notions such as no-signalling, parameter independence etc. which play a key rôle in quantum foundations. Arrow's theorem is a no-go theorem of a similar flavour to results such as Bell's theorem. A central assumption is **IIA**, which is analogous to the various forms of independence which appear as hypotheses of the results in quantum foundations. In particular, the functorial treatment we have developed in the present paper has common features with the rôle of presheaves in the sheaf-theoretic account of quantum non-locality and contextuality given in [1].

It must be said that, although some degree of commonality has been exposed by the present account, the arguments are substantially different. Nevertheless, the use of the categorical language to put results from such different settings in a common framework is suggestive, and may prove fruitful in exploring the rôle of various forms of independence. It will also be interesting to relate this to the logics of dependence and independence being developed by Jouko Väänänen and his colleagues.

Altogether, although modest in its scope, we hope the present paper may help to suggest some further possibilities for elucidating the general structure of no-go results, and of the notions of independence which play a pervasive part in these results.

Bibliography

[1] S. Abramsky and A. Brandenburger. The sheaf-theoretic structure of non-locality and contextuality. *New Journal of Physics*, 13(11):113036, 2011.

[2] K.J. Arrow. A difficulty in the concept of social welfare. *The Journal of Political Economy*, 58(4):328–346, 1950.

[3] K.J. Arrow. *Social choice and individual values*. Yale University Press, 1963. Second edition.

[4] E.S. Bainbridge, P.J. Freyd, A. Scedrov, and P.J. Scott. Functorial polymorphism. *Theoretical Computer Science*, 70(1):35–64, 1990.

[5] J.H. Blau. A direct proof of Arrow's theorem. *Econometrica*, 40:61–67, 1972.

[6] A. Gibbard. Manipulation of voting schemes: a general result. *Econometrica: Journal of the Econometric Society*, pages 587–601, 1973.

[7] E. Grädel and J. Väänänen. Dependence, Independence, and Incomplete Information. In *Proceedings of 15th International Conference on Database Theory, ICDT 2012*. ACM, 2012.

[8] W. Hodges. *A shorter model theory*. Cambridge University Press, 1997.

[9] A.P. Kirman and D. Sondermann. Arrow's theorem, many agents, and invisible dictators. *Journal of Economic Theory*, 5(2):267–277, 1972.

[10] M.A. Satterthwaite. Strategy-proofness and Arrow's conditions: Existence and correspondence theorems for voting procedures and social welfare functions. *Journal of economic theory*, 10(2):187–217, 1975.

[11] R. Sikorski. *Boolean algebras*. Springer, 1969.

[12] A.D. Taylor. *Social choice and the mathematics of manipulation*, volume 6. Cambridge University Press, 2005.

[13] J. Väänänen. *Dependence logic: a new approach to independence friendly logic*, volume 70. Cambridge University Press, 2007.

[14] P. Wadler. Theorems for free! In *Proceedings of the fourth international conference on Functional programming languages and computer architecture*, pages 347–359. ACM, 1989.

John T. Baldwin

How Big Should the Monster Model Be?

To Jouko Väänänen on his 60th birthday

Abstract: We compare the role of set theory in formulating notions of 'monster model' or 'universe' in the work of Shelah and Grothendieck.

Keywords: monster model, Grothendieck universe, simple theory, bounded orbit, saturation and omission

MSC: 03C45, 14A99

John T. Baldwin: Department of Mathematics, Statistics, and Computer Science, University University of Illinois at Chicago, e-mail:jbaldwin@uic.edu

The 1970's witnessed two attempts to organize large areas of mathematics by introducing a structure in which all of that mathematics took place. Both of these programs explicitly invoke large cardinal hypotheses; advocates of both programs deny that these hypotheses are necessary. There were a few years ago no extant metatheorems justifying these conclusions (but see McLarty [24]). By a metatheorem, I mean a way to tell by looking at a particular proposition whether the purported use of large cardinals is necessary. The natural expectation is that these criteria would be syntactic. But the possibility of geometric criteria arise in [21] and the general model theoretic development suggests other 'contentual criteria' (Section 4.) We have been unable to develop any reasonable such theorem in the model theoretic case. This paper just explores the territory.

As we will see, in both cases, the role of replacement as an intermediary between Zermelo Set Theory and large cardinals arises in the analysis. This work was stimulated by some questions raised by Newelski in [27] concerning the 'absoluteness' of the existence of a bounded orbit and the discussions at about the same time concerning the possible use of large cardinals in Wiles' proof of Fermat's last theorem.

On the one hand Grothendieck wanted to provide a framework in which to organize large areas of algebraic geometry. For this he invented the notion of a universe, a large enough set to be closed under the usual algebraic operations. He explicitly developed cohomology theory using the existence of (a proper class of) universes[1].

At about the same time, Shelah introduced his version of a 'universal domain', later dubbed a *monster model*: $\mathfrak{C}$ is a saturated model of T with cardinality $\overline{\kappa}$ a strongly inaccessible cardinal. He writes in [30], 'The assumption on $\overline{\kappa}$ does not, in fact, add any axiom of set theory as a hypothesis to our theorems.' This is a rather unclearly stated metatheorem. 'Which theorems?' An acceptable answer in the short run is, 'the theorems in this book'. And, apparently, the statement is true in that sense. We seek here to clarify what theorems are intended and which are not.

We give a brief comparison of the algebraic geometry side (in Section 1.2) relying heavily on [23, 24, 21] and the model theoretic issues (in Subsection1.3). Section 2 provides the connection between monster models and Hanf numbers. In Section 3 we examine the model theoretic question more carefully in three examples and pose some specific problems. We conclude with speculations about the ways one might sort out the properties which do not depend on a long universe in Section 4

In addition to correspondence with McLarty, we want to thank Maryanthe Mariallis and Enrique Casanovas for helpful discussions.

1 The two programs

1.1 Set Theoretic Background

We discuss four different set theories. The standard background theory for model theory is ZFC. While there has been some work lately on the necessary uses of set theory in model theory [33, 15, 3] they have focused on the role of the axiom of choice[2] or on principles near the GCH[3] and mostly on infinitary model theory. Our subject here is first order model theory and the role of axioms having to do with the length of the universe.

---

1 The issue we address here of the size of the universe is distinct from the organization via sheaves addressed in [20].

2 See Shelah's proof in [33] that for countable theories Morley's categoricity theorem is provable in ZF.

3 In particular many results in infinitary logic apparently require $2^{\kappa} < 2^{\kappa^+}$, at least for $\kappa < \aleph_\omega$. See [2] for background and more references.

We consider in Subsection 3.1 whether even relatively small cardinals such as $\beth_{\omega_1}$ are necessary in model theory. In the section on foundations of cohomology we discuss ZC (Zermelo Set Theory), i.e. ZFC without either replacement or foundation, but with the full axiom of separation. Maclane set theory (i.e. bounded Zermelo or MC in McLarty) weakens ZC still more by postulating only separation for Δ_0-formulas (bounded quantification).

In the other direction, we consider whether certain uses of inaccessible cardinals are as transparently unnecessary as is usually claimed by model theorists. For this it is useful to extend from ZFC to NBG, (Von Neumann, Bernays, Gödel) set theory. This is a *conservative* extension of ZFC, which distinguishes between classes and sets. See [23] for a more detailed account.

1.2 Grothendieck Universes and Wiles proof

In formulating the general theory of cohomology Grothendieck developed the concept of a universe – a collection of sets large enough to be closed under any operation that arose. Grothendieck proved that the existence of a single universe is equivalent over ZFC to the existence of a strongly inaccessible cardinal [16, Vol. I, p. 196]. More precisely, U is the set V_α of all sets with rank below α for some uncountable strongly inaccessible cardinal.

McLarty summarised the general situation in [23]:

> Large cardinals as such were neither interesting nor problematic to Grothendieck and this paper shares his view. For him they were merely legitimate means to something else. He wanted to organize explicit calculational arithmetic into a geometric conceptual order. He found ways to do this in *cohomology* and used them to produce calculations which had eluded a decade of top mathematicians pursuing the Weil conjectures [29]. He thereby produced the basis of most current algebraic geometry and not only the parts bearing on arithmetic. His cohomology rests on universes but weaker foundations also suffice at the loss of some of the desired conceptual order.

As written the great applications of cohomology theory (e.g. Wiles and Faltings) implicitly rely on universes. Most number theorists regard the applications as requiring much less than their ‘on their face’ strength and in particular believe the large cardinal appeals are ‘easily eliminable’. An animated discussion on Math Overflow [37] and earlier on FOM[4] emphasizes this belief. This discussion is perhaps best summed up by the amusing remark of Pete Clark, ‘If a mathematician

4 Search ‘universes’ or Wiles on the Fom archive.

of the caliber of Y.I. Manin made a point of asking in public whether the proof of the Weil conjectures depends in some essential way on inaccessible cardinals, is this not a sign that "Of course not; don't be stupid" may not be the most helpful reply?'
There are in fact two issues. McLarty[23] writes: 'Wiles's proof uses hard arithmetic some of which is on its face one or two orders above PA, and it uses functorial organizing tools some of which are on their face stronger than ZFC.'

There are two current programs for verifying in detail the intuition that the formal requirements for Wiles proof of Fermat's last theorem can be substantially reduced. On the one hand, McLarty's current work ([24, 25]) aims to reduce the 'on their face' strength of the results in cohomology from large cardinal hypotheses to finite order Peano. On the other hand Macintyre aims to reduce the 'on their face' strength of results in hard arithmetic[21] to Peano. These programs may be complementary or a full implementation of Macintyre's might avoid the first.

McLarty ([24]) reduces

1. ' all of SGA[5]' to Bounded Zermelo plus a Universe.
2. "'the currently existing applications" to Bounded Zermelo itself, thus the consistency strength of simple type theory.'

The Grothendieck duality theorem and others like it become theorem schema.

The essential insight of the McLarty's papers on cohomology is the role of replacement in giving strength to the universe hypothesis. In [24], a *ZC-universe* is defined to be a transitive set U modeling *ZC* such that every subset of an element of U is itself an element of U. He remarks that any V_α for α a limit ordinal is provable in *ZFC* to be a *ZC*-universe. McLarty then asserts the essential use of replacement in the original Grothendieck formulation is to prove: For an arbitrary ring R every module over R embeds in an injective R-module and thus injective resolutions exist for all R-modules. But he gives a proof in a system with the proof theoretic strength of finite order arithmetic[6] that every sheaf of modules on any small site has an infinite resolution. (Section 3.6 of [24].)

Macintyre [21] dismisses with little comment the worries about the use of 'large-structure' tools in Wiles proof. He begins his appendix [21], 'At present, all roads to a proof of Fermat's Last Theorem (henceforward FLT) pass through some version of a Modularity Theorem (generically MT) about elliptic curves defined over Q... A casual look at the literature may suggest that in the formulation of MT

5 SGA refers to a sequence of Grothendiecks works.

6 In [25], McLarty reduces the strength to second order arithmetic for certain cohomology theories.

(or in some of the arguments proving whatever version of MT is required) there is essential appeal to higher-order quantification, over one of the following'. He then lists such objects as $\mathbb{C}$, modular forms, Galois representations ...and summarises that a superficial formulation of MT would be Π^1_m for some small m. But he continues, 'I hope nevertheless that the present account will convince all except professional sceptics that MT is really Π^0_1.' There then follows a 13 page highly technical sketch of an argument for the proposition that MT can be expressed by a sentence in Π^0_1 along with a less-detailed strategy for proving MT in PA.

Macintyre's complexity analysis is in traditional proof theoretic terms. But his remark that 'genus' is more a useful geometric classification of curves than the syntactic notion of degree suggests that other criteria may be relevant.

From the standpoint of this paper the McLarty's approach is not really a metatheorem but a statement that there was only one essential use of replacement and it can be eliminated. In contrast, Macintyre argues that 'apparent second order quantification' can be replaced by first order quantification. But the argument requires deep understanding of the number theory for each replacement in a large number of situations. Again, there is no general theorem that this type of result is provable in PA. A battery of techniques is displayed for translating the statements to Π^0_1 and reducing the proof theoretic strength of the axioms.

1.3 Monster models in Model theory

Many model theory papers begin 'We work in a big saturated model' or slightly more formally, 'We are working in a saturated model of cardinality κ for sufficiently large κ (a monster model).'

What does *sufficiently* mean? In every case I know such a declaration is not intended to convey a reliance on the existence of large cardinals. Rather, in Marker's phrase, it is a declaration of laziness, 'If the stakes were high enough I could write down a ZFC proof'. As we note below, in standard cases the author isn't being very lazy; but the route to formalizing a metatheorem expressing this intuition does not seem clear.

This was not a problem for the early history of classification theory. Work focussed on stable theories. And a stable theory has a saturated model in every λ with $\lambda^{\kappa(T)} = \lambda$, where $\kappa(T)$ is an invariant that is less than $|T|^+$. Thus, there is a plentiful supply of monster models. But recently model theory has moved to the investigation of unstable theories and these issues become more acute, as we discuss in Subsection 3.2. We will see that the difficulty is not just the lack of saturated models but lack of the control of structure provided by stability theory.

The fundamental unit of study is a particular first order theory. The need is for a monster model of the theory T. If M is a κ-saturated model of T, then every model N of T with cardinality at most κ is elementarily embedded in M and every type over a set of size $< \kappa$ is realized in M. So every configuration of size less than κ that could occur in any model of T occurs in M. Then general theorems are asserted to hold for each theory.

In fact, the requirement that the monster model be saturated in its own cardinality is excessive. A more refined version of the 'monster model hypothesis' asserts: Any first order model theoretic properties of sets of size less than kappa can be proved in a κ-saturated strongly κ-homogenous model M (any two isomorphic submodels of card less than κ are conjugate by an automorphism of M). Such a model exists (provably in ZFC) in some κ' not too much bigger than κ. See Hodges [18] or my monograph on categoricity [2] for the refined version. (Hodges' big model condition is ostensibly stronger and slightly more complicated to state; but existence is also provable in ZFC.) Buechler [7], Shelah [30], and Marker [22] expound the harmless nature of the fully saturated version. Ziegler [38] adopts a class approach that could be formulated in Gödel Bernays set theory. And we will follow that approach below.

In order to clarify the problem, we will address several specific problems where some issues arise in calculating the size of the necessary 'monster'.

2 Hanf Numbers and Monster models

In this section we expand a bit on the arguments for the eliminability of large cardinal hypotheses in uses of the monster model. Then we connect the properties of a class monster model with the calculability of certain Hanf functions.

Buechler[7] argues that the apparent reliance can be removed by a sequence of applications of the same proof. To prove model theoretic statements about structures of size at most κ, use a κ-monster. If κ increases, choose a larger monster. Note that the size of the monster was not used in the argument.

So, for example, to compute the spectrum function of a first order theory via the strategy of classification theory, theories are divided into categories by properties (stability class, DOP, OTOP, depth) which have no dependence on the size of the model. Then for each class P a function f_P is defined such that for $\kappa < \rho$ which is the size of a given choice of monster model $f_P(\kappa)$ is (or is at least an upper bound

7 I paraphrase an argument that Buechler says holds 'with few exceptions' on page 70 of [7].

for) the number of models in κ. This function works for all κ by just redoing the argument for a larger ρ as κ grows.

We work in Von Neumann, Bernays, Gödel set theory NBG, a *conservative* extension of ZFC, which admits classes as objects.

Definition 2.1. A *monster model* is a class model $\mathbb{M}$ which is a union of κ-saturated models for arbitrarily large κ.

This definition (from [38]) is quite different from the usual usage in model theory. We connect it with more standard usage by defining the notion of a κ-monster which formalizes monster set models as certain kinds of special models [18, 9, 34].

Definition 2.2. 1. A structure M of infinite cardinality κ is *special* if M is the union of an elementary chain $\langle M_\lambda : \lambda < \kappa, \lambda \text{ a cardinal} \rangle$, where each M_λ is λ^+-saturated.
2. A structure M is strongly κ-homogeneous if for every A contained in M with $|A| < \kappa$, every elementary embedding of A into M can be extended to an automorphism of M.
3. A κ-*monster* model $\mathbb{C}_\kappa$ is a special model of cardinality $\mu = \beth_{\kappa^+}(\kappa)$.

Fact 2.3. *A κ-monster is unique up to isomorphism, μ^+-universal and strongly κ^+-homogenous.*

Now the natural conjecture is:

Conjecture 2.4. For any property P, the class monster $\mathbb{M}$ satisfies P if and only if all sufficiently large κ-monsters $\mathbb{C}_\kappa$ satisfy P.

The main problem is to specify what is meant by a property. A too generous definition is ‘a class in NBG’. But the issue is to refine this notion. And all we actually give here are some specific examples that should be considered in making a definition.

This conjecture would follow if ‘all sufficiently large κ-monsters $\mathbb{C}_\kappa$ satisfy P.’ were replaced by a ‘uniform proof’ that ‘for all sufficiently large κ-monsters that $\mathbb{C}_\kappa$ satisfies P.’ This is the strategy that works successfully for the spectrum problem. But I don’t see how to get this claim in general; we examine a specific problem where a uniform argument is not apparent in Subsection 3.2.

Finding such a uniform argument seems related to Hanf numbers. Hanf [17] introduced the following extremely general and soft argument. $P(\boldsymbol{K}, \lambda)$ ranges over such properties as: $\boldsymbol{K}$ has a model in cardinality λ, $\boldsymbol{K}$ is categorical in λ, or the type q is omitted in some model of $\boldsymbol{K}$ of cardinality λ. We will see some more novel examples below.

Theorem 2.5 (Hanf). *Fix a set of classes* $\mathbf{K}$ *of a given kind (e.g. the classes of models defined by sentences of* $L_{\mu,\nu}$ *for some fixed* μ, ν *of a given similarity type). For any property* $P(\mathbf{K}, \lambda)$ *there is a cardinal* κ *such that if* $P(\mathbf{K}, \lambda)$ *holds for some* $\lambda > \kappa$ *then* $P(\mathbf{K}, \lambda)$ *holds for arbitrarily large* λ.

Proof. Let

$$\mu_{\mathbf{K}} = \sup\{\lambda : P(\mathbf{K}, \lambda) \text{ holds }\}$$

where $\mu_{\mathbf{K}} = \infty$ if there is no bound on the cardinality of models of $\mathbf{K}$ satisfying P. Then

$$\kappa = \sup\{\mu_{\mathbf{K}} : \mu_{\mathbf{K}} < \infty\}.$$

$\square_{2.5}$

Definition 2.6. P is *downward closed* if there is a κ_0 such that if $P(\mathbf{K}, \lambda)$ holds with $\lambda > \kappa_0$, then $P(\mathbf{K}, \mu)$ holds if $\kappa_0 < \mu \leq \lambda$.

The following is obvious.

Theorem 2.7. *If a property* P *is downward closed then for any* κ *there is a cardinal* μ *such that for any class of models* $\mathbf{K}$ *with vocabulary*[8] *of size* κ, *if some model in* $\mathbf{K}$ *with property* P *has cardinality greater than* μ, *then there is a model in* $\mathbf{K}$ *with property* P *in all cardinals greater than* μ.

That is, if each of a collection of classes is downward closed for a property P there is a Hanf Number for P in the following stronger sense.

Definition 2.8 (Hanf Numbers). The *Hanf number* for P, among classes $\mathbf{K}$ with vocabulary of cardinal κ, is μ if: if there is a model in $\mathbf{K}$ with cardinality $> \mu$ that has property P, then there is a model with property P in all cardinals greater than μ. In this situation μ is the Hanf number of P (for classes with vocabulary of cardinal κ).

Definition 2.9.
1. A function f (a class-function from cardinals to cardinals) is *strongly calculable* if f can (provably in ZFC) be defined in terms of cardinal addition, multiplication, exponentiation, and iteration of the $\beth$ function.
2. A function f is *calculable* if it is (provably in ZFC) eventually dominated by a strongly calculable function. If not, it is *incalculable*.

Here are several examples of properties of first order T where Hanf numbers may or may not be calculable.
1. T has a model in κ.

8 This gets a bit more technical; see page 32 of [2].

2. T has a saturated (or special) model in κ.
3. T has a model that is a group with a bounded orbit in the sense of Subsection 3.2.
4. The Hanf number for omission and saturation (Subsection 3.3).

For 1), the Hanf number is the cardinality of the vocabulary, so it is calculable. For 2) the Hanf number is the first stability cardinal for stable theories and again this is calculable. But for unstable theories there is considerable not yet determinative research on the existence of saturated models so the Hanf number has not been calculated. See, e.g., [31, 13]. Note that 2) is not downward closed. We explore some cases of 3) where the Hanf number is calculated and some where it remains an open question in Subsection 3.2. And we note in Subsection 3.3 that the Hanf number for case 4) is incalculable in general but it is for superstable T.

If Conjecture 2.4 holds, the natural size for a monster model for studying a property P is the Hanf number of P. Unfortunately, as our discussion in Subsection 3.2 shows, the equivalence in Conjecture 2.4 is not obvious.

Most examples in the literature of Hanf numbers are variants on the Morley's omitting types theorem and the Hanf number[9] is $\beth_{\omega_1}$. There are more complicated examples in [32, 2].

3 Three Examples

3.1 Is replacement needed?

One of the fundamental tools of model theory constructs indiscernibles realizing types from a prescribed set.

Theorem 3.1. *Let M be a big saturated model. For every* large enough *set $I \subset M$, there exists an infinite sequence of order indiscernibles $J \subset M$ such that for every finite* $\mathbf{b} \in J$ *there is an* $\boldsymbol{a} \in I$ *with* $\mathrm{tp}(\mathbf{b}/\emptyset) = \mathrm{tp}(\boldsymbol{a}/\emptyset)$.

The crux here is the requirement that the complete types of the sequences in J are types realized in I. With no requirement on the types appearing in J, only Ramsey's theorem and compactness is needed in the standard Ehrenfeucht-Mostowski proof.

9 This is for countable vocabularies; for a vocabulary with cardinality κ, the relevant Hanf number $\beth_{(2^\kappa)^+}$.

We can guarantee this result only if $|M| \geq \beth_{\omega_1}$. This example makes the question, 'What set theory is used in model theory?' a little sharper. Friedman proved [14] that Borel determinacy required the existence of $\beth_{\omega_1}$. Are there such examples of necessary uses of replacement in first order model theory? Morley [26] showed both that $\beth_{\omega_1}$ sufficed for the cardinality of I and that it was necessary. But this necessity argument itself uses replacement.

In some sense Theorem 3.1 and Hanf numbers for omitting types require the existence of $\beth_{\omega_1}$ even to be stated. Those notions are about size or about 'logics'. But here is a theorem clearly stated in ZC, but for which known proofs use replacement. Byunghan Kim[19] proved:

Theorem 3.2 (Kim). *For a simple first order theory non-forking is equivalent to non-dividing.*

The usual easily applicable descriptions of simple theories involve uncountable objects. But definitions of simple, non-forking, and non-dividing are equivalent in ZC to statements about countable sets of formulas. Indeed we quote below such formulations which were given as the definitions in Casanovas' recent exposition [8]. Nevertheless, the argument for Kim's theorem employs Morley's technique for omitting types; that is: The standard argument uses the Erdos-Rado theorem on cardinals less than $\beth_{\omega_1}$.

Our goal here is simply to state this proposition clearly enough to show that it is properly formulated without any use of replacement. For this, we simply repeat the basic definitions from [8] where the exact result we are after is given a short complete proof. We work in a complete first order theory in a countable vocabulary.

Definition 3.3. Let $\mathbf{a}_i$ be a sequence of finite tuples in a model of a first order theory T. A set of formulas $X = \{\phi(\mathbf{x}, \mathbf{a}_i) : i < \omega\}$ is k-inconsistent if every k element subset of X is inconsistent.

With this notion in hand we can define forking and dividing.

Definition 3.4. Let $A \cup \{\mathbf{a}\} \cup \{\mathbf{a}_j : j < n\}$ be a subset of a model of T.

1. The formula $\phi(\mathbf{x}, \mathbf{a})$ k-divides over A if there is an infinite set $I = \{\mathbf{a}_i : i < \omega\}$ such that $\{\phi(\mathbf{x}, \mathbf{a}_i) : i < \omega\}$ is k-inconsistent and all the $\mathbf{a}_i$ realize $\mathrm{tp}(\mathbf{a}/A)$. ϕ *divides* if it k-divides for some k.
2. The formula $\pi(\mathbf{x}, \mathbf{a})$ forks over A if for some finite set of formulas $\psi_j(\mathbf{x}, \mathbf{a}_j)$ with $j < n$, $\pi(\mathbf{x}, \mathbf{a}) \vdash \bigvee_{j<n} \psi_j(\mathbf{x}, \mathbf{a}_j)$ and each $\psi_j(\mathbf{x}, \mathbf{a}_j)$ divides over A.

Definition 3.5. The formula $\phi(\mathbf{x}, \mathbf{y})$ has the tree property with respect to $k < \omega$ if there is a tree $(\mathbf{a}_s : s \in \omega^{<\omega})$ (in some model of T) such that for all $\eta \in \omega^{\omega}$, the branch $\{\phi(\mathbf{x}, \boldsymbol{a}_{\eta\restriction n}) : n < \omega\}$ is consistent and for all $s \in \omega^{<\omega}$, the family of siblings $\{\phi(\mathbf{x}, \boldsymbol{a}_{s^\frown i}) : i < \omega\}$ is k-inconsistent.

Definition 3.6. T is *simple* if there is no formula $\phi(\mathbf{x}, \mathbf{y})$ which has the tree property in T.

There is a direct proof of the following result in [8].

Theorem 3.7. *Let T be a simple theory. A partial type $\pi(\mathbf{x}, \boldsymbol{a})$ divides over A if and only it forks over A.*

None of the arguments given in [8] directly invoke replacement. But Lemma 1.1 of that paper, which is applied at a crucial point, is a variant of the standard:

Theorem 3.8. *[Morley omitting types theorem] Let T be a τ-theory, Γ a set of partial τ-types (in finitely many variables) over $\emptyset$ and $\mu = (2^{|\tau|})^+$, Suppose M_α for $\alpha < \mu$ are a sequence of τ-structures, each omitting all types in Γ and satisfying T such that $|M_\alpha| > \beth_\alpha$.*

Then, in a model M of T omitting all types in Γ, there is a countable sequence I of order indiscernibles such that for every finite sequence $\boldsymbol{a} \in I$, $\mathrm{tp}(\boldsymbol{a}/\emptyset)$ is realized in each M_α.

The crucial message of Morley's theorem as opposed to the standard Ehrenfeucht Mostowski argument which takes place in ZC is the requirement that the finite types that will be realized in J are specified in advance to come from the M_α. I know of no proof of Theorem 3.7 that does not rely on replacement to invoke Erdos-Rado via Morley's omitting types theorem.

Question 3.9. Is Theorem 3.7 provable in ZC?

Remark 3.10. As we'll see later in other examples, restricting to stable theories often reduces the set theoretic strength needed for a result. In particular, Shelah proved the equivalence of forking and dividing for stable theories without any reliance on replacement. The strength of stationarity in stable theories allows the construction of the required indiscernibles using only Ramsey's theorem (as in Ehrenfreucht-Mostowski). See Section V.3, in particular Theorem V.3.9 of [4].

Recently, Chernikov and Kaplan [10] proved a slight weakening of the theorem for a much broader class of theories. They prove that for types over models[10] in

---

10 Examples of unstable theories where types over sets can fork without dividing are well known.

NTP2 theories; these[11] include all simple and all NIP theories. But they still rely on 'Kim's Lemma' which uses replacement.

Building on work of Adler [1], Vasey [35] has proved Theorem 3.7 is provable in ZC.

3.2 Bounded orbits of types

Newelski in [27] raised questions about how to calculate whether the action of a monster model of a first order theory of groups on its Stone space admitted a bounded orbit. Also in [27], he reduced the problem to the calculation of the Hanf number for saturation and omission, which makes general model theoretic sense. In Section 3.3, we recall that this Hanf number is not calculable. In this section, we work only with the Hanf number for bounded orbits in an attempt to see if it might be calculable (for arbitrary theories of groups).

Assumption 3.11. T is a first order theory extending the theory of groups with countable vocabulary τ. Each model of T acts on itself and therefore on its Stone space by left translation. For $p \in S(M)$, $p \mapsto gp = \{\phi(g^{-1}x, \mathbf{a}) : \phi(x, \mathbf{a}) \in p\}$.

Notation 3.12. Consider $M \prec M'$, which are groups. Let $p \in S(M')$, and let the M'-orbits of p be represented by p_α ($\alpha < \mu$). Consider the restrictions $q_\alpha = p_\alpha \restriction M$ and let $P1, P2$ denote the following conditions.
P1: Each q_α extends uniquely to some type over M' in the orbit of p.
P2: Every p_α restricts to some q_α.

The following definition is adapted from [27].

Definition 3.13. Let $\mathbb{C}$ be a κ-monster, $M \prec \mathbb{C}$, $q \in S(M)$, $\mathcal{O} = Mq$. That is, $\mathcal{O}$ is the orbit of q under left translation by M.

1. $\mathcal{O}$ is $\mathbb{C}$-*bounded if there is* $p \in S(\mathbb{C})$ *with P1 and P2 and* $q = p \restriction M$.
2. $\mathcal{O}$ *is* ∞-bounded *if it is* $\mathbb{C}'$*-bounded for all* κ*-monsters* $\mathbb{C}' \succ M$ *(for any* $\kappa > |M|$*).*
3. *T has an ∞-bounded orbit if for some $M \models T$, there is some ∞-bounded orbit in $S(M)$.*
4. *For a fixed countable theory T, let μ_T be the least cardinality of an ∞-bounded orbit (for some κ-monster modeling T) if such a model exists.*
5. *Let* $\mu_{\mathrm{bd}} = \sup\{\mu_T : T$ has a model with an ∞-bounded orbit $\}$.

11 T has TP2 if for some formula $\phi(\mathbf{x}, \mathbf{y})$ and some $k < \omega$ there is an array $\{a_i^j : i, j < \omega\}$ such that each row $\phi(x, \mathbf{a}_i^j)$ is k-inconsistent and for every $\eta : \omega \to \omega$, each type $\phi(\mathbf{x}, \mathbf{a}_i^\eta(i))$ is consistent.

Now Newelski[27] asked[12].

> If we consider the existence of a bounded orbit in $S(\mathcal{C})$ as a property of T, then this property should not depend on the choice of $\mathcal{C}_{\overline{\kappa}}$ provided $\overline{\kappa}$ is large enough. In particular, if a type $p \in S(\mathcal{C})$ has bounded orbit and $\mathcal{C}' = \mathcal{C}'_{\kappa} \succ \mathcal{C}_{\kappa'}$ is a larger monster model (for some $\kappa' > \kappa$, then there should exist a type $p' \in S(\mathcal{C}')$ extending p with the $G^{\mathcal{C}'}$ orbit of the same size as that of the $G^{\mathcal{C}'}$ -orbit of p. It is unclear to me, how large $\overline{\kappa}$ should be for $\mathcal{C}$ to have this property.

In this section we give several other formulations of this question.

Question 3.14. 1. Find a bound on μ_{bd} in terms of $\beth$-numbers.
2. Find an example where there is an ∞-bounded orbit of size $> 2^{2^{\aleph_0}}$ in a countable theory.
3. Is there a bound on the size of an ∞-bounded orbit in a countable theory of a group.

By Remark 1.12 of [27] the Hanf number for μ_{bd} is no more than $H(N_\lambda)$, (i.e. $H(P^\lambda_N)$) as discussed in Section 3.3. But we showed in [6] that this number is incalculable. However it is only an upper bound for the Hanf number for μ_{bd}.

We can't answer any of these questions; we try to place them in a broader context and address related questions. To begin with, instead of trying to bound the size of the bounded orbit, one could bound the size of models which have a bounded orbit. If there was no such bound, one would like to think that the monster model had a bounded orbit in the sense of Definition 3.17.

Pillay and Conversano [11, 12] for o-minimal and even NIP theories and Newelski [28] for more general classes defined in terms of the action of groups have calculated the Hanf number. But our question here is:

Question 3.15. Is there a calculation of the Hanf number for bounded orbits that works for every first order theory of groups? Or are some kind of stability conditions necessary to get a bound.

Note that the key aspect of ∞-bounded is the property of extending a κ-monster to a κ'-monster preserving the size of a small orbit. It was implicit in [27] that such an extension theorem holds in stable theories. For context, we include a proof.

Fact 3.16. *If T is stable and $\mathbb{C}$ is a κ-monster of T with a type $p \in S(\mathbb{C})$ with bounded orbit of cardinality $\lambda < \mu$ (with μ as in Definition 2.2.3), then in any κ'-monster $\mathbb{C}' \succ \mathbb{C}$, there is a type with orbit of cardinality λ.*

12 Newelski's κ-monsters were not unique, complicating the issue even further.

Proof. Let $p' \in S(\mathbb{C}')$ be a nonforking extension of $p = p_0 \in S(\mathbb{C})$. Fix $M_0 \prec \mathbb{C}$ that is $\kappa(T)^+$-saturated and $\langle q_\alpha \in S(M_0) : \alpha < \lambda\rangle$ which are the conjugates of $p \restriction M_0$. Each q_α is a nonforking extension of a stationary type over some $X_\alpha \subset M_0$ with $|X_\alpha| < \kappa(T)$. If every $\mathbb{C}'$-conjugate of p' is a nonforking extension of some q_α, we are finished. Now suppose some $m \in \mathbb{C}'$ conjugates p' so that $mp' \restriction M_0 \neq q_\alpha$ for all $\alpha < \mu$. Then mX_0 is not contained in M_0. But there exists a subset Y of M_0 with $|Y| = \kappa(T)$ such that $r = \mathrm{tp}(mX_0/M_0)$ is based on Y. Now by $|M_0|$-saturation of $\mathbb{C}$, construct in $\mathbb{C}$ a nonforking sequence $m_\beta Y_\beta$ for $\beta < |M_0|$ of realizations of r. Each Y_β is the base of a conjugate of $p' \restriction M_0$ contrary to hypothesis. $\square_{3.16}$

Newelski's formulation focuses on whether a particular orbit remains bounded as the ambient monster changes. To connect with the true monster we consider a variant– which monsters have bounded (i.e. small) orbits?

Definition 3.17 (A model has a Bounded orbit). 1. Let $\mathbb{M}$ be the monster model of T. We say $\mathbb{M}$ has a *bounded orbit* if there is a $p \in S(\mathbb{M})$ such that $\mathbb{M}p = \{ap : a \in \mathbb{M}\}$ is a set.
2. Let $\mathbb{C} = \mathbb{C}_\kappa$ be a κ-monster model of T. We say $\mathbb{C}_\kappa$ has a *bounded orbit* if there is a $p \in S(\mathbb{C})$ such that, with $\mathbb{C}p$ denoting $\{ap : a \in \mathbb{C}\}$, $|\mathbb{C}p| < |\mathbb{C}|$.

Question 3.18. When does $\mathbb{M}$ have a bounded orbit? That is, can we define a cardinal κ and a *property of set models* such that $\mathbb{M}$ has a bounded orbit if and only if the set monster $\mathbb{C}_\kappa$ has the property.

This is the formulation of Conjecture 2.4 in this context. We will see it is problematic.

Notation 3.19. We will write $\hat{M}$ for a model which may be either a κ-monster $\mathbb{C}_\kappa$ or the true monster $\mathbb{M}$. The saturation hypothesis may not always be used.

For the next few paragraphs we analyze the relation between the orbits of a type and its restrictions. Note that for any *set or class* model M and $p \in S(M)$, the cardinality of the orbit of p is the index of $\mathrm{stb}_M(p)$ in M. (The *stabilizer of* p, $\mathrm{stb}_M(p)$, is the subgroup of $a \in M$ such that $ap = p$.)

Claim 3.20. *Let $M \prec N \prec \hat{M}$ and $\hat{p} \in S(\hat{M})$.*
1. *If $a \in \mathrm{stb}_{\hat{M}}(\hat{p})$ and $a \in M$ then $a \in \mathrm{stb}_M(\hat{p} \restriction N)$. In particular, for any $M \prec \hat{M}$, $(\mathrm{stb}_{\hat{M}}(\hat{p}) \cap M) \subseteq \mathrm{stb}_M(\hat{p} \restriction M)$.*
2. *So if M contains representatives of all cosets of* $\mathrm{stb}_{\hat{M}}(\hat{p})$,

$$|\hat{M}/\mathrm{stb}_{\hat{M}}(\hat{p}))| = |M/\mathrm{stb}_{\hat{M}}(\hat{p}) \cap M| \geq |M/\mathrm{stb}_M(\hat{p} \restriction M)|.$$

Proof. For any $\phi(x, \mathbf{c}) \in \hat{p} \restriction N$, $\phi(x, a^{-1}\mathbf{c}) \in \mathrm{stb}_{\hat{M}}(\hat{p})$ and $a^{-1}\mathbf{c} \in N$ so $\phi(x, a^{-1}\mathbf{c}) \in \mathrm{stb}_N(\hat{p} \restriction N)$. For the equality in the second assertion, note that for any $b \in M$ there is an a_i in the set of representatives such that $b^{-1}a_i \in \mathrm{stb}_{\hat{M}}(\hat{b})$ but also in M. $\square_{3.20}$

We establish downward monotonicity for having a bounded orbit.

Claim 3.21. *Let $M \prec N \prec \hat{M}$ and $\hat{p} \in \hat{M}$. If $\hat{M}$ has a bounded orbit of size μ then for every κ with $\mu < \kappa < |\hat{M}|$ and every model M of size κ, there is a model M' of size κ with $M \prec M'$ such that M' has an orbit of size $\leq \mu$.*

Proof. Fix a model M of size κ. Let $\hat{p} \in S(\hat{M})$ have a bounded orbit of cardinality μ, i.e. $\mathrm{stb}_{\hat{M}}(\hat{p})$ has μ cosets with representatives $\langle a_i : i < \mu\rangle$. Let N be a submodel containing the a_i. Choose M' as an elementary extension of N and M of cardinality κ. By Claim 3.20, the orbit of $\hat{p}\restriction M'$ has size at most μ. $\square_{3.21}$

We show every model is contained in a model of the same cardinality where the containment in Claim 3.20.1 becomes equality.

Claim 3.22. *Fix $\hat{p} \in S(\hat{M})$. For any M, there is an M' with $M \prec M'$ and $|M| = |M'|$ such that*

$$\mathrm{stb}_{\hat{M}}(\hat{p}) \cap M' = \mathrm{stb}_{M'}(\hat{p}\restriction M').$$

Proof. By Claim 3.20 $\mathrm{stb}_{\hat{M}}(\hat{p}) \cap M' \subseteq \mathrm{stb}_M(\hat{p}\restriction M)$. Now let $M_1 = M^*$ be a model of size $|M|$ such that for every $a \in \mathrm{stb}_M(\hat{p}\restriction M) - \mathrm{stb}_{\hat{M}}(\hat{p}) \cap M$, there is $\mathbf{c} \in M^*$ with $\phi(x, \mathbf{c}) \in \hat{p}$, $\phi(a^{-1}x, \mathbf{c}) \notin \hat{p}$. Thus, every such a is not in $\mathrm{stb}_{M^*}(\hat{p}\restriction M^*)$. Now if $M_{i+1} = M_i^*$, then $M' = \bigcup_{i<\omega} M_i$ satisfies the desired condition, as by Claim 3.20 $\mathrm{stb}_{\hat{M}}(\hat{p}) \cap M' \subseteq \mathrm{stb}_{M'}(\hat{p}\restriction M')$. $\square_{3.22}$

With these technical lemmas we can relate the size of the orbit a fixed type over the (class or set) monster to the size of the orbits of its restrictions.

Lemma 3.23. *Fix a type $\hat{p} \in S(\mathbb{M})$. The following are equivalent.*
1. *The orbit of $\hat{p}$ is bounded.*
2. *Let $\mu = |\mathbb{M}\hat{p}|$. For every M with $|M| \geq \mu$, there exists an M' containing M with $|M'| = |M|$ such that $\hat{p}\restriction M'$ has cardinality μ.*
3. *Let $\mu = |\mathbb{M}\hat{p}|$. For every M with $|M| \geq \mu$, there is a κ-monster $\mathbb{C}_\kappa$ with $M \prec \mathbb{C}_\kappa$ and the restriction of $\hat{p}$ to the κ-monster $\mathbb{C}_\kappa$ has an orbit of cardinality less than κ.*

Proof. 1) implies 2) is Claim 3.21 with $\hat{M}$ taken as $\mathbb{M}$. For 2) implies 1), we prove the converse. If $\hat{p} \in S(\mathbb{M})$ is unbounded then for every κ, we can choose M of cardinality κ so that κ elements of M are in distinct cosets of $\mathrm{stb}_{\mathbb{M}}(\hat{p})$. By Lemma 3.22, we can extend M to M' with $\mathrm{stb}_{\mathbb{M}}(\hat{p}) \cap M' = \mathrm{stb}_{M'}(\hat{p}\restriction M')$ and the index of this subgroup is at least κ.

3) To prove 3) from 1) make the same construction as in proving 3) but continue for κ^+ steps with the $\alpha + 1$'st structure saturated over its predecessor.

3) implies 1). Build a chain $\mathbb{C}_\kappa$ for arbitrarily large κ satisfying the condition in iii). Since $\hat{p}$ is fixed for the entire construction $\hat{p}$ is unbounded. $\square_{3.23}$

Remark 3.24. Note that condition 3) of Lemma 3.23 depends on the particular embedding of M_κ in the monster. That is, if we take a different copy of M_κ, there will be a conjugate of $\hat{p}$ under the automorphism group of $\mathbb{M}$ whose restriction to this M_κ has a bounded orbit.

Thus 3) implies:

4) For arbitrarily large κ, there is a $\hat{p}' \in S(\mathbb{M})$ whose restriction to the κ-monster M_κ has an orbit of cardinality less than κ.

But 4) is ostensibly weaker; the extension keeps changing.

Question 3.25. Is it possible that there is a type $p \in S(M)$ that is bounded, say with orbit size μ and for arbitrarily large κ there are extensions p_κ over set models M_κ which all have orbit size μ. But there is no increasing chain of such types?

We can rephrase this analysis with the following implication.

Corollary 3.26. *From Lemma 3.23 and Remark 3.24, we know 1) implies 2):*
1. *$\mathbb{M}$ has a bounded orbit.*
2. *For arbitrarily large κ, the κ-monster M_κ has an orbit of cardinality less than κ.*

Question 3.27. But does the converse hold?

If the converse held, then we would know that the true monster does not have a bounded orbit, exactly if there is an upper bound on the cardinalities of set monsters with bounded orbits. And then we could reduce the existence of a true monster with bounded orbits to the following ZFC question.

Question 3.28. Find an *explicit description* of a function $f : \mathrm{On} \to \mathrm{On}$ such that if T has cardinality σ and the $f(\sigma)$-monster has a bounded orbit then $\mathbb{M}$ has a bounded orbit. More precisely, is there such an f which is *calculable*?

3.3 Saturation and omission

In this section, we state what seems to be a natural model theoretic property. It arose from Newelski's consideration of the issues in Section 3.2. But the Hanf number for the property is not calculable.

Newelski reduced the existence of bounded orbits to the question. When is a there a triple of a theory in vocabulary τ, a sub-vocabulary τ_1 and a τ-type p such that there is a model M with $M \restriction \tau_1$-saturated but M omits p.

Let us state the property more formally.

Definition 3.29. We say $M_1 \models \boldsymbol{t}$ where $\boldsymbol{t} = (T, T_1, p)$ is a triple of two theories and a type. The theories are in vocabularies $\tau \subset \tau_1$ with $|\tau_1| \leq \lambda$, $T \subseteq T_1$ and p is a τ_1-type over the empty set if M_1 is a model of T_1 which omits p, but $M_1 \restriction \tau$ is saturated.

Let $\boldsymbol{N}_\lambda$ denote[13] the set of $\boldsymbol{t}$ with $\tau_1 = \lambda$ and $\boldsymbol{N}_\lambda^{ss}$ the subset of those where the theory T is superstable. Then $H(\boldsymbol{N}_\lambda)$ denotes the Hanf number of $\boldsymbol{N}_\lambda$, $H(\boldsymbol{N}_\lambda)$ is least so that if some $\boldsymbol{t} \in \boldsymbol{N}_\lambda$ has a model of cardinality $H(\boldsymbol{N}_\lambda)$ it has arbitrarily large models.

To avoid difficulties about finding saturated models, we assume in this subsection:

Assumption 3.30. Assume the collection of λ with $\lambda^{<\lambda} = \lambda$ is a proper class.

This assumption follows from the generalized continuum hypothesis or from 'There are a proper class of strongly inaccessible cardinals'.

Under the mild set theoretic hypotheses of 3.30, we showed in [6] that $H(\boldsymbol{N}_\lambda)$ equals the Löwenheim number of second order logic, which is incalculable[36]. In [5] we restrict the question by requiring that the theory T satisfy stability conditions. The distinction between superstable and stable is immense. For the superstable case the number is easily calculable in terms of Beth numbers.

Theorem 3.31. *Let* $|\tau| = \lambda$.

1. *If T is superstable*

$$H(\boldsymbol{N}_\lambda^{ss}) < H(L_{(2^\lambda)^+,\omega}) \leq \beth_{(2^{(2^\lambda)^+})^+}.$$

2. *In general,*

$$H(\boldsymbol{N}_\lambda) \geq \ell^2(L^{II})$$

where $\ell^2(L^{II})$ denotes the Löwenheim number of second order logic. Thus $H(\boldsymbol{N}_\lambda)$ is a large cardinal if such exist.

For the superstable case, we have stated a calculable bound. For stable theories, work in progress (with Shelah) purports to show the upper bound $H(L_{(2^\omega)^+,\omega_1})$, which still depends on the choice of set theory, although it is lower than the general case.

---

13 Thus, 'there is an $M \in \boldsymbol{N}_\lambda$ with cardinality κ' replaces the more cumbersome notation in [6], '$P_N^\lambda(\boldsymbol{K_t}, \kappa)$ holds'.

4 What kind of conditions imply only weak set theory is needed to prove a proposition?

The usual answers to this question are in terms of syntactic properties of the properties under consideration. Properties of infinitary logic are inherently more complicated that properties of first order logic. Or as Macintyre phrased it, various properties are phrased with second order quantification.

None of the three examples in Section 3 seem *a priori* to have any real connection with stability. Even in the first case, forking and dividing can be defined for arbitrary theories and one can ask, 'does forking equal dividing'. But we have three propositions where there is apparent (in the first case), definite (in the third) and (open in the second) dependence on set theory to obtain the proposition for an arbitrary theory[14]. But in each case a stability hypothesis gives a positive solution to the problem. In stable theories forking is the same as dividing. In stable theories, there are bounded orbits. In superstable theories, the Hanf number of N_λ is calculable.

Note that in these examples the dependence of the result on the length of the set theoretic universe varies depending on the stability of the underlying theory. In contrast, calculating the spectrum function of a first order theory has no such dependence (although the proof uses the stability hierarchy).

Can we find a way to characterize such conditions as stability or o-minimality or the geometric conditions such as genus mentioned by Macintyre so as to account for lesser set theoretic strength necessary to study 'tame' mathematics? We see here that the cardinalities of the objects studied are not the defining characteristic of tame.

Acknowledgement: I thank the John Templeton Foundation for its support through Project #13152, Myriad Aspects of Infinity, hosted during 2009-2011 at the Centre de Recerca Matematica, Bellaterra, Spain. I thank Rutgers University for support of visits where some of this work was done.

14 Strictly speaking to apply this comment to the case of NTP2 theories we should only consider types over models, since we know there are non-simple theories for which forking (over arbitrary (countable) sets) does not equal dividing.

Bibliography

[1] H. Adler. Thorn-forking as local forking. *Journal of Mathematical Logic*, pages 21–38, 2009. preprint.

[2] John T. Baldwin. *Categoricity*. Number 51 in University Lecture Notes. American Mathematical Society, Providence, USA, 2009. www.math.uic.edu/˜ jbaldwin.

[3] John T. Baldwin. Amalgamation, absoluteness, and categoricity. In Arai, Feng, Kim, Wu, and Yang, editors, *Proceedings of the 11th Asian Logic Conference, 2009*, pages 22–50. World Sci. Publ., Hackensack, NJ, 2012.

[4] J.T. Baldwin. *Fundamentals of Stability Theory*. Springer-Verlag, 1988.

[5] J.T. Baldwin and S. Shelah. A Hanf number for saturation and omission II. to appear Fundamenta Mathematica, preprint:http://www.math.uic.edu/~jbaldwin/pub/shnew22.

[6] J.T. Baldwin and S. Shelah. A Hanf number for saturation and omission. *Fund. Math.*, 213:255–270, 2011. preprint:http://www.math.uic.edu/~jbaldwin/pub/shnew22.

[7] Steven Buechler. *Essential Stability Theory*. Springer-Verlag, 1991.

[8] Enrique Casanovas. The number of types in simple theories. *Annals of Pure and Applied Logic*, 98:69–86, 1999.

[9] C.C. Chang and H.J Keisler. *Model theory*. North-Holland, 1973. 3rd edition 1990.

[10] A. Chernikov and I. Kaplan. Forking and dividing in ntp2 theories. *Journal of Symbolic Logic*, 77:1–20, 2012.

[11] A. Conversano and A. Pillay. Connected components of definable groups and o-minimality i. preprint.

[12] A. Conversano and A. Pillay. Connected components of definable groups and o-minimality II. preprint.

[13] Džamonja+Shelah. On $\triangleleft^*$-maximality. *Annals Pure and Applied Logic*, 125:119–158, 2004. item 692 in Shelah archive.

[14] Harvey Friedman. Higher set theory and mathematical practice. *Ann. Math.Log.*, pages 325–357, 1971.

[15] Sy-David Friedman and Martin Koerwien. On absoluteness of categoricity in aecs. preprint, 200x.

[16] A Grothendieck. Elements de géometrie algébriques (rédigés avec la collaboration de J. Dieudonné), IV Etudes Locale des Schémas et des Morphismes de Schémas. *Publ Math I.H.E.S.*, 28:5–255, 1966.

[17] William Hanf. Models of languages with infinitely long expressions. In *Abstracts of Contributed papers from the First Logic, Methodology and Philosopy of Science Congress, Vol.1*, page 24. Stanford University, 1960.

[18] W. Hodges. *Model Theory*. Cambridge University Press, 1993.

[19] Byunghan Kim. Forking in simple theories. *Journal of the London Mathematical Society*, 57:257–267, 1998.

[20] Angus J. Macintyre. Model theory: Geometrical and set-theoretic aspects and prospects. *Bulletin of Symbolic Logic*, 9:197 – 212, 2003.

[21] Angus J. Macintyre. The impact of Gödel's incompleteness theorems on mathematics. In Baaz et al, editor, *Kurt Gödel and the Foundations of Mathematics*, pages 3–26. Cambridge University Press, 2011.

[22] D. Marker. *Model Theory: An introduction*. Springer-Verlag, 2002.

[23] C. Mclarty. What does it take to prove Fermat's last theorem? Grothendieck and the logic of number theory. *Bulletin of Symbolic Logic*, 16:359–377, 2010.

[24] C. McLarty. A finite order arithmetic foundation for cohomology. preprint:http://arxiv.org/pdf/1102.1773v3.pdf, 2012.

[25] C. McLarty. Zariski cohomology in second order arithmetic. preprint:http://arxiv.org/pdf/1207.0276v2.pdf, 2012.

[26] M. Morley. Omitting classes of elements. In Addison, Henkin, and Tarski, editors, *The Theory of Models*, pages 265–273. North-Holland, Amsterdam, 1965.

[27] Ludomir Newelski. Bounded orbits and measures on a group. *Israel Journal of Mathematics*, 187:209–229, 2012.

[28] Ludomir Newelski. Bounded orbits and strongly generic sets. preprint, 2012.

[29] B. Osserman. The weil conjectures. In T. Gowers, editor, *The Princeton Companion to Mathematics*, pages 729–732. Princeton University Press, 2008.

[30] S. Shelah. *Classification Theory and the Number of Nonisomorphic Models*. North-Holland, 1978.

[31] S. Shelah. Simple unstable theories. *Annals of Mathematical Logic*, 19:177–203, 1980.

[32] S. Shelah. Categoricity for abstract classes with amalgamation. *Annals of Pure and Applied Logic*, 98:261–294, 1999. paper 394. Consult Shelah for post-publication revisions.

[33] S. Shelah. Model theory without choice: categoricity. *Journal of Symbolic Logic*, 74:361–401, 2009. paper 840.

[34] Katrin Tent and Martin Ziegler. *A course in Model Theory*. Lecture Notes in Logic. Cambridge University Press, 2012.

[35] Sebastien Vasey. Indiscernible extraction and Morley sequences. preprint.

[36] Jouko Väänänen. Abstract logic and set theory II: Large cardinals. *Journal of Symbolic Logic*, 47:335–346, 1982.

[37] various contributors. Inaccessible cardinals and Andrew Wiles's proof. mathoverflow:http://mathoverflow.net/questions/35746/inaccessible-cardinals-and-andrew-wiless-proof, 2012.

[38] M. Ziegler. Introduction to stability theory and Morley rank. In E. Bouscaren, editor, *Model Theory and Algebraic Geometry : An Introduction to E. Hrushovski's Proof of the Geometric Mordell-Lang Conjecture*, pages 19–44. Springer-Verlag, 1999.

John P. Burgess

Modal Logic in the Modal Sense of Modality

for Jouko Väänänen

Abstract: One reason is explained why the pursuit of modal logic's original aim of evaluating the formal validity of arguments turning on necessity and possibility in a 'metaphysical' sense must cross borders and draw the logician into 'metaphilosophy'.

John P. Burgess: Department of Philosophy, Princeton University, Princeton, NJ 08544-1006, USA, e-mail: jburgess@princeton.edu

1 Modal Logic & Border-Crossing

Originally 'geometry' named an enterprise aiming to describe the forms of physical space. Today geometry is the study of a vast range of mathematical structures, closely enough related to structures once thought to model physical space that they can usefully be studied by the same methods or adaptations thereof. Some of these structures have important applications, as with the use of phase space in statistical mechanics; but as this example shows, the applications may have little to do with the original aims of the subject. The part of geometry devoted to the original aim is now but a small corner of a large field, and it is the more marginal because it involves a crossing of disciplinary borders. It does so because it is no longer tenable to view geometry as determining in advance the form of the arena in which physical forces act, before electromagnetic and gravitational theory, say, get to work describing those forces. On the contrary, the pursuit of what were the original aims of geometry is now inseparable from electromagnetic and gravitational theory themselves (which to begin with have taught us that we must not speak of physical space, but rather of physical spacetime).

Originally 'logic' named an enterprise aiming to prescribe the norms of deductive argumentation in philosophy and elsewhere. Today logic is the study of a vast range of formal systems, closely enough related to systems once thought to model deductive argumentation that they can usefully be studied by the same methods or adaptations thereof. Some of theses systems have important applications, as with the use of boolean logic in circuit design; but as this example shows, the applications may have little to do with the original aims of the subject. The part of logic devoted to the original aim is now but a small corner of a large field, and it is the more marginal because it involves a crossing of disciplinary borders. It must do so because it is no longer uncontroversial to assume that logic can serve as an organon, determining in advance the norms according to which substantive philosophical or other issues should be debated, before philosophers or others get to work debating. On the contrary, the pursuit of what were the original aims of logic is now for many inseparable from philosophy generally, and philosophy of language especially.

As it is with logic in general, so it is with modal logic in particular. But what is meant by 'modal' here? Mood-and-modality is the linguistic category pertaining to the expression of irreality of one kind or another, insofar as this is done grammatically rather than lexically, which is to say, insofar as it is expressed by inflections of the verb, or by use of special modal auxiliary verbs. Linguists distinguish deontic, epistemic, and dynamic modality. All are of interest in philosophy, but most of all the third, which more or less amounts to what philosophers call 'metaphysical' modality. Its key notions are expressible in various ways, but chiefly by use of the modal auxiliaries 'could' and 'would' and 'might'. For those key notions are the necessary in the sense of *what is and would have been (could not have failed to be) no matter what*, and the possible in the sense of *what is or isn't but potentially might (or could) have been*. Henceforth 'necessary' and 'possible', and with them 'modality', when used without distinguishing epithet, should be understood as pertaining to this kind rather than any other. It will here be our default sense of modality. Something very like this sense – or anyhow, something much more like it than like any other sense in play today – seems to have been assumed in the first work on modal logic, Aristotle's modal syllogistic. And therefore without much exaggeration or distortion we can say that to articulate the logic of this notion of modality was modal logic's original aim.

Modal logic today is an enormously broader enterprise than it was in Aristotle's time. Much of modal logic seems to have been developed for its purely mathematical interest, since the number of formal systems far exceeds the number of conceivable intuitive interpretations. But there are some significant external applications, as with the use of dynamic logic in theoretical computer science. The model-theoretic machinery developed to bring order out of the chaos of competing systems is indeed found invoked today in many areas of logic and across the border from pure logic in theoretical computer science and linguistics. Very little of modal logic is devoted to the attempt to determine what are the valid forms of argument involving necessity and possibility in *any* sense, and vanishingly little is devoted to the original aim of modal logic, that of dealing with the logic of modality in our default sense.

In fact, apart from some scolding by critics of modern modal logic about the dangers of 'Aristotelian essentialism' – which were met by firm denials from modal logicians that their logic harbored any 'essentialist commitments' – until 1970 the very topic of modality in our default sense was lost sight of, not only in logic, but in philosophy as well. Today what were the original aims of modal logic have begun to be pursued again, though the pursuit may not as yet have gotten very far. As with geometry and with logic in general, pursuit of the original aims will require crossing borders into another domain.

In the remarks that follow I will indicate how, in particular, one must enter into the realm of metaphysics, and beyond that into the realm of 'metaphilosophy', where the issues are about *what the issues are about* when debating metaphysics. One must confront a notorious 'modal paradox', and consider not only various views about how to deal with the paradox, but also various meta-views about what 'dealing with' a paradox can mean.

2 Modality Lost

But how did the neglect, until a few decades ago, of modality, in what I have claimed is the default sense of the term, come about? Well, not all at once, but in stages. The 'metaphysical' notion of *what would have been no matter what* (the necessary) was conflated with the epistemological notion of *what independently of sense-experience can be known to be* (the *a priori*), which in turn was identified with the semantical notion of *what is true by virtue of meaning* (the analytic), which in turn was reduced to a mere product of human convention. And what motivated these reductions? Mainly, I would suggest, efforts at 'demystification'. Let me outline the story.

The mystery of modality, for early modern philosophers, was how we can have any knowledge of it. Here is how the question arises. We think that when things are some way, in some cases they could have been otherwise, and in other cases they couldn't. That is the modal distinction between the contingent and the necessary. And we not only think that there is such a distinction, we also think that we know some examples. For instance, though there has been no female U. S. president, there could have been, whereas the number 29 not only has no nontrivial divisors but couldn't have had any. Thus we may speculate counterfactually about 'What if Hilary Clinton had been president in September, 2001?' but won't speculate counterfactually about 'What if 5 had been a divisor of 29?' And what is called the question of modal epistemology is just this: How do we know that the examples in question *are* examples of that of which they are supposed to be examples?

And why should this question be considered a difficult problem, a kind of mystery? Well, that is because, on the one hand, when we ask about most other items of purported knowledge how it is we can know them, sense-experience seems to be the source, or anyhow the chief source of our knowledge, but, on the other hand, sense-experience seems able only to provide knowledge about what is or isn't, not what could have been or couldn't have been. How do we bridge the gap between 'is' and 'could'? The classic statement of the problem was given by Immanuel Kant, early (p.3) in the introduction to the second or B edition of his first critique, *The Critique of Pure Reason*: 'Experience teaches us that a thing is so, but not that it cannot be otherwise.'

Note that this formulation allows that experience *can* teach us that a necessary truth is true; what it is not supposed to be able to teach is that it is necessary. For example, experience with what they call 'manipulatives' in pre-school, trying to arrange 29 blocks in various patterns, may convince one that 29 is a prime number, one that has no nontrivial factorization; but granted that it doesn't *actually* have one, how do we know that it *couldn't possibly* have had one, and is necessarily a prime number? How can experience teach us that *extra* fact?

The problem becomes more vivid if one adopts the language that was once used by G. W. Leibniz, and much later repopularized by Saul Kripke in his famous work on model theory for formal modal systems, the usage according to which the necessary is that which is 'true in all possible worlds'. In these terms the problem is that the senses only show us *this* world, the world we live in, the *actual* world as it is called, whereas when we claim to know about what could or couldn't have been, we are claiming knowledge of what is going on in some or all *other* worlds. For *that* kind of knowledge, it seems, we would need a kind of sixth sense, or extrasensory perception, or nonperceptual mode of apprehension, to see beyond the world in which we live to these various other worlds.

That is the problem. Let me now rapidly review some of the highlights of the history of attempted solution, beginning with Kant himself. Kant concludes, in the very next sentence after the one I quoted earlier, that our knowledge of necessity must be what he calls *a priori* knowledge or knowledge that is 'prior to' or before or *independent of* experience, rather than what he calls *a posteriori* knowledge or knowledge that is 'posterior to' or after or *dependant on* experience. And so the problem of the origin of our knowledge of necessity becomes for Kant the problem of the origin of our *a priori* knowledge.

Well, that is not *quite* the right way to describe Kant's position, since there is one special class of cases where Kant thinks it isn't really so hard to understand how we can have *a priori* knowledge. He doesn't think *all* of our *a priori* knowledge is mysterious, but only most of it. He distinguishes what he calls *analytic* from what he calls *synthetic* judgments, and holds that *a priori* knowledge of the former is unproblematic, since it is not really knowledge of external objects, but only knowledge of the content of our own concepts, a form of self-knowledge.

We can generate any number of examples of analytic truths by the following three-step process. First, take a simple logical truth of the form 'Anything that is both an A and a B is a B', for instance, 'Anyone who is both a man and unmarried is unmarried'. Second, find a synonym C for the phrase 'thing that is both an A and a B', for instance, 'bachelor' for 'one who is both a man and unmarried'. Third, substitute the shorter synonym for the longer phrase in the original logical truth to get the truth 'Any C is a B', or in our example, the truth 'Any bachelor is unmarried'. Our knowledge of such a truth seems unproblematic because it seems to reduce to our knowledge of the meanings of our own words.

So the problem for Kant is not exactly how knowledge *a priori* is possible, but more precisely how *synthetic* knowledge *a priori* is possible. Kant thought we do have examples of such knowledge, and indeed the example I have been using, that 29 is a prime number, is one of them. Arithmetic, according to Kant, was supposed to be synthetic *a priori*, and geometry, too – all of pure mathematics. In his *Prolegomena to Any Future Metaphysics*, Kant listed 'How is pure mathematics possible?' as the first question for metaphysics, for the branch of philosophy concerned with space, time, substance, cause, and other grand general concepts – including modality.

Kant offered an elaborate explanation of how synthetic *a priori* knowledge is supposed to be possible, an explanation reducing it to a form of self-knowledge, but later philosophers questioned whether there really were any examples of the synthetic *a priori*. Geometry, so far as it is about the physical space in which we live and move – and that was the original conception, and the one still prevailing in Kant's day – came to be seen as, not synthetic *a priori*, but rather *a posteriori*. (This development was one of the main causes of the extended usage, alluded to earlier, of 'geometry' to cover a vast range of mathematical investigations.) The great mathematician Carl Friedrich Gauß had already come to suspect that geometry is *a posteriori*, like the rest of physics, by some time in the first half of the nineteenth century. Since the time of Albert Einstein in the early twentieth century the *a posteriori* character of physical geometry has been the received view (whence the need for border-crossing from mathematics into physics if one is to pursue the original aim of geometry).

As for arithmetic, the great logician Gottlob Frege in the late nineteenth century claimed that it was not synthetic *a priori*, but analytic – of the same status as 'Any bachelor is unmarried', except that to obtain something like '29 is a prime number' one needs to substitute synonyms in a logical truth of a form much more complicated than 'Anything that is both an A and a B is a B'. This view was subsequently adopted by many philosophers in the analytic tradition of which

Frege was a forerunner, whether or not they immersed themselves in the details of Frege's program for the reduction of arithmetic to logic.

Once Kant's synthetic *a priori* has been rejected, the question of how we have knowledge of necessity reduces to the question of how we have knowledge of analyticity, which in turn resolves into a pair of questions: On the one hand, how do we have knowledge of synonymy, which is to say, how do we have knowledge of meaning? On the other hand how do we have knowledge of logical truths? As to the first question, presumably we acquire knowledge, explicit or implicit, conscious or unconscious, of meaning as we learn to speak, by the time we are able to ask the question whether *this* is a synonym of *that*, we have the answer. But what about knowledge of logic? That question didn't loom large in Kant's day, when only a very rudimentary logic existed, but after Frege vastly expanded the realm of logic – only by doing so could he find any prospect of reducing arithmetic to logic – the question loomed larger.

Many philosophers, however, convinced themselves that knowledge of logic also reduces to knowledge of meaning, namely, of the meanings of logical particles, words like 'not' and 'and' and 'or' and 'all' and 'some'. To be sure, there are infinitely many logical truths, in Frege's expanded logic. But they all follow from or are generated by a finite list of logical rules, and philosophers were tempted to identify knowledge of the meanings of logical particles with knowledge of rules for using them: Knowing the meaning of 'or', for instance, would be knowing that 'A or B' follows from A and follows from B, and that anything that follows both from A and from B follows from 'A or B'. So in the end, knowledge of necessity reduces to conscious or unconscious knowledge of explicit or implicit semantical rules or linguistics conventions or whatever.

On this picture we know that five isn't a divisor of 29 in any possible world, not because we have some kind of ESP that permits us to engage in telepathic remote viewing of other possible worlds while remaining in this one, but rather because, just as we have implicitly or unconsciously learned conventions or rules that tell us no one who is married may properly be described as a 'bachelor', so also we have implicitly or unconsciously learned conventions or rules that tell us, or from which it follows, that one may not describe any world as a world in which 5 is a 'divisor' of 29.

For instance, a world in which whenever one lays out five rows of six blocks and then counts the total, one finds 29, is not properly described as a world in which five times six is 29. Rather it is a world where, by some curious physical principle, whenever 30 blocks are brought together in five rows of six, one of them disappears; or where, by some curious psychological principle, whenever we try to count 30 blocks arranged in five rows of six, we always miss one; or something of the sort.

Such is the sort of picture that had become the received wisdom in philosophy departments in the English speaking world by the middle decades of the last century. For instance, A. J. Ayer, the notorious logical positivist, and P. F. Strawson, the notorious ordinary-language philosopher, disagreed with each other across a whole range of issues, and for many mid-century analytic philosophers such disagreements were considered the main issues in philosophy (though some observers would speak of the 'narcissism of small differences' here). But both sides put forward a picture quite like the one I have just drawn, Ayer in his *Language Truth and Logic* (Golancz) of 1936 and Strawson in his *Introduction to Logical Theory* (Methuen) of 1952.

And people like Ayer and Strawson in the 1920s through 1960s would sometimes go on to speak as if linguistic convention were the source not only of our knowledge of modality, but of modality itself, and go on further to speak of the source of language lying in *ourselves*. Individually, as children growing up in a linguistic community, or foreigners seeking to enter one, we must consciously or unconsciously learn the explicit or implicit rules of the communal language as something with a source outside us to which we must conform. But by contrast, collectively, *as* a speech community, we do not so much *learn* as *create* the language with its rules. And so if the origin of modality, of necessity and its distinction from contingency, lies in language, it therefore lies in a creation of ours, and so in us. 'We, the makers and users of language' are the ground and source and origin of necessity. Well, this is not a literal quotation from any one philosophical writer of the last century, but a pastiche of paraphrases of several.

3 Modality Regained

Such was the climate of opinion before January, 1970, when Kripke delivered his Princeton lectures on 'Naming and Necessity'. Since then it has become clear that there are several things wrong with the conclusion that we, the makers and users of language, are the originators of necessity, and the line of thought from Kant onwards that led to this sort of conclusion, and that incidentally led to necessity, as something distinct from aprioriness or analyticity, being submerged and lost sight of. I won't try to say in advance just how *many* main things are wrong with the line of thought I will be criticizing, lest I end up like Cardinal Ximenes in the Monty Python sketch about the Spanish Inquisition, and find myself repeatedly having to revise my estimate upwards. I'll just mention – following Kripke, though not very closely or strictly – several difficulties one by one; we can count them up afterwards if we wish.

For one thing, there is a difficulty to be found right at the beginning of the whole line of argument, with Kant, and indeed at the very first step, Kant's inference from the premise that experience does not teach us that a thing cannot be otherwise to the conclusion that our knowledge of necessity is *a priori*. Strawson was less dismissive of Kant than was Ayer, but he famously described one argument of Kant's as a '*non sequitur* of numbing grossness'. I don't know whether the argument to which this phrase was first applied really deserved words so harsh, but Kant's argument from 'experience isn't sufficient to teach us that a thing couldn't have been otherwise' to 'experience isn't needed to learn that a thing couldn't have been otherwise' does seem to call for a pretty severe rebuke. It's like arguing from 'sunshine isn't sufficient to grow corn (rainfall is also needed)' to 'sunshine isn't needed to grow corn (rainfall is sufficient)'. It really is a *non sequitur* of numbing grossness.

Kripke's achievement was not just to point out this logical lapse on the part of a famous philosopher – finding logical lapses in the works of famous philosophers is something it is all too easy to do, I'm afraid, and one doesn't become famous just for doing *that* – but rather, his achievement was to offer plausible counterexamples to Kant's claim, plausible cases of *a posteriori* knowledge of necessity. One famous example, perhaps the most celebrated, is that the Queen – or to call her by name rather than title, Elizabeth Windsor – could not have been the daughter of Harry Truman. Suppose a tabloid publishes a story according to which she *is* the daughter of Harry Truman, smuggled into the royal family as an infant. The story could be disproved by a chain of documents and testimony leading back to persons who moved in royal circles in the 1920s. Or it could be directly disproved by DNA testing. But all of these methods of disproof require sense-experience. The tabloid's sensational conspiracy theory can't be disproved by so-called 'armchair methods', just sitting and thinking. The story is false, but its falsehood is an item of *a posteriori*, not *a priori* knowledge.

Nonetheless, according to Kripke, the fact that Elizabeth Windsor is not the daughter of Harry Truman is a necessary fact. She not only *actually isn't* but *couldn't possibly have been* the daughter of anyone other than her actual father George Windsor, and in particular, not of Harry Truman. A 'possible world' or counterfactual situation in which a daughter of Harry Truman was smuggled into the royal family as an infant and later became Queen, would be a situation in which a person other than Elizabeth Windsor became Queen (and presumably along the way came to be *called* 'Elizabeth Windsor'). It is not a situation in which Elizabeth Windsor (the person we actually call 'Elizabeth Windsor') had a different father.

Such examples of *a posteriori* necessities show that what philosophers of the period of Ayer and Strawson tended to claim about our knowledge of necessary truths is wrong: it does *not* just reduce to knowledge of linguistic rules. In fact, one must keep carefully distinct the age-old notion of *necessity* from the modern notions of *aprioriness* and *analyticity*.

4 Necessity and Convention Reconsidered

The purely negative conclusion that the primæval notion necessity is *not* to be equated with eighteenth-through-twenty-first-century notions of aprioriness or analyticity is what Kripke left us with at the end of his 1970 Princeton lectures. By two years later, however, when the transcript of the audiotape of those lectures was being published, he had something a bit more positive to say, and he said it in some addenda to the lectures proper. There he suggests that in the background, behind each of his examples of *a posteriori* necessity, there is some *a priori* principle according to which whatever is so about certain matters is so of necessity.

On this picture, we know *a priori* that whatever the facts are about whether one number divides another, they couldn't have been otherwise. Experience with manipulatives may teach us that five *doesn't* divide 29, and combining what experience thus teaches us with our background *a priori* knowledge, we can conclude that five *couldn't have* divided 29. Similarly, we know *a priori* that, whoever a given person's parents were, someone with different parents wouldn't count as *that* person. We know *a posteriori* that the Queen *isn't* a daughter of Harry Truman, and combining this *a posteriori* knowledge with our *a priori* knowledge, we can conclude that the Queen *couldn't have been* a daughter of Harry Truman.

Kripke describes this *a priori* knowledge in the background as 'discoverable by philosophical analysis', which is near enough for present purposes to calling it 'analytic'. Thus the view appears to be that our knowledge of necessity reduces to something like a combination of *a posteriori* knowledge gained through sense-experience with analytic knowledge of semantic rules or linguistic conventions or whatever. In that case, the formulation that our knowledge of necessity derives from our knowledge of linguistic rules would be defensible after all, provided it is understood as meaning only this, that though our knowledge of the *truth* of a necessary truth may depend on sense-experience, our knowledge of the *necessity* of that truth, given that it is true, derives from knowledge of linguistic rules. The main point is that the epistemology of modality is still demystified: there is no need for ESP. And perhaps that is what the logical positivists and ordinary language philosophers really cared about. (Unfortunately for them, Kripke has found *another* problem with the line of thought leading to the conclusion that our knowledge of necessity reduces to knowledge of linguistic rules, a 'skeptical paradox' about the notion of 'rule' itself; but this is a topic I must leave aside here for lack of space among other reasons.)

The point we have reached is this: We have learned that we must be very careful about how we understand such a formulation as 'Our knowledge of necessity derives from knowledge of linguistic rules'; but it has not really been shown that such formulations are wholly off the mark. There remains, however, to be evaluated the step from the intermediate conclusion 'knowledge of modal distinctions derives knowledge of linguistic rules' – presumably rules about using modal auxiliary verbs, such as 'would' and 'might' – to the further conclusion that modal distinctions *themselves* derive from linguistic rules, and the final conclusion that the origin of necessity lies in *us*, the makers and users of language.

One immediate difficulty with these further steps is that the conclusion just enunciated, along with all talk about the 'origin' of necessity, is simply nonsense: that the problem of the 'origin' of necessity is not a mystery but a muddle. For if one takes the word 'origin' at all literally, then to speak of the origin of necessity is to speak of a time when necessity came into being, and before which there was none. And *that* certainly seems to be nonsense. The number 29 not only is today but *always has been* necessarily prime. Even if we substitute 'ground' for 'origin', to get rid of the temporal connotations, in asking after the ground of something we still seem to be asking what is it without which that thing would not have been, whereas the necessary as opposed to the contingent is precisely that which would have been *no matter what* as opposed to that which might not have been *if only*. So we seem to be asking, 'What is it, without which that which could not have failed to be would have failed to be?' And surely such a question is nonsensical.

As for *our* role in the matter, as makers and users of language, certainly any suggestion that if only *we* had acted differently, if only we had adopted different semantical rules or linguistic conventions, different things would have been necessary, involves a kind of confusion. Suppose, for instance, we had adopted the conventions of a base-thirteen numeration system rather than our actual decimal or base-ten system. Then the number that we wrote as a two followed by a nine, and perhaps even pronounced 'twenty-nine', would indeed have been a number divisible by five, since it would have been two times thirteen plus nine or 35 that we would have been calling 'twenty-nine', and that number is five times seven. But our adopting a different numeration system would not have changed any facts about the number 29, except the fact that we *call* it 'twenty-nine' and not 'twenty-three' as we might if we used base thirteen.

So in speaking of ourselves as the originators of modality, the philosophers of the mid-twentieth-century will have been speaking nonsense if they were speaking literally. But surely they, or the best of them, had the wit to see this, and so we must take them to be speaking figuratively, even if they make no particularly strenuous efforts to warn against literalism or to spell out in terms that *can* be taken literally what they express figuratively by speaking of ourselves, the makers and users of language, as being the creators of necessity. Let us grant this, and let us allow ourselves to go on speaking figuratively for a bit. Even so, the formulation according to which we are the creators of necessity is open to at least one further challenge.

Even granting that what is needed to get from the kind of knowledge sense-experience can give us to knowledge of the necessity of necessary truths is knowledge of linguistic rules, and even granting also (despite 'skeptical paradoxes' about 'rules') that we do have the requisite kind of knowledge of linguistic rules, still it is premature to conclude, even speaking figuratively, that necessity itself derives from linguistic rules. Another interpretation of the situation remains possible, and logical positivists and ordinary language philosophers have given no real argument against this rival interpretation, but merely assumed that the rival view is too 'metaphysical', in a pejorative sense, to be taken seriously by analytic philosophers. The real problem or mystery of the so-called 'origin of necessity' lies in the clash between these two interpretations.

5 Realism and Pragmatism about Modality

Let me now try to describe what the two conflicting views about the nature of modality amount to, before turning to consider the bearing of their conflict on other issues concerning modality, and ultimately on modal *logic*. I will give first the view that people like Ayer and Strawson want to *reject*, and then the view they want to *accept*.

Here is the first view. Though the fact is not mentioned in *Genesis*, the first thing God said on the first day of creation was 'Let there be necessity'. And there was necessity. And God saw necessity, that it was good. And God divided necessity from contingency. And only then did He say 'Let there be light'. Several days later, Adam and Eve were introducing names for the animals into their language, and during a break between the fish and the birds, introduced also into their language modal auxiliary verbs, or devices that would be translated into English using modal auxiliary verbs, and rules for their use, rules according to which it can be said of some things that they 'could' have been otherwise, and of other things that they 'could not'. In so doing they were merely putting labels on a distinction that was no more their creation than were the fishes of the sea or the beasts of the field or the birds of the air.

And here is the rival view. The failure of *Genesis* to mention any command 'Let there be necessity' is to be explained simply by the fact that no such command was issued. We have no reason to suppose that the language in which God speaks to the angels contains modal auxiliary verbs or any equivalent device. (I am speaking of 'metaphysical' modality here as always, not of deontic modality of the kind involved in 'Thou shalt not…') Sometime after the Tower of Babel some tribes found that their purposes would be better served by introducing into their language certain modal auxiliary verbs, and fixing certain rules for their use. When we say that *this* is necessary while *that* is contingent, we are applying such rules, rules that are products of human, not divine intelligence.

I have been allowing myself here to use theological language in describing the distinction I want to draw. That would have been the natural way for seventeenth or eighteenth century philosophers, who nearly all were or professed to be theists or deists, to discuss the matter. For many today, such language cannot be literally accepted, and if it is only taken metaphorically, then I have not really done better than those who speak figuratively and frame the question as that of whether the 'origin' of necessity lies outside us or within us. So let me drop the theological language, and try again, without it, to describe the division between two views that I have in mind.

Well, here the first view: Ultimately reality as it is in itself, independently of our attempts to conceptualize and comprehend it, contains both facts about what is, and superfacts about what not only is but had to have been. Our modal usages, for instance, the distinction between the simple indicative 'is' and the construction 'had to have been', simply *reflect* this fundamental distinction in the world, a distinction that is and from the beginning always was there, independently of us and our concerns.

And here is the second view: We have reasons, connected with our various purposes in life, to use certain words, including 'would' and 'might', in certain ways, and thereby to make certain distinctions. The distinction between those things in the world that *would have been* no matter what and those that *might have failed to be* if only is a *projection* of the distinctions made in our language. Our saying there were necessities there before us is a *retroactive* application to the pre-human world of a way of speaking invented and created by human beings in order to solve human problems.

Well, that's my second try. With it I have gotten rid of theology, but unfortunately I have *not* gotten rid of all metaphors. The key remaining metaphor is the *optical* one: reflection *vs* projection. Perhaps I should give up the attempt to get rid of all metaphors, and admit that the two views I am discussing are not so much philosophical *theses* or *doctrines* as 'metaphilosophical' *attitudes* or *orientations*: a stance that finds the 'reflection' metaphor congenial, and the stance that finds the 'projection' metaphor congenial. But let me try a third time to describe the distinction between the two outlooks in literal terms, avoiding optics as well as theology.

To begin with, both sides grant that there is a correspondence or parallelism between two items. On the one hand, there are facts about the contrast between what is necessary and what is contingent, for instance, between 29 being a prime number and 29 being the number of years it takes for Saturn to orbit the sun. On the other hand, there are facts about our usage of modal auxiliary verbs such as 'would' and 'might', and these include, for instance, the fact that we have no use for questions of the form 'Would 29 still have been a prime number if such-and-such?' but may have use for questions of the form 'Would 29 still have been the number of years it takes for Saturn to orbit the sun if such-and-such?' The difference between the two sides concerns the order of explanation of the relation between the two parallel ranges of facts.

And what do I mean by *that*? Well, both sides grant that '29 is necessarily prime', for instance, is a proper thing to say, but they differ in the explanation *why* it is a proper thing to say. Asked why, the first side will say that ultimately it is simply because 29 *is* necessarily prime. That makes the proposition that 29 is necessarily prime true, and since the sentence '29 is necessarily prime' expresses that proposition, it is true also, and a proper thing to say. The second side will say instead that '29 is necessarily prime' is a proper thing to say because there is a rule of our language according to which it is a proper thing to say. This formulation of the difference between the two sides gets rid of metaphor, though it does put an awful lot of weight on the perhaps fragile 'why' and 'because'.

Note that the adherents of the second view need not *deny* that 29 is necessarily prime. On the contrary, having said that the sentence '29 is necessarily prime' is, per rules of our language, a proper thing to say, they will go on to say it. Nor need the adherents of the first view deny that recognition of the propriety of saying '29 is necessarily prime' is enshrined in a rule of our language. The adherents of the first view need not even deny that proximately, as individuals, we learn that '29 is necessarily prime' is a proper thing to say by picking up the pertinent rule in the course of learning our language. But the adherents of the first view will maintain that the rule itself is only proper because collectively, as the creators of the language, we or our remote answers have, in setting up the rule, managed to achieve correspondence with a pre-existing fact, or rather, a pre-existing superfact, the superfact that 29 is necessarily prime. The difference between the two views is, as I said, in the order of explanation.

I will want labels for the two sides, or 'metaphilosophical' stances, and rather than invent new ones, I will simply take two of the most overworked terms in the philosophical lexicon and give them one more job to do. I will call the reflection view 'realism' about modality, and the projection view 'pragmatism'. That at least will be easy to remember, since 'realism' and 'reflection' begin with the same first two letters, as do 'pragmatism' and 'projection'. The realist/pragmatist distinction has bearing across a range of issues and problems, and above all it has bearing on the meta-issue of which issues are significant. For the two sides will, or ought to, recognize quite different questions as the central unsolved problems in the theory of modality.

For those on the realist side, the old problem of the *ultimate* source of our knowledge of modality remains, even if it is granted that the *proximate* source lies in knowledge of linguistic conventions. For knowledge of linguistic conventions constitutes knowledge of a reality independent of us only insofar as our linguistic conventions reflect, at least to some degree, such an ultimate reality. So for the realist the problem remains of explaining how such degree of correspondence as there is between distinctions in language and distinctions in the world comes about. If the distinction in the world is something primary and independent, and not a mere projection of the distinction in language, then how the distinction in language comes to be even imperfectly aligned with the distinction in the world remains to be explained. For it cannot be said that we have faculties responsive to modal facts independent of us – not in any sense of 'responsive' implying that if the facts had been different, then our language would have been different, since *modal* facts *couldn't have been* different. What then *is* the explanation? This is the problem of the epistemology of modality as it confronts the realist, and addressing it is or ought to be at the top of the realist agenda.

As for the pragmatist side, a chief argument of thinkers from Kant to Ayer and Strawson and beyond for their anti-realist stance has been precisely that if the distinction we perceive in reality is taken to be merely a projection of a distinction created by ourselves, then the epistemological problem dissolves. That seems more like a reason for *hoping* the Kantian or Ayerite or Strawsonian view is the right one, than for *believing* that it is; but in any case, even supposing the pragmatist view is the right one, and the problems of the epistemology of modality are dissolved, still the pragmatist side has an important unanswered question of its own to address. The pragmatist account, as I formulated it earlier, begins by saying that we have certain reasons, connected with our various purposes in life, to use certain words, including 'would' and 'might', in certain ways, and thereby to make certain distinctions. What the pragmatist owes us is an account of what these purposes are, and how the rules of our language help us to achieve them. Addressing *that* issue is or ought to be at the top of the pragmatists' to-do list.

While the positivist Ayer dismisses all metaphysics, the ordinary-language philosopher Strawson distinguishes *good* metaphysics, which he calls 'descriptive', from *bad* metaphysics, which he calls 'revisionary', but which I will call 'transcendental' (without intending any specifically Kantian connotations). Descriptive metaphysics aims to provide an explicit account of our 'conceptual scheme', of the most general categories of commonsense thought, as embodied in ordinary language. Transcendental metaphysics aims to get beyond or behind all merely human conceptual schemes and representations to ultimate reality as it is in itself, an aim that Ayer and Strawson agree is infeasible and probably unintelligible. The descriptive/transcendental divide in metaphysics is a paradigmatically 'metaphilosophical' issue, one about *what philosophy is about*. Realists about modality are paradigmatic transcendental metaphysicians. Pragmatists must in the first instance be descriptive metaphysicians, since we must to begin with understand much better than we currently do how our modal distinctions work and what work they do for us, before proposing any revisions or reforms. And so the difference between realists and pragmatists goes beyond the question of what issue should come first on the philosopher's agenda, being as it is an issue about *what philosophical agendas are about*.

6 The Varied Sorrows of Modal Logic

So much for generalities about the realist/pragmatist distinction. Before taking up the issue of the bearing of the division on modal *logic*, I must return to my earlier historical sketch, and consider what was going on, in the period of Ayer and Strawson and Carnap, and more generally in the pre-1970 period, in the domain of modal logic.

The subject began, as I have already had occasion to mention, with logic's founder, Aristotle. It was actively pursued through antiquity and the middle ages and in a more desultory fashion on into the early modern period as far as the nineteenth century. But Frege, the founder of truly modern logic, swept away such fragments of modal logic as had still been left in Kant and still remained in his own day. As he explains early (§4) in his *Begriffsschrift*, the founding work of truly modern logic, his concept-writing is to contain no symbols for modalities, on the grounds that modal distinctions only indicate something about how an asserted statement is *known* and contribute nothing to the content of the statement asserted.

Apparently Frege is here falling in with the tendency to identify necessity and aprioriness, which we have seen to be a mistake. A better reason for his leaving modal distinctions aside would have been that the task of developing a logic capable, as the traditional syllogsistic was not, of analyzing mathematical arguments was enough to keep him fully occupied. Distinctions of mood, like those of tense, play no role in mathematics, which deals with facts that are necessary, as well as timeless. Developing a logic for such notions, which do figure in philosophical as opposed to mathematical argumentation, could well be left for the future, after mathematical reasoning has been taken care of.

Modality continued to be ignored by Frege's heirs: Russell, for one, and certainly Hilbert and his school. Eventually the eminent twentieth-century logicians Kurt Gödel and Alfred Tarski were to take an interest in modal systems, but in both cases involvement with modal logic was a sideline in a career devoted mainly to other matters, and the systems considered were ones that had already been developed by other, lesser twentieth-century figures, above all C. I. Lewis.

As is well-known, Lewis's main aim was to develop in the object language of a formal system a theory of matters that orthodox logicians express only in the metalanguage. His primary focus was on what he called 'strict implication'. Close reading shows that by this he did not mean formal logical deducibility or consequence in the strictest sense, but something more like analytic deducibility or consequence. (Thus 'Jones is unmarried' would be for Lewis 'strictly implied' not only by 'Jones is a man and Jones is unmarried', but also by 'Jones is a bachelor'.) It is perhaps not going too far to suggest that Lewis's enterprise, for which the label 'endometalogic' might be appropriate, came to be called 'modal' logic only because the modal notion of necessity had by Lewis's day become conflated and confounded with the notion of analyticity.

Now the chief difference between treating implication or entailment in the metalanguage and representing it by an operator in the object language is that such operators can be iterated and nested and embedded, whereas, if 'implies' or 'entails' appear only in metalinguistic formulations and only in application to object-language formulas, then there can be no question or such iteration or nesting or embedding. Lewis thought he had 'intuitions' directly about 'strict implication', and in devising a formal system he would consult these to decide which axioms involving iteration and nesting and embedding were acceptable and which not.

Sometimes he changed his mind. Does *p*'s strictly implying *q* strictly imply that *q*'s strictly implying *r* strictly implies *p*'s strictly implying *r*? Or can we only say that *p*'s strictly implying *q* and *q*'s strictly implying *r* together strictly imply *p*'s strictly implying *r*? Lewis leaned first towards the first position, then towards the second, and so were born the systems known as **S3** and **S2**. These were soon joined by others, even before the mathematically-minded, who claimed no intuitions about strict implication, took over and produced dozens and scores of variants.

Lewis rather belatedly introduced the diamond symbol for possibility, and allowed that '*p* strictly implies *q*' could be analyzed as 'it is not the case that it is possible that *p* and not *q*'. But he never used the box symbol for necessity, and never took the step of analyzing 'possibly' as 'not necessarily not' or '*p* strictly implies *q*' as 'necessarily, if *p* then *q*' (with the 'material' sense of 'if': it is not the case that *p* and not *q*). Formal systems with necessity as primitive derive from Gödel rather than Lewis.

When bringing in 'possibly', Lewis insists that it must be understood as what he calls 'logical' possibility, as not involving a contradiction (really meaning by this *analytical* possibility or not involving a contradiction-in-terms). Had he considered 'necessarily', he inevitably would have insisted on a 'logical', which is to say *analytical*, interpretation of it as well. His position was fully in accord with the tendency in his day to identify the necessary and the analytic.

Carnap, a pioneer of quantified modal logic, and one of the few modal logicians after Lewis to say much about the meaning of modal formulas, explicitly presented his key notion as an explication alike of 'Leibniz' notion of necessity and Kant's notion of analyticity'. And this provoked objections from W. V. Quine, who saw that quantification into the context 'the statement "----------" is analytic' is nonsense. Protesting against defenders of modal logic, who in his view were not answering his objections but changing the subject, he insisted that he was concerned about the status of quantification into modal contexts 'in the quantificational sense of quantification and the modal sense of modality'; and by the latter he meant reading the box symbol in formal systems as 'it is analytic that'. For that is what the box was supposed to mean according to the writers Quine was most concerned to criticize, above all Carnap. But for us the 'modal sense of modality' should rather mean our default 'metaphysical' sense.

If Quine failed to convince, it was perhaps because many of his opponents, though officially committed to explaining necessity as analyticity, were unconsciously influenced by thoughts about necessity in the original, default, 'metaphysical' sense. The general pre-1970 failure to recognize both notions explicitly and to distinguish between them naturally made it impossible to appreciate that different logics may be appropriate to the different notions. That is part of what I meant by saying that the original aim of modal logic, to deal with 'metaphysical' modality, became submerged and lost sight of.

As for the dispute between Quine and the defenders of quantified modal logic, by hindsight, belatedly distinguishing notions formerly conflated, Quine seems to be right about the meaninglessness of 'There is someone *x* such that the statement "*x* did not defeat Nixon and win the 1968 U. S. presidential election" is not analytic', while *of course* it makes sense to say that there is someone such *it might have happened* that he defeated Nixon and won the 1968 U. S. presidential election. Indeed this not only makes sense to say, but is true: For *Hubert Humphrey* is someone such that it might have happened that he defeated Nixon and won the 1968 U. S. presidential election.

But we had to wait for Kripke (from whose 1970 lectures the foregoing example derives) before it was made plain that reading the box as 'it is analytic that...' (the official Ludovician explanation, to the extent that there was one) and reading the box as 'it could not have failed to be the case that...' (a reading which doubtless influenced the 'intuitions' of some) are two different and incompatible readings. Before then there was much arguing at cross-purposes, and much confused groping guided now by one reading of the box and now by another, without recognizing the switch.

It is said that when Cauchy lectured at the Académie des Sciences on the distinction between convergent and divergent series, Laplace rushed home to check the series in his *Mécanique Céleste*. But when Kripke lectured in Princeton on the distinction between 'logical' and 'metaphysical' necessity – or in his preferred terminology, simply between analyticity and necessity *tout court et sans phrase* – modal logicians did not rush home to check which formal system was the right one for which notion of necessity. For by 1970 the idea that there might be a 'right' modal logic – which had been a serious issue between Aristotle and his student Theophrastus, and a serious issue for Lewis as he compared **S2** and **S3** – had come to seem quaint, so far had the subject drifted from its origins.

7 Realism, Pragmatism, and the Modal Paradox

To this day the literature on the question which is the right modal logic for 'metaphysical' modality remains thin. The candidate systems most often considered have been **S5** and **S4**, though I have seen (as referee) a soon-to-be-published (I hope) paper making an ingenious case for McKinsey's **S4.1**. And there is a well-known argument, developed by Nathan Salmon and others, for the system **T**, or rather *against* the axiom that distinguishes **S4** from **T**: the putative law that if p is necessary, it is necessarily necessary (and dually, if p is possibly possible, it is possible).

The objection goes by various names, but is perhaps most often called simply 'the modal paradox'. To illustrate how issues of modal logic, in the original or old-fashioned sense of the search for the right system of logic for modality properly so called, are inseparable from issues of philosophy and even 'metaphilosophy', of metaphysics and even 'metametaphysics', I would like to describe the example, and indicate how the realist/pragmatist division bears on it.

The paradox goes back to some of the discussion of *a posteriori* necessities in 'Naming and Necessity', and in particular to some examples pertaining to the necessity of origins. I have in mind here under that head now not the necessity, spoken of earlier, of the genealogy of human beings, whether of royal or noble or commoner rank. Rather, the issue I wish to consider is over the original composition of inanimate artifacts.

In a much-discussed passage in his third 'Naming and Necessity' lecture and attendant footnotes, Kripke suggests that the wooden table before him could not have originally been made of wholly different pieces of wood – really one should say wood and hardware, I suppose. In so saying he is apparently tacitly granting that the table could have been made of *slightly* different pieces. Perhaps a different screw could have been used at a certain joint, for instance.

The same principle would seem to apply to, say, a ship made of a thousand planks. There is an intuition X to the effect that the same ship could have been made of 999 of the same planks, arranged the same way, plus a replacement for plank #473; but at the same time there is an intuition Y to the effect that a ship made of a thousand different planks would have been a different ship. The puzzle is that these two intuitions, X and Y, seem to conflict. Let me describe the conflict in Leibniziano-Kripkean 'possible worlds' language.

Here in the actual world A we have a ship, let us name it the good ship *Theseus*, made of 1000 planks. Our first intuition X is that the same ship could have been made of 999 of these planks plus a replacement for plank #473. In possible-worlds terms that means there is another world B where the same good ship *Theseus* exists with all but one plank the same as in our world A, and only plank #473 different. But then in world B one has a good ship *Theseus* made of 1000 planks, and by the same sort of intuition, there must another world C where the same good ship *Theseus* exists with all but one plank the same as in the world B, but with plank #692 different. That means for us back in world A there is another world C where the good ship *Theseus* exists with all but *two* planks the same as in our world A, but with planks #473 and #692 different, so one could have two planks different and still have the same ship. The same sort of considerations can then be used to argue that one could have three planks different, or four, or five, or all 1000. But that is contrary to our other intuition Y.

The modal paradox resembles well-known paradoxes of vagueness, such as the *heap* and the *bald one*, for which proposed solutions are a dime a dozen – except that here what seems to be vague is the relation of *identity*. And the idea that 'is the very same thing as' could be vague is for many a far more troubling idea than the idea that 'heap' or 'bald' is vague. Indeed, according to many, it is an outright incoherent idea. Salmon has suggested that the solution here, the way to avoid treating identity as vague, is to recognize that though world B is a possibility for us in world A, and world C is a possibility for those in world B, still C is not a possibility for us in world A, but only a 'possible possibility'.

There are reasons, however, to doubt the cogency of this suggestion. For the modal puzzle seems entirely parallel to the famous temporal puzzle discussed by Plutarch and Thomas Hobbes. According to this puzzle, in the year A, the hero Theseus returned to Athens after slaying the Minotaur, and his ship was left as a monument in the harbor at Piraeus. In the year B, one plank, beginning to rot, was replaced, and in the year C another plank, and so on, until in the end none of the original planks were left. The puzzle is that we have an intuition that changing one plank will leave us with the same ship but that if we change them all we have only a replica, and not the original ship.

No one thinks the solution to this classical puzzle is to maintain that though time B may be later than time A, and time C later than time B, still time C is not later than year A, that it is at time A a time that 'is going to be going to come' but not a time that 'is going to come'. Parity of reasoning suggests that if we are suspicious of the idea of a future future that is not a present future, we should be equally suspicious of the idea of a possible possibility that is not an actual possibility.

The two puzzles or paradoxes can be developed in a way that makes the parallelism between the modal and temporal cases more strikingly apparent. We have a thousand planks laid out in a warehouse, and instructions for assembling them into a ship, indicating which plank is to go where, and the following standing instruction. Each day we are to look out for a signal at noon, and if it comes, we are to assemble the ship. If the signal does not come, we are to inspect the planks, remove whichever one seems least sound and burn it, replacing it with the top plank in a reserve pile stored in a secure place. Once the signal does come and a ship is assembled, we are to inspect it each day, and replace whichever plank seems least sound, in the same manner.

In both versions of the puzzle, the signal comes the first day, and the ship is assembled. In the temporal version, we compare the ship constructed on the first day with the ship in the harbor on the second day, the ship in the harbor on the third day, and so on through a thousand daily changes of a plank, until the last day. In the modal version we consider ship constructed on the first day with the ship that would have been constructed if the signal had not arrived until the second day, with the ship that would have been constructed if the signal had not arrived until the third day, and so on. I leave it to the reader to think through the details.

The parallelism between the temporal and modal cases, and the clear untenability in the temporal case of the suggestion that 'later than' is not transitive, to me make Salmon's solution to the modal paradox is uninviting. Now Salmon's is not the only purported solution on offer, but I don't think any of the others on offer is any better, nor have I a better one of my own to suggest. My reason for mentioning this problem about the necessity of origins is not to showcase some new purported solution, but rather to underscore the significance of the difference between the realist and pragmatist views, thus indicating how that division may ultimately be relevant to the choice of the right modal logic, and specifically to the decision whether it includes the law that *possibly possible* implies *possible*. The difference between the realist and pragmatist views is not over which solution to the modal paradox is best, but over how confident one can be that there *is* any solution, and ultimately over how significant the puzzle is.

From a the standpoint of a genuine realist, presumably there just *has to be* some solution, even if it is one requiring us to adopt a radically revisionist metaphysics. Since the problem is that the vagueness of *identity* seems to be intolerable, in a way other kinds of vagueness generally are not, the most obvious solution would be to get rid of the object, the ship, whose identity seems to be vague. This might be done by adopting a revisionist metaphysics in which such extended, composite objects as ships don't really exist at all, but only atoms in the void, or in more modern terms, leptons and quarks, or strings or something, some of which sometimes swarm in ship-like formation.

For the pragmatist, however, it is all only a matter of whether a certain object in a certain contemplated counterfactual situation would count as the same object as a certain other object in a certain other contemplated counterfactual situation. And that, in turn, is only really a question of whether the two *ought to* count as one according to some rules ultimately created by *us*, as makers and users of language. I produced early on a famous quotation from Kant. Let me now produce an even more famous one, from his 'Idea for a Universal History': 'Out of the crooked timber of humanity no straight thing was ever made.' Given the imperfection, the 'crookedness' of every creation of ourselves, it may well be that our rules and conventions yield in certain cases conflicting answers, or no answers; and in such a case the puzzle can have no solution. On the pragmatist view it is always well to be aware our imperfections, and the puzzle of the origin of necessity is of significance insofar as it helps highlight one of those imperfections – but not because the puzzle demands a solution. For the pragmatist, until we are shown that some practical decision of genuine importance would hinge crucially on how we answer the question 'Would it be the same ship or not?' we may be content to leave the question unanswered. Moreover, if the question ever did prove to be, in some specific case, of practical importance, then according to the pragmatist view we could then settle it by *fiat*, making a new rule for ourselves, since it is rules we have made to which the answers to all questions about modality are ultimately to be traced.

In sum, for those who take a realist attitude towards the question of the origin of modality, there will be a real problem of principle about the necessity of origin. By contrast, to those who take a pragmatist attitude there will only be the potential for a practical problem, a potential that is thus far apparently unrealized, and that if it ever were realized could always be solved by fiat. On the realist view it seems that the question of which modal logic is the right one must remain in suspense until the right solution to the modal paradox has been found. On the pragmatist view, there simply may *be* no answer to the question of which modal logic is right, since there may be none to the question of which solution to the paradox is right. Clearly, then, pursuit the right modal logic for 'metaphysical' modality, which was arguably modal logic's original aim, will take the logician deep into philosophy and beyond it into 'metametaphysics'.

Note

Portions of this paper derive from the second annual Kripke Lecture, which I presented at the CUNY graduate center in fall, 2012, and reprised a week later at a conference on modality at Princeton hosted by Boris Kment, and at the University of Connecticut in spring, 2013, using the title 'The Origin of Necessity and the Necessity of Origin' for the first two presentations, and 'Realism and Pragmatism about Modality' for the third, somewhat revised version. I am grateful to comments from the audience on all three occasions, which have led to revisions that I hope are improvements. The lecture material in turn was partly a summary, but partly an elaboration going beyond, my recent book *Saul Kripke: Puzzles and Mysteries* (Polity Press, 2013). In particular, I refer the reader to that book (especially the first sections of the Introduction and of chapter 2) for fuller treatment of the material in §§ 3, 4, and 6 above (and for discussion in chapter 5 of the 'skeptical paradox' about 'rules', which I have barely alluded to here), and especially for bibliographical references connected with this material. Portions of §6 trace back to an old unpublished paper of mine, 'The Varied Sorrows of Modality' (whose title is a phrase of Quine's), and I remain grateful to my colleagues Gilbert Harman and the late David Lewis for comments on that draft. Finally, I am indebted to Teresa Robertson for discussions of 'the modal paradox', and refer the reader to her *Stanford Encyclopedia of Philosophy* entry 'Essential vs. Accidental Properties' and its linked supplement 'Arguments for Origin Essentialism' for references to and discussion of the work of Salmon mentioned in §7 above, and the large body of related work I have not had space to discuss.

Xavier Caicedo

Lindström's Theorem for Positive Logics, a Topological View

Abstract: The topology induced on classes of first order structures by the elementary classes is characterized among finer topologies by intrinsic non-boolean properties. This yields a topological version of Lindström's first theorem which applies to certain extensions of first order logic lacking classical negation. In this setting, we examine the possibility of extending the equilibrium game semantics of imperfect information logic on finite models to infinite models.

Keywords: abstract model theory, negationless logics, elementary topology, imperfect information logic, game semantics

MSC: 03C95, 03C52, 91A05, 91A40

Xavier Caicedo: Departamento de Matemáticas, Universidad de los Andes, AA 4976, Bogotá, Colombia

1 Introduction

Closure under classical negation is an essential feature in the usual proofs of the celebrated Lindström's theorems characterizing first order logic [21] and many other results in abstract model theory. We provide in this paper an intrinsic characterization of the elementary topology, actually a topological version of Lindström's first theorem, from which negationless forms of this result follow. It is well known that for each first-order vocabulary τ the class of first-order structures of type τ becomes a topological space by taking as a basis the elementary classes $Mod(\varphi)$, φ a first-order sentence, so that model theoretic properties as compactness correspond to genuine topological properties. We show that among finer topologies this is the unique regular compact topology for which the class of countable structures is dense, and the forgetful and renaming functors relating distinct vocabularies are continuous. Since any model theoretic logic gives rise similarly to a natural topology, we obtain versions of Lindström's theorem which apply to certain extensions of first-order logic lacking classical negation.

One of the most interesting such extensions is imperfect information logic in its various versions (cf. [23], [30]). In this context, we discuss the problem of extending to infinite models the [0,1]-valued equilibrium semantics of these logics for finite models introduced in [28], [15]. We show that under natural conditions this is only possible for essentially first order fragments of imperfect information logic.

Model theoretic logics without negation have been considered by García-Matos in [17]. Topological ideas have been present in classical model theory from its beginnings, mainly via the study of Polish spaces of enumerated countable models [11] or spaces of types [24], [25]. Working directly with spaces of models appears first in Fraïssé's beautiful proof of countable compactness of first order logic [13]. This approach has been exploited in the context of model theoretic logics by Mundici in [26] and the author in [5], [6], [7].

We refer the reader to [10], [2], [32] for unexplained concepts in model theory, abstract model theory or topology, respectively.

2 The elementary topology

Given a first-order vocabulary τ, $L_{\omega\omega}(\tau)$ is the set of first-order sentences of type τ. The *elementary topology* on the class St_τ of first-order structures type τ is obtained by taking the family of *elementary classes*

$$Mod(\varphi) = \{M : M \models \varphi\}, \quad \varphi \in L_{\omega\omega}(\tau)$$

as an open basis. Due to the presence of classical negation, this family is also a closed basis and thus the closed classes of St_τ are the first-order axiomatizable classes $Mod(T)$, $T \subseteq L_{\omega\omega}(\tau)$. Possible foundational problems due to the fact that the topology is a class of classes may be settled observing that it is indexed by a set, namely the set of theories of type τ.

The main facts of model theory are reflected by the topological properties of these spaces. Thus, the *downward Löwenheim-Skolem theorem* for sentences amounts to topological density of the subclass of countable structures. *Łoś theorem* on ultraproducts grants that *U-limits* exist for any ultrafilter U, condition well known to be equivalent to topological compactness, and to model theoretic compactness in this case.

These spaces are not Hausdorff or T_1, but having a clopen basis they are *regular*; that is, closed classes and exterior points may be separated by disjoint open classes. All properties or regular compact spaces are then available: normality, complete regularity, uniformizability, the Baire property, etc.

Many model theoretic properties are related to the continuity of natural operations between classes of structures. The following operations are readily seen to be continuous and play an important role in abstract model theory:

- *Reducts*, $\pi_{\sigma\tau} : St_\sigma \to St_\tau$, $\pi_{\sigma\tau}(M) = M \upharpoonright \tau$, which forgets the interpretation of the symbols in $\sigma \setminus \tau$ when $\tau \subseteq \sigma$.

- *Renamings*, $\hat{\alpha} : St_\sigma \to St_\tau$ that change the names of the interpretations according to a bijection of vocabularies $\alpha : \sigma \to \tau$ respecting kind and arity of the symbols. More precisely, $(\alpha R)^{\hat{\alpha}(M)} = R^M$ for any $R \in \sigma$.

Not only these, but a host of operations are continuous for the elementary topology: first-order interpretations, cartesian products (as operations on several arguments), any operation with Σ^1_1 graph (cf. [6]), any operation enjoying the *uniform reduction property* in the sense of Feferman and Vaught [12]. Consider, for example:

- *Restrictions*, $r_{\tau P} : St_\tau \to St_\tau$, where $P \in \tau$ is a monadic predicate symbol, sending M to the substructure $M \upharpoonright P^M$ with universe P^M, when this set is a subuniverse of M. If τ has function symbols, this is a partial operation with closed domain. It is continuous because $r_P^{-1}(Mod_\tau(\varphi)) = Mod_\tau(\varphi^P)$ where φ^P results of relativizing the quantifiers of φ to P.

- *Disjoint sums*, which assign to a family of structures $(M_i)_{i\in I} \in \Pi_i St_{\tau_i}$ the structure of type $\oplus_i \tau_i := \cup_i \{P_i\} \cup \underline{\tau_i}$:

$$\Sigma_{i\in I} M_i = (\cup_i \underline{A_i}, \underline{M_i})_{i\in I},$$

where the $\underline{M_i}$ are mutually disjoint renamings of the M_i with disjoint vocabularies $\underline{\tau_i}$ and universes $\underline{A_i}$ which interpret the predicates P_i. The resulting map $S : \Pi_i St_{\tau_i} \to St_{\oplus_i \tau_i}$ is continuous due to Feferman-Vaught theorem [12] which grants for each sentence $\varphi \in L_{\omega\omega}(\oplus_i \tau_i)$ sentences $\psi_j \in L_{\omega\omega}(\tau_{i_j})$, $j = 1, ..., k$, and a boolean condition $B(p_1, ..., p_k)$ such that $\Sigma_{i\in I} M_i \models \varphi$ if and only if $B(M_{i_1} \models \psi_1, ..., M_{i_k} \models \psi_k)$ holds. Taking a disjunctive normal form of B, this is seen to mean that $(M_i)_{i\in I}$ belongs to a union of basic classes of the product topology on $\Pi_i St_{\tau_i}$. Hence, $S^{-1}(Mod(\varphi))$ is open.

3 Topological regularity

As noticed before, a topological space is regular if closed sets and exterior points may be separated by open sets. It is *normal* if disjoint closed sets may be separated by disjoint open sets. We do not include the Hausdorff property in our definitions. Thus, normality does not imply regularity here. However, a regular compact space is normal. Actually, a regular Lindelöf space is already normal (cf. [32], exc. 32.4).

Consider the following equivalence relation in a space X:

$$x \equiv y \Leftrightarrow cl\{x\} = cl\{y\}$$

where cl denotes topological adherence. Clearly, $x \equiv y$ if and only if x and y belong to the same closed (open) subsets (of a given basis). Let $X_{/\equiv}$ be the quotient space and $\eta : X \to X_{/\equiv}$ the natural projection. Then $X_{/\equiv}$ is T_0 by construction but not necessarily Hausdorff. The following claims are easily verified:

a) *$\eta : X \to X_{/\equiv}$ induces an isomorphism between the respective lattices of Borel subsets of X and $X_{/\equiv}$. In particular, it is open and closed, preserves disjointedness, preserves and reflects compactness and normality.*

b) *The assignment $X \longmapsto X_{/\equiv}$ is functorial,* because $\equiv$ is preserved by continuous functions and thus any continuous map $f : X \to Y$ induces a continuous assignment $f_{/\equiv} : X_{/\equiv} \to Y_{/\equiv}$ which commutes with composition.

c) *$X \longmapsto X_{/\equiv}$ preserves products;* that is, $(\Pi_i X_i)_{/\equiv}$ is canonically homeomorphic to $\Pi_i(X_{i/\equiv})$ with the product topology (monomorphisms are not preserved).

d) *If X is regular, the equivalence class of x is $cl\{x\}$* (this may fail in the non regular case).

e) *If X is regular, $X_{/\equiv}$ is Hausdorff*: if $x \not\equiv y$ then $x \notin cl\{y\}$ by (d); thus there are disjoint open sets U, V in X such that $x \in U$, $cl\{y\} \subseteq V$, and their images under η provide an open separation of ηx and ηy in $X_{/\equiv}$ by (a).

f) *If K_1 and K_2 are disjoint compact subsets of a regular topological space X that can not be separated by open sets there exist $x_i \in K_i, i = 1, 2,$ such that $x_1 \equiv x_2$.* Indeed, ηK_1 and ηK_2 are compact in $X_{/\equiv}$ by continuity and thus closed because $X_{/\equiv}$ is Hausdorff by (e) above. They can not be disjoint; otherwise, they would be separated by open sets whose inverse images would separate K_1 and K_2. Pick $\eta x = \eta y \in \eta K_1 \cap \eta K_2$ with $x \in K_1$, $y \in K_2$.

Clearly, for the elementary topology on St_τ, the relation $\equiv$ coincides with elementary equivalence of structures and $St_{\tau/\equiv}$ is homeomorphic to the Stone space of complete theories.

As noticed before, the spaces St_τ with the elementary topology are uniformizable. It is easy to check that the family of relations $\equiv_{n,\tau_0}$ (τ_0-elementary equivalence of structures up to quantifier rank n), with $n \in \omega$ and finite $\tau_0 \subseteq \tau$, forms a uniformity basis.

4 An intrinsic characterization of the elementary topology

In this section, $x, y, \ldots$ will denote first-order structures in St_τ, $x \approx y$ will denote isomorphism.

$x \sim_{n,\tau} y$ means that there is a sequence $\emptyset \neq I_0 \subseteq .... \subseteq I_n$ of sets of τ-partial isomorphism of finite domain so that, for $i < j \leq n$, $f \in I_i$ and $a \in x$ (respectively, $b \in y$), there is $g \in I_j$ such that $g \supseteq f$ and $a \in Dom(g)$ (respectively, $b \in \mathrm{Im}(g)$). The later is called the *extension property*.

$x \sim_\tau y$ means the above holds for an infinite chain $\emptyset \neq I_0 \subseteq .... \subseteq I_n \subseteq ...$

Fraïssé's characterization of elementary equivalence [13] says that for finite relational vocabularies: $x \equiv_{n,\tau} y$ if and only if $x \sim_{n,\tau} y$. To have it available for vocabularies containing function symbols add the complexity of terms in atomic formulas to the quantifier rank.

It is well known that for countable $x, y : x \sim_\tau y$ implies $x \approx y$.

Given a vocabulary τ let τ^* be a disjoint renaming of τ. If $x, y \in St_\tau$ have the same power, let y^* be an isomorphic copy of y sharing the universe with x and renamed to be of type τ^*. In this context, (x, y^*) will denote the $\tau \cup \tau^*$-structure that results of expanding x with the relations of y^*.

Lemma 4.1. *There is a vocabulary $\tau^+ \supseteq \tau \cup \tau^*$ such that for each finite vocabulary $\tau_0 \subseteq \tau$ there is a sequence of elementary classes $\Delta_1 \supseteq \Delta_2 \supseteq \Delta_3 \supseteq ....$ in St_{τ^+} such that if $\pi = \pi_{\tau^+,\tau\cup\tau^*}$ then (1) $\pi(\Delta_n) = \{(x, y^*) : |x| = |y| \geq \omega, x \equiv_{n,\tau_0} y\}$, (2) $\pi(\bigcap_n \Delta_n) = \{(x, y^*) : |x| = |y| \geq \omega, x \sim_{\tau_0} y\}$. Moreover, $\bigcap_n \Delta_n$ is the reduct of an elementary class.*

Proof. Let Δ be the class of structures $(x, y^*, <, a, I)$ where $<$ is a discrete linear order with minimum but no maximum and I codes for each $c \leq a$ a family $I_c = \{I(c, i, -, -)\}_{i \in x}$ of partial τ_0-τ_0^*–isomorphisms from x into y^*, such that for $c < c' \leq a$: $I_c \subseteq I_{c'}$ and the extension property holds. Describe this by a first-order sentence θ_Δ of type $\tau^+ \supseteq \tau_0 \cup \tau_0^*$ and set $\Delta_n = Mod_L(\theta_\Delta \wedge \exists^{\geq n} x(x \leq a)\}$. Then condition (1) in the Lemma is granted by Fraïssé's characterization and the fact that x being infinite has room to code all partial isomorphisms of finite domain, and condition (2) is granted because $(x, y^*, <, a, I) \in \bigcap_n \Delta_n$ if and only if $<$ contains an infinite increasing ω-chain below a, a Σ_1^1condition. □

A topology on St_τ is *invariant* if its open (closed) classes are closed under isomorphic structures. In the following we will always assume this property of the topologies considered. Of course, it is superfluous if we identify isomorphic structures.

Theorem 4.2. *Let Γ_τ be a regular compact topology finer than the elementary topology on each class St_τ such that the countable structures are dense in St_τ and reducts and renamings are continuous for these topologies. Then Γ_τ is the elementary topology for all τ.*

Proof. We first show that any pair of disjoint closed classes C_1, C_2 of Γ_τ may be separated by an elementary class. Assume this is not the case. Since the C_i are compact in the topology Γ_τ then they are compact for the elementary topology and, by regularity of the latter, there exist $x_i \in C_i$ such that $x_1 \equiv x_2$ in $L_{\omega\omega}(\tau)$ by claim (f) in the previous section. The x_i must be infinite, otherwise they would be isomorphic contradicting the disjointedness of the C_i. By normality of Γ_τ, there are towers $U_i \subseteq C_i' \subseteq U_i' \subseteq C_i''$, $i = 1, 2$, separating the C_i with U_i, U_i' open and C_i', C_i'' closed in Γ_τ and disjoint. Let I be a first-order sentence of type $\tau' \supseteq \tau$ such that $(z, ..) \models I \Leftrightarrow z$ is infinite, and let π be the corresponding reduct operation. For fixed $n \in \omega$ and finite $\tau_0 \subseteq \tau$ let t be a first-order sentence describing the common $\equiv_{n,\tau_0}$-equivalence class of x_1, x_2. As

$$(x_i, ..) \in Mod_{\tau'}(I) \cap \pi^{-1}Mod(t) \cap \pi^{-1}U_i, \quad i = 1, 2$$

and this class is open in $\Gamma_{\tau'}$ by continuity of π, then by the density hypothesis there are countable $x_i \in U_i$, $i = 1, 2$, such that $x_1 \equiv_{n,\tau_0} x_2$. Thus for some expansion of (x_1, x_2^*),

$$(x_1, x_2^*, ..) \in \Delta_{n,\tau_0} \cap \pi_1^{-1}(C_1') \cap (\rho\pi_2)^{-1}(C_2'), \tag{1}$$

where Δ_{n,τ_0} is the class of Lemma 4.1, π_1, π_2 are reducts, and ρ is a renaming:

$$\begin{aligned} &\pi_1(x_1, x_2^*, ...) = x_1 && \pi_1 : St_{\tau^+} \to St_{\tau\cup\tau^*} \to St_\tau \\ &\pi_2(x_1, x_2^*, ...) = x_2^* && \pi_2 : St_{\tau^+} \to St_{\tau\cup\tau^*} \to St_{\tau^*} \\ &\rho(x_2^*) = x_2 && \rho : St_{\tau^*} \to St_\tau. \end{aligned}$$

Since the classes (1) are closed by continuity of the above functors then $\bigcap_n \Delta_{n,\tau_0} \cap \pi_1^{-1}(C_1') \cap (\rho\pi_2)^{-1}(C_2')$ is non-emtpy by compactness of Γ_{τ^+}. But $\bigcap_n \Delta_{n,\tau_0} = \pi(V)$ with V elementary of type $\tau^{++} \supseteq \tau^+$. Then

$$V \cap \pi^{-1}\pi_1^{-1}(U_1') \cap \pi^{-1}(\rho\pi_2)^{-1}(U_2') \neq \emptyset$$

is open of $\Gamma_{L^{++}}$ and by the density condition it must contain a countable structure $(x_1, x_2^*, .., ...)$. Thus $(x_1, x_2^*, ..) \in \bigcap_n \Delta_{n,\tau_0}$, with $x_i \in U_i' \subseteq C_i''$. It follows that $x_1 \sim_{\tau_0} x_2$ and thus $x_1 \upharpoonright \tau_0 \approx x_2 \upharpoonright \tau_0$. Let δ_{τ_0} be a first-order sentence of type $\tau \cup \tau^* \cup \{h\}$ such that: $(x, y^*, h) \models \delta_{\tau_0} \Leftrightarrow h : x \upharpoonright \tau_0 \approx y \upharpoonright \tau_0$. By compactness,

$$\Big(\bigcap_{\tau_0 \subseteq_{fin} \tau} Mod_{\tau\cup\tau^*\cup\{f\}}(\delta_{\tau_0})\Big) \cap \pi_1^{-1}(C_1'') \cap (\rho\pi_2)^{-1}(C_2'') \neq \emptyset$$

and we have $h : x_1 \approx x_2$, $x_i \in C_i''$, contradicting the disjointedness of the C_i''. Finally, if C is a closed class of Γ_τ and $x \notin C$, $cl_{\Gamma_\tau}\{x\}$ is disjoint from C by regularity of Γ_τ. Then $cl_{\Gamma_\tau}\{x\}$ and C may be separated by open classes of the elementary topology, which implies C is closed in this topology. □

If we ask continuity of further operations, we may trade the compactness hypothesis in Theorem 4.2 for normality or the Lindelöf property, due to the following topological version of Th. 15 in [5].

Theorem 4.3. *Let Γ_τ be a regular normal (or Lindelöf) topology on each St_τ such that reducts, renamings, restrictions, and disjoint sums are continuous. Then Γ_τ is compact.*

Proof. By hypothesis, the disjoint sum embedding $S : \Pi_i St_{\tau_i} \to St_{\oplus_i \tau_i}$ is continuous for the Tychonoff product of the topologies Γ_{τ_i} in the domain and the topology $\Gamma_{\oplus_i \tau_i}$ in the co-domain. On the other hand, the map $D : St_{\oplus_i \tau_i} \to \Pi_i St_{\tau_i}$:

$$M \longmapsto (renam_{\underline{\tau_i},\tau_i}(M \restriction P_i^M \restriction \underline{\tau}_i))_i$$

decomposing a structure into a family of restricted renamed reducts is also continuous for these topologies since each projection $St_{\oplus_i \tau_i} \to St_{\tau_i}$ is continuous by hypothesis. Clearly, $D \circ S$ is the identity on $\Pi_i St_{\tau_i}$ (module isomorphism) and by the functorial properties of $X \longmapsto X_{/\equiv}$, $D_{/\equiv} \circ S_{/\equiv}$ is the identity in $(\Pi_i St_{\tau_i})_{/\equiv}$. Therefore, $S_{/\equiv}$ is injective and its image C in $St_{\oplus_i \tau_i/\equiv}$ is a retract of this space. Since the latter space is Hausdorff by regularity of $\Gamma_{\oplus_i \tau_i}$ and retracts of Hausdorff spaces are closed, then C is closed. As $St_{\oplus_i \tau}$ is normal by hypothesis, so is $St_{\oplus_i \tau_i/\equiv}$ and thus C is normal also. Therefore, $\Pi_i (St_{\tau_i/\equiv}) \approx (\Pi_i St_{\tau_i})_{/\equiv} \approx C$ is normal. In particular, all powers $(St_{\tau/\equiv})^I$ are normal. By Noble's theorem stating that a Hausdorff space X is compact if and only if the power space X^κ is normal for all κ ([27], Corollary 2.2), we conclude that $St_{\tau/\equiv}$ and thus St_τ is compact for Γ_τ. Finally, recall that regular Lindelöf spaces are normal. □

5 Lindström's theorem revisited

A model theoretic logic is a pair $(L, \models)$ where L is an assignment $\tau \longmapsto L_\tau$ from first-order vocabularies to classes of "sentences", and $\models$ is a relation in $\cup_\tau S_\tau \times L_\tau$ satisfying Lindström's axioms (cf. [2], II. Definition1.1.1):

- *Isomorphism*: If $M \approx N$ then $M \models \varphi$ iff $N \models \varphi$.
- *Reduct*: If $\sigma \subseteq \tau$ then $L_\sigma \subseteq L_\tau$ and for any $\varphi \in L_\sigma$, $M \in St_\tau : M \restriction \sigma \models \varphi$ iff $M \models \varphi$.
- *Renaming*: A bijection $\alpha : \sigma \to \tau$ respecting kind and arity induces a map $t : L_\sigma \to L_\tau$, so that $M \models \varphi$ iff $\hat{\alpha} M \models t\varphi$, where $\hat{\alpha} : St_\sigma \to St_\tau$ renames the structures according to α.

Our results will hold under relaxed versions of the reduct and renaming axioms where the inclusion $L_\sigma \subseteq L_\tau$ and the substitution map $t : L_\sigma \to L_\tau$ are replaced with maps $L_\sigma \to P(L_\tau)$ from symbols to theories.

Define $L \leq_w L'$ (L' is a *weak extension* of L) if each sentence of L_τ is equivalent to a theory of L'_τ. Then, L is *equivalent in theories* to L' if $L \leq_w L'$ and $L' \leq_w L$.

Closure of L under boolean connectives $\neg, \vee, \wedge$, or relativization is defined as usual. The same for compactness. As noticed in [17], the downward Löwenheim-Skolem theorem splits for negationless logics in several non equivalent versions:

- LSk_1: If any sentence has a model it has a countable model (the familiar version).
- LSk_2: Any sentence true in all countable models is true in all models.
- LSk_3: Two sentence equivalent in all countable models are equivalent (this property implies the other two).

Σ^1_1 (Second order existential logic) satisfies the first but not the second version, and the opposite is true of Π^1_1 (second order universal logic) because $\{\mathbb{R}\}$ is Π^1_1-definable. The fragment $L_{\omega\omega}(Q_0^+)$ of $L_{\omega\omega}(Q_0)$ obtained by closing $L_{\omega\omega}$ under the quantifier Q_0 (there are infinitely many...) and $\wedge, \vee, \exists, \forall$, but not $\neg$, satisfies all three versions.

Definition. For any model-theoretic logic L let $\Gamma_\tau(L)$ the topology on St_τ obtained by taking the classes $Mod(\theta)$, $\theta \in L_\tau$, as a sub-basis of closed classes.

This topology is invariant, and reducts and renamings become automatically continuous due to the (relaxed) Lindström's axioms. Model theoretic compactness of L is equivalent to topological compactness of $\Gamma_\tau(L)$ by Alexander sub-basis lemma (cf [32]). Closure under connectives refers to properties of the sub-basis; for example, closure under disjunctions grants it is a closed basis. LSk_2 corresponds in the later case to topological density of the countable structures. LSk_1, instead, is basis dependent if the logic is not closed under negations. Finally, $L \leq_w L'$ means that $\Gamma_\tau(L')$ is finer that $\Gamma_\tau(L)$.

Call a logic *regular* if the topology it induces is regular[1]. In a regular compact logic closed under disjunctions and conjunctions, LSk_1 implies LSk_2: if $M \notin Mod(T)$, then by regularity M belongs to a closed class C contained in $Mod(T)^c$. By compactness there are finitely many closed basics C_i such that $C \subseteq \cap_i C_i \subseteq Mod(\varphi)^c$. By the closure conditions $\cap_i C_i$ contains a countable model if LSk_1 holds. Thus, $Mod(\varphi)^c$ has a countable model and LSk_2 holds.

The next version of Lindström's theorem follows immediately from Theorem 4.2 and the previous observations.

Theorem 5.1. (Positive Lindström's theorem) *Any regular compact logic L extending weakly $L_{\omega\omega}$, closed under disjunctions and satisfying LSk_2, is equivalent in theories to $L_{\omega\omega}$. If L is also closed under conjunctions, the results holds under LSk_1.*

1 No to be confused with the notion of regular logic in [2].

Trivial examples of logics satisfying the hypothesis of the first part of this theorem are $th^{\kappa}_{\omega\omega,\tau} = \{T : T \text{ is a first-order theory of type } \tau, |T| \leq \kappa\}$. From these, only $th^{\omega}_{\omega\omega,\tau}$ satisfies LSk_1. This shows that regularity and compactness do not grant finite dependence nor any bound on the dependence number.

Regularity can not be dropped from the hypothesis of Theorem 5.1. For example, $L_{\omega\omega}(Q_0^+)$ satisfy LSk_2 and LSk_1, and it is compact by an ultraproduct argument (see next section), but the class $Mod(\exists x Q_0 y(y < x))$ is not first-order axiomatizable because its complement is not closed under ultrapowers. We conclude that this logic can not be regular.

Similarly, Σ^1_1 can not be regular because it satisfies the other hypothesis of Theorem 5.1 (with LSk_1) but its sentences are not all reducible to first-order theories. However, its topology is normal due to Robinson's consistency lemma. This shows that we can not replace regularity with normality in any of our previous results.

Topological regularity of $\Gamma_\tau(L)$ does not have a simple model theoretic description, but in compact logics closed under $\vee$ and $\wedge$, it implies the following weak form of negation: *for any* $\varphi \in L_\tau$ *there is* $\{\theta_i\}_{i\in I} \subseteq L_\tau$ *such that* $M \not\models \varphi$ *if and only if* $M \models \vee_i \theta_i$. The second claim of Theorem 5.1, but not the first, may be shown utilizing this hypothesis instead of regularity.

Lindström's original theorem (for a wider notion of extension) is an immediate corollary of Theorem 5.1.

Theorem 5.2. *A compact weak extension of first-order logic closed under boolean connectives and satisfying* LSk_1 *is equivalent (in sentences) to* $L_{\omega\omega}$.

Proof. Closure of L under boolean connectives means that the closed sub-basis of $\Gamma_\tau(L)$ is actually a clopen basis and thus the space is regular. By Theorem 5.1, for any $\theta \in L_\tau$ the complementary classes $Mod(\theta)$ and $Mod(\neg\theta)$ are equivalent to first-order theories, hence they are elementary. □

Versions of Theorems 5.1 and 5.2 in terms of $[\omega_1, \infty]$-compactness instead of compactness may be obtained from Theorem 4.3 if we add closure under relativizations and the uniform reduction property for disjoint sums.

6 Compactness and ultraproducts

The relation between compactness and ultraproducts survives in negationless logics because it is a purely topological convergence phenomenon. Recall that given an ultrafilter U over I, a family $\{a_i\}_{i\in I}$ *U-converges* to an element x in a space X, in symbols $\{a_i\}_{i\in I} \to_U x$, if and only if $\{i \in I : a_i \in V\} \in U$ for any open (sub-basic) neighborhood V of x. It is well known that a space is compact iff any I-family of X has an U-limit.

Given a logic L, convergence of structures, say $\{M_i\}_{i\in I} \to_U M$, expressed in terms of open sub-basics of $\Gamma_\tau(L)$ means: $M \not\models \varphi \Rightarrow \{i \in I : M_i \not\models \varphi\} \in U$; that is

$$\text{for all } \varphi \in L_\tau : \{i \in I : M_i \models \varphi\} \in U \Rightarrow M \models \varphi. \tag{1}$$

Therefore, a logic is compact if and only if for any family of structures $\{M_i\}_{i\in I}$ of the same type and ultrafilter U over I there exists M such that (1) holds. Thus, any logic L satisfying the following *one-way ultraproduct property* must be compact:

$$\{i \in P : M_i \models \varphi\} \in U \Rightarrow \Pi_i M_i/_U \models \varphi. \tag{2}$$

Example. Let $\mathcal{K}$ be a family of Lindström quantifiers closed under ultraproducts (cardinality quantifiers Q_α, Magidor-Malitz quantifiers, the Hartig quantifier, etc.), then the logic $L_{\omega\omega}(\mathcal{K}^+)$ obtained by closing $L_{\omega\omega}$ under the quantifiers in $\mathcal{K}$ and $\wedge, \vee, \exists, \forall$ may be shown to satisfy (2) by a simple induction.

Reciprocally, compactness of L implies that for any family $\{M_i\}_{i\in I}$ of structures and any ultrafilter U over I there is an extension $M \geq \Pi_i M_i/_U$ such that for any $\varphi \in L_{\tau\cup\{c,d,..\}}$, $f, g, .. \in \Pi_i M_i$

$$\{i \in P : M_i \models \varphi[f(i), g(i), ..]\} \in U \Rightarrow M \models \varphi[f_{/U}, g_{/U}, ..], \tag{3}$$

because the family of expansions $M_i^+ = (M_i, f(i))_{f\in\Pi_i M_i}$, where $f(i)$ interprets c_f, must have a U-limit $(M, a_f)_{f\in\Pi_i M_i}$, and applying (3) to atomic sentences yields an embedding $f/_U \mapsto a_f : \Pi_i M_i/_U \leq M$.

In this vein, the topological proof given in [7] of Shelah's characterization of $[\lambda, \kappa]$-compact boolean closed logics in terms of ultraproducts, [22], may be readily adapted to negationless logics. Regularity of the logic is not required.

Recall that a logic L is $[\lambda, \kappa]$*-compact* if, for any family of theories $\{T_\alpha\}_{\alpha<\kappa}$ in L_τ, if $\cup_{j\in S} T_{\alpha_j}$ is satisfiable for all $S \subseteq \kappa$ of power less than λ then $\cup_{\alpha<\kappa} T_\alpha$ is satisfiable. An ultrafilter U is (λ, κ)*-regular* if there is $F \subseteq U$ of power κ such that $\cap J = \varnothing$ for any $J \subseteq F$ of power λ.

Theorem 6.1. (Positive abstract compactness theorem) *Let L be a logic L closed under disjunctions. If there is a (λ, κ)-regular ultrafilter U over $\kappa^{<\lambda}$ such that for any family $\{M_i\}_{i\in I}$ of structures there is an extension $M \geq \Pi_i M_i/_U$ such that for any $\varphi \in L_{\tau\cup\{c,d,..\}}$ and $f, g, .. \in \Pi_i M_i$ (3) holds, then L is $[\lambda, \kappa]$-compact. If L is closed under relativizations the reciprocal holds.*

7 Imperfect information logic as [0,1]-valued logic

One of the most interesting logics lacking classical negation is *imperfect information logic* in its various versions. These derive from independence friendly logic *IF*, introduced originally by Hintikka as a "friendly" way of expressing Henkin quantifiers [19]. Its syntax extends that of $L_{\omega\omega}$ allowing for each finite set of variables Y and formulas φ, ψ, expressions $\exists x_{/Y}\varphi$, $\forall x_{/Y}\varphi$, $\varphi \vee_{/Y} \psi$, $\varphi \wedge_{/Y} \psi$, where the decoration $_{/Y}$ expresses independence with respect to the variables in Y, as in the usual definition of uniform continuity of a function:

$$\forall x\ \forall \varepsilon > 0\ \exists \delta_{/x}\ \forall y\ (|x-y| < \delta \rightarrow |fx - fy| < \varepsilon)$$

"*there is δ not depending on x*"

The meaning of $\forall x_{/Y}\varphi$, $\varphi \vee_{/Y} \psi$, $\varphi \wedge_{/Y} \psi$ is more clearly explained by the game semantics governing this logic [23]. To each sentence of φ and structure M of the same type is associated a two-players game $G(\varphi, M)$ which is played from the root down along a branch of the syntactical tree of φ. Calling E and A the two players, they make their moves at each node depending on its label as follows:

$\vee_{/Y}$: E chooses the left or right descendant (subformula)
$\wedge_{/Y}$: A chooses the left or right descendant
$\exists x_{/Y}$: E assigns a value $a \in M$ to the variable x
$\forall x_{/Y}$: A assigns a value $a \in M$ to the variable x
$\neg$: interchange roles and proceed to next node

Each move is made in ignorance of the value previously assigned to the variables in Y. When a terminal node is reached, E wins if the atomic formula labeling the node is true for the valuation of variables constructed by the players along the way; otherwise A wins (if the roles have been reversed by the last negation reached, it is the opposite).

A sentence φ is said to be *true* if E has a winning strategy in this game, *false* if A has a winning strategy. We write, respectively, $M \models^{+} \varphi$ and $M \models^{-} \varphi$ (positive and negative satisfaction).

For first order sentences (where $_{/\varnothing}$ is the only decoration) the game is determined and thus one of the players has a winning strategy according to the classical truth value. This may be seen observing that the existence of a winning strategy for E (for A) corresponds to the satisfaction of a particular (dual) Skolem normal form of φ. For the new sentences this is a game of imperfect information which may be undetermined. It may happen that $M \not\models^{+} \varphi$ and $M \not\models^{-} \varphi$ and thus φ does not have a truth value in the model M.

Game semantics of IF behaves so differently from ordinary semantics that it raises several non-trivial issues. The valid formulas of the form $\forall x \exists y \forall z \exists w_{/x} \varphi$ (φ first-order) form already a non arithmetical set. Validity is not even Σ_2 definable in set theory (Väänänen [29]). A serious study was possible only after streamlining its syntax and Hodges discovery that it has a compositional semantics [18], contradicting Hintikka's previous claims. Väänänen has introduced an essentially equivalent but conceptually different dual version based on the notion of *dependence* between variables, expressed at the atomic level [30]. Some strengthenings have been considered also by Abramsky and Väänänen [1]. We will consider a variant IF^* which eliminates all restrictions on the use of the quantifier slashes and where $\vee, \wedge$ are not slashed (cf. [8]), but our observations below are translatable to dependence logic.

Since a winning strategy is a choice of appropriate Skolem functions making true certain first-order condition, the disjoint classes

$$Mod^+(\varphi) = \{M : M \models^+ \varphi\}, \;\; Mod^-(\varphi) = Mod^+(\neg\varphi) = \{M : M \models^- \varphi\}$$

are Σ^1_1. Therefore, IF^* has the compactness property and the Löwenheim-Skolem property with respect to $\models^+$. Moreover, in finite models it defines exactly the NP classes. Burgess [4] has observed that for any pair of disjoint Σ^1_1-classes K_1, K_2 the union of which contains all one element structures there is an IF^* sentence φ such that $Mod^+(\varphi) = K_1, Mod^-(\varphi) = K_2$.

The semantics of IF^* may be seen as a partial two-valued semantics taking the truth value 1 in $Mod^+(\varphi)$, 0 in $Mod^-(\varphi)$, and no value out of these classes. It is natural to ask if it is possible to interpolate in a sensible manner a [0,1]-valued semantic assigning a value strictly between 0 and 1 to the structures where the game is undetermined. An intriguing positive answer in the realm of finite models is given by *equilibrium semantics*, suggested by M. Ajtai as explained in [3], and studied by Sevenster and Sandu [28] for IF and by Galliani [15] for Väänänen's dependence logic.

Given a particular run of $G(\varphi, M)$ in which the players use strategies σ, τ, the payoff of the run for E is $u(\sigma, \tau) = 1$ if she wins, 0 otherwise. For A the payoff is $1 - u(\sigma, \tau)$. Probabilistic distributions on sets of strategies are called *mixed strategies*. Given a pair of mixed strategies p, q on the respective sets of strategies S_1, S_2 of each player, the expected value of the random variable $u(\sigma, \tau)$, assuming the players choose their strategies independently, is:

$$u^*(p, q) = \Sigma_{\sigma,\tau} p(\sigma)q(\tau)u(\sigma, \tau).$$

Since this is a constant sum game, the theory of games [31] grants that if M is finite there exists an *equilibrium pair* of mixed strategies (p_0, q_0) such that $u^*(p', q_0) \leq$

$u^*(p_0, q_0) \leq u^*(p_0, q')$ for all p', q' so that the players will not improve their payoff by changing them. Obviously all such pairs give the same value which (less obviously) is identical to

$$u^*(p_0, q_0) = \max_p \min_q u^*(p, q) = \min_q \max_p u^*(p, q). \tag{1}$$

Equilibrium semantics takes this *equilibrium value* as the true value of φ in the model M. We will call it φ^M. If $M \models^+ \varphi$ and σ is a winning strategy for E then any pair (σ, τ) is an equilibrium pair and $\varphi^M = 1$; similarly, $\varphi^M = 0$ if $M \models^- \varphi$.

For example, $\varphi^M = 1 - \frac{1}{|M|}$ for the sentence $\forall x \exists y/x(x \neq y)$ (cf. [3]).

This semantics may be seen to satisfy on finite models:

C1. $\varphi^M = 1$ if and only if $M \models^+ \varphi$, $\varphi^M = 0$ if and only if $M \models^- \varphi$
C2. $\neg\varphi^M = 1 - \varphi^M$
C3. $(\varphi \vee \psi)^M, = \max\{\varphi^M, \psi^M\}$
C4. For any rationals $r, \varepsilon \in [0, 1]$, $\varepsilon < 1$, and any sentence $\varphi \in IF^*$ there is $\varphi_{r,\varepsilon} \in IF^*$ such that $\varphi^M \geq r$ iff $\varphi^M_{r,\varepsilon} \geq \varepsilon$.

C2 is straightforward, C1 and C4 are shown in [28], and C3 may be proven by game theoretic considerations. It is an open question whether C4 holds for $\varepsilon = 1$; that is, there is φ_r such that

$$\varphi^M \geq r \text{ iff } \varphi^M_r = 1. \tag{C4*}$$

Equivalently, whether the complexity of the query $\varphi^M \geq r$ with respect to the size of M is in NP. Sevenster and Sandu [28] have shown that $\Sigma^1_2 \cap \Pi^1_2$ is an upper bound for the complexity of these queries. Thus, a collapse of the polynomial hierarchy to NP would make (C4*) true. Finding a counterexample seems a though job because it would separate NP from $\Sigma^1_2 \cap \Pi^1_2$, solving an outstanding problem in algorithmic complexity theory.

Our main question is whether it is possible to have a [0,1]-valued semantic of IF^* in infinite models satisfying C1 to C4 and preserving useful model theoretic properties of IF^*, as compactness and the downward Löwenheim-Skolem property.

The first candidate is equilibrium semantics itself. To extend equilibrium semantics to infinite models we must consider pairs of probabilistic measures p, q on S_1, S_2 for which $u(\sigma, \tau)$ is measurable in the product measure, so that the expected payoff

$$u^*(p, q) = \iint_{(\sigma,\tau)\in S_1 \times S_2} u(\sigma, \tau) dp dq$$

exists. If equilibrium pairs exist, the equilibrium value is given by the generalization of (1): $\sup_p \inf_q u^*(p,q) = \inf_q \sup_p u^*(p,q)$. However, the results may be wild. The existence of equilibrium pairs may depend on the cardinality of the model and the sup-inf identity may already fail for the humblest sentences of IF^*:

- $\forall x \exists y_{/x}(x \neq y)$ *does not have mixed equilibrium pairs in the set* $\mathbb{N}$, *but* $\sup_p \inf_q u^*(p,q) = \inf_q \sup_p u^*(p,q) = 1$.
- $\forall x \exists y_{/x}(x \neq y)$ *has mixed equilibrium pairs with value 1 in the set* $\mathbb{R}$; *in fact, any pair of atomless probabilistic measures is an equilibrium pair.*
- *For* $\forall x \exists y_{/x}(x \leq y)$ *in* $(\mathbb{N}, \leq)$: $\sup_p \inf_q u^*(p,q) < \inf_q \sup_p u^*(p,q)$.

Not withstanding this grim panorama, there are genuine IF^* sentences having mixed equilibrium pairs for all models.

Example. Consider the sentence $Inf : \exists u \forall x \exists y_{/u} \exists z_{/ux}(z = x \wedge y \neq u)$ for which $Mod^+(Inf) = \{A : A \text{ infinite}\}$ and $Mod^-(Inf) = \{A : |A| = 1\}$. It has mixed equilibrium pairs with value 1 in all the infinite models and a well defined equilibrium value $Inf^M < 1$ in each finite model M.

Call an IF^* sentence *total* if it has equilibrium pairs for all models. This family of sentences has pleasant properties and is worth studying, but it has a serious flaw for our purposes. Notice that C1 is precisely the condition we would not like to abandon if we wish to extend faithfully game semantics.

Fact 1. *Equilibrium semantics on total sentences satisfy* C2, C3, C4 *in all models but not necessarily* C1.

Proof. C2 is straightforward, C3 may be proved by game theoretic considerations and the proof in [28] of C4 may be seen to hold for infinite structures. Now, there is a sentence $incf \in IF^*$ such that $Mod^+(incf) = \{M : M \not\cong (\mathbb{R},+,\cdot,0,1,<)\}$ and $Mod^-(incf) = \emptyset$ because being an incomplete ordered field is a Σ_1^1-condition. The formula $incf \vee \forall x \exists y_{/x}(x \neq y)$ has equilibrium value 1 in all models because in structures non isomorphic to $\mathbb{R}$ there is a pure winning strategy for E choosing the left formula of the disjunction, and we have seen that the right formula has equilibrium value 1 in $\mathbb{R}$. However, $\mathbb{R} \not\models^+ incf \vee \forall x \exists y_{/x}(x \neq y)$, violating C1. □

We could restrict further the formulas of IF^* or we could search for [0,1]-valued semantics on infinite models not necessarily related to equilibrium values. An obvious possibility in the second course of action is three-valued semantics:

$$M \longmapsto \varphi^M = \begin{cases} 1 & \text{if } M \models^+ \varphi \\ 0 & \text{if } M \models^- \varphi \\ \frac{1}{2} & \text{otherwise} \end{cases}$$

Indeed, it satisfies C1-C4* in infinite models. The straightforward proofs are left to the reader. But this is not the end of our search because this semantics does no satisfy compactness nor the Löwenheim-Skolem property. In fact, no [0,1]-valued semantics $M \mapsto \varphi^M$ satisfies the later properties and C1 at the same time in all formulas of IF^*. To see this we must specify first what we mean by these properties. Define for any $r \in [0,1] \cap \mathbb{Q}$:

$$Mod_{\geq r}(\varphi) = \{M : \varphi^M \geq r\},\ Mod_{\leq r}(\varphi) = \{M : \varphi^M \leq r\}.$$

Compactness: If $\{Mod_{\geq(\leq) r_i}(\varphi_i)\}_{i\in I}$ is finitely satisfiable then it is satisfiable.
Löwenheim-Skolem property: If $Mod_{\geq(\leq) r}(\varphi)$ is satisfiable then it has a countable model.

We will assume that any [0,1]-valued semantics satisfies the natural generalization of Lindström's axioms:

C0 (a) $\varphi^M = \varphi^N$ if $M \approx N$; (b) $\varphi^M = \varphi^{M\upharpoonright\tau}$ if $\tau \subseteq \mu$; (c) $\varphi^M = \alpha\varphi^{\hat{\alpha}M}$ if α is a renaming.

Fact 2: *There is no* [0,1]*-valued semantics on all formulas of IF* satisfying simultaneously* C1 *and compactness or the Löwenheim-Skolem property.*

Proof. Consider a semantics satisfying C0, C1. Then for the sentence *incf* introduced in the previous example we must have $\{(M,a) : M \approx \mathbb{R}\} = Mod_{\leq r}(incf)$ for some rational $r < 1$ due to C1 and C0 (a),(b). But this class is a counterexample to compactness and the Löwenheim-Skolem property. □

Thus, we are bound to consider proper fragments of IF^*. In this context, it is worth noticing that the fragment of perfect recall sentences satisfies C1-C4 and C4* plus compactness and the Löwenheim-Skolem property for the three-valued semantics. The next result shows the limitations if we wish more than that.

Theorem 7.1. *Let L be an extension of $L_{\omega\omega}$ closed syntactically under* $\neg, \vee$, *with a* [0,1]*-valued semantics extending the* {0,1}*-valued semantics of $L_{\omega\omega}$. Suppose L satisfies* C2-C4, *compactness, and the Löwenheim-Skolem property. Then for each $\varphi \in L$ and rational (real) $r \in [0,1]$ there is a first order theory T_r such that; $\varphi^M \geq r$ iff $M \models T_r$.*

Proof. Consider the topology on St_L having for a sub-basis of closed classes $Mod_{\geq r}(\varphi)$, $\varphi \in L$, $r \in [0,1] \cap \mathbb{Q}$. By C2 it includes the classes $Mod_{\leq r}(\varphi)$ and due to C3, C4, it is closed under unions: $Mod_{\geq r_1}(\varphi_1) \cup Mod_{\geq r_2}(\varphi_2) = Mod_{\geq\varepsilon}(\varphi_{1r_1}) \cup Mod_{\geq\varepsilon}(\varphi_{2r_2}) = Mod_{\geq\varepsilon}(\varphi_{1r_1} \vee \varphi_{2r_2})$; hence, it is a closed basis. The topology is regular because $M \notin Mod_{\geq r}(\varphi)$ implies $\varphi^M < s < r$ for some rational s and thus the open classes $Mod_{\geq s}(\varphi)^c$ and $Mod_{\leq s}(\varphi)^c$ separate M and $Mod_{\geq r}(\varphi)$. Countable structures are dense because $\varphi^M < r$ implies $\varphi^M \leq s < r$. Finally, C0 grants the continuity of reducts and renamings. Therefore, this is the elementary topology by Theorem 4.2. □

Corollary. *If L is a fragment of IF^* with a [0,1]-valued semantics satisfying C1-C4, compactness, and the Löwenheim-Skolem property, then $Mod^+(\varphi)$ and $Mod^-(\varphi)$ are recursively first-order axiomatizable for any $\varphi \in L$.*

Proof. $Mod^+(\varphi) = Mod_{\geq 1}(\varphi) = Mod(T_1)$ by $C1$ and the previous theorem. Moreover, the first-order consequences of T_1 coincide with the consequences of a Σ_1^1-sentence. □

Assuming C4* a simpler proof of the corollary may be obtained because $Mod^+(\varphi)$ being Σ_1^1 is closed under ultraproducts, and its complement $Mod^+(\varphi)^c = \cup_{r<1} Mod_{\leq r}(\varphi) = \cup_{r<1} Mod^+((\neg\varphi)_r)$ is closed under ultrapowers; hence, it is first order axiomatizable (cf. [10]). Moreover, under C4* all the T_r of Theorem 7.1 result recursively axiomatizable.

8 Final remarks

Further Lindström's characterizations of first order logic also hold for regular logics, the proofs being essentially topological with minimal model theoretic contents. Thus, any regular weak extension of $L_{\omega\omega}$ satisfying compactness and the (countable) Tarski's chain property, or having relativizations and satisfying the uncountable omitting types theorem, is equivalent in theories to $L_{\omega\omega}$. Many other model theoretic results known for boolean logics are essentially topological and may also be lifted to logics without classical negation.

The case of equilibrium semantics for IF^* logic shows that [0,1]-valued semantics makes perfect sense for classical structures, and topological regularity appears naturally in this setting. Theorem 7.1 is actually a Lindström's theorem for [0,1]-valued logics on classical structures. This viewpoint may be translated to the realm of [0,1]-valued structures to obtain Lindström's theorems and other results mentioned in this paper for extensions of continuous logic and Lukasiewicz-Pavelka logic. These results will appear elsewhere (cf. [9]).

Bibliography

[1] S. Abramsky and J. Väänänen. From IF to BI: A tale of dependence and separation. *Synthese*, 167(2):207-230, 2008.

[2] J. Barwise and S. Feferman (eds.), *Model-Theoretic Logics*. Springer Verlag, 1985.

[3] A. Blass and Y. Gurevich. Henkin quantifiers and complete problems. *Annals of Pure and Applied Logic*, 32(1):1-16, 1986.
[4] J. P. Burgess. A remark on henkin sentences and their contraries. *Notre Dame Journal of Formal Logic*, 44(3):185-188, 2003.
[5] X. Caicedo. Compactness and normality in abstract logics. *Annals of Pure and Applied Logic*, 59:33-43, 1993.
[6] X. Caicedo. Continuous operations on spaces of structures. In *Quantifiers: models, logics and computation*, volume II, M. Krynicki, M. Mostowski, and L. W. Szczerba (eds.), pages 263-296, Kluwer Acad. Publ., 1995.
[7] X. Caicedo. The abstract compactness theorem revisited. In *Logic and Foundations of Mathematics*, Cantini, Casari, Minari (eds.), pages 131-141, Kluwer Acad. Publ., 1999.
[8] X. Caicedo, F. Dechesne, and T. M. V. Janssen. Equivalence and quantier rules for logic with imperfect information. *Logic Journal of IGPL*, 17(1):91-129, 2009.
[9] X. Caicedo, J. Iovino, Omitting uncountable types, and the strength of [0,1]-valued logics. *Annals of Pure and Applied Logic*, 165:1169–1200, 2014.
[10] C.C. Chang and J. Keisler. *Model theory*, 2nd. ed. North Holland, 1973.
[11] A. Ehrenfeucht and A. Mostowski. A topological space of models of first-order theories. *Bull. de L'Academie Polonaise des Sciences*. Série des sciences math., astr. et phys. IX: 369-373, 1961.
[12] S. Feferman and R.L. Vaught. The first-order properties of products of algebraic systems. *Fundam. Math.*, 47:57–103, 1959.
[13] R. Fraïssé. *Cours de logique mathematique, Tome 2, Theorie des modeles*, 2e edition. Gauthier-Villars, Paris, 1972.
[14] R. Fagin. Generalized 1rst-order spectra and polynomial-time recognizable sets. In *Complexity of Computation*, volume 7, R. M. Karp (ed.), pages 43-73, SIAM-AMS Proceedings, 1974.
[15] P. Galliani. Game Values and Equilibria for Undetermined Sentences of Dependence Logic, MSc thesis, ILLC Publications, MoL-2008-08, 2008.
[16] P. Galliani and A. L. Mann. Lottery Semantics: A Compositional Semantics for Probabilistic First-Order Logic with Imperfect Information. *Studia Logica*, 101:293–322, 2013.
[17] M. García-Matos. Abstract model theory without negation. Ph.D. Dissertation, University of Helsinki, 2005.
[18] W. Hodges. Compositional semantics for a language of imperfect information. *Logic Journal of the IGPL*, 5(4):539–563, 1997.
[19] J. Hintikka and G- Sandu, G.. Game-theoretical semantics. In *Handbook of Logic & Language*, J. van Benthem and A. ter Meulen (eds.), pages 361–410, Elsevier Science, 1997.
[20] J. Kontinen and J. Väänänen. On definability in dependence logic. *Journal of Logic, Language and Information*, 18(3):317–332, 2009.
[21] P. Lindström. On extensions of elementary logic. *Theoria*, 35: 1-11, 1969.
[22] A. Makowski and S. Shelah. Positive results in abstract model theory. *Ann. of Pure and Applied Logic*, 25(3):263-300, 1983.
[23] A. L. Mann, G. Sandu and M. Sevenster. *Independence-Friendly Logic: A Game-Theoretic Approach*. London Mathematical Society Lecture Note Series, 2011.
[24] M. Morley. Categoricity in Power. *Transactions of the A.M.S.*, 114:514-538, 1965.
[25] M. Morley. Applications of topology to $L_{\omega_1\omega}$. *Proceedings of the Tarski Symposium, A.M.S. Proceed. Pure Math.*, 25:234-240, 1974.

[26] D. Mundici. Inverse topological systems and compactness in abstract model theory. *Journal of Symbolic Logic*, 51:785-794, 1986.
[27] N. Noble. Products with closed projections, II. *Trans. Amer. Math. Soc.*, 160:169-183, 1971.
[28] M. Sevenster & G. Sandu. Equilibrium semantics of languages of imperfect information. *Annals of Pure and Applied Logic*, 161:618-631, 2010.
[29] J. Väänänen. On the semantics of informational independence. *Logic Journal of the IGPL* 10(3):339-352, 2002.
[30] J. Väänänen. *Dependence logic: A New Approach to Independence Friendly Logic*. London Mathematical Society Student Texts, Vol. 70. Cambridge University Press, 2007.
[31] J. von Neumann and O. Morgenstern. *Theory of Games and Economic Behavior*. Princeton University Press, 1947.
[32] S. Willard. *General Topology*. Addisson-Wesley, 1970.

Zoé Chatzidakis

Model Theory of Fields With Operators – a Survey

For Jouko, on the occasion of his 60th birthday

Abstract: This paper presents a survey of the model theory of fields with operators, such as derivations, automorphisms, or the λ-functions of separably closed fields.

Keywords: fields with operators

MSC: 03C60, 12L12, 12H99

Zoé Chatzidakis: CNRS (UMR 8553), Ecole Normale Supérieure

1 Introduction

The model theory of fields with operators has proven to be very useful in applications of model theory to problems outside logic. Differentially closed fields and separably closed fields were instrumental in Hrushovski's proof of the conjecture of Mordell-Lang ([39]), and in work of Hrushovski and Pillay on counting the number of transcendental points on certain varieties ([44]). Let me also mention Ax's work ([1]) on connections between transcendence theory and differential algebra, which has been recently elaborated on by Bertrand and Pillay ([6]), and by Kowalski ([51]) in positive characteristic. The model theory of difference fields also provided various achievements: Hrushovski's new proof of the conjecture of Manin-Mumford ([40]), Scanlon's approach to p-adic abc conjectures and proximity questions ([89], [90], [91]), and his solution to Denis' conjecture ([92]); more recently, applications to algebraic dynamics by Hrushovski and the author ([21], [22], [16]), and by Medvedev and Scanlon ([61]). Model theory also provides some insight on the Galois theory of systems of differential (or difference) equations and of Picard-Vessiot extensions, or of strongly normal extensions of Kolchin, see e.g. [74].

The aim of this article is not to present these applications, but to give a survey of what is known of the model theory of these enriched fields. We will discuss such issues as existence of model companions and their various axiomatisations, elimination of quantifiers and decidability; we will also investigate stability theoretic properties and mention some open problems.

Another notable omission of this survey is that of exponential fields, which have seen many extraordinary developments in the past twenty years: in the context of the field of real numbers, or in the context of the field of complex numbers. I do not consider myself an expert in this subject, parts of which are still in full progress.

2 Basic results

2.1 Differential fields of characteristic 0

Recall that a *differential field* is a field K endowed with a derivation D, i.e., an additive map $D : K \to K$ satisfying the Leibniz rule $D(xy) = xD(y) + D(x)y$. We study them in the language of rings augmented by a function symbol for the derivation, $\{+, -. \cdot .0, 1, D\}$.

The existence of a model completion (commonly called DCF_0) of the theory of differential fields of characteristic 0 was first shown by Robinson ([84]); hence the theory eliminates quantifiers and is complete ([83]). The models of DCF_0 are called *differentially closed fields*. Since every derivation of a field extends to its algebraic closure, a differentially closed field is algebraically closed as a field. Blum gave a simple axiomatisation of DCF_0 in her PhD thesis ([8], see also [9]), and showed that DCF_0 is ω-stable. A consequence of this result is the existence and uniqueness of the differential closure of a differential field: any totally transcendental theory has a unique prime model (Shelah). Poizat showed ([79]) that DCF_0 eliminates imaginaries.

2.2 Several derivations

These results were extended to several commuting derivations by McGrail ([59]): the theory $\mathrm{DCF}_{0,n}$ of differentially closed fields of characteristic 0 with n commuting derivations is complete, eliminates quantifiers (in the language of rings augmented by n function symbols for the derivations), is ω-stable of rank ω^n and eliminates imaginaries.

The case of fields with non-commuting derivations but which have a finite dimensional commutator Lie subalgebra was first studied by Yaffe ([104]); Singer ([95]) then showed how to reduce the study of their existentially closed models to the case of of models of $DCF_{0,n}$. The case of arbitrary non-commuting derivations is a consequence of a result of Moosa and Scanlon, and will be discussed later.

2.3 Geometric axiomatisation

The axiomatisation given by Blum is very simple and uses differential polynomials in one variable. If $F(X) \in K[X]$, $X = (X_0, \ldots, X_n)$, involves non-trivially the variable X_n, then the associated differential polynomial $f(X) = F(X, DX, \ldots, D^n X)$ is said to have *order n*. Besides the usual axioms for differential fields of characteristic 0, Blum adds the scheme of axioms for algebraically closed fields, and the following scheme of axioms: for every pair of differential polynomials f and g, with the order of f strictly greater than the order of g, the axiom

$$\exists x\ f(x) = 0 \wedge g(x) \neq 0.$$

[Note that these axioms imply that the derivation is non-trivial: $\exists x\ Dx = 1$.]

A different kind of axiomatisation of DCF_0 (so called *geometric axiomatisation*) was given by Pierce and Pillay ([72]). Its formulation uses a very limited amount of differential algebra, but nevertheless allows one to describe several variable differential varieties. It contains the usual axioms for differential fields of characteristic 0; for the remaining axioms we need first a definition.

Let $f_1, \ldots, f_m \in K[X]$, $X = (X_1, \ldots, X_n)$. Suppose that they generate a prime ideal of $K[X]$ and let $V \subset K^n$ be the irreducible affine variety defined by the equations $f_1(x) = \cdots = f_m(x) = 0$. Define the *twisted tangent bundle of* V, $\tau(V)$, by the following equations:

$$(x, y) \in \tau(V) \iff \bigwedge_{i=1}^{m} f_i(x) = 0 \wedge \sum_{j=1}^{n} \frac{\partial f_i}{\partial X_j}(x) y_j + f_i^D(x) = 0.$$

Here f_i^D denotes the polynomial obtained by applying D to the coefficienst of f_i. Note that if $a \in V(K)$, then $(a, Da) \in \tau(V)$, and for $a \in V$, the algebraic set $\tau(V)(a) = \{b \mid (a, b) \in \tau(V)\}$ coincides with the algebraic set defined by "$y - D(a)$ belongs to the tangent space of V at a". There is a natural onto map $\pi : \tau(V) \to V$. The geometric axiomatisation of DCF_0 is then obtained by adding to the theory of algebraically closed differential fields the scheme of axioms:

For every irreducible varieties $V \subset K^n$ and $W \subset \tau(V)$ such that $\pi(W)$ is Zariski dense in V, there is a point a such that $(a, D(a)) \in W$.

[It is well-known that the polynomials $f_1, \ldots, f_n$ generating a prime ideal is a first-order property of the coefficients of the f_i's.] Several attempts were made to find geometric axiomatisations for $DCF_{0,n}$, see [71], [86]; the problem is however much harder than in the one derivation case, and the axiomatisations which were obtained are in my opinion less elegant.

2.4 Other theories of differential fields of characteristic 0

All the fields discussed above are algebraically closed. One can however impose conditions on the pure field, and ask that it be existentially closed (as a differential field) with respect to those conditions. This was done by Singer ([94]) for ordered differential fields (thus obtaining a notion of closed ordered differential field), and by Tressl for the class of differential fields which are *large*, and whose theory in the language of rings is model-complete[1]. More work on closed ordered differential fields was done by Brihaye, Michaux, Point, and Rivière ([64], [78], [80], [81], [82]). The "uniform" axiomatisation of Tressl was generalised by Guzy in [34].

Itai and Hrushovski ([42]) showed the existence of many model-complete theories of differential fields of characteristic 0 which are superstable, and are not DCF_0. Chatzidakis and Hrushovski ([19]) studied the limit theory of the differential fields $(\mathbb{F}_p(t)^s, d/dt)$ as p goes to ∞, and showed that it interprets arithmetic[2].

One can also consider valued fields with operators. Valued differential fields have been studied extensively by Guzy, Michaux, Point and Rivière in [13], [30], [31], [32], [33], [35], [36]. Valued difference fields were studied by Hrushovski ([41]), by Durhan and Van den Dries ([2], [3]), by Bélair, Macintyre and Scanlon ([4]), by Pal ([70]), ...

2.5 More on DCF_0

Shelah ([93]) showed that differential closures are not necessarily minimal: it is possible to find a differential field K, with differential closure $\hat{K}$ containing a strictly smaller K-isomorphic copy of $\hat{K}$. (This result was also proved independently by Rosenlicht and by Kolchin). In the same paper Shelah also showed that there are 2^κ non-isomorphic differentially closed fields of cardinality κ when κ is uncountable. The question on the number of non-isomorphic countable dif-

1 A field K is *large* if it is existentially closed in the field of power series $K((t))$.

2 $\mathbb{F}_p(t)^s$ denotes the separable closure of $\mathbb{F}_p(t)$.

ferentially closed fields stayed open for a long time, and was finally solved by Hrushovski and Sokolovich, who showed there are $2^{\aleph_0}$ many of them ([45], see [57] for a published proof).

2.6 Differential fields of positive characteristic

Let K be a differential field of positive characteristic p, with derivation D. The Frobenius map $x \mapsto x^p$ is then a ring embedding, and we denote the image of K by K^p. It is a subfield of K, and D vanishes on K^p. Thus, differential fields of positive characteristic have very large subfields of constants. We say that K is *differentially perfect*, or *strict*, if K^p coincides with the constant subfield, i.e.:

$$\forall x\, Dx = 0 \rightarrow \exists y\, y^p = x.$$

Existentially closed differential fields of characteristic $p > 0$ are called *differentially closed*. Their theory, denoted DCF_p, was axiomatised by Wood ([100]), who showed that it is complete and model complete, and eliminates quantifiers in the language of differential rings augmented by a function symbol r which is interpreted as the inverse of the Frobenius map $x \mapsto x^p$: one adds the axiom $\forall x\, (Dx \neq 0 \wedge r(x) = 0) \vee r(x)^p = x$. Note that this axiom implies that the field is differentially perfect, and thus DCF_p is the model completion of the theory of differentially perfect differential fields of characteristic p. Wood and Shelah showed that DCF_p is stable not superstable, and that every differential field K has a differential closure (see [102], where Wood also gives Shelah's proof that the differential closure is unique). The underlying subfield of a differentially closed field of positive characteristic is of course separably closed, and of infinite degree of imperfection. I think one does not know a natural language in which differentially closed fields of positive characteristic eliminate imaginaries. See also the next section on separably closed fields.

2.7 Separably closed fields of positive characteristic

Separably closed fields are fields K with the following property: if $f(X) \in K[X]$ is such that $f'(X)$ is not identically 0 (such polynomials are called separable), then $f(X)$ has a zero in K. If the characteristic of the field is 0, then all polynomials are separable. In positive characteristic p, non-separable polynomials are of the form $f(X) = g(X^p)$, where $g \in K[X]$.

Let us fix a field K of characteristic $p > 0$. Then K^p is a subfield of K, and K is a K^p-vector space. A *p-basis of* K is a set $B \subset K$ such that $K = K^p[B]$, and B is minimal such. The size of B (an integer or ∞) is called the *degree of imperfection* of K (also called its *Ershov invariant*). If $|B| = e \in \mathbb{N}$, then $[K : K^p] = p^e$. For more details on p-bases, see [10].

Ershov ([28]) showed that given a prime p and $e \in \mathbb{N} \cup \{\infty\}$, the theory $\mathrm{SCF}_{p,e}$ of separably closed fields of characteristic p and degree of imperfection e is complete (and therefore decidable). He proved it through a model completion result: expand the language by adding n-ary predicates Q_n, $n \in \mathbb{N}$, and add to the theory of fields of characteristic p an axiom saying that $Q_n(x_1, \ldots, x_n)$ holds if and only if the elements $x_1, \ldots, x_n$ are *p-independent*, i.e., if $x_i \notin K^p[x_1, \ldots, \hat{x}_i, \ldots, x_n]$ for every $1 \leq i \leq n$. Consider the theory T_e obtained from the theory of fields of characteristic p by adding the defining axioms for the predicates Q_n and the axiom $\forall x_1, \ldots, x_{e+1} \neg Q_{e+1}(x_1, \ldots, x_{e+1})$ [actually, only adding to the language $Q_1, \ldots, Q_{e+1}$ would suffice]. Then the model companion of T_e is precisely $T_e \cup \mathrm{SCF}_{p,e}$.

Let K be a field of positive characteristic $p > 0$. For each positive integer n, fix an enumeration $m_i(\bar{x})$, $i < p^n$, of the *p-monomials* in $\bar{x} = (x_1, \ldots, x_n)$, i.e., of the monomials $x_1^{i_1} x_2^{i_2} \ldots x_n^{i_n}$ with $0 \leq i_1, \ldots, i_n < p$. We then define $n+1$-ary functions $\lambda_i(\bar{x}, y) : K^{n+1} \to K$ by $\lambda_{i,n}(\bar{a}, b) = 0$ if either the elements of $\bar{a}$ are not p-independent in K, or if $b \notin K^p[\bar{a}]$; if $b \in K^p[\bar{a}]$ and the elements of $\bar{a}$ are p-independent, then the elements $\lambda_{i,n}(\bar{a}, b)$ ($i < p^n$) of K are uniquely determined by the equation

$$b = \sum_{i<p^n} \lambda_{i,n}(\bar{a}, b)^p m_i(\bar{a}).$$

As far as I know, these functions first appear in a paper of Srour which shows that the theory of separably closed fields eliminates quantifiers in the expanded language (Lemma 3, [96]). In this paper, he also studies prime models, and describes independence. In case the degree of imperfection of the field is $e \in \mathbb{N}$, one can also fix a p-basis $\bar{b} = \{b_1, \ldots, b_e\}$, add to the language of rings e new constant symbols (for the elements of $\bar{b}$) and the functions $\lambda_{i,e}(\bar{b},\ )$ ($i < p^e$); one then adds to the theory of separably closeed fields an axiom saying that $b_1, \ldots, b_e$ form a p-basis, as well as the defining axioms of the functions $\lambda_{i,e}(\bar{b},\ )$. Delon shows in [26] that separably closed fields eliminate quantifiers and imaginaries in this extended language.

No natural language in which separably closed fields of infinite Ershov invariant eliminate imaginaries is known, this has been an open problem for a long time. Note that in the finite invariant case, it was necessary to add to the language new symbols of constants for the elements of a p-basis, but of course one cannot do that in the infinite invariant case.

2.8 Hasse-Schmidt derivations

We fix a field K of characteristic $p > 0$. An *iterative Hasse-Schmidt derivation* on K is a stack of operators $D_i : K \to K$, $i \in \mathbb{N}$, which are additive, and satisfy

- $D_0(x) = x$,
- $D_n(xy) = \sum_{i+j=n} D_i(x)D_j(y)$,
- $D_n D_m(x) = \binom{n+m}{m} D_{n+m}$.

Note that D_0 is the identity map, that D_1 is just an ordinary derivation, that for $i < p$, D_i is a rational multiple of $D_1^{(i)}$ (the i-th iterate of D_1), and that $D_1^{(p)} = 0$. One can also show that D_{p^n} satisfies $D_{p^n}(x^{p^n}) = D_1(x)^{p^n}$ for $n \geq 1$, so that D_{p^n} defines a derivation on K^{p^n}. The theory of fields with a Hasse-Schmidt derivation is axiomatisable (in the language $\{+, -, \cdot, 0, 1, D_n\ (n \in \mathbb{N})\}$), and has a model companion (see Ziegler [105]) which is obtained from the original theory by adjoining to it $\mathrm{SCF}_{p,1}$ together with the axiom expressing strictness (i.e., $\ker D_1 = K^p$). The resulting theory $\mathrm{SCH}_{p,1}$ eliminates imaginaries and quantifiers. Note that we did not have to add constant symbols for the elements of a p-basis to eliminate imaginaries.

The results can be extended to arbitrary finite invariant e, by adding e stacks of commuting operators to get the theory $\mathrm{SCH}_{p,e}$ of e commuting iterative Hasse-Schmidt derivations. Kowalski ([48]) shows that the theory of existentially closed differential fields with e commuting derivations admits a geometric axiomatisation.

Bi-interpretability with SCF_e.
Let K be a separably closed field of characteristic p and finite degree of imperfection e, fix a p-basis $b_1, \ldots, b_e$ of K. One can then define e Hasse-Schmidt derivations $\mathcal{D}^i$, $i = 1, \ldots, e$ on K by setting:
- $D_0^i = id$ for all i;
- $D_1^i(b_j) = 1$ if $j = i$, 0 otherwise;
- $D_{p^n}^i(b_j) = 0$ for $n \geq 1$ and all i, j.

These conditions, together with the axioms for commuting iterative Hasse-Schmidt derivations, define uniquely the stacks $\mathcal{D}^i$ on K for $i = 1, \ldots, e$. Ziegler ([105]) shows that this expansion makes K into a model of $\mathrm{SCH}_{p,e}$, and that the above operators are definable in the language of rings augmented by constants for the elements of the p-basis $\{b_1, \ldots, b_e\}$.

Note also that there are many ways of expanding K, but any one of them will make K into a model of $\mathrm{SCH}_{p,e}$. The elimination of quantifiers in the language with the

Hasse-Schmidt derivations implies that the λ-functions $\lambda_i(x)$ over a given p-basis b are terms in (x, b) (in the language of $\mathrm{SCH}_{p,e}$).

2.9 Derivations of the Frobenius

Let K be a field of characteristic $p > 0$, and $q = p^n$ a power of p. A derivation of the q-Frobenius on K is an additive endomorphism of K which satisfies a twisted Leibnitz rule

$$\delta(ab) = a^q\delta(b) + b^q\delta(a).$$

When $q = 1$, this is an ordinary derivation on K. Kowalski shows in [49] that the theory of fields endowed with a derivation of the q-Frobenius has a model companion, which is complete, eliminates quantifiers in the language

$$\{+, -, \cdot, \delta, \lambda_{i,m}\ m \in \mathbb{N}, i < p^m\},$$

is stable not superstable. He gives a geometric axiomatisation of the theory, after introducing the appropriate version of the twisted tangent bundle. Strictness of K is defined as with ordinary differential fields: K is strict iff $\ker\delta = K^p$. It is unknown whether it suffices to add the inverse of the Frobenius map to the language of differential rings to get quantifier elimination. Kowalski's paper leaves open several other questions, some of them quite intriguing.

2.10 Difference fields

A difference field is a field K with a distinguished endomorphism, usually denoted by σ. Macintyre showed that the theory of difference fields has a model companion, ACFA ([56]). Its models are inversive difference fields, i.e., the endomorphism σ is onto. It admits a geometric axiomatisation: one says that σ is an automorphism of the algebraically closed difference field σ, and one takes the following scheme of axioms:

For every irreducible varieties U and V, with $V \subset U \times U^\sigma$, such that V projects dominantly onto U and U^σ, there is a tuple a such that $(a, \sigma(a)) \in V$.

Here U^σ denotes the variety image of U under σ. We now list some properties of ACFA: its completions are obtained by describing the behaviour of σ on the algebraic closure of the prime field; ACFA is decidable; it does not eliminate quantifiers in the language of difference fields, the reason being that an automorphism of a field E may have non-isomorphic extensions to the algebraic closure of E. Macintyre gives in [56] a language in which ACFA eliminates quantifiers.

Chatzidakis and Hrushovski start an in-depth study of ACFA in [18]. Among other things, they show that all completions of ACFA eliminate imaginaries, are unstable, but are supersimple and of SU-rank ω.

A field which plays an important role is the *fixed field*, $\mathrm{Fix}(\sigma) = \{a \in K \mid \sigma(a) = a\}$. In characteristic $p > 0$, if $q = p^n$ for some $n \in \mathbb{Z}$, we denote by Frob_q the map $x \mapsto x^q$. Then the automorphisms $\tau = \sigma^n \mathrm{Frob}^m$ are also definable ($n \in \mathbb{N}, m \in \mathbb{Z}$), as are their fixed fields. These fields are pseudo-finite, and therefore unstable (they have the independence property); but they are *stably embedded*, i.e., every K-definable subset of a cartesian power of $\mathrm{Fix}(\tau)$ is definable with parameters from $\mathrm{Fix}(\tau)$. One can also show that if K is a model of ACFA, then so is the difference field (K, τ).

2.11 Mixing operators: the commutative case

We already talked about several commuting derivations. The other examples all involve an endomorphism or automorphism. The difficulty in all these cases is to show that the class of existentially closed models is elementary. Once it is done, results by Chatzidakis and Pillay ([25]) give much information on the behaviour of the models, definable sets, algebraic closure, independence, etc. Usually, the proofs done for ACFA generalise almost instantly. Below we will describe what is known.

Separably closed fields

The first new example is probably the theory of separably closed fields of fixed degree of imperfection (in the language with the λ-functions) with a distinguished endomorphism σ. The model companion SCFA of this theory exists ([15]), is simple but not supersimple nor stable, and eliminates imaginaries if the degree of imperfection is finite. As is the case with ACFA, it does not eliminate quantifiers.

Difference differential fields

The next example was investigated by Bustamante-Medina in [14]. Consider the theory of differential fields of characteristic 0 endowed with an endomorphisms σ. It admits a model companion, denoted DCFA (this was first proved by Hrushovski in the late 90's; for an axiomatisation, see [14]). The theory is supersimple of SU-rank ω^2, eliminates imaginaries. There are three important subfields: the fixed field, the field of constants, and their intersection. Recently, the results of Bustamante-Medina were generalised by León Sánchez, to $DCFA_n$, the theory of existentially closed models of the theory of fields of characteristic 0 with n commuting derivations and an endomorphism σ of the differential field. It appears in [87], and uses the techniques of [86].

Commuting automorphisms

Hrushovski showed in the 90's that the theory of fields with two commuting automorphisms does not have a model companion. The example appears in a paper of Kikyo ([47]), I will give a slightly different one here. We let $\mathcal{K}$ be the class of all difference fields with commuting automorphisms σ, τ.

We let K be an algebraically closed field containing $\mathbb{Q}$, on which τ is the identity, and let σ be any automorphism of K such that if ω is a primitive cubic root of 1, then $\sigma(\omega) = \omega^2$, and such that K has an element b_n with $\langle\sigma\rangle$-orbit of size exactly n for each $n > 2$, and an element b with infinite $\langle\sigma\rangle$-orbit.

Consider the following formula:

$$\varphi(x, y, z) := y^3 = x + z \wedge \tau(y) = \omega y \wedge \sigma(x) = \tau(x) = x.$$

Consider the field $K(t)$, t an indeterminate, and define $\sigma(t) = \tau(t) = t$. Because the elements $\sigma^i(b)$, $b \in \mathbb{Z}$, are all distinct, the algebraic extensions of K obtained by adjoining to $K(t)$ a cubic root α_i of $t + \sigma^i(b)$ are linearly disjoint over $K(t)$. So, setting L to be the algebraic extension of $K(t)$ generated by the α_i's, $i \in \mathbb{Z}$, and defining $\sigma(\alpha_i) = \alpha_{i+1}$, and $\tau(\alpha_i) = \sigma^i(\omega)\alpha_i$ for all i, the automorphisms σ and τ of L commute, and (t, α_0) is a solution of $\varphi(x, y, b)$.

Assume now that $\varphi(a, \alpha, b_n)$ holds in some difference field in $\mathcal{K}$ extending K. Then $\sigma^n(\alpha) = \omega^j\alpha$ for some j. On the other hand, the fact that σ and τ commute implies

$$\tau\sigma^n(\alpha) = \tau(\omega)^j\tau(\alpha) = \omega^{j+1}\alpha = \sigma^n(\omega\alpha) = \omega^{2^n+j}\alpha,$$

so that necessarily $2^n \equiv 1$ modulo 3, i.e., n is even. So, when n is odd, the formula $\varphi(x, y, b_n)$ has no solution in any extension of K in $\mathcal{K}$. [Note that if n is even, then the construction done for b can be easily adapted to yield a solution of the formula $\varphi(x, y, b_n)$.]

Take a non-principal ultrapower M of K on the set I of odd integers. If $b^* = (b_n)_{n\in I}/\mathcal{U}$, then the $\langle\sigma\rangle$-orbit of b^* is infinite, so that the formula $\varphi(x, y, b^*)$ has a solution in some extension $M' \in \mathcal{K}$ of M. However, any formula $\psi(z)$ satisfied by b^* will be satisfied by infinitely many b_i's, for which we know that the formula $\varphi(x, y, b_i)$ has no solution. This shows that the model companion of the theory of fields with two commuting automorphisms does not exist.

Action of $\mathbb{Q}$ on a field. A contrario, Medvedev shows in [60] that the theory of fields with an action of $(\mathbb{Q}, +)$ has a model companion (in the language of rings augmented by function symbols σ_q, $q \in \mathbb{Q}$). She also generalises many of the results true for ACFA to this context.

2.12 A remark

Note that the example showing that the theory of commuting automorphisms does not have a model companion does not provide an answer to the following two questions:

Let T be a theory which eliminates quantifiers in a language $\mathcal{L}$, consider the class $\mathcal{K}'$ of models of T with a distinguished automorphism σ. The questions are whether the class of existentially closed elements of $\mathcal{K}'$ is an elementary class, when

(i) T is a complete theory extending ACFA, or

(ii) T is a complete theory of pseudo-finite field.

The reason is that it is difficult to show that a given system of difference equations has a solution in a member of $\mathcal{K}'$, since the underlying field of this extension must be a model of ACFA or a pseudo-finite field.
For instance, take the difference field $K(t)$ described above. We want to show that the formula $\varphi(x, y, b)$ has a solution in some extension of $K(t)$ in $\mathcal{K}'$. Consider the field $L = K(t, \alpha_i)_{i \in \mathbb{Z}}$.

In case (i), the satisfaction of $\varphi(x, y, b)$ is equivalent to the fact that (L, σ, τ) embeds in a member of $\mathcal{K}'$; hence we need to extend σ and τ to the algebraic closure L^{alg} of L, in such a fashion that σ and τ commute. This seems hard to achieve.

In case (ii), let L_0 be the subfield of L fixed by τ; then σ restricts to an automorphism of L_0, and we consider the difference field (L_0, σ), which we wish to extend to a member of $\mathcal{K}'$. The theory of pseudo-finite fields reduces the problem to the following: extend σ_0 and τ to the algebraic closure L^{alg} of L_0 in such a fashion that σ_0 normalises $\langle \tau \rangle$. Again, this seems hard to achieve.

2.13 A general framework for fields with operators

The following is a particular case of a setup introduced in [68] by Moosa and Scanlon, and which permits to treat some of the above considered cases in a uniform fashion. Consider a $\mathbb{Q}$-algebra D which is finite-dimensional over $\mathbb{Q}$ and has basis $\{\varepsilon_0, \dots, \varepsilon_{\ell-1}\}$ (as a $\mathbb{Q}$-vector space). We assume that the projection π of D onto the first coordinate is a $\mathbb{Q}$-algebra homomorphism. We call ψ the isomorphism $D \to \mathbb{Q}^\ell$ associated to the basis. If R is a $\mathbb{Q}$-algebra, one sets $\mathcal{D}(R) = D \otimes_{\mathbb{Q}} R$.

A $\mathcal{D}$-ring is a $\mathbb{Q}$-algebra R, together with a sequence $\partial = (\partial_1, \dots, \partial_{\ell-1})$ of operators on R, such that the map

$$e : R \to \mathcal{D}(R), \qquad x \mapsto x\varepsilon_0 + \partial_1(x)\varepsilon_1 + \cdots + \partial_{\ell-1}(x)\varepsilon_{\ell-1}$$

is a homomorphism of $\mathbb{Q}$-algebras.

The class of $\mathcal{D}$-rings is axiomatisable in the language of rings augmented by symbols for the operators. There is a unique $\mathcal{D}$-ring structure on $\mathbb{Q}$, obtained by forcing $1 = \varepsilon_0 + \partial_1(1)\varepsilon_1 + \cdots + \partial_{\ell-1}\varepsilon_{\ell-1}$.

Some examples of $\mathcal{D}$-rings:

1. Differential rings: Let $\mathcal{D}(R) = R[\eta]/(\eta^2)$, with R-basis $\{1, \eta\}$.
2. Rings with higher order derivations: let $\mathcal{D} = R[\eta]/(\eta^n)$, with R-basis $\{1, \eta, \dots, \eta^{n-1}\}$. This is a generalisation of ordinary differential rings, since no iterativity condition is imposed.
3. Difference rings: Let $\mathcal{D}(R) = R \times R$, with R-basis $\{(1,0), (0,1)\}$.
4. Rings with several (higher order) derivations and endomorphisms: take the tensor product (over R) of various rings as above. The derivations do not commute.
5. D-rings: $\mathcal{D}(R) = R^2$, with multiplication defined by

   $$(x_1, y_1) \cdot (x_2, y_2) = (x_1x_2, x_1y_2 + x_2y_1 + y_1y_2c).$$

   If $c = 0$, this is just an ordinary derivation. If $c \neq 0$, the map $\sigma(x) = x + c\partial_1(x)$ defines a ring endomorphism of R, and (because $c \in \mathbb{Q}$) ∂_1 is definable from σ. These structures were considered by Scanlon in [88].
6. Derivations of an endomorphism σ: $\mathcal{D}(R) = R \times R[\eta]/(\eta^2)$, with R-basis $\{(1,0), (0,1), (1,\eta)\}$. Then the endomorphism σ and the operator δ satisfy $\delta(xy) = \sigma(x)\delta(y) + \sigma(y)\delta(x)$.
7. Taking fibered products of finite dimensional $\mathbb{Q}$-algebras.

The main result of [68] is that under certain conditions on the algebra $\mathcal{D}$, the theory of $\mathcal{D}$-rings which are integral domains has a model companion, $\mathcal{D}$-CF_0. The axiomatisation is geometric, once one has introduced the appropriate twisted tangent bundle. They then describe completions of the theory, independence, types, and show that they eliminate imaginaries. The theory is simple, and sometimes supersimple, stable, or ω-stable.

3 Stability theoretic notions

Basically the fields we study fall within four (= two times two) classes. The operators considered are variations of derivations or of automorphisms. If some automorphism occurs, the theory of the enriched field will be simple and unstable. If only derivations occur, then it will be stable. One can then refine this analysis. If the underlying field is algebraically closed then we actually get ω-stability in the case of commuting derivations, and supersimplicity when at most one automorphism is present. In the case of an endomorphism of an algebraically closed field, or of an automorphism of a separably closed field, or of several non-commuting automorphisms (of a field or differential field) we only obtain simplicity.

3.1 Algebraic closure, independence

In all cases where the underlying field is algebraically closed or separably closed, model-theoretic algebraic closure and independence are extremely easy to describe, once one knows algebraic closure and independence in ACF. Basically, to get the model-theoretic algebraic closure, one closes under all obvious definable functions, and then takes the (relative, field-theoretic) algebraic closure. The independence of algebraically closed sets is then the one induced by ACF (or possibly in the case of infinite degree of imperfection, SCF):

$$A \perp_C B \text{ if and only } \mathrm{acl}(AC) \text{ and } \mathrm{acl}(BC) \text{ are independent over } \mathrm{acl}(C).$$

3.2 Zilber's conjecture – the trichotomy

Let T be a stable theory, M a model of T, and D a strongly minimal definable subset of M. Then acl defines a pregeometry on D, i.e., it satisfies the Steinitz exchange property: if the elements a, b of D and the set $C \subset D$ are such that $a \in \mathrm{acl}(Cb) \setminus \mathrm{acl}(C)$, then $b \in \mathrm{acl}(Ca)$. Quotienting $S = D \setminus (\mathrm{acl}(\emptyset) \cap D)$ by the equivalence relation $a \sim b \iff \mathrm{acl}(a) = \mathrm{acl}(b)$, one obtains a geometry cl on the set $N := S/\sim$, which is of one of the following kinds:

- degenerate or trivial: for any $A \subset N$, $\mathrm{cl}(A) = \bigcup_{a \in A} \mathrm{cl}(\{a\})$;
- modular non-trivial: there are $a_1, \ldots, a_n, b$ with $b \in \mathrm{cl}(a_1, \ldots, a_n) \setminus \bigcup_i \mathrm{cl}(\{a_i\})$; if A, B are closed and have non-empty intersection C, then $A \perp_C B$.
- non-modular: there are closed sets A and B which have non-empty intersection C and are not independent over C.

Zilber's conjecture stated that any geometry originating from a strongly minimal set, coincides with the geometry of one of the following structures:

- (Trivial) An infinite set with no-structure.
- (Modular non-trivial): A vector space $(V, +, -, 0, \alpha)_{\alpha \in F}$ over a fixed division ring F.
- (Non-modular) An algebraically closed field $(F, +, \cdot, c)_{c \in F_0}$ with constants for a subfield F_0.

The conjecture was disproved by Hrushovski in [38] (see also [37]). Let us say that a (stable) structure M *satisfies the trichotomy* if whenever D is a strongly minimal set definable in M, then the pregeometry associated to D satisfies Zilber's conjecture. It turns out that many of the stable fields with operators do satisfy the trichotomy: this was proved by Hrushovski and Sokolovich for DCF$_0$ ([45]); to speak about the other examples we first need a discussion.

The concept of satisfying the trichotomy can be generalised in two fashions. First, in the stable case, one can allow ∞-definable sets, as long as they are of U-rank 1, i.e., as long as acl defines a pregeometry. Then the trichotomy is stated as above. In the unstable case, some care needs to be taken, and the trichotomy becomes a *dichotomy* between the first two cases (*modular case*) on one hand, and the third case (*field case*) on the other. We still work with a definable (or ∞-definable) set D, on which acl defines a pregeometry, but which is not necessarily strongly minimal. We also assume that there is a good independence notion - e.g., that the theory is simple. We say that D is *modular* iff whenever a, b are tuples of elements of D, then they are independent over the intersection of their algebraic closures in D^{eq}. The dichotomy then states:

If D is not modular, then some type p of rank 1 which is realised in D is non-orthogonal to a field.

Recall that a type p of rank 1 over a set A of parameters is non-orthogonal to an ∞-definable set S iff there is some B containing A, a tuple c in S such that $c \perp_A B$, and $\mathrm{acl}(Bc)$ contains a realisation a of p which is not in $\mathrm{acl}(B)$.

3.3 Two generalisations of modularity

1 - The definition of modularity we gave generalises easily to types of arbitrary rank, and in that case it is called *one-basedness*: a (∞-)A-definable set S is *one-based* iff whenever $B \supset A$ and a is a tuple of elements of S, then a and B are independent over the intersection of their algebraic closures in M^{eq}.

Recall that a type p over A is *regular* iff it is orthogonal to any of its forking extensions: if C is a set, a and b are realisations of p, such that $a \perp_A C$ and $b \not\perp_A C$, then $a \perp_C b$. Forking makes it possible to define a pregeometry on the set of realisations of p, i.e., one replaces acl by "forking". A regular type is called modular iff its associated geometry is modular; and non-modular otherwise.

3.4 Properties of one-basedness

The defining property of one-basedness is, in my opinion, quite remarkable. It is extremely desirable in case of a group structure, because of the following

Theorem (Hrushovski-Pillay, [43]) *Let G be a one-based $\mathcal{L}$-structure, which is stable and of finite U-rank, and assume that it is a group (maybe with additional structure). Then for every n, every definable subset of G^n is a finite Boolean combination of cosets of definable subgroups of G^n. Furthermore, these subgroups are defined over* $\mathrm{acl}(\emptyset)$, *and in particular there are at most $|\mathcal{L}| + \aleph_0$ many of them.*

Note a consequence of this result: If G satisfies the assumptions of the theorem, then G has a subgroup G_0 of finite index which is commutative. This comes from looking at the graph of multiplication in G^3, or at the graph of the inverse map.

3.5 Some other structures which satisfy the dichotomy

Separably closed fields of finite degree of imperfection.
If K is a saturated separably closed field, then no definable subset of K is ranked by the U-rank. But there are many ∞-definable sets D which have U-rank 1, for instant the perfect core $k = \bigcap_n K^{p^n}$ of K. Note that the field k is algebraically closed. Moreover, the induced structure on k is that of a pure field: if $S \subset K^n$ is definable in K, then $S \cap k^n$ is definable in the pure field k. Assume that the degree of imperfection of K is finite. Hrushovski showed in [39] that non-modular *thin*[3] types of U-rank 1 are non-orthogonal to the field k; this result was later extended by Delon in [27] to all U-rank 1 non-modular types.

These results extend to fields with commuting iterative Hasse-Schmidt derivations. I do not know if they extend to any of the following characteristic p fields: separably closed fields of infinite degree of imperfection, differentially closed

3 i.e., those such that given a realisation a over a model K, the field $\mathrm{dcl}(Ka)$ has finite transcendence degree over K.

fields of positive characteristic, or to existentially closed fields with a derivation of the q-Frobenius.

Differentially closed fields with several commuting derivations.
Moosa, Pillay and Scanlon study in [69] the modularity of regular types in $DCF_{n,0}$. They show that a regular non-modular type is non-orthogonal to the generic of a definable subgroup of $\mathbb{G}_a$. This has for consequence the dichotomy for types of rank 1: a type of rank 1 which is non-modular, is non-orthogonal to the generic type of the field of absolute constants, defined by $D_1x = D_2x = \cdots D_nx = 0$.

The hope is that the result extends to regular types of higher rank: that regular non-modular types are non-orthogonal to a field of constants, i.e., a subfield defined by a conjunction of conditions of the form $\sum a_iD_i(x) = 0$. A partial result in that direction was obtained by Süer ([97]).

Difference fields.
Chatzidakis and Hrushovski, later joined by Peterzil, show in [18], [24] that models of ACFA satisfy the dichotomy: a type of SU-rank 1 which is not modular, is non-orthogonal to a fixed field, i.e., to $\mathrm{Fix}(\sigma)$ if the characteristic is 0; or to some $\mathrm{Fix}(\sigma^n\mathrm{Frob}^m)$ if the characteristic is positive.

Because the theory ACFA is unstable, other issues connected to stability or instability appear. Most types are not stationary. In characteristic 0, a type which is foreign to $\mathrm{Fix}(\sigma)$ will be stable stably embedded. So, the unique source of unstability is $\mathrm{Fix}(\sigma)$. In positive characteristic this is not the case: there are modular subgroups of $\mathbb{G}_a$ which are unstable.

So, in characteristic 0, if G is a definable group which is one-based, then the result of Hrushovski-Pillay on stable one-based groups of finite U-rank will apply, and give that any definable subset of G^n is a Boolean combination of cosets of definable subgroups. In characteristic $p > 0$, we cannot apply the result, however, in case G is a definable subgroup of an algebraic group H (so that the group law is quantifier-free definable), then every quantifier-free definable subset of G^n is a finite Boolean combination of cosets of quantifier-free definable subgroups of G^n. This partial result suffices for most applications. If H is a semi-abelian variety, then G is also stable stably embedded, and the full result holds, see [23].

$\mathcal{D}$-fields.
Moosa and Scanlon show in [68] that the dichotomy theorem holds for finite-dimensional types of SU-rank 1, i.e., types over a set A of tuples a such that the

transcendence degree over acl(A) of acl(Aa) is finite. This brings me to another remark.

3.6 Some words on proofs of the dichotomy

With the exception of ACFA in characteristic 0, the early results of the dichotomy for theories of enriched fields[4] used a very technical tool, Zariski geometries. Zariski geometries were introduced by Hrushovski and Zilber in [46] and provide a set of axioms for a geometry on a structure, which, if satisfied, imply that the structure satisfies the trichotomy. The original setting is intended for strongly minimal sets. In [39] a version for sets of realisations of minimal types appears, see also [58] and [106]. This was unsatisfactory in many respects, as checking the axioms usually is very long and involved.

A new way of showing the dichotomy: the CBP
Inspired by results of Fujiki and Campana on compact complex manifolds, Pillay introduced in [75] a property of types of canonical bases of types of finite rank, and which implies the dichotomy. Later, Pillay and Ziegler ([77]) somewhat generalised this property to a wider context. This property is called the *Canonical Base Property*, *CBP* for short. Pillay and Ziegler then proceed to show the CBP for certain algebraic examples, in a very algebraic and elementary (and ingenious) fashion. They show it for the following theories and families of types:
- In characteristic 0, types of finite rank in the theories DCF_0 and ACFA.
- In separably closed fields of finite invariant, the very thin types[5] have the CBP.

By the above mentioned result of Pillay, their result gives in particular a direct proof of the dichotomy for DCF_0 and $ACFA_0$. Their proof can be (and was) adapted to the so-called finite-dimensional types of DCFA and of $DCF_{0,n}$, $\mathcal{D}$-CF_0, $DCFA_n$, see [14] and [68], [87]. In the case of $DCF_{0,n}$, and later of DCFA, the method of arc-spaces introduced in [69] yields the full dichotomy. It is at the moment still open for arbitrary rank 1 types in $\mathcal{D}$-CF_0.

Let me also mention that the dichotomy holds for types of transformally algebraic elements in SCFA and in SCFE: one reduces the problem to the case of

4 that is: for DCF_0 in [45], for minimal types in separably closed fields of finite degree of imperfection [27], for ACFA in positive characteristic [24].
5 i.e., those such that given a realisation a over a model K, the structure $K\langle a\rangle$ obtained by closing $K(a)$ under the p-th powers of the λ-functions, has a finite separating transcendence basis over K.

ACFA. Here SCFE denotes the model companion of the theory of difference fields with σ not surjective. It can also be described as the limit theory of the structures $(\mathbb{F}_p(t)^s, x \mapsto x^{p^n})$ when p^n goes to ∞.

3.7 A few words on the CBP

The CBP has other interesting consequences, which can be found in [77], in [99], in [65] and in [16]. It is still open for minimal types in SCF_e, and more generally in most fields of positive characteristic (maybe with additional structure) which are not algebraically closed. Chatzidakis shows in [16] that types of finite rank in existentially closed difference fields of characteristic $p > 0$ have the CBP.

3.8 Definable groups

A classical result says that a group definable in an algebraically closed field K is isomorphic to $G(K)$ for some algebraic group G.
The intended generalisation to a field with operators K is:
$(*)$ *If H is definable in K, then there is a group homomorphism $f : H' \to G(K)$, where H' is a definable subgroup of H of finite index, G is an algebraic group defined over K, and* $\ker(f)$ *is finite.*

This generalisation holds for most theories we described above (and maybe all). They appear in papers by Messmer, and by Bouscaren and Delon on separably closed fields of finite invariant ([62], [11]), by Benoist [5] on fields with Hasse-Schmidt derivations, by Pillay on differential groups ([73], [76]), by Kowalski and Pillay ([52], [53], [54], [55]), by Bustamante-Medina [14], Freitag [29], ... (to name a few).

In a recent paper [7], Blossier, Martin-Pizarro and Wagner introduce the notion of a theory being one-based over another, and describe definable groups in models of the first one as almost embeddable in definable groups in the smaller language of the second one. An application of their result gives then that $(*)$ holds in all the theories we have so far considered, provided that

- the theory eliminates imaginaries,
- independence is given by ACF-independence of the algebraic closures in the sense of the theory.

Thus it seems to me that among the theories I introduced above, the only theories which do not satisfy the hypotheses of the result of Blossier, Martin-Pizarro and Wagner, are those of fields of positive characteristic and which have infinite degree of imperfection.

One should note that this is not quite the optimal result in case of differential fields, since there $(*)$ can be strengthened to conclude that f is injective.

Acknowledgement: The author was partially supported by ANR-09-BLAN-0047.

Bibliography

[1] James Ax, On Schanuel's conjectures, Ann. of Math. (2) 93 (1971), 252 – 268.
[2] Salih Azgin, Valued fields with contractive automorphism and Kaplansky fields, J. Algebra 324 (2010), no. 10, 2757 – 2785.
[3] Salih Azgin, Lou van den Dries, Elementary theory of valued fields with a valuation-preserving automorphism. J. Inst. Math. Jussieu 10 (2011), no. 1, 1 – 35.
[4] Luc Bélair, Angus Macintyre, Thomas Scanlon, Model theory of the Frobenius on the Witt vectors, American Journal of Mathematics, Volume 129, Number 3, June 2007, 665 – 721.
[5] Franck Benoist, Théorie des modèles des corps munis d'une dérivation de Hasse, Ph.D. thesis, Université Denis Diderot - Paris VII, 2005.
[6] Daniel Bertrand, Anand Pillay, A Lindemann-Weierstrass theorem for semi-abelian varieties over function fields,J. Amer. Math. Soc. 23 (2010), no. 2, 491 – 533.
[7] Thomas Blossier, Amador Martin Pizarro et Frank O. Wagner, Géométries relatives, preprint 2010, hal-00514393.
[8] Lenore Blum, Generalized Algebraic Structures: A Model Theoretical Approach, PhD thesis, MIT, 1968.
[9] Lenore Blum, Differentially closed fields: A model-theoretic tour, in *Contributions to Algebra* (H. Bass, P. Cassidy, and J. Kovacic, Eds.), Academic Press, San Diego, 1977.
[10] N. Bourbaki, XI, *Algèbre Chapître 5, Corps commutatifs*, Hermann, Paris 1959.
[11] E. Bouscaren, F. Delon, Groups definable in separably closed fields, Trans. Amer. Math. Soc. 354 (2002), no. 3, 945–966.
[12] E. Bouscaren, F. Delon, Minimal groups in separably closed fields, J. Symbolic Logic 67 (2002), no. 1, 239–259.
[13] Thomas Brihaye, Christian Michaux, Cédric Rivière, Cell decomposition and dimension function in the theory of closed ordered differential fields, Ann. Pure Appl. Logic 159 (2009), no. 1-2, 111 – 128.
[14] Ronald F. Bustamante Medina, Model Theory of Differentially Closed Fields with a Generic Automorphism, PhD thesis, Paris 2005. Available at http://arxiv.org/abs/0907.3959.
[15] Zoé Chatzidakis, Generic automorphisms of separably closed fields, Illinois J. Math. 45 (2001), no. 3, 693 – 733.
[16] Zoé Chatzidakis, A note on canonical bases and one-based types in supersimple theories, Confluentes Math., 04 No 3, 1250004 (2012).

[17] Z. Chatzidakis, G. Cherlin, S. Shelah, G. Srour and C. Wood, Orthogonality of types in separably closed fields, in: *Classification theory, Proc. of the 1985 Chicago Conference*, Springer Lecture Notes 1292.

[18] Zoé Chatzidakis, Ehud Hrushovski, Model theory of difference fields, Trans. Amer. Math. Soc. 351 (1999), 2997 – 3071.

[19] Zoé Chatzidakis, Ehud Hrushovski, Asymptotic theories of differential fields, Illinois J. of Mathematics 47 Nr 3 (2003), 593 – 618.

[20] Zoé Chatzidakis, Ehud Hrushovski, Model theory of endomorphisms of separably closed fields, J. of Algebra 281 (2004), no. 2, 567 – 603.

[21] Zoé Chatzidakis, Ehud Hrushovski, Difference fields and descent in algebraic dynamics, I, Journal of the IMJ, 7 (2008) No 4, 653 – 686.

[22] Zoé Chatzidakis, Ehud Hrushovski, Difference fields and descent in algebraic dynamics, II, Journal of the IMJ, 7 (2008) No 4, 687 – 704.

[23] Zoé Chatzidakis, Ehud Hrushovski, On subgroups of semi-abelian varieties defined by difference equations, preprint 2012. Available on arxiv 1112.0920.

[24] Zoé Chatzidakis, Ehud Hrushovski, Yaacov Peterzil, The model theory of difference fields, II: periodic ideals and the trichotomy in all characteristics, Proceedings of the London Math. Society (3) 85 (2002), 257 – 311.

[25] Z. Chatzidakis, A. Pillay, Generic structures and simple theories, Ann. Pure Applied Logic 95 (1998), 71 – 92.

[26] Françoise Delon, Idéaux et types sur les corps séparablement clos, Supplément au Bull. de la S.M.F, Mémoire 33, Tome 116, 1988.

[27] Françoise Delon, Separably closed fields, in: *Model Theory and Algebraic Geometry*, E. Bouscaren ed., Springer Verlag, Lecture Note 1696, Berlin 1998.

[28] Ju. L. Ershov, Fields with a solvable theory (Eng. transl.), Sov. Math. Doklady 8 (1967), 575 – 576.

[29] James Freitag, Indecomposability for differential algebraic groups, available on arxiv 1106.0695.

[30] Nicolas Guzy, Note sur les corps différentiellement clos valués, C. R. Math. Acad. Sci. Paris 341 (2005), no. 10, 593 – 596.

[31] Nicolas Guzy, Quelques remarques sur les corps D-valués, C. R. Math. Acad. Sci. Paris 343 (2006), no. 11-12, 689 – 694.

[32] Nicolas Guzy, Valued fields with K commuting derivations, Comm. Algebra 34 (2006), no. 12, 4269 – 4289.

[33] Nicolas Guzy, 0-D-valued fields, J. Symbolic Logic 71 (2006), no. 2, 639 – 660.

[34] Nicolas Guzy, Some elements of Lie-differential algebra and a uniform companion for large Lie-differential fields, Ann. Pure Appl. Logic 150 (2007), no. 1-3, 66 – 78.

[35] Nicolas Guzy, Françoise Point, Topological differential fields, Ann. Pure Appl. Logic 161 (2010), no. 4, 570 – 598.

[36] Nicolas Guzy, Cédric Rivière, Geometrical axiomatization for model complete theories of differential topological fields, Notre Dame J. Formal Logic 47 (2006), no. 3, 331 – 341.

[37] Ehud Hrushovski, Strongly minimal expansions of algebraically closed fields, Israel J. Math, 79, 129 – 151 (1992).

[38] Ehud Hrushovski, A new strongly minimal set, in *Stability in model theory, III (Trento, 1991)*, Ann. Pure Appl. Logic 62 (1993), no. 2, 147 – 166.

[39] Ehud Hrushovski, The Mordell-Lang conjecture for function fields, J. of the Amer. Math. Soc. Vol 9 Nr 3 (1996), 667 – 690.

[40] Ehud Hrushovski, The Manin-Mumford conjecture and the model theory of difference fields, Ann. Pure Appl. Logic 112 (2001), no. 1, 43 – 115.
[41] Ehud Hrushovski, The elementary theory of the Frobenius automorphism, preprint (1996, revised in 2004, ... 2013). Available at http://www.math.huji.ac.il/~ehud/.
[42] E. Hrushovski, M. Itai, On model complete differential fields, Trans. Amer. Math. Soc. 355 (2003), no. 11, 4267 – 4296.
[43] U. Hrushovski, A. Pillay, Weakly normal groups, in: *Logic Colloquium 85*, North Holland 1987, 233 – 244.
[44] Ehud Hrushovski, Anand Pillay, Effective bounds for the number of transcendental points on subvarieties of semi-abelian varieties, Amer. J. Math. 122 (2000), no. 3, 439 – 450.
[45] Ehud Hrushovski, Željko Sokolović, Minimal subsets of differentially closed fields, preprint 1994. Accepted by Trans. Amer. Math. Soc..
[46] Ehud Hrushovski, Boris Zilber, Zariski geometries, J. of the Amer. Math. Soc. Vol 9 Nr 1 (1996), 1 – 56.
[47] Hirotaka Kikyo, On generic predicates and automorphisms, Kokyuroku of RIMS 1390 (2004), 1 – 8. Available at http://www.kurims.kyoto-u.ac.jp/~kyodo/kokyuroku/contents/1390.html.
[48] Piotr Kowalski, Geometric axioms for existentially closed Hasse fields, Annals Pure Appl. Logic 135 (2005), 286 – 302.
[49] Piotr Kowalski, Derivations of the Frobenius map, J. Symbolic Logic 70 (1) (2005) 99 – 110.
[50] Piotr Kowalski, Jet operators on fields, J. Algebra 289 (2005), no. 1, 312 – 319.
[51] Piotr Kowalski, Schanuel property for additive power series, Israel J. Math. 190 (2012), 349 – 363.
[52] Piotr Kowalski, Anand Pillay, Pro-algebraic and differential algebraic groups on affine spaces, American Journal of Math., 122, (2000), 213 – 222.
[53] Piotr Kowalski, A. Pillay, A note on groups definable in difference fields, Proceedings AMS, 130, (2001), 205-212.
[54] Piotr Kowalski, Anand Pillay, Quantifier elimination for algebraic D-groups, Trans. Amer. Math. Soc. 358 (2006), no. 1, 167 – 181.
[55] Piotr Kowalski, Anand Pillay, On algebraic σ-groups, Trans. Amer. Math. Soc. 359 (2007), no. 3, 1325–1337.
[56] Angus Macintyre, Generic automorphisms of fields, Ann. Pure Appl. Logic 88 Nr 2-3 (1997), 165 – 180.
[57] David Marker, Model theory of differential fields, in *Model Theory of fields* (D. Marker, M. Messmer, A. Pillay Eds.), Lecture Notes in Logic Nr 5, Springer Verlag 1996, 38 – 113.
[58] David Marker, Zariski Geometries, in: *Model Theory and Algebraic Geometry*, E. Bouscaren ed., Springer Verlag, Lecture Note 1696, Berlin 1998, 107 – 128.
[59] Tracey McGrail, The model theory of differential fields with finitely many commuting derivations, J. Symbolic Logic, vol. 65 (2000), no.2, 885 – 913.
[60] Alice Medvedev, $\mathbb{Q}$ACFA, preprint (2012).
[61] Alice Medvedev, Thomas Scanlon, Invariant varieties for polynomial dynamical systems, Annals of Mathematics (2) 179 (2014) no. 1, 81 – 177.
[62] Margit Messmer, Groups and fields interpretable in separably closed fields, Trans. Amer. Math. Soc. 344 (1994), 361 – 377.

[63] Margit Messmer and Carol Wood, Separably closed fields with higher derivations, Journal of Symbolic Logic 60 (1995), 898 – 910.
[64] Christian Michaux, Cédric Rivière, Quelques remarques concernant la théorie des corps ordonnés différentiellement clos, Bull. Belg. Math. Soc. Simon Stevin 12 (2005), no. 3, 341 – 348.
[65] Rahim Moosa, Anand Pillay, On canonical bases and internality criteria, Illinois J. Math. 52 (2008), no. 3, 901 – 917.
[66] Rahim Moosa and Thomas Scanlon, Jets and prolongations spaces, Journal of the Institute of Mathematics Jussieu 9 (2010), no. 2, 391 – 430.
[67] Rahim Moosa and Thomas Scanlon, Generalized Hasse-Schmidt varieties and their jet spaces, Proc. Lond. Math. Soc. (3) 103 (2011), no. 2, 197 – 234.
[68] Rahim Moosa and Thomas Scanlon, Model theory of fields with free operators in characteristic zero, preprint 2012, available on arXiv 1212.5838.
[69] Rahim Moosa, Anand Pillay and Thomas Scanlon, Differential arcs and regular types in differential fields, J. Reine Angew. Math. 620 (2008), 35 – 54.
[70] Koushik Pal, Multiplicative valued difference fields, J. Symbolic Logic 77 (2012), no. 2, 545 – 579.
[71] David Pierce, Fields with several commuting derivations, J. Symbolic Logic 79 (2014) no. 1, 1 – 19.
[72] David Pierce, Anand Pillay, A note on the axioms for differentially closed fields of characteristic zero, J. Algebra 204 (1998), no. 1, 108 – 115.
[73] Anand Pillay, Some foundational questions concerning differential algebraic groups, Pacific J. 179 (1997), no 1, 179 – 200.
[74] Anand Pillay, Differential Galois Theory I, Illinois Journal of Math. 42 (1998), 678 – 699.
[75] Anand Pillay, Model-theoretic consequences of a theorem of Campana and Fujiki, Fundamenta Mathematicae, 174 (2002), 187 – 192.
[76] Anand Pillay, Algebraic D-groups and differential Galois theory, Pacific J. Math. 216 (2004), no. 2, 343 – 360.
[77] Anand Pillay, Martin Ziegler, Jet spaces of varieties over differential and difference fields, Selecta Math. (N.S.) 9 (2003), no. 4, 579 – 599.
[78] Françoise Point, Ensembles définissables dans les corps ordonnés différentiellement clos, C. R. Math. Acad. Sci. Paris 349 (2011), no. 17-18, 929 – 933.
[79] Bruno Poizat, Une théorie de Galois imaginaire, J. of Symb. Logic vol. 48 (1984) no. 4, 1151 – 1170.
[80] Cédric Rivière, The model theory of m-ordered differential fields, Math. Log. Q. 52 (2006), no. 4, 331 – 339.
[81] Cédric Rivière, The theory of closed ordered differential fields with m commuting derivations, C. R. Math. Acad. Sci. Paris 343 (2006), no. 3, 151 – 154.
[82] Cédric Rivière, Further notes on cell decomposition in closed ordered differential fields, Ann. Pure Appl. Logic 159 (2009), no. 1-2, 100 – 110.
[83] Abraham Robinson, Relative model-completeness and the elimination of quantifiers, Dialectica Volume 12, Issue 3-4 (1958), pages 394 – 407.
[84] Abraham Robinson, On the concept of a differentially closed field, Bulletin of the Research Council of Israel, 1959, 113 – 128.
[85] A. Robinson, *Introduction to Model Theory and to the Metamathematics of Algebra*, North-Holland, Amsterdam, 1974.

[86] Omar León Sánchez, Geometric Axioms for Differentially Closed Fields with Several Commuting Derivations, Journal of Algebra, Vol. 362 (2012), 107 – 116.

[87] Omar León Sánchez, PhD Thesis, U. of Waterloo, Canada, 2013.

[88] Thomas Scanlon, *A model complete theory of valued D-fields*, Journal of Symbolic Logic, 65 , no. 4 (2000), 1758 - 1784.

[89] Thomas Scanlon, The abc-theorem for commutative algebraic groups in characteristic p, International Mathematics Research Notices 1997, no. 18, 881 – 898.

[90] Thomas Scanlon, p-adic distance from torsion points of semi-abelian varieties, J. Reine Angew. Math. 499 (1998), 225 – 236.

[91] Thomas Scanlon, The conjecture of Tate and Voloch on p-adic proximity to torsion, International Mathematics Research Notices, 1999, no. 17, 909 – 914.

[92] Thomas Scanlon, Diophantine geometry of the torsion of a Drinfeld module, J. Number Theory 97 (2002), no. 1, 10 – 25.

[93] Saharon Shelah, Differentially closed fields, Israel J. of Math. 16 (1973), 314 – 328.

[94] Michael F. Singer, The model theory of ordered differential fields. J. Symbolic Logic 43 (1978), no. 1, 82 – 91.

[95] Michael F. Singer, Model theory of partial differential fields: from commuting to noncommuting derivations, Proc. Amer. Math. Soc. 135 (2007), no. 6, 1929 – 1934.

[96] G. Srour, The independence relation in separably closed fields, J. Symbolic Logic 51 (1986), no. 3, 715 – 725.

[97] Sonat Süer, On subgroups of the additive group in differentially closed fields, J. Symbolic Logic 77 (2012), no. 2, 369 – 391.

[98] Marcus Tressl, The uniform companion for large differential fields of characteristic 0, Trans. Amer. Math. Soc., vol.357 (2005), no.10, 3933 – 3951.

[99] Frank O. Wagner, Some remarks on one-basedness, J. of Symb. Logic 69 Nr 1 (2004), 34 – 38.

[100] Carol Wood, The model theory of differential fields of characteristic $p > O$, Proc. Amer. Math. Soc. 40 (1973), 577 – 584.

[101] Carol Wood, Prime model extensions for differential fields of characteristic $p \neq 0$, J. Symbolic Logic 39 (1974), 469 – 477.

[102] Carol Wood, The model theory of differential fields revisited, Israel Journal of Mathematics 25 (1976) 331 – 352.

[103] Carol Wood, Notes on the stability of separably closed fields, J. Symbolic Logic 44 (1979), 412 – 416.

[104] Yoav Yaffe, Model completion of Lie differential fields, Ann. Pure Appl. Logic, vol. 107 (2001), no.1-3, 49 – 86.

[105] Martin Ziegler, Separably closed fields with Hasse derivations, Journal of Symbolic Logic 68 (2003), 311 – 318.

[106] Boris Zilber, *Zariski Geometries. Geometry from the logician's point of view*, London Mathematical Society Lecture Note Series, 360. Cambridge Univ. Press, Cambridge, 2010.

Carlos Augusto Di Prisco

Some Aspects of the Ramsey Theory of Real Numbers

Abstract: We examine several variations of the Ramsey property for collections of infinite sets or equences of natural numbers. Particular enphasis is put on the existence of homogeneous infinite products of finte sets of natural numbers. Parametrizations of some of these partition properties are also considered, and then used to obtain independence results.

Keywords: partition properties, partition calculus, Ramsey sets

MSC: 03E02, 03E05

Carlos Augusto Di Prisco: Instituto Venezolano de Investigaciones científicas, Caracas; Universidad de Los Andes, Bogotá, email: cdiprisc@ivic.gob.ve

1 Introduction

The theory of partitions as an area of combinatorial set theory originated in the 1930's with F. P. Ramsey's famous theorem [25]. Some earlier results of Schur and Van der Waerden have some of the same combinatorial flavor, but Ramsey's theorem became the initial point of a whole area of study (see [14]). In the 1950's Erdös and Rado gave shape to the theory and developed it in several different directions with their partition calculus [12].

Ramsey's Theorem asserts that given positive integers n and k, for every partition of $\mathbb{N}^{[n]} = \{a \subseteq \mathbb{N} : |a| = n\}$ into k many pieces, there is an infinite $H \subseteq \mathbb{N}$ such that $H^{[n]}$, the collection of all of its n-element subsets, is contained in one piece. Such a set H is said to be homogeneous for the partition. A finitary version of Ramsey's theorem is stated as follows, given positive integers n, k, m, there is a positive integer N such that for every set A of N elements and every partition of $A^{[n]} = \{a \subseteq A : |a| = n\}$ into k many pieces, there is a subset $H \subseteq A$ with $|H| = m$, for which $H^{[n]}$ is contained in one of the pieces (i.e. H is homogeneous for the partition).

Ramsey proved his theorem as a technical tool to answer a question about decidability of first order logic [25]. Many different applications of the theorem have been found, and the multiple ways it has been generalized or adapted to other contexts constitute a rich theory with applications in other areas of mathematics.

The books [14] and [23] provide a good sample of the degree of development of Ramsey Theory. Several other developments in this area have started from metamathematical considerations. In these other cases as well, the solution of a metamathematical problem led to the development of concepts and theories of a purely combinatorial character. The converse direction has also been fruitful, as some natural questions about partitions have required a metamathematical analysis, for example to establish consistency results.

The obvious infinite dimensional generalization of Ramsey's theorem to partitions of the set $\mathbb{N}^{[\infty]}$ of infinite subsets of $\mathbb{N}$, instead of the collection of subsets of a specified finite size, is false. Nevertheless, partitions of $\mathbb{N}^{[\infty]}$ into a finite number of "topologically simple" pieces always admit an infinite homogeneous set, i.e. an infinite set $H \subseteq \mathbb{N}$ such that $H^{[\infty]} = \{A \subseteq H : A \text{ infinite}\}$ is contained in one of the pieces.

Since we can identify subsets of $\mathbb{N}$ with their characteristic functions, the set $\mathbb{N}^{[\infty]}$ corresponds to a subset of the Cantor space $2^{\mathbb{N}}$, which, with the inherited topology, is homeomorphic to the set of irrational numbers, and also to the Baire space, the set $\mathbb{N}^{\infty}$ of all infinite sequences of natural numbers endowed with the product topology. In this sense, we are dealing with partitions of the set of real numbers, considering these numbers represented by infinite subsets of $\mathbb{N}$, or infinite sequences of natural numbers.

Partition properties are frequently stated in terms of colorings. A k-coloring of a set S is simply a function $c : S \to K$ where K is a set of size k. Clearly, every k-coloring of S determines a partition of S into k pieces. For example, Ramsey's Theorem states that for every $n, k \in \mathbb{N}$ and every k-coloring of $\mathbb{N}^{[n]}$, there is $H \in \mathbb{N}^{[\infty]}$ such that $H^{[n]}$ is monochromatic.

We will consider partitions (or colorings) of $\mathbb{N}^{[\infty]}$ the set of infinite subsets of $\mathbb{N}$, and partitions of infinite products of various structures, finite or infinite, and the existence of different types of monochromatic sets for them. Different types of monochromatic sets usually give rise to corresponding partition properties of different strengths. We will be interested in the interrelationship between them, looking both into metamathematical and purely combinatorial aspects.

Some quite old questions of the theory remain open, like the question about the necessity of inaccessible cardinals for the consistency of the Ramsey property $\omega \to (\omega)^{\omega}$, or if this partition property follows from AD.

We use a very standard notation. $\mathbb{N}$ is the set of natural numbers, which is identified with ω, the first infinite ordinal. $\mathbb{N}^{\infty}$ is the set of infinite sequences of natural numbers; the topological space obtained giving it the product topology (obtained by giving $\mathbb{N}$ the discrete topology) is called the Baire space. $\mathbb{N}^{<\infty} = \bigcup_{n=0}^{\infty} \mathbb{N}^n$ is the set of finite sequences of natural numbers. Given a set A and $n \in \mathbb{N}$, $A^{[n]}$ is the collection of those subsets of A which have exactly n elements. The collection of finite subsets of $\mathbb{N}$ is denoted by $\mathbb{N}^{[<\infty]}$. For every infinite $A \subseteq \mathbb{N}$, we use $A^{[\infty]}$ to denote the collection of infinite subsets of A, accordingly, $\mathbb{N}^{[\infty]}$ is the set of infinite subsets of $\mathbb{N}$. If $A \in \mathbb{N}^{[\infty]}$ and $a \in \mathbb{N}^{[<\infty]}$, then $A/a = \{n \in A : \max a < n\}$.

If $s \in \mathbb{N}^{[<\infty]}$ and $n \in \mathbb{N}$, we write $n < s$ to express that $n < \min(s)$.

We will use the letters $n, m, k, l, \ldots$ to denote natural numbers, and A, B, C, H, X, $Y, \ldots$ to denote infinite sets of natural numbers. The letters $a, b, c, \ldots$ will be used to denote finite sets of natural numbers, and $s, t, r, \ldots$ to denote finite sequences.

2 Ramsey's Theorem

In this section we start presenting Ramsey's theorem, which can be viewed as the origin of the branch of combinatorics known today as Ramsey Theory. Then, we consider some extensions or generalizations.

Theorem 2.1. *(Ramsey [25]) For every $n, k \in \mathbb{N}$, and every $c : [\mathbb{N}]^n \to k$, there is $H \in [\mathbb{N}]^\infty$ such that c is constant on $[H]^\infty$.*

In the arrow notation of Erdös and Rado, this is expressed by

$$\omega \to (\omega)^n_k.$$

For a proof of Ramsey's theorem the reader can consult [14] or [30].

It should be clear that the statement of Ramsey's theorem for $k = \omega$ is false, just consider the function $c : [\mathbb{N}]^n \to \omega$ given by $c(a) = min(a)$.

Other functions like $c(a) = \{min(a), max(a)\}$, give counterexamples. The following theorem of Erdös and Rado classifies all the counterexamples.

Definition 2.2. Let $a = \{x_1, \ldots, x_r\}$ be a set of size r, and $X \subseteq \{1, \ldots, r\}$. Then $a \restriction X = \{x_i : i \in X\}$. A partition $c : [B]^r \to Ord$ of the collection of r-element subsets of B is canonical on B if there is a set $X \subseteq \{1, \ldots, r\}$ such that for every $a, b \in [B]^r$,

$$c(a) = c(b) \text{ if and only if } a \restriction X = b \restriction X.$$

The set X is unique if it exists.

Theorem 2.3. *([11]) For every positive integer r, and for every partition $f : [\mathbb{N}]^r \to Ord$, there is a set $B \in [\mathbb{N}]^\infty$, such that f is canonical on B.*

The sets of the form $[\mathbb{N}]^r$ with r a positive integer are the uniform families of finite rank. A result of Pudlak and Rödl ([24]) generalizes Theorem 2.3 to the realm of all uniform families.

For the convenience of the reader we include the following definition.

By induction on the countable ordinals we define what is an α-uniform family of finite sets of natural numbers.

Definition 2.4. A family $\mathcal{F} \subseteq [\mathbb{N}]^{<\omega}$ is α-uniform on $A \in [\mathbb{N}]^{\infty}$ if

1. $\alpha = 0$ and $\mathcal{F} = \{\emptyset\}$,
2. $\alpha = \beta + 1$ and for every $n \in \omega$, the collection
$$\mathcal{F}_{(n)} = \{t \in \mathbb{N}^{[<\infty]} : n < t, \{n\} \cup t \in \mathcal{F}\}$$
is β-uniform on A/n,
3. α is a limit ordinal and there exists a strictly increasing sequence $\{\alpha_n : n \in \omega\}$ of ordinals with limit α and for every $n \in \omega$,
$$\mathcal{F}_{(n)} = \{t : n < t, \{n\} \cup t \in \mathcal{F}\}$$
is α_n-uniform on A/n.

We say that $\mathcal{F}$ is uniform on A if it is α-uniform on A for some $\alpha < \omega_1$.

Observe that $\{\emptyset\}$ is the only 0-uniform family. And in general, for every $k \in \omega$, $\mathbb{N}^{[k]}$ is the only k-uniform family. The collection of all the sets of natural numbers of the form $\{n_1, n_2, \dots, n_k\}$ such that $n_1 < n_2 < \cdots < n_k$ y $k = n_1 + 1$ is an ω-uniform family called the Schreier family. More generally, if $\{m_i : i \in \omega\}$ is a strictly increasing sequence of natural numbers, the collection $\{\{n_1, n_2, \dots, n_k\} : k = m_{n_1}\}$ is an ω-uniform family.

And given $\{m_i : i \in \omega\}$ a strictly increasing sequence of natural numbers, the family $\{\{n_1, \dots, n_k\} : k > i,\ k = m_{n_i}\}$ is $\omega + i$-uniform

Recall that a family $\mathcal{F}$ of finite sets of natural numbers is a barrier on a set $M \in \mathbb{N}^{[\infty]}$ if for every $s, t \in \mathcal{F}$, $s \not\subseteq t$ if $s \neq t$, and every infinite subset of M has an initial segment in $\mathcal{F}$ [22].

It is easy to show by induction on the countable ordinals that every uniform family is a barrier. Also, every barrier is uniform on some infinite set.

Definition 2.5. A function $\phi : \mathcal{S} \to \omega$ is a canonical partition of $\mathcal{S}$ on X if $\mathcal{S}$ is a uniform family on X and there is a uniform family $\mathcal{T}$ and an application $f : \mathcal{S} \to \mathcal{T}$ such that

(a) $f(s) \subseteq s$ for every $s \in \mathcal{S}$,
(b) for every $s, t \in \mathcal{S}$, $\phi(s) = \phi(t)$ if and only if $f(s) = f(t)$.

Clause (b) is equivalent to

(b') there is a one-to-one function $\psi : \mathcal{T} \to \mathbb{N}$ such that for every $s \in \mathcal{B}$, $\phi(s) = \psi(f(s))$.

If $\mathcal{F}$ is a family of finite sets of natural numbers and $A \in \mathbb{N}^{[\infty]}$, we write $\mathcal{F} \upharpoonright A$ to denote the subfamily $\mathcal{F} \cap \mathcal{P}(A)$.

Theorem 2.6. *[24] For every uniform barrier $\mathcal{S}$ on a set $X \in [\mathbb{N}]^\infty$ and every function $\phi : \mathcal{S} \to \mathbb{N}$, there is $Y \in X^{[\infty]}$ such that $\phi \restriction (\mathcal{S} \restriction Y)$ is a canonical partition of $\mathcal{S} \restriction Y$ on Y.*

To end this section let us consider partitions of $[\mathbb{N}]^\infty$. A straightforward generalization of Ramsey's theorem to this kind of partition is prevented by the axiom of choice. This is so since using this axiom one proves the existence of a partition of $[\mathbb{N}]^\infty$ into two parts neither of them containing a subset of the form $[A]^\infty$. Moreover, an argument of [19] shows that the existence of a non principal ultrafilter on $\mathbb{N}$ is enough to give a counterexample.

The topology induced on $[\mathbb{N}]^\infty$ by the product topology of $2^{\mathbb{N}}$ allows us to view this in a different form. Using the axiom of choice one can prove the existence of a totally imperfect subset of $[\mathbb{N}]^\infty$, that is, a subset of $[\mathbb{N}]^\infty$ that together with its complement meets every perfect subset of $[\mathbb{N}]^\infty$ ([1]). And since for every $A \in [\mathbb{N}]^\infty$ the set $[A]^\infty$ is perfect, a totally imperfect set determines a partition as above. Nevertheless, for partitions into definable portions there are homogeneous sets.

A subset $\mathcal{A}$ of $[\mathbb{N}]^\infty$ is Ramsey if there is $A \in [\mathbb{N}]^\infty$ such that $[A]^\infty \subseteq \mathcal{A}$ or $[A]^\infty \cap \mathcal{A} = \emptyset$.

Theorem 2.7. *(Galvin and Prikry, [13]) Every Borel subset of $[\mathbb{N}]^\infty$ is Ramsey.*

Theorem 2.8. *(Silver, [27]) Every analytic subset of $[\mathbb{N}]^\infty$ is Ramsey.*

It is consistent with $ZF + DC$ that every subset of $[\mathbb{N}]^\infty$ is Ramsey, relative to the consistency of $ZFC+\exists$ inaccessible cardinal. In Solovay's model [28], where every set of real numbers is Lebesgue measurable, every subset of $[\mathbb{N}]^\infty$ is Ramsey [20].

Using the arrow notation, the statement "every subset of $[\mathbb{N}]^\infty$ is Ramsey" is expressed by

$$\omega \to (\omega)^\omega.$$

In the next sections we will examine several variants of this property.

3 Sublattices of $\mathcal{P}(\mathbb{N})$

We consider now a partition property defined using of a different type of homogeneity for partitions of $\mathbb{N}^{[\infty]}$. Instead of requiring for every coloring $c : \mathbb{N}^{[\infty]} \to \{0, 1\}$ the existence of an infinite set B with all of its infinite subsets of the same color, we only require that all subsets of B containing a fixed subset A have the same color.

The partition symbol

$$\omega \to ((\omega))^\omega$$

means that for every coloring $c : \mathbb{N}^{[\infty]} \to \{0,1\}$ there is a pair (A, B) with $A \subseteq B \in \mathbb{N}^{[\infty]}$ and $B \setminus A \in \mathbb{N}^{[\infty]}$ such that

$$[A, B] = \{X : A \subseteq X \subseteq B\}$$

is monochromatic (we put no requirements on the size of A). Such sets of the form $[A, B]$ are called "doughnuts" in [7] and [6].

We write

$$\omega \to (\text{perfect})^{\omega}$$

to express that for every coloring $c : \mathbb{N}^{[\infty]} \to \{0,1\}$ there is a perfect monochromatic set. Notice that this is equivalent to saying that there are no Bernstein sets. A Bernstein set (or totally imperfect set) is a subset S of $\mathbb{N}^{[\infty]}$ such that S and its complement meet every perfect set.

Since every "doughnut" $[A, B]$ is a perfect subset of $\mathbb{N}^{[\infty]}$, it follows immediately from the definitions that

$$\omega \to (\omega)^{\omega} \text{ implies } \omega \to ((\omega))^{\omega} \text{ implies } \omega \to (\text{perfect})^{\omega}.$$

Using the argument of [19] mentioned above, it can be shown that a non-principal ultrafilter on $\mathbb{N}$ provides a counterexample for the property $\omega \to ((\omega))^{\omega}$, but the property $\omega \to (\text{perfect})^{\omega}$ is consistent with the existence of such ultrafilters (see [3]).

The consistency of $(ZF+DC+\omega \to ((\omega))^{\omega})$ follows just from the consistency of ZFC, with no hypothesis involving inaccessible cardinals. This is so since $\omega \to_{Baire} ((\omega))^{\omega}$ holds (see [7]), and the consistency of "every subset of $\mathbb{N}^{\infty}$ has the Baire property" has been established by Shelah assuming just the consistency of ZFC [26].

J. Brendle, L. Halbeisen and B. Lowe have studied the property $\omega \to ((\omega))^{\omega}$ restricted to Σ^1_2 sets, Δ^1_2 sets, and projective sets in general (see, for example, [2], and [15], where it is shown that $\omega \to_{\Sigma^1_2} ((\omega))^{\omega}$ does not imply $\omega \to_{\Sigma^1_2} (\omega)^{\omega}$).

The exact relationship between the properties $\omega \to (\omega)^{\omega}$ and $\omega \to ((\omega))^{\omega}$ was revealed in [4]. In section 7 we will explain this and also the relation with a partition property of polarized type.

4 Homogeneity in terms of products of finite sets

In this section we consider partitions of the Baire space $\mathbb{N}^{\infty} = \mathbb{N} \times \mathbb{N} \times \ldots$, the set of infinite sequences of natural numbers with the product topology obtained when $\mathbb{N}$ is considered as a discrete space. We will also consider partitions of some subspaces of the Baire space given by infinite products of finite sets. For such partitions we ask for the existence of homogeneous sets of a very particular form, namely, products of finite sets.

Given

$$c : \mathbb{N}^{\infty} \to \{1, 2\},$$

we want a monochromatic product $\prod_{i=0}^{\infty} H_i$, with $H_i \subseteq \mathbb{N}$ of some specified size for every i. ([5]).

We use the partition symbol

$$\begin{pmatrix} \omega \\ \omega \\ \vdots \end{pmatrix} \to \begin{pmatrix} m_0 \\ m_1 \\ \vdots \end{pmatrix}$$

to express that for every coloring $c : \mathbb{N}^{\infty} \to 2$, there is a monochromatic product $\prod_{i=0}^{\infty} H_i$ with $H_i \subseteq \mathbb{N}$ and $|H_i| = m_i$ for every i.

Notice that here we do not require the sequence $\{H_i\}_{i=0}^{\infty}$ to be increasing, but as shown in [5] this is not essential.

It is easy to realize that we cannot require H_0 and H_1 to be both infinite: let $c : \mathbb{N}^{\infty} \to \{0, 1\}$ be defined by $c(x) = 0$ if and only if $x(0) < x(1)$. Clearly, a product $\prod_{i=0}^{\infty} H_i$ with $H_i \subseteq \mathbb{N}$ and with H_0 and H_1 both infinite cannot be monochromatic for c. In the same fashion a counterexample can be given if we require infinite sets in any two specified (fixed) coordinates. If no coordinates are specified, it is possible to get for every $c : \mathbb{N}^{\infty} \to \{0, 1\}$ a monochromatic product $\prod_{i=0}^{\infty} H_i$ with at least two infinite factors, the position of the infinite factors depending on the partition. This follows from results of G. Moran and D. Strauss [21]. They prove that for every $k \in \mathbb{N}$ and every partition of $\mathbb{N}^{\infty}$ into two pieces, there is a monochromatic product $\prod_{i=0}^{\infty} H_i$ with k factors that are infinite (in fact, each of them is the whole set $\mathbb{N}$) and the rest of the factors are singletons. It is not known if we can also require the finite factors to be non-trivial, having at least two elements each.

Question 4.1. Is the following statement consistent? (We mean consistent with *ZF*, provided that *ZF* is consistent). For every partition $c : \mathbb{N}^{\infty} \to \{1, 2\}$, there is a homogeneous product $\prod_{i=0}^{\infty} H_i$ such that for every i, $|H_i| > 1$ and there are $m, n \in \mathbb{N}$, $m \neq n$ with $|H_m| = |H_n| = \aleph_0$. (see [7]).

Henle proved in [16] that $\omega \to (\omega)^{\omega}$ implies that every partition of $\mathbb{N}^{\infty}$ into two pieces admits a homogeneous product $\prod_{i=0}^{\infty} H_i$ such that for every i, $|H_i| > 1$ and H_0 is infinite.

Identifying infinite subsets of $\mathbb{N}$ with their increasing enumeration, it can be easily verified that $\omega \to (\omega)^\omega$ implies $\begin{pmatrix}\omega\\ \omega\\ \vdots\end{pmatrix} \to \begin{pmatrix}m_0\\ m_1\\ \vdots\end{pmatrix}$ for every sequence $\{m_i\}_{i=0}^\infty$ of positive integers.

The exact relationship between the partition properties $\omega \to (\omega)^\omega$ and $\begin{pmatrix}\omega\\ \omega\\ \vdots\end{pmatrix} \to \begin{pmatrix}m_0\\ m_1\\ \vdots\end{pmatrix}$ (for a given sequence $\{m_i\}_{i=0}^\infty$ of positive integers) turned out to be an interesting problem. The two partition properties are not equivalent since polarized partition property is consistent with the existence of non principal ultrafilters on $\mathbb{N}$ (see [10]). To prove this, a theory of partitions of products of finite sets and their parametrized versions was developed. This theory, of intrinsic combinatorial interest, is another example of a combinatorial development motivated by a metamathematical question (see [8, 9, 29]).

We will deal now with partitions of subspaces of $\mathbb{N}^\infty$ of the form $\prod_{i=0}^\infty n_i$, where $\{n_i\}_{i=0}^\infty$ is a sequence of positive integers and, as usual, we identify each n with the set $\{0, 1, \dots, n-1\}$. In other words, we will consider partitions of the set of infinite sequences of positive integers bounded by some sequence $\{n_i\}_{i=0}^\infty$.

Given a sequence $\{n_i\}_{i=0}^\infty$, invoking the axiom of choice we can define a coloring $c : \prod_{i=0}^\infty n_i \to \{0, 1\}$ such that for no sequence $\{H_i\}_{i=0}^\infty$ with $H_i \subseteq n_i$ and $|H_i| = 2$ for every i, the product $\prod_{i=0}^\infty H_i$ is monochromatic. Restricting the class of colorings to be considered, some positive results are obtained. For example, there is a sequence $\{n_i\}_{i=0}^\infty$ such that every Borel coloring $c : \prod_{i=0}^\infty n_i \to \{0, 1\}$ admits a monochromatic product of pairs ([18, 8]). An interesting connection appeared here between these partition relations restricted to definable classes of colorings and the Grzegorczyk hierarchy of primitive recursive functions (see [8] and [30]).

In more general terms, for every sequence $\{m_i\}_{i=0}^\infty$ of positive integers, there is a sequence $\{n_i\}_{i=0}^\infty$ of positive integers such that for every Souslin-measurable coloring $c : \prod_{i=0}^\infty n_i \to \{0, 1\}$ there is a monochromatic product $\prod_{i=0}^\infty H_i$ where for every i, $H_i \subseteq n_i$ has size m_i.

Partitions of infinite products of finite sets were considered in [5] and [17], with particular interest in the question of the consistency of the existence some sequence $\{n_i\}_{i=0}^\infty$ such that for every partition of $\prod_{i=0}^\infty n_i$ into two pieces there is a monochromatic product of pairs.

We describe now, for each sequence $\{m_i\}_{i=0}^\infty$, a class of products of finite sets and a class of colorings which admit homogeneous products whose factors have sizes determined by $\{m_i\}_{i=0}^\infty$. First we need some definitions. Given a sequence $\vec{n} = \{n_i\}_{i=0}^\infty$ of natural numbers, a sequence $\vec{H} = \{H_i\}_{i=0}^\infty$ of finite sets of natural numbers is said to be of type $\vec{n}$, or a $\vec{n}$-sequence, if for every i, $|H_i| = n_i$. We also say in this case that the product $\prod_{i=0}^{\infty} H_i$ is an $\vec{n}$-product, or (n_i)-product.

Whenever $\{H_i\}_{i=0}^{\infty}$ and $\{J_i\}_{i=0}^{\infty}$ are sequences of finite sets of natural numbers we say that $\prod_{i=0}^{\infty} J_i$ is a subproduct of $\prod_{i=0}^{\infty} H_i$ if $J_i \subseteq H_i$ for every i. We write $\vec{J} \leq_k \vec{H}$ if $J_i \subseteq H_i$ for every $i \in \mathbb{N}$ and $J_i = H_i$ for $i < k$.

Given sequences of positive integers $\vec{m} = \{m_i\}_{i=0}^{\infty}$ and $\vec{n} = \{n_i\}_{i=0}^{\infty}$,

$$(\vec{n}) \to (\vec{m})$$

expresses that for every partition $c : \prod_{i=0}^{\infty} H_i \to \{0,1\}$ such that $|H_i| = n_i$ for all i, there is a sequence $\{J_i\}_{i=0}^{\infty}$ with $J_i \subseteq H_i$, and $|J_i| = m_i$ for all i, such that c is constant on $\prod_{i=0}^{\infty} J_i$. More concisely, for every 2-coloring of a product of type $\vec{n}$, there is a monochromatic subproduct of type $\vec{m}$. As before, if we restrict ourselves to colorings measurable with respect to a certain σ-field $\mathcal{C}$ of subsets of $\mathbb{N}^{[\infty]}$, we use the notation

$$(\vec{n}) \to_{\mathcal{C}} (\vec{m})$$

for the corresponding property.

Notice that for sequences $\vec{m} = \{m_i\}_{i=0}^{\infty}$ and $\vec{n} = \{n_i\}_{i=0}^{\infty}$ of positive integers, $(\vec{n}) \to (\vec{m})$ does not follow from $\omega \to (\omega)^{\omega}$, as

$$\begin{pmatrix} \omega \\ \omega \\ \vdots \end{pmatrix} \to \begin{pmatrix} m_0 \\ m_1 \\ \vdots \end{pmatrix}$$

does.

For partitions of finite products we use the notation

$$\begin{pmatrix} n_0 \\ n_1 \\ \vdots \\ n_k \end{pmatrix} \to \begin{pmatrix} m_0 \\ m_1 \\ \vdots \\ m_k \end{pmatrix}$$

which means that for every partition $c : \prod_{i=0}^{k} H_i \to \{0,1\}$ such that $|H_i| = n_i$ for all $i \leq k$, there is a sequence $\{J_i\}_{i=0}^{k}$ with $J_i \subseteq H_i$, and $|J_i| = m_i$ for all $i \leq k$, such that c is constant on $\prod_{i=0}^{k} J_i$.

Definition 4.2. Let $S : \mathbb{N}^{<\infty} \to \mathbb{N}$ be defined as follows

$$S(m_0) = 2m_0 - 1$$

$$S(m_0, \ldots, m_{i+1}) = 2(m_{i+1} - 1)\left[\prod_{k=0}^{i} \binom{m_k}{S(m_0, \ldots, m_k)}\right] + 1.$$

The function S has the following property which can be verified by induction.

Lemma 4.3. *Let $(m_i) \in \mathbb{N}^\infty$ and let $n_i = S(m_0, \ldots, m_i)$ for all i. Then for every k*

$$\begin{pmatrix} n_0 \\ n_1 \\ \vdots \\ n_k \end{pmatrix} \to \begin{pmatrix} m_0 \\ m_1 \\ \vdots \\ m_k \end{pmatrix}.$$

It follows that we can get monochromatic subproducts of type $\vec{m}$ for every continuous 2-coloring of $\prod_{i=0}^{\infty} S(m_o, \ldots, m_i)$ (note that a continuous coloring gives a partition into clopen pieces).

The existence of monochromatic $\vec{m}$-subproducts for semicontinuous 2-colorings, i.e. partitions into a closed set and its complement, also follows from the Lemma (see [8]). To extend this result to a wider class of colorings, for example to Borel-measurable colorings, iterates of the function S are defined in order to carry out certain diagonalization arguments.

The iterates of the function S are defined recursively as follows.

$$\begin{aligned} &S^{(0)}(m_0, \ldots, m_i) = S(m_0, \ldots, m_i), \\ &S^{(p+1)}(m_0, \ldots, m_i) = \\ &S(S^{(p)}(m_0), S^{(p)}(m_0, m_1), \ldots, S^{(p)}(m_0, \ldots, m_i)). \end{aligned}$$

For $M \subseteq \mathbb{N}$, define $S_M : \mathbb{N}^{<\infty} \to \mathbb{N}$ by

$$\begin{aligned} S_M(m_0) &= S(m_0), \\ S_M(m_0, \ldots, m_i) &= S^{(p)}(m_0, \ldots, m_i). \end{aligned}$$

where

$$p = |\bigcup_{k \in M \cap i} \prod_{j=0}^{\infty} S_M(m_0, \ldots, m_j)|.$$

For every sequence $\{m_i\}_{i=0}^{\infty}$ of positive integers, $S_M((m_i))$ denotes the infinite sequence whose i-th term is $S_M(m_0, \ldots, m_i)$ (analogously for $S^{(p)}((m_i))$).

We get the following lemma.

Lemma 4.4. *Let $\{m_i\}_{i=0}^{\infty}$ be a sequence of positive integers, $M \subseteq \mathbb{N}$, $k \in M$, and $N = M \setminus \{k\}$. For every $S_M((m_i))$-product $\prod_i H_i$ and every closed set $\mathcal{X} \subseteq \prod_i H_i$, there is an $S_N((m_i))$-subproduct $\prod_i J_i$ of $\prod_i H_i$ such that for every $x, y \in \prod_i J_i$, if $x \upharpoonright k = y \upharpoonright k$, then $x \in \mathcal{X} \leftrightarrow y \in \mathcal{X}$.*

When this happens we say that $\mathcal{X} \cap \prod_i J_i$ depends only on coordinates $< k$.

Given $n \in \mathbb{N}$, and two sequences (J_i) and (H_i) we use

$$(J_i) \leq_n (H_i)$$

to express that $J_i = H_i$ for every $i < n$ and $J_i \subseteq H_i$ for every $i \geq n$.

A pair (A, B) of infinite subsets of $\mathbb{N}$ is said to be regular if $A \subseteq B$ and $B \setminus A$ is infinite. If (A, B) and $(C.D)$ are regular pairs,

$$(C, D) \leq (A, B)$$

if $A \subseteq C \subseteq D \subseteq B$ with both $C \setminus A$ and $B \setminus D$ infinite. And for $n \in \mathbb{N}$,

$$(C, D) \leq_n (A, B)$$

if $(C, D) \leq (A, B)$ and also $A \cap n = C \cap n$ and $D \cap n = B \cap n$.

Definition 4.5. We say that a set $\mathcal{X} \subseteq \mathbb{N}^\infty$ is (m_i)-Ramsey if for every regular pair (A, B), every $n \in B \setminus A$, and every S_B-product $\prod_i H_i$, there is a regular pair $(C, D) \leq_n (A, B)$ and and an S_D-product $\prod_i J_i \leq_n \prod_i H_i$ such that $\mathcal{X} \cap \prod_i J_i$ depends only on coordinates $< n$.

$\mathcal{X} \subseteq \mathbb{N}^\infty$ is (m_i)-Ramsey null if for every regular pair (A, B), every $n \in B \setminus A$, and every S_B-product $\prod_i H_i$, there is a regular pair $(C, D) \leq_n (A, B)$ and and an S_D-product $\prod_i J_i \leq_n \prod_i H_i$ such that $\mathcal{X} \cap \prod_i J_i = \emptyset$.

For every sequence (m_i), the ideal of (m_i)-Ramsey null sets is a σ-ideal, and the collection of (m_i)-Ramsey sets is a σ-field of subsets of $\mathbb{N}^\infty$.

Every closed subset of $\mathbb{N}^\infty$ is (m_i)-Ramsey, and therefore all Borel subsets of $\mathbb{N}^\infty$ are (m_i)-Ramsey. Furthermore, all analytic subsets of $\mathbb{N}^\infty$ are (m_i)-Ramsey since the field of (m_i)-Ramsey sets if closed under the Souslin operation.

We obtain thus, the following.

Theorem 4.6. *For every sequence $\vec{m} = \{m_i\}_{i=0}^\infty$, there is a sequence $\vec{n} = \{n_i\}_{i=0}^\infty$ such that for any Souslin-measurable*

$$c : \prod_{i=0}^{\infty} n_i \to \{0, 1\}$$

there exist $H_i \subseteq N_i$ with $|H_i| = m_i$ for all i, such that c is constant on $\prod_{i=0}^\infty H_i$.

These results were obtained in [8] and [10], and in this final form in [29] and [30].

As it was mentioned before, $\omega \to (\omega)^\omega$ implies the polarized partition relation

$$\begin{pmatrix} \omega \\ \omega \\ \vdots \end{pmatrix} \to \begin{pmatrix} m_0 \\ m_1 \\ \vdots \end{pmatrix}$$

for every sequence $\{m_i\}_{i=0}^{\infty}$, but these two partition relations are not equivalent. To show this, a parametrization of $\omega \to (\omega)^{\omega}$ by infinite products of finite sets was developed in [9] and [10]. We give the basic facts about this parametrization in the next section.

5 The Ramsey property parametrized by products of finite sets.

The following lemma from [9] is a more uniform version of Lemma 4.3. Besides being interesting in its own right, it is useful to parametrize the partition property $\omega \to (\omega)^{\omega}$ with partition properties of the form $(\vec{n}) \to (\vec{m})$.

Lemma 5.1. *There is $R : \mathbb{N}^{<\infty} \to \mathbb{N}$ such that for every infinite sequence $\vec{m} = \{m_i\}_{i=0}^{\infty}$ of positive integers and every coloring*

$$c : \bigcup_{k} \prod_{i<k} R(m_0, \dots, m_i) \to \{0, 1\},$$

there exist $H_i \subseteq R(m_0, \dots, m_i)$, $|H_i| = m_i$ for all i, and an infinite $A \subseteq \mathbb{N}$, such that c is constant on $\bigcup_{k \in A} \prod_{i<k} H_i$.

We cannot hope to get even more uniformity, since it is easy to define a partition for which there is no $\vec{m}$-sequence $\vec{H}$ such that for every k, the subproduct $\prod_{i=0}^{k} H_i$ monochromatic.

The original proof of lemma 5.1 given in [9] uses the hypothesis “all Σ_2^1-sets are Ramsey”. Todorcevic gave afterwards a proof which avoids this hypothesis using the following Ramsey uniformization theorem (see [30]).

Theorem 5.2. *Let X be a Polish space and R a coanalytic subset of the product $\mathbb{N}^{[\infty]} \times X$ with the property that for all $M \in \mathbb{N}^{[\infty]}$ there is $x \in X$ such that $R(M, x)$ holds. Then there is an infinite $M \in \mathbb{N}^{[\infty]}$ and a continuous function $F : [M]^{\infty} \to X$ such that $R(N, F(N))$ holds for every $N \in [M]^{\infty}$.*

For our purposes below, it is convenient to take such an R as in 5.1 that majorizes the function S of Lemma 4.3. We also can assume that $R(m_0, \dots, m_i)$ majorizes each of its arguments and is monotonically increasing in each of them.

The iterates of the function R given by Lemma 5.1 are defined as those for S in the previous section.

$$R^{(0)}(m_0, ..., m_i) = R(m_0, ..., m_i),$$
$$R^{(p+1)}(m_0, ..., m_i) =$$
$$R(R^{(p)}(m_0), R^{(p)}(m_0, m_1), ..., R^{(p)}(m_0, ..., m_i)).$$

For $M \subseteq \mathbb{N}$, define $R_M : \mathbb{N}^{<\infty} \to \mathbb{N}$ recursively by

$$R_M(m_0) = R(m_0),$$
$$R_M(m_0, ..., m_i) = R^{(p)}(m_0, ..., m_i).$$

where

$$p = |\bigcup_{k \in M \cap i} \prod_{j=0}^{\infty} R_M(m_0, \dots, m_j)|.$$

For every sequence $\{m_i\}_{i=0}^{\infty}$ of positive integers, $S_M((m_i))$ denotes the infinite sequence whose i-th term is $S_M(m_0, \dots, m_i)$ (analogously for $S^{(p)}((m_i))$). And as before, for every sequence $\{m_i\}_{i=0}^{\infty}$ of positive integers, $R_M((m_i))$ denotes the infinite sequence whose i-th term is $R_M(m_0, \dots, m_i)$ (analogously for $R^{(p)}((m_i))$).

A field of subsets of $\mathbb{N}^{\infty} \times \mathbb{N}^{[\infty]}$ is defined as follows:

Definition 5.3. A subset $\mathcal{X} \subseteq \mathbb{N}^{\infty} \times \mathbb{N}^{[\infty]}$ is (m_i)-para-Ramsey if for every regular pair (N, M), every $R_M((m_i))$-sequence $\vec{H}$, every $n \in M \setminus N$, every $A \in \mathbb{N}^{[\infty]}$, and every $a \in \mathbb{N}^{[<\infty]}$, there exist a regular pair $(Q, P) \leq_n (N, M)$ an $R_P((m_i))$-sequence $\vec{J}$ such that $\vec{J} \leq_n \vec{H}$, an integer $k \geq n$, and an infinite $B \subseteq A$ such that $\forall s \in \prod_{i<k} J_i$

$$[s, \vec{J}] \times [a, B] \subseteq \mathcal{X} \text{ or } [s, \vec{J}] \times [a, B] \cap \mathcal{X} = \emptyset.$$

Here, $[s, \vec{J}]$ is the collection of elements of $\prod_{i=0}^{\infty} J_i$ which extend s.

A rather involved argument gives the following results (see [10] or [30]).

Lemma 5.4. *Every open subset of the product space $\mathcal{X} \subseteq \mathbb{N}^{\infty} \times \mathbb{N}^{[\infty]}$ is (m_i)-para-Ramsey.*

Theorem 5.5. *([10]) For every sequence $\vec{m}$, the field of (m_i)-para-Ramsey subsets of $\mathcal{X} \subseteq \mathbb{N}^{\infty} \times \mathbb{N}^{[\infty]}$ is a σ-field of subsets of $\mathbb{N}^{\infty} \times \mathbb{N}^{[\infty]}$ containing the open sets and closed under Souslin's operation.*

The proof of this theorem is much harder than the proof of the corresponding result for (m_i)-Ramsey subsets of $\mathbb{N}^{\infty}$. To show that open subsets of $\mathbb{N}^{\infty} \times \mathbb{N}^{[\infty]}$ are in (m_i)-para-Ramsey, the notion of barrier of Nash-Williams is used, and a combinatorial forcing is developed along the lines of [13].

As for the case of the field of (m_i)-Ramsey subsets of $\mathbb{N}^\infty$, the argument of [10] to show that for every (m_i) the field of (m_i)-para-Ramsey subsets of $\mathcal{X} \subseteq \mathbb{N}^\infty \times \mathbb{N}^{[\infty]}$ is closed under Souslin's operation uses metamathematical tools. There is a more recent combinatorial proof due to S. Todorcevic (see [30]).

Corollary 5.6. *For every sequence (m_i), every analytic subset of $\mathbb{N}^\infty \times \mathbb{N}^{[\infty]}$ is (m_i)-para-Ramsey.*

It can be verified ([9] 5.2) that for every sequence $\{m_i\}_{i=0}^\infty$ there is a sequence $\{n_i\}_{i=0}^\infty$ such that given a set $\mathcal{X} \subseteq \prod_{i=0}^\infty n_i \times \mathbb{N}^{[\infty]}$ which is (m_i)-para-Ramsey, there is an (m_i)-sequence $\{H_i\}_{i=0}^\infty$ and an infinite set H such that

$$\prod_{i=0}^\infty H_i \times H^{[\infty]} \subseteq \mathcal{X} \text{ or } \prod_{i=0}^\infty H_i \times H^{[\infty]} \cap \mathcal{X} = \emptyset.$$

Thus, we obtain the following.

Corollary 5.7. *For every sequence $\{m_i\}_{i=0}^\infty$, there is a sequence $\{n_i\}_{i=0}^\infty$ such that for every analytic $\mathcal{X} \subseteq \prod_{i=0}^\infty n_i \times \mathbb{N}^{[\infty]}$, there is an (m_i)-sequence $\{H_i\}_{i=0}^\infty$ and an infinite set H such that*

$$\prod_{i=0}^\infty H_i \times H^{[\infty]} \subseteq \mathcal{X} \textit{ or } \prod_{i=0}^\infty H_i \times H^{[\infty]} \cap \mathcal{X} = \emptyset.$$

6 Independence of the Ramsey property from the polarized property.

We say that a model M is a Solovay model over a ground model V if $M = L(\mathbb{R})$, the class of sets constructible from $\mathbb{R}$, where $\mathbb{R}$ is the set of reals in a generic extension of V obtained using the Levy order to collapse an inaccessible cardinal of V to $(\omega_1)^M$.

The following theorem establishes that in a Solovay model, the fields of (m_i)-Ramsey sets and of (m_i)-para-Ramsey sets defined in the previous sections include all subsets of the corresponding spaces. Since in a Solovay model all subsets of $\mathbb{N}^{[\infty]}$ are Ramsey, Lemma 5.1 holds there.

Theorem 6.1. *([10]) In every Solovay model, for every sequence $\{m_i\}_{i=0}^\infty$ of positive integers, every subset of $\mathbb{N}^\infty$ is (m_i)-Ramsey, and every subset of $\mathbb{N}^\infty \times \mathbb{N}^{[\infty]}$ is (m_i)-para-Ramsey.*

The main ingredients of the proof are several well known properties of Solovay models, in particular the fact that in a Solovay model every subset of $\mathbb{N}^\infty$ or of $\mathbb{N}^\infty \times \mathbb{N}^{[\infty]}$ is the union of $\aleph_1$ analytic sets.

It follows that in a Solovay model, for every sequence $\{m_i\}_{i=0}^\infty$, there is a sequence $\{n_i\}_{i=0}^\infty$ such that $(\vec{n}) \to (\vec{m})$. The parametrized version also holds.

Corollary 6.2. *In every Solovay model, for every sequence $\{m_i\}_{i=0}^\infty$ there exists a sequence $\{n_i\}_{i=0}^\infty$ such that for every coloring*

$$c : \prod_{i=0}^\infty n_i \times \mathbb{N}^{[\infty]} \to \{0,1\}$$

there exist $H_i \subseteq n_i$, $|H_i| = m_i$ for every i, and an infinite $H \subseteq \mathbb{N}$ such that $(\prod_{i=0}^\infty H_i) \times H^{[\infty]}$ is monochromatic.

Notice that this implies that in every Solovay model, for every sequence $\{m_i\}_{i=0}^\infty$ and for every coloring

$$c : \mathbb{N}^\infty \times \mathbb{N}^{[\infty]} \to \{0,1\}$$

there exist $H_i \subseteq n_i$, $|H_i| = m_i$ for every i, and an infinite $H \subseteq \mathbb{N}$ such that $(\prod_{i=0}^\infty H_i) \times H^{[\infty]}$ is monochromatic.

The consistency of this parametrized partition property can be used to show that the property $\omega \to (\omega)^\omega$ is not implied by the polarized partition property

$$\begin{pmatrix} \omega \\ \omega \\ \vdots \end{pmatrix} \to \begin{pmatrix} m_0 \\ m_1 \\ \vdots \end{pmatrix}.$$

This is achieved starting from a Solovay model $L(\mathbb{R})$, and adding a generic selective ultrafilter U to obtain the generic extension $L(\mathbb{R})[U]$. The parametrized partition property in $L(\mathbb{R})$ is then used to show that this generic extension satisfies the polarized partition property

$$\begin{pmatrix} \omega \\ \omega \\ \vdots \end{pmatrix} \to \begin{pmatrix} m_0 \\ m_1 \\ \vdots \end{pmatrix}.$$

Since in the presence of non principal ultrafilters on $\mathbb{N}$ there are non Ramsey subsets of $\mathbb{N}^{[\infty]}$, the property $\omega \to (\omega)^\omega$ does not hold in the generic extension $L(\mathbb{R})[U]$. Therefore, the property

$$\begin{pmatrix} \omega \\ \omega \\ \vdots \end{pmatrix} \to \begin{pmatrix} m_0 \\ m_1 \\ \vdots \end{pmatrix}$$

does not imply $\omega \to (\omega)^\omega$.

A natural question arises about the relationship between the polarized partition property

$$\begin{pmatrix} \omega \\ \omega \\ \vdots \end{pmatrix} \to \begin{pmatrix} m_0 \\ m_1 \\ \vdots \end{pmatrix}$$

and the property $\omega \to (\text{perfect})^\omega$.

It is clear that $\omega \to (\text{perfect})^\omega$ follows from

$$\begin{pmatrix} \omega \\ \omega \\ \vdots \end{pmatrix} \to \begin{pmatrix} 2 \\ 2 \\ \vdots \end{pmatrix},$$

since every infinite product of pairs is perfect. Are these two partition properties equivalent?

We address this question in the next section.

7 Independence of the sublattice property from the polarized property.

Let $L[\{a_\xi : \xi < \omega_1\}]$ be a generic extension of L obtained by adding $\aleph_1$-many Cohen generic reals to L, and let $N_1 = L(\{a_\xi : \xi < \omega_1\})$ computed in the generic extension. This is the inner model of the generic extension containing all the Cohen generic reals as elements but not their collection.

We will see below that in the model N_1 the property $\omega \to ((\omega))^\omega$ holds, but not the polarized partition relation

$$\begin{pmatrix} \omega \\ \omega \\ \vdots \end{pmatrix} \to \begin{pmatrix} 2 \\ 2 \\ \vdots \end{pmatrix}.$$

On the other hand, in [10] it is established that the polarized partition relation

$$\begin{pmatrix} \omega \\ \omega \\ \vdots \end{pmatrix} \to \begin{pmatrix} 2 \\ 2 \\ \vdots \end{pmatrix}$$

does not imply $\omega \to ((\omega))^\omega$ (see Section 6). So, none of these properties implies the other.

For $s, t \in 2^{<\omega}$, we write $s \sqsubseteq t$ to express that s is an initial segment of t, and $s \sqsubset t$ when s is a proper initial segment of t. As usual, $T \subseteq 2^{<\omega}$ is a tree if $t \in T$ and $s \sqsubset t$, implies $s \in T$. Given a tree $T \subseteq 2^{<\omega}$, we say T is *uniform* if for every $s, t \in T$, of the same length,

$$s \frown 0 \in T \longleftrightarrow t \frown 0 \in T \text{ and } s \frown 1 \in T \longleftrightarrow t \frown 1 \in T.$$

A *branch* of a tree $T \subseteq 2^{<\omega}$ is an infinite sequence $c \in 2^\omega$ such that for every n, $c \restriction n \in T$. We denote by $[T]$ the set of branches of T. A tree $T \subseteq 2^{<\omega}$ is *perfect* if for every $t \in T$ there is $s \in T$ such that $t \sqsubseteq s$ for which $s \frown 0 \in T$ and $s \frown 1 \in T$.

It is clear that every perfect uniform tree $T \subseteq 2^{<\omega}$ determines a sublattice $[A, B]$. Namely, $A = \{n \in \omega : \forall c \in [T](c(n) = 1)\}$ and $B = \{n \in \omega : \exists c \in [T](c(n) = 1)\}$

We will show that $\omega \to ((\omega))^\omega$ holds in N_1. For this we prove a more general result.

Theorem 7.1. *Let M be a transitive model of ZFC, let $M[\{a_\xi : \xi < \omega_1\}]$ be the generic extension obtained using the forcing notion $\mathbb{C}_{\omega_1}$ which adds ω_1 Cohen reals, and let $N = L(\{a_\xi : \xi < \omega_1\})$ computed in the generic extension $M[\{a_\xi : \xi < \omega_1\}]$. In N the property $\omega \to ((\omega))^\omega$ holds.*

Proof. Notice that if $y \in M[\{a_\xi : \xi < \omega_1\}]$ is Cohen generic over M, then the forcing $\mathbb{C}_{\omega_1}$ can be factored as $\mathbb{C} \times \mathbb{C}_{\omega_1}$, where $\mathbb{C}$ is Cohen forcing. So,

$$M[\{a_\xi : \xi < \omega_1\}] = M[y][G'],$$

where G' is $\mathbb{C}_{\omega_1}$-generic over $M[y]$.

We will prove that if $\mathcal{A} \subseteq \mathbb{N}^{[\infty]}$ in N, there is a sublattice $[K, H] \in N$ such that $[K, H] \subseteq \mathcal{A}$ or $[K, H] \cap \mathcal{A} = \emptyset$.

Let $\mathcal{A} \subseteq \mathbb{N}^{[\infty]}$ be such that $\mathcal{A} \in N$, and let $\alpha = rank_N(\mathcal{A})$. There is a formula $\Phi(z, a_{\delta_0}, \dots, a_{\delta_{k-1}})$ that defines $\mathcal{A}$, and by the factoring mentioned at the begining of the proof, we can assume that the parameters a_{δ_i} $(i < k)$ of the formula are in the base model M.

Let $\mathbb{P}$ be the forcing of finite uniform subtrees of $2^{<\omega}$ ordered by end extension, thus, $T_2 \le T_1$ if and only if $T_1 \subseteq T_2$ and $T_2 \restriction \text{height}(T_1) = T_1$, where height$(T_1)$ is the length of any branch of T_1 (all branches of a finite uniform tree have the same length).

Since $\mathbb{P}$ is countable and atomless, it is equivalent to Cohen forcing, therefore there are $\mathbb{P}$-generic filters over M in $M[\{a_\xi : \xi < \omega_1\}]$. A $\mathbb{P}$-generic over M is a perfect uniform tree, thus it is a sublattice $[A, B]$. It can be shown by a standard argument that every element of $[A, B]$ is Cohen generic over M.

We will see that for some $[A', B'] \subseteq [A, B]$, $[A', B'] \subseteq \mathcal{A}$ or $[A', B'] \cap \mathcal{A} = \emptyset$. Take $x \in [A, B]$. If $M[\{a_\xi : \xi < \omega_1\}] \models x \in \mathcal{A}$, then in $M[x]$, by the weak homogeneity of the partial order $\mathbb{C}_{\omega_1}$, we have that

$$\emptyset \Vdash_{\mathbb{C}_{\omega_1}} \check{x} \in \mathcal{A},$$

i.e., the empty condition forces $\check{x} \in \mathcal{A}$, in the forcing $\mathbb{C}_{\omega_1}$, where $\check{x}$ is the canonical name for x as an element of the base model $M[x]$.

Therefore, since x is Cohen generic over M, in M there is a Cohen condition p such that

$$p \Vdash \text{“}\emptyset \Vdash_{\mathbb{C}_{\omega_1}} \check{\Gamma} \in \mathcal{A}\text{”},$$

where Γ is the canonical name for the $\mathbb{C}$-generic, and where we admit a slight abuse of notation to express $\check{\Gamma}$ as the canonical name for the object named by Γ.

Consider $[A, B] \restriction p$, i.e. the sublattice $[A', B'] = \{y \in [A, B] : p \sqsubset y\}$. Notice that A' differs from A just in a finite set, and the same for B' and B. For every $y \in [A', B']$, since $p \Vdash \text{“}\emptyset \Vdash_{\mathbb{C}_{\omega_1}} \check{\Gamma} \in \mathcal{A}\text{”}$, we have

$$M[y] \models \text{“}\emptyset \Vdash_{\mathbb{C}_{\omega_1}} \check{y} \in \mathcal{A}\text{”}.$$

Since the forcing that extends $M[y]$ to the full extension $M[\{a_\xi : \xi < \omega_1\}]$ is (isomorphic to) $\mathbb{C}_{\omega_1}$, we have that

$$M[\{a_\xi : \xi < \omega_1\}] = M[y][G'] \models y \in \mathcal{A}.$$

We proceed in the same fashion if $M[\{a_\xi : \xi < \omega_1\}] \models x \notin \mathcal{A}$. In the first case we obtain that $[A', B'] \subseteq \mathcal{A}$, and in the second, we get $[A', B'] \cap \mathcal{A} = \emptyset$.

It remains to be shown that there is such a homogeneous sublattice $[A, B]$ in the model N. We can code finite uniform subtrees of $2^{<\omega}$ by finite sequences of 0's and 1's, i.e. by elements of $2^{<\omega}$, in a canonical way. Dense subsets of the partial order $\mathbb{P}$ will correspond to dense subsets of the Cohen partial order. This means that any Cohen generic in the extension codes a generic sublattice. Therefore, we will have generic sublattices in the model N. □

Notice that from this proof it follows that in the model N, every $\mathcal{A} \subseteq \mathbb{N}^{[\infty]}$ is Silver measurable, since we can reproduce this argument below any arbitrary $[C, D]$.

This argument is similar to the one used in [2] for projective partitions.

Corollary 7.2. *In the model $N = L(\{a_\xi : \xi < \omega_1\})$ of the previous theorem, every set $\mathcal{A} \subseteq \mathbb{N}^{[\infty]}$ contains or is disjoint from a perfect subset of $\mathbb{N}^{[\infty]}$, i.e., there are no Bernstein sets in this model.*

Definition 7.3. Let M be a transitive model of ZFC, and $M[G]$ a generic extension. A real $x \in \mathbb{N}^{\infty}$ is eventually different if for every $y \in \mathbb{N}^{\infty} \cap M$ there is $k \in \omega$ such that for every $i \geq k$ $x(i) \neq y(i)$.

Let us observe that if M is a transitive model of ZFC, and $M[g]$ is a Cohen extension, then the Cohen generic real g is not eventually different. This is so because given $y \in M$ and $k \in \omega$, the set $D_k = \{p : \exists i \geq k(p(i) = y(i))\}$ is dense. Moreover, for every $y \in M$ and $k \in \omega$, the set $E_k = \{p : \exists i \geq k(p(i) = y(i) \text{ and } p(i+1) = y(i+1))\}$ is also dense, and therefore, for every $y \in M$, $g(i) = y(i)$ and $g(i+1) = y(i+1)$ for infinitely many i.

Thus, in a Cohen extension $M[g]$ no eventually different reals are added. To see this, let $r \in \omega^\omega$ such that $r \in M[g] \setminus M$, and let $\dot{r}$ be a $\mathbb{C}$-name for r in M. Enumerate the conditions of Cohen forcing $\mathbb{C}$ by $p_0, p_1, p_2, p_3, \ldots, p_n, \ldots$. Define $x \in \omega^\omega$ as follows:

$$x(n) := min\{k : \exists q \leq p_n(q \Vdash \dot{r}(\check{n}) = \check{k})\}.$$

Then, $x \in M$, and we will see that $\{i \in \omega : r(i) = x(i)\}$ is infinite. Suppose not, then $\exists n_0 \in \omega[\forall n \geq n_0(r(n) \neq x(n))]$. By the Forcing Theorem there is p such that $p \Vdash \forall n \geq \check{d}(\dot{r}(n) \neq \check{x}(n))$. Pick $m > d$ such that $p_m \leq p$. If $x(m) = k$, then (by definition) there is $q \leq p_m[q \Vdash \dot{r}(\check{m}) = \check{k}]$. Therefore $q \Vdash \dot{r}(\check{m}) = \check{x}(\check{m})$, which is a contradiction.

Once we know that $\forall x \in M[g]\ \exists y \in M$ such that $\{i \in \omega : x(i) = y(i)\}$ is infinite, it is easy to see that for all $x \in M[g]$ there is $y \in M$ such that $\{i \in \omega : x(i) = y(i) \wedge x(i+1) = y(i+1)\}$ is infinite: Let $r_1 \in \omega^\omega$ be such that $r_1 \in M[g] \setminus M$. Then consider the real r_2 which codifies the consecutive pairs of r_1. Apply to r_2 the previous result to find in M a real r_3 such that $\{i \in \omega : r_2(i) = r_3(i)\}$ is infinite. Decoding from r_3 we obtain a real $r_3{}^*$ in M such that the set $\{i \in \omega : r_1(i) = r_3{}^*(i) \wedge r_1(i+1) = r_3{}^*(i+1)\}$ is infinite.

Finally observe that if $M[G]$ is obtained adding ω_1 many Cohen reals, then no eventually different reals are added.

An argument due to J. Brendle that we sketch below gives that the polarized partition relation

$$\begin{pmatrix} \omega \\ \omega \\ \vdots \end{pmatrix} \to \begin{pmatrix} 2 \\ 2 \\ \vdots \end{pmatrix}$$

fails in N_1.

Remember that N_1 is the model obtained adding ω_1 many Cohen reals to L, and then computing $L(\{a_\xi : \xi < \omega_1\})$ in the generic extension.

Since $L[G]$ does not contain eventually different reals,

$$\forall x \in L[G] \exists y \in L\, \{i \in \omega : x(i) = y(i)\} \text{ is infinite.}$$

Moreover, for every $x \in L[G]$, there is $y \in L$ such that the set

$$\{i \in \omega : x(i) = y(i) \text{ and } x(i+1) = y(i+1)\}$$

is infinite, as we have observed in the paragraphs preceding the statement of the theorem.

For every x, there is a first $y \in L$, in the Σ^1_2 ordering of the reals in L, for which $\{i \in \omega : x(i) = y(i) \text{ and } x(i+1) = y(i+1)\}$ is infinite. We call it $f(x)$. Let i_x be the first i such that $x(i) = f(x)(i)$.

Define $c : \mathbb{N}^\infty \to 2$ by $c(x) = 0$ if and only if i_x is even.

We show that no product of pairs is monochromatic for c. In $L[G]$, let $\{H_i\}_{i\in\omega}$ be a sequence of pairs of integers, and let $x \in \Pi_{i\in\omega} H_i$. Notice that we can assume that $x(i_x) = f(x)(i_x)$ and $x(i_x + 1) = f(x)(i_x + 1)$, since we can modify a finite number of values from x and obtain another member of the product with the same $f(x)$ satisfying this requirement.

Now, modifying the i_x-th value of x we obtain $x' \in \Pi_{i\in\omega} H_i$ such that $c(x) \neq c(x')$.

The coloring c is Δ^1_2, since $c^{-1}\{0\}$ and $c^{-1}\{1\}$ are both $\boldsymbol{\Pi}^1_2$. To see this, $c(x) = 0$ if and only if the first i such that $x(i) = f(x)(i)$ is even; and the function f is defined by $\boldsymbol{\Pi}^1_2$ formula: given $x, z = f(x)$ if and only if $\forall y(y \leq_L z \Rightarrow \{i : z(i) = x(i) \text{ and } z(i+1) = x(i+1)\}$ is finite).

We have that $M[G]$ satisfies that for every sequence $\{H_i\}_{i\in\omega}$ of pairs of natural numbers, there are $x, y \in \Pi_{i\in\omega} H_i$ such that $c(x) \neq c(y)$.

Since c is definable, it is in $N_1 = L(\{a_\xi : \xi < \omega_1\})$, and in this model the same holds. Thus the polarized partition property does not hold in N_1.

Summarizing, we state the next theorem.

Theorem 7.4. *The Bernstein property $\omega \to (\text{perfect})^\omega$ does not imply*

$$\begin{pmatrix} \omega \\ \omega \\ \vdots \end{pmatrix} \to \begin{pmatrix} 2 \\ 2 \\ \vdots \end{pmatrix}.$$

None of the properties $\omega \to ((\omega))^\omega$ and

$$\begin{pmatrix} \omega \\ \omega \\ \vdots \end{pmatrix} \to \begin{pmatrix} 2 \\ 2 \\ \vdots \end{pmatrix}$$

implies the other.

Proof. The model N_1 considered above satisfies $\omega \to ((\omega))^\omega$ (and therefore $\omega \to (\text{perfect})^\omega$) and not the polarized partition property. In [10] it is shown that a selective ultrafilter can be added generically to a Solovay model $L(\mathbb{R})$ to obtain a model $L(\mathbb{R})[U]$ in which the polarized partition property holds but not $\omega \to ((\omega))^\omega$. □

Clearly,

$$\begin{pmatrix}\omega\\ \omega\\ \vdots\end{pmatrix} \to \begin{pmatrix}2\\ 2\\ \vdots\end{pmatrix} \text{ implies } \omega \to (\text{perfect})^{\omega},$$

since a product of the form $\prod_{i\in\omega} H_i$, with $\{i \in \omega : |H_i| \geq 2\}$ infinite, is a perfect set. But the previous results show that the reverse implication does not hold, considering that

$$\omega \to ((\omega))^{\omega} \text{ implies } \omega \to (\text{perfect})^{\omega},$$

since any $[K, H]$ is a perfect subset of $\mathbb{N}^{[\infty]}$.

We have then that the only implications that hold between these properties are the ones shown in the following diagram.

$$\omega \to (\omega)^{\omega}$$

$$\swarrow \qquad \searrow$$

$$\omega \to ((\omega))^{\omega} \qquad \begin{pmatrix}\omega\\ \omega\\ \vdots\end{pmatrix} \to \begin{pmatrix}m_0\\ m_1\\ \vdots\end{pmatrix}$$

$$\searrow \qquad \swarrow$$

$$\omega \to (\text{perfect})^{\omega}.$$

Bibliography

[1] Bernstein, F., Zur Theorie der Trigonometrischen Reiche. *Berichte über die Verhandlungen der Königlich Sächsischen Gesellschaft der Wissenschaften zu Leipzig Mathematisch-Physiche Klasse* 60 (1908) 325-338.

[2] Brendle, J., L. Halbeisen and B. Löwe, Silver measurability and its relations to other regularity properties. Math. Proc. Cambridge Philos. Soc., 138 (2005) 135-149.

[3] Di Prisco, C. A., Partition properties and perfect sets. *Notas de Lógica Matemática*, Universidad Nacional del Sur, Bahía Blanca, Argentina, 38 (1993)119-127.

[4] Di Prisco, C. A. and F. Galindo, Perfect set properties in models of ZF. *Fundamenta Mathematicae* 208 (2010) 249-262.

[5] Di Prisco, C. A. and J. Henle, Partitions of products. *Journal of Symbolic Logic* 58 (1993) 860-871.

[6] Di Prisco, C. A. and J. Henle, Partitions of the reals and choice. In: Models, algebras and proofs (X. Caicedo and C. H. Montenegro, Eds.) Marcel Dekker, Inc. 1999, 13-23.

[7] Di Prisco, C. A. and J. Henle, Doughnuts, floating ordinals, square brackets, and ultraflitters. *Journal of Symbolic Logic* 65 (2000) 461-473.

[8] Di Prisco, C. A., J. Llopis and S. Todorcevic, Borel partitions of products of finite sets and the Ackermann function. *Journal of Combinatorial Theory*, Series A, 93 (2001) 333-349.

[9] Di Prisco, C. A., J. Llopis y S. Todorcevic, Parametrized partitions of products of finite sets. *Combinatorica* (to appear).

[10] Di Prisco, C.A. and S. Todorcevic, Souslin partitions of products of finite sets. *Advances in Mathematics*, 176 (2003) 145-173.

[11] Erdös, P. and R. Rado, A combinatorial theorem. *J. London Math. Soc.* 25 (1950) 249-255.

[12] Erdös, P. and R. Rado, A partition calculus in set theory, *Bulletin of the Amer. Math. Soc.* 62 (1956), 427-489. (Abstract 68T-368).

[13] Galvin, F. and K. Prikry, Borel sets and Ramsey's theorem, *Journal of Symbolic Logic* 38 (1973), 193-198.

[14] Graham, R. L., B. L. Rotschild and J. H. Spencer, Ramsey Theory. John Wiley and sons, 1990.

[15] Halbeisen, L., Making doughnuts of Cohen reals. *Mathematical Logic Quarterly* 49 (2003) 173–178.

[16] Henle, J., The consistency of one fixed omega, *Journal of Symbolic Logic* 60 (1995) 172-177.

[17] Llopis, J., A note in polarized partitions, *Notas de Lógica Matemática*, Universidad Nacional del Sur, Bahía Blanca, Argentina, 39 (1994) 89-94.

[18] Llopis, J. and S. Todorcevic, Borel partitions on products of finite sets, *Acta Científica Venezolana* 47 (1996) 85-88.

[19] Mathias, A. R. D., A remark on rare filters, in: Infinite and finite sets (A. Hajnal. R. Rado and Vera Sós, Eds.). Colloquia Mathematica Societatis Janos Bolyai 10, , North Holland, 1975.

[20] Mathias, A.R.D., Happy Families. *Annals of Pure and Applied Logic* 12 (1977) 59-111.

[21] Moran, G. and D. Strauss, Countable partitions of product spaces, *Mathematika* 27 (1980) 213-224.

[22] Nash-Williams, C.St. J. A., On well quasi-ordering transfinite sequences. *Proceedings of the Cambridge Philosophical Society* 61 (1965) 33-39.

[23] Nešetřil, J. and V. Rödl, Eds., Mathematics of Ramsey Theory. Springer Verlag, 1990.

[24] Pudlák, P. V. Rödl, Partition theorems for systems of finte subsets of the integers. *Discrete Mathematics* 39 (1982) 67-73.

[25] Ramsey, F. P. , On a problem of formal logic. *Proc. of the London Mathematical Society, Ser. 2* 30, Part 4 (1928) 338–384.

[26] Shelah, S., Can you take Solovay's inaccessible away? *Israel Journal of mathematics* 48 (1984) 1-47.

[27] Silver, J., Every analytic set is Ramsey, *Journal of Symbolic Logic* 35 (1970), 60-64.

[28] Solovay, R., A model of set theory where every set of reals is Lebesgue measurable. *Annals of Mathematics* 92 (1970) 1-56.

[29] Todorcevic, S., A new quantitative analysis of some basic principles in the theory of functions of a real variable. *Bull. de l'Acad. Serbe des Sci.* vol. CXXII, No, 26 (2001) 133-144.

[30] Todorcevic, S., Introduction to Ramsey Spaces. Annals of Mathematical Studies No. 174. Princeton University Press, 2010.

Mirna Džamonja

The Singular World of Singular Cardinals

Abstract: The article uses two examples to explore the statement that, contrary to the common wisdom, the properties of singular cardinals are actually more intuitive than those of the regular ones.

Keywords: singular cardinals, SCH, combinatorics

MSC: 03E05

Mirna Džamonja: School of Mathematics, University of East Anglia, Norwich, NR4 7TJ, UK, email: h020@uea.ac.uk

1 Introduction

Infinite cardinals can be regular or singular. Regular cardinals and especially successors of regular cardinals, tend to lend themselves to easier and better understood combinatorial methods and hence are often considered as being in some sense easier. For example, Todd Eisworth in his Handbook of Set Theory article [6] artfully exposes difficulties one has in dealing with the combinatorics of the successors of singulars and explains on a number of examples why the methods used at a successor of a regular most often cannot work when dealing with the successor of a singular. Indeed, it is known that in many situations, dealing with singular cardinals and their successors has to involve techniques beyond combinatorics and forcing, and it notably requires large cardinals. This is true even for such seemingly elementary properties as the calculation of the size of the power set of the cardinal κ as a function of the size of the power sets of the cardinals below, which is basically the content of the famous Singular Cardinal Hypothesis and which has lead to some of the deepest results throughout set theory. In fact, the common wisdom and the thesis of [6] are that if the universe is close to L then the singular cardinals and their successors are “manageable”, and the opposite is true in the models obtained by using strong enough large cardinal hypothesis.

We shall explore the antithesis, which is that (a) in some situations singular cardinals are more manageable than the regular ones and (b) in some models obtained from large cardinals the successors of singulars actually behave quite close to how they do in L, even if we do our best to mess up L by changing the cardinal arithmetic drastically. Our exposition will involve two examples, one for each one of (a) and (b), which we now explain.

2 Tree embeddings

In this part of the paper we discuss an issue that we investigated with Väänänen in [5], inspired by work that he had initiated in a number of earlier papers, concentrating there on $\aleph_1$ and other regular cardinals. In contrast, we worked with κ a singular cardinal of countable cofinality and obtained a surprisingly different situation.

Consider rooted trees of height and cardinality κ, which we call κ-*Trees* [1]. We are interested in κ-Trees which in addition do not have κ-branches, which we call *bounded*. We studied the natural notion of reduction, which is simply a strict-order preserving function from one tree to another. The existence of a reduction from T to T' is denoted by $T \leq T'$. Furthermore, we write $T < T'$ if both $T \leq T'$ and $T' \not\leq T$ hold. Considerations of tree reductions have arisen in the context of model theory and descriptive set theory, as we now explain. Any undefined or unreferenced notion can be found in [5], where in particular we give ample references to EF games and chain models.

A relational structure of size λ can be considered as an element of 2^λ. To understand the classification of such models up to isomorphism one uses the Ehrenfeucht-Fraïssé (EF) games, especially in the countable case. There is an older method of classification, using the fact that for two fixed countable models the set of all isomorphisms between them is $F_{\delta\sigma}$, and hence the set of pairs of models that are non-isomorphic is co-analytic. In fact the set,

$$\{(A, B) : A, B \text{ countable models and } \exists f : A \cong B\}$$

is the same set as the set of pairs (A, B) of countable models for which II has a winning strategy in the EF game. This analysis made it possible to attach to each pair (A, B) of non-isomorphic countable models a rank, called Scott watershed $S(A, B)$, which in this case is an ordinal $\alpha < \omega_1$.

The rank can be thought of as a clock of the EF game in the following sense: during the EF game the Nonisomorphism player I has to go down this clock at every stage, starting at α itself, and a condition of winning for I is that he has not run out of time before the lack of an isomorphism has been exposed. This game

1 This is different than the usual κ-*trees*, which are also required to have levels of size $< \kappa$.

is called the dynamic EF game of rank $\alpha + 1$ and denoted by $EFD_{\alpha+1}$. The fact that $S(A, B) = \alpha + 1$ means that II wins EFD_α (and hence EFD_β for any $\beta < \alpha$), while I wins $EFD_{\alpha+1}$ (and hence EFD_β for any $\beta > \alpha$).

For uncountable models, say of size $\aleph_1$, one can generalise the Ehrenfeucht-Fraïssé Theorem by considering games of length ω_1. One is then tempted to find the corresponding notion of the Scott watershed. It turns out that it is no longer enough to use the ordinals, as the game is transfinite. The right notion this time is that of bounded $\aleph_1$-Trees, the class of which we shall denote by $\mathcal{T}_{\aleph_1}$. If $T \leq T'$ then it is easier for II to win EFD_T than $EFD_{T'}$. Respectively, if $T \leq T'$ then it is easier for I to win $EFD_{T'}$ than EFD_T. Finally, if II wins EFD_T and I wins $EFD_{T'}$, then one can prove $T < T'$. In [7] it was shown by Hyttinen and Väänänen that the analogue of the Scott watershed exists also in the uncountable case. Naturally, the importance of $(\mathcal{T}_{\aleph_1}, \leq)$ in this context led to a systematic study of its structural properties and a development of the descriptive set theory of the space 2^{ω_1} based on $\mathcal{T}_{\aleph_1}$. The theory of $(\mathcal{T}_\lambda, \leq)$ for λ successor of regular is also quite known, although it is not completely parallel to that of $(\mathcal{T}_{\aleph_1}, \leq)$.

When moving to the singular cardinals in [5], the surprise was that new possibilities opened up. Firstly, it is possible to make links between chain models and the infinitary logic $L_{\kappa\kappa}$. For κ singular of countable cofinality, one can develop model theory of $L_{\kappa\kappa}$ based on the concept of a *chain model*, as was done by C. Karp and her successors. A chain model is an ordinary model A equipped with a presentation of A as a union of a chain $A_0 \subseteq A_1 \ldots \subseteq A_n \subseteq \ldots$ for $n < \omega$. The chain is not assumed to be elementary. Let us denote such a system as (A_n). The point of chain models is the following modification of the truth definition of $L_{\kappa\kappa}$

$$(A_n) \models \exists \bar{x}\varphi(\bar{x}) \iff \text{there are } n < \omega \text{ and } \bar{a} \in A_n \text{ with } A \models \varphi(\bar{a}), \tag{1}$$

where $\bar{x}$ is a sequence of length $< \kappa$. If we restrict to chain models, the model theory of $L_{\kappa\kappa}$ is very much like that of $L_{\omega_1\omega}$. For example, one can prove the Completeness Theorem using consistency properties, and one can also prove undefinability of well order, Craig Interpolation Theorem, Beth Definability Theorem, etc. (None of these theorems is true for the classical $L_{\kappa\kappa}$ logic). In [5] we extended Scott's analysis of countable models to chain models of size κ. In particular, we considered versions of EF game for (chain) models of a singular cardinality κ of countable cofinality and discovered that the relevant clock trees of these games are bounded κ-Trees.

In fact the main point of this is that κ-Trees for κ as above have properties that make them rather similar to ordinals. The reason for this is that there is a natural notion of rank. Using this notion we can for example show that the universality number of bounded κ-Trees under reduction is just κ^+, and that within each rank in $[1, \kappa^+)$ the universality number is just ω. This is in sharp contrast with the situation of λ-Trees where λ is a regular cardinal. For example for $\lambda = \aleph_1$ Mekler and Väänänen [10] have established that the universality number for bounded λ-Trees under reduction cannot be computed in ZFC, and the consistency of this number being equal to 1 for $\lambda = \aleph_1$ is not known.

3 Universal graphs

In this section we shall discuss an embedding question which comes from an even more familiar object than trees, namely graphs. Given a cardinal κ, we are interested to know what is the smallest size of a family of graphs of size κ which embeds every graph of size κ as an induced subgraph. This is known as the universality number. For $\kappa = \aleph_0$ this number is 1, as the random graph is a universal countable graph. For uncountable κ the situation becomes sensitive to the axioms of set theory. Namely, by the classical results in model theory on the existence of saturated and special models (see [1]), in the presence of GCH there is a universal graph on every infinite κ, and in fact $\kappa = 2^{<\kappa}$ suffices. On the other hand, a result of Shelah mentioned in [11] and described in [8], is that adding $\aleph_2$ Cohen reals to a model of CH makes the universality number of graphs at $\aleph_1$ equal to $\aleph_2$. This is an easy proof, in fact Shelah says in the abstract of [11] "The consistency of the non-existence of a universal graph of power $\aleph_1$ is trivial. Add $\aleph_2$ generic Cohen reals.", and instead he concentrates on a much more complex proof of the consistency of the existence of a universal graph at $\aleph_1$ and the negation of CH (which in fact was not right in [11]. Shelah corrected his proof in [12] and Mekler gave a different proof in [9]). Other successors of regulars behave in a similar way, although neither Mekler's nor Shelah's proof seem to carry over from $\aleph_1$ to larger successors of regulars. Namely, in [4] Džamonja and Shelah obtained the consistency of the universality number of graphs at κ^+ for an arbitrary large regular κ being equal κ^{++} while 2^κ is as large as desired. The negative consistency results directly translate to other successors of regulars and even to a class of them, and to let the reader appreciate the way Cohen's forcing is used, we give a rendition of that argument here. (The proof of Theorem 3.1 presented in [8] is less formal.)

Theorem 3.1 (Shelah, see [8]). *Suppose that $\kappa^{<\kappa} = \kappa$ and let $\mathbb{P}$ be the forcing to add λ many, with $\mathrm{cf}(\lambda) \geq \kappa^{++}$ and $\lambda \geq 2^{\kappa^+}$, Cohen subsets to κ. Then the universality number for graphs on κ^+ in the extension by $\mathbb{P}$ is λ.*

Proof. First notice that in the extension we have that $2^\kappa = 2^{\kappa^+} = \lambda$, and hence the total number of graphs on κ^+ is at most λ, so the universality number for graphs on κ^+ is also $\leq \lambda$. Now we show that it is $\geq \lambda$. Suppose to the contrary, that $\{H_\gamma : \gamma <$

$\gamma^*\}$ for some $\gamma^* < \lambda$ in the extension are graphs with universe κ^+ that are universal for graphs on κ^+. By standard arguments about the factoring of the Cohen forcing and using $\mathrm{cf}(\lambda) \geq \kappa^{++}$, we may assume that all graphs H_γ are from the ground model. Let $\langle A^j_i : i \in [\kappa, \kappa^+), j < \kappa^{++}\rangle$ be a 1-1 enumeration of the first κ^{++} Cohen subsets of κ added by $\mathbb{P}$, where the indexing is chosen for the convenience in the argument to follow. For each $j < \kappa^{++}$ we define in the extension a graph G_j on κ^+ by letting for $\alpha < i < \kappa^+$ there be an edge between α and i iff $\alpha < \kappa \leq i$ and $\alpha \in A^j_i$. For each j let h_j be an embedding of G_j to some H_{γ_j}. Note that there is a club C of κ^{++} such that for all $j \in C$ of cofinality κ^+, $h_j \restriction [\kappa+1)$ is in $V[A^k_i : i \in [\kappa, \kappa^+), k < k^*]$ for some $k^* < j$. Then for any $j \in C$ we have

$$A^j_\kappa = \{\alpha < \kappa : (\alpha, \kappa) \text{ are an edge in } G_j\} =$$

$$\{\alpha < \kappa : (h_j \restriction [\kappa + 1)(\alpha), h_j \restriction [\kappa + 1)(\kappa)) \text{ are an edge in } H_{\gamma_j}\},$$

which is an object in $V[A^k_i : i \in [\kappa, \kappa^+), k < k^*]$, a contradiction. $\bigstar_{3.1}$ □

Using a standard argument about Easton forcing we can see that it is equally easy to get negative universality results for graphs at a class of regular cardinals:

Theorem 3.2. *Suppose that the ground model V satisfies GCH and $\mathfrak{C}$ is a class of regular cardinals in V, while F is a non-decreasing function on $\mathfrak{C}$ satisfying that for each $\kappa \in \mathfrak{C}$ we have $\mathrm{cf}(F(\kappa)) \geq \kappa^{++}$. Let $\mathbb{P}$ be Easton's forcing to add $F(\kappa)$ Cohen subsets to κ for each $\kappa \in \mathfrak{C}$. Then for each $\kappa \in \mathfrak{C}$ the universality number for graphs on κ^+ in the extension by $\mathbb{P}$ is $F(\kappa)$.*

Proof. Recall that the forcing notion $\mathbb{P}$ is constructed as follows: for each $\kappa \in \mathfrak{C}$ we have the forcing P_κ which adds $F(\kappa)$ Cohen subsets to κ, using functions of size $< \kappa$ from $\kappa \times F(\kappa)$ to $\{0, 1\}$ with the extension given by the extension of functions. Then $\mathbb{P}$ is the Easton product of $\{P_\kappa : \kappa \in \mathfrak{C}\}$, which means that each condition $p \in \mathbb{P}$ is an element of $\Pi_{\kappa \in \mathfrak{C}} P_\kappa$ with support $\mathrm{spt}(p)$ satisfying $|\mathrm{spt}(p) \cap \theta| < \theta$ for every regular cardinal θ. Denoting by p_κ the projection of condition p on the coordinate κ, each condition in $\mathbb{P}$ can be viewed as a function on triples (κ, α, β) where $p(\kappa, \alpha, \beta)$ is defined as $p_\kappa(\alpha, \beta)$. Then standard arguments show that for every regular θ the forcing breaks into $\mathbb{P}^{\leq\theta} \times \mathbb{P}^{>\theta}$ where $\mathbb{P}^{\leq\theta} = \{p \restriction (\kappa, \alpha, \beta) : \kappa \leq \theta\}$ and $\mathbb{P}^{>\theta} = \{p \restriction (\kappa, \alpha, \beta) : \kappa > \theta\}$, and that $\mathbb{P}^{>\theta}$ is θ-closed while $\mathbb{P}^{\leq\theta}$ satisfies the θ^+-chain condition.

Now let $\kappa \in \mathfrak{C}$ and suppose for a contradiction that in the extension by $\mathbb{P}$ we have a universal family $\{H_\gamma : \gamma < \gamma^*\}$ of graphs on κ^+ for some $\gamma^* < F(\kappa)$. Since $\mathbb{P}^{>\kappa^+}$ is κ^+-closed, it does not add any new subsets to κ^+, and hence the universal family is added by $\mathbb{P}^{\leq\kappa^+}$. We shall once more use an argument about factoring, in that for every $\theta < F(\kappa)$ we can consider $\mathbb{P}^{\leq\kappa^+}$ as the product

$$\begin{aligned} & Q^{\leq\theta} = \{p \in \mathbb{P}^{\leq\kappa^+} : (\kappa, \alpha, \beta) \in \mathrm{dom}(p) \implies \beta \leq \theta\} \\ \times \quad & Q^{>\theta} = \{p \in \mathbb{P}^{\leq\kappa^+} : (\kappa, \alpha, \beta) \in \mathrm{dom}(p) \implies \beta > \theta\}. \end{aligned}$$

Since $\mathrm{cf}(F(\kappa)) \geq \kappa^{++}$, there is some $\theta < F(\kappa)$ such that all graphs H_γ are added by $Q^{\leq\theta}$. Now we can basically repeat the argument from the proof of Theorem 3.1: let $\langle A^j_i : i \in [\kappa, \kappa^+), j < F(\kappa)\rangle$ be a 1-1 enumeration of the Cohen subsets of κ added by $Q^{>\theta}$ and for each $j < \kappa^{++}$ we define in the extension a graph G_j on κ^+ by letting for $\alpha < i < \kappa^+$ there be an edge between α and i iff $\alpha < \kappa \leq i$ and $\alpha \in A^j_i$. For each j let h_j be an embedding of G_j to some H_{γ_j}. Note that there is a club C of $F(\kappa)$ such that for all $j \in C$ of cofinality $\geq \kappa^+$, $h_j \restriction [\kappa + 1)$ is in $V[A^k_i : i \in [\kappa, \kappa^+), k < k^*]$ for some $k^* < j$. Then for every $j \in C$

$$\begin{gathered} A^j_\kappa = \{\alpha < \kappa : (\alpha, \kappa) \text{ are an edge in} G_j\} = \\ \{\alpha < \kappa : (h_j \restriction [\kappa + 1)(\alpha), h_j \restriction [\kappa + 1)(\kappa)) \text{ are an edge in } H_{\gamma_j}\}, \end{gathered}$$

which is an object in $V^{Q^{\leq\theta}}[A^k_i : i \in [\kappa, \kappa^+), k < k^*]$, a contradiction. $\bigstar_{3.2}$ □

Things change at the successor of a singular! Positive results analogous to the Džamonja-Shelah [4] were obtained for κ of countable cofinality by Džamonja and Shelah in [3] and for arbitrary cofinality by Cummings, Džamonja, Magidor, Morgan and Shelah in [2]. We quote that general result:

Theorem 3.3 (Cummings et al. [2]). *If κ is a supercompact cardinal, $\lambda < \kappa$ is a regular cardinal and Θ is a cardinal with $\mathrm{cf}(\Theta) \geq \kappa^{++}$ and $\kappa^{+3} \leq \Theta$ there is a forcing extension in which $\mathrm{cf}(\kappa) = \lambda$, $2^\kappa = 2^{\kappa^+} = \Theta$ and there is a universal family of graphs on κ^+ of size κ^{++}.*

However, it is not known in Theorem 3.3 if the universality number is exactly κ^{++}. For all we know, in that model there could be a universal graph on κ^+. Worse, we do not know how to imitate the negative universality presented in Theorems 3.1 and 3.2 above. We do not know how to obtain a model in which the relevant instances of GCH fail and the universality number of graphs is 2^λ for λ the successor of a singular. We have given Shelah's argument about the Cohen forcing in gory detail to invite the reader to think if anything like this can be produced at a singular κ. Initial results by Cummings and Magidor (private communication) indicate that a naive generalization might not be possible. Perhaps there is some sort of singular cardinal hypothesis-like behaviour here. Perhaps we cannot just monkey around with the universality number for graphs at the successor of singular even when we are basically as far from L as we can possibly be, at least as far as the power set function is concerned?

4 Conclusion

We have discussed two problems where the intuition of the singular cardinal being in a sense more difficult than a regular one, seems to be completely false. In fact, the truth seems to be that although the properties of the singular cardinals are *harder to discover*, once we have done that difficult discovery, these properties are actually *nicer* than their analogues at the regular cardinals. Some other results but the ones presented here can be viewed with this idea in mind, for example does not the whole story of the Singular Cardinal Hypothesis including the celebrated theorem of Shelah $[(\forall n < \omega)2^{\aleph_n} < \aleph_\omega] \implies 2^{\aleph_\omega} < \aleph_{\omega_4}$, does not this story say that the singulars are in fact more intuitive than the regulars? Erdös has said something to the extent of the infinite being the easy part, and the finite the difficult one. If the infinite is the limit of the finite, a singular cardinal is a limit of the successors of regulars, and maybe it is at such limits that the unruly universe of set theory wishes to express its more tame behaviour. It seems possible that by investigating finer combinatorics than that expressed by the power set function we may find combinatorial versions of SCH which are just outright true.

Acknowledgement: The author thanks EPSRC for their grant EP/I00498.

Bibliography

[1] Chen Chung Chang and H. Jeremy Keisler. *Model Theory*. North Holland, Amsterdam, third edition, 1990.

[2] James Cummings, Mirna Džamonja, Menachem Magidor, Charles Morgan, and Saharon Shelah. A framework for forcing constructions at successors of singular cardinals. Submitted, 2013.

[3] Mirna Džamonja and Saharon Shelah. Universal graphs at the successor of a singular cardinal. *J.Symbol. Logic*, 68:366–387, 2003.

[4] Mirna Džamonja and Saharon Shelah. On the existence of universal models. *Arch. Math. Logic*, 43(7):901–936, 2004.

[5] Mirna Džamonja and Jouko Väänänen. Chain models, trees of singular cardinality and dynamic EF games. *Journal of Mathematical Logic*, 11(1):61–85, 2011.

[6] Todd Eisworth. Successors of singular cardinals. In Matthew Foreman and Akihiro Kanamori, editors, *Handbook of Set Theory*, volume 2, pages 1229–1350. Springer, Dordrecht, Heidelberg, London, New York, 2010.

[7] Tapani Hyttinen and Jouko Väänänen. On Scott and Karp trees of uncountable models. *J. Symbolic Logic*, 55(3):897–908, 1990.

[8] Menachem Kojman and Saharon Shelah. Nonexistence of universal orders in many cardinals. *J. Symbolic Logic*, 57(3):875–891, 1992.

[9] A. Mekler. Universal structures in power $\aleph_1$. *Journal of Symbolic Logic*, 55(2):466–477, 1990.
[10] A. Mekler and J. Väänänen. Trees and Π^1_1-subsets of ${}^{\omega_1}\omega_1$. *The Journal of Symbolic Logic*, 58(3):1052–1070, 1993.
[11] Saharon Shelah. On universal graphs without instances of CH. *Annals of Pure and Applied Logic*, 26(1):75 – 87, 1984.
[12] Saharon Shelah. Universal graphs without instances of CH: revisited. *Israel Journal of Mathematics*, 70(1):69 – 81, 1990.

Curtis Franks

Logical Nihilism

Abstract: Traditional debates about the scope and nature of logic do not do justice to the details of its maturation. In asking whether certain inferential practices are properly logical or more aptly viewed as part of the special sciences, for example, we ignore how logic has been shaped by developments in extra-logical culture. A tour of some of logic's modern contours showcases how the most interesting, and most distinctively logical, features of the discipline emerge precisely when we let go of our expectations of logical correctness. We are left wondering what role that notion of correctness is supposed to serve and invited to question its coherence.

Keywords: admissibility, intuitionistic logic, philosophy, proof theory

MSC: 01, 03, 08

Curtis Franks: Department of Philosophy, University of Notre Dame, 100 Malloy Hall, Notre Dame, IN 46556, e-mail: cfranks@nd.edu

"This seems to abolish logic, but does not do so."
—L. Wittgenstein, *Philosophical Investigations* §242

1 Introduction

The idea that there may be more than one correct logic has recently attracted considerable interest. This cannot be explained by the mere fact that several distinct logical systems have their scientific uses, for no one denies that the "logic" of classical mathematics differs from the "logics" of rational decision, of resource conscious database theory, and of effective problem solving. Those known as "logical monists" maintain that the panoply of logical systems applicable in their various domains says nothing against their basic tenet that a single relation of logical consequence is either violated by or manifest in each such system. "Logical pluralists" do not counter this by pointing again at the numerous logical systems, for they agree that for all their interest many of these indeed fail to trace any relation of logical consequence. They claim, instead, that no one logical consequence relation is privileged over all others, that several such relations abound.

Interesting as this debate may be, I intend to draw into question the point on which monists and pluralists appear to agree and on which their entire discussion pivots: the idea that one thing a logical investigation might do is adhere to a relation of consequence that is "out there in the world," legislating norms of rational inference, or persisting some other wise independently of our logical investigations themselves. My opinion is that fixing our sights on such a relation saddles logic with a burden that it cannot comfortably bear, and that logic, in the vigor and profundity that it displays nowadays, does and ought to command our interest precisely because of its disregard for norms of correctness.

I shall not argue for the thesis that there are no correct logics. Although I do find attempts from our history to paint a convincing picture of a relation of logical consequence that attains among propositions (or sentences, or whatever) dubious, I should not know how to cast general doubt on the very idea of such a relation. By "drawing this point into question" I mean only to invite reflection about what work the notion of a correct logic is supposed to be doing, why the debate about the number of logical consequence relations is supposed to matter to a logician, and whether the actual details of logic as it has developed might be difficult to appreciate if our attention is overburdened by questions about the correctness of logical principles. Rather than issue any argumentative blows, I propose merely to lead the reader around a bit until his or her taste for a correct logic sours.

2 The law of excluded middle

Surely the most notorious bone of contention in the discussion of logical correctness is the law of excluded middle, $\phi \vee \neg\phi$. Is this law logically valid, so that we know that its instances, "Shakespeare either wrote all those plays or he didn't," "Either the continuum hypothesis is true or it is not true," "He's either bald or he isn't," etc., each are true in advance of any further information about the world?

One hardly needs to mention that hundreds of spirited disavowals and defenses of *lem* have been issued in the last century. Many of these have even been authored by expert logicians. But let us turn our backs to these ideological matters and consider briefly some of what we have learned about *lem* quite independently of any question about its correctness.

The simplest setting for this is the propositional calculus.[1] The intuitionistic propositional calculus (IPC) differs from the classical propositional calculus (CPC) precisely in its rejection of *lem*. For a concrete and standard formalization of IPC one may take a typical Hilbert-style axiomatization of CPC and erase the single axiom for double negation elimination, $\vdash \neg\neg\phi \supset \phi$, leaving the rules of inference as before.

In fact, one of the first things observed about *lem* is its equivalence with *dne*, which is most easily seen in the setting of natural deduction. Standard natural deduction presentations of CPC have a rule allowing one to infer any formula ϕ from the single premise $\neg\neg\phi$. It is easy to derive in such a system the formula $\phi \vee \neg\phi$. It is similarly easy to show that if we modify this calculus by disallowing *dne* but now introduce a new rule that allows one to write down any instance of *lem* in any context, ϕ will be derivable in all contexts in which $\neg\neg\phi$ is derivable.

In *1934–35*, Gentzen observed to his own surprise[2] that the sequent calculus presentation of CPC admits an even more elegant modification into a presentation of IPC than the one just described. One simply disallows multiple-clause succedents and leaves the calculus otherwise unchanged. Thus *lem* and with it the entire distinction between intuitionistic and classical logic is subsumed into the background structure of the logical calculus. All the inference rules governing the logical particles ($\wedge, \vee, \supset, \neg$) and all the explicit rules of structural reasoning (*identity*, *cut*, *weakening*, *exchange*, and *contraction*) are invariant under this transformation. Thus it appears that a duly chosen logical calculus allows a precise analysis of what had been thought of as a radical disagreement about the nature of logic. When classicists and intuitionists are seen to admit precisely the same inference rules, their disagreement appears in some ways quite minor, if more global than first suspected.

---

1 I emphasize that this really is a matter of perspicuity. One should not think that the phenomena described below are artifacts of peculiar features of propositional logic. They are nearly all consequences of decisions about *lem* that are invariant across a wide spectrum of logics. Consider: the structural subsumption of *lem* applies also to the predicate calculus; the admissible propositional rules of Heyting Arithmetic (and of Heyting Arithmetic with Markov's principle) are exactly those of IPC (*Visser 1999*); the disjunction property holds for Heyting Arithmetic (*Kleene 1945*) and for intuitionistic Zermelo–Fraenkel set theory (*Myhill 1973*).

2 Gentzen described the fact that *lem* prescribes uses of logical particles other than those given by their introduction and elimination rules as "troublesome." The way that in the sequent calculus the logical rules are quarantined from the distinction between classical and intuitionistic logic he called "seemingly magical." He wrote, "I myself was completely surprised by this property . . . when first formulating that calculus" *Gentzen 1938* (259).

Exactly how minor, on closer inspection, is the difference between these superficially similar calculi CPC and IPC? Not very. Even before Gentzen's profound analysis, Gödel (*1932*) observed that IPC satisfied the "disjunction property": formulas such as $\phi \vee \psi$ are provable only if either ϕ or ψ is as well. At first sight this might appear to be no more than a restatement of the intuitionist's rejection of *lem*. After all, that rejection was motivated by the idea that instances of *lem* are unwarranted when neither of their disjuncts can be independently established. But, one might think, if any disjunction is warranted in the absence of independent verification of one of its disjuncts, those like $\phi \vee \neg\phi$ are, so rejecting *lem* should lead to something like the disjunction property. This reasoning strikes me as worthy of further elaboration and attention, but it should be unconvincing as it stands. For one thing, the formal rejection of *lem* only bars one from helping oneself to its instances whenever one wishes. The gap between this modest restriction and the inability ever to infer any disjunction of the form $\phi \vee \neg\phi$ at all, unless from a record of that inference one could effectively construct a proof either of ϕ or of $\neg\phi$, is a broad one. More, there are infinitely many formulas ϕ unprovable in IPC, such that IPC+ϕ is consistent but insufficient to prove *lem*. Examples like $\neg p \vee \neg\neg p$ may well fuel suspicions that even in the shadow of global distrust of *lem*, some disjunctions $\phi \vee \psi$ are more plausible than any instance of *lem* even in the absence of resources sufficient to derive ϕ or ψ.

Thus the disjunction property is a non-trivial consequence of the invalidity of *lem*. In fact, this situation exemplifies a recurring phenomenon in logic, wherein from the assumption of a special case of some general hypothesis, that hypothesis follows in its full generality. This often happens even when, as in the present case, the general phenomenon does not appear, even in hindsight, to be a "logical consequence" of its instance in any absolute sense. The disjunction property has further consequences of its own, however, to which we can profitably turn.

In the approach to semantics known as inferentialism, the meaning of a logical particle should be identified with the conditions under which one is justified in reasoning one's way to a statement governed by that particle. From this point of view, which is given expression already in some of Gentzen's remarks, and owing to the separation in sequent calculus of *lem* from the logical rules, the meanings of the familiar logical particles might be said not to differ in intuitionistic and classical logic.

However, the disjunction property gives rise to an alternative interpretation of the logical particles of IPC in which each theorem refers back to the notion of provability in IPC itself. For if $\phi \vee \psi$ is provable only if one of ϕ and ψ is as well, then a candidate and interesting reading of the sentence $\vdash_{\text{IPC}} \phi \vee \psi$ is "Either $\vdash_{\text{IPC}} \phi$ or $\vdash_{\text{IPC}} \psi$." Expanding on this idea, one might suggest that the provability of a conjunction "means" that each of its conjunctions is provable, that the provability of a conditional, $\phi \supset \psi$, "means" that given a proof of ϕ one can construct a proof of ψ, and that the provability of $\neg\phi$ "means" that a contradiction can be proved in the event that a proof of ϕ is produced.

As we shall see shortly, this so far informal interpretation of intuitionistic logic is riddled with ambiguities. All the same, some reflection should bring home the idea that some disambiguation of this reading is a possible way to understand the theorems of IPC. When one compares the situation with CPC, where conditional truth comes so cheap and the disjunction property fails badly, one can only conclude that the difference between these calculi is in some sense great after all, greater even than the debate over the validity of *lem* alone first suggests. When one then recalls the earlier observation that these calculi can be presented so that their formal differences are slight and their rules identical, the observation that radical differences in meaning result from so subtle a change in form is striking.

If we are to take seriously the idea that theorems of IPC refer back to IPC provability, then some care must be taken in making this interpretation precise. For if the provability of a conditional, $\phi \supset \psi$, "means" that $\vdash_{\text{IPC}} \psi$ in the event that $\vdash_{\text{IPC}} \phi$, then one should expect $\vdash_{\text{IPC}} \phi \supset \psi$ in every situation in which the set of theorems of IPC is closed under the rule "from ϕ, infer ψ." However, the disjunction property implies that these expectations will not be met.

To see this, consider the Kreisel–Putnam rule, "From $\neg\phi \supset (\psi \vee \chi)$, infer $(\neg\phi \supset \psi) \vee (\neg\phi \supset \chi)$." The only derivations of $\neg\phi \supset (\psi \vee \chi)$ in natural deduction are proofs whose last inferences are instances of $\supset$-*elim* (modus ponens), $\wedge$-*elim*, *dne*, or $\supset$-*intro*. It is easy to see that a proof ending in $\supset$-*elim* or $\wedge$-*elim* cannot be the *only* way to prove this formula, that in fact any such proof can be normalized into a proof of the same formula whose last inference is an instance of one of the other two rules. If, further, we consider the prospects of this formula being a theorem of IPC, then *dne* is no longer a rule, and we may conclude that any proof necessarily contains a subproof of $\psi \vee \chi$ from the assumption $\neg\phi$ (to allow for $\supset$-*intro*). What might this subproof look like? Once more, *dne* is not an option, so again by insisting that the proof is normalized (so that it doesn't end needlessly and awkwardly with $\supset$-*elim* or $\wedge$-*elim*) we ensure that its last step is an instance of $\vee$-*elim*. But this means that an initial segment of this subproof is a derivation in IPC either of ψ or of χ from the assumption $\neg\phi$ (it is here that the disjunction property rears its head), and in each case it is clear how to build a proof of $(\neg\phi \supset \psi) \vee (\neg\phi \supset \chi)$. Putting this all together, we see that whenever $\neg\phi \supset (\psi \vee \chi)$ is a theorem of IPC, so too is $(\neg\phi \supset \psi) \vee (\neg\phi \supset \chi)$.

When a logical system's theorems are closed under a rule of inference, we say that the rule is "admissible" for that logic. The above argument established that the Kreisel–Putnam rule is admissible for IPC. One might expect that the rule is also "derivable," that $\vdash_{\text{IPC}} (\neg\phi \supset (\psi \vee \chi)) \supset ((\neg\phi \supset \psi) \vee (\neg\phi \supset \chi))$. However, it is not (*Harrop 1960*). This situation is dis-analogous to that of classical logic, where all admissible rules are derivable so that the distinction between admissibility and derivability vanishes. In the parlance, we say that CPC is "structurally complete" but that IPC is not.

In fact much of the logical complexity of IPC can be understood as a residue of its structural incompleteness. For the space of intuitionistically valid formulas is far more easily navigated than the space of its admissible rules: although it is decidable whether or not a rule is admissible in IPC, there is no finite basis of rules that generates them all (*Rybakov 1997*).[3]

What ought one make of the structural incompleteness of IPC? One thing that can definitely be said is that reading the expression $\phi \supset \psi$ as "There is a procedure for transforming a proof of ϕ into a proof of ψ" is problematic for both the classicist and the intuitionist, but for different reasons. This reading is wrong for the classicist, because the idea of procedurality simply does not enter into the conditions of classical validity. By contrast, procedures of proof transformation are central for the intuitionist. However, we now know that there *are* procedures for transforming a IPC-proof of ϕ into a IPC-proof of ψ in cases where $\phi \supset \psi$ is not a theorem of IPC. So at best one could say that IPC is incomplete with respect to this semantics, and more plausibly one should say that this reading of $\vdash_{\mathrm{IPC}} \phi \supset \psi$ is erroneous.

Thus we see a sense in which the phenomenon of structural completeness is related to a sort of semantic completeness: a structurally incomplete logic will be incomplete with respect to the most naive procedural reading of its connectives. It also happens that structural completeness bears a precise relation to the phenomenon of Post-completeness, the situation in which any addition made to the set of theorems of some logic will trivialize the logic by making all formulas in its signature provable. To state this relationship, we refer to a notion of "saturation." For a logical calculus L whose formulas form the set S, let $\mathrm{Sb}(X)$ be the set of substitution instances of formulas in $X \subseteq S$ and let $\mathrm{Cn}_L(X)$ be the set of formulas ϕ such that $X \vdash_L \phi$. L is saturated if for every $X \subseteq S$ $\mathrm{Cn}_L(X) = \mathrm{Cn}_L(\mathrm{Sb}(X))$ for every $X \subseteq S$. By a *1973* theorem of Tokarz, a Post-complete calculus is structurally complete if, and only if, it is saturated.

For these and perhaps other reasons many authors have felt that the presence of non-derivable, admissible rules is a deficiency of systems like IPC. The very term "structural incompleteness" suggests that something is missing from IPC because

3 Generally, the set of admissible rules has higher complexity than the set of derivable rules for the same logic. In IPC admissibility is CONEXP-complete whereas derivability is PSPACE-complete. A striking example from *Chagrov 1992* is a decidbale modal logic for which admissibility is undecidable.

correct inferences about provability in this logic are not represented as theorems in IPC. Rybakov (*1997*), for example, suggests that "there is a sense in which a derivation inside a [structurally incomplete] logical system corresponds to conscious reasoning [and] a derivation using [its] admissible rules corresponds to subconscious reasoning." He faults such systems for having rules that are "valid in reality" yet "invalid from the viewpoint of the deductive system itself" (10–11). Structurally complete systems, by contrast, are "self contained" in the sense that they have the "very desirable property" of being conscious of all the rules that are reliable tools for discovering their own theorems (476). It seems to me that this attitude derives from wanting to preserve the naive procedural understanding of the logical connectives. The situation ought rather, I counter, lead one to appreciate the subtlety of procedurality exhibited in intuitionism. For the logical lesson to be learned is that in the absence of *lem* the context of inference takes on a new role. Thus $\phi \supset \psi$ means that given any background of assumptions from which ϕ is provable, a proof of ϕ can be transformed into a proof of ψ under those same assumptions, and this understanding does not reduce, as it does with logics insensitive to context, to the idea that any proof of ϕ can be transformed into a proof of ψ. This irreducibility strikes me as a "very desirable property" for many purposes. I should like to know more about the conditions that lead to it.

From this point of view, it is natural to ask whether there are logics that, unlike classical logic, admit a constructive interpretation but, like classical logic, are not sensitive in this way to context. Perhaps the constructive nature of IPC derives from its context-sensitivity. Surprisingly, Jankov's logic, IPC+$\neg\phi \vee \neg\neg\phi$, appears to undermine any hope of establishing a connection between these phenomena. Consider the Medvedev lattice of degrees of solvability. The setting is Baire space ω^ω (the set of functions from ω to ω) and the problem of producing an element of a given subset of this space. By convention, such subsets are called mass problems, and their elements are called solutions. One says that one mass problem reduces to another if there is an effective procedure for transforming solutions of the second into solutions of the first. If one defines the lattice $\mathfrak{M}$ of degrees of reducibility of mass problems, it happens that under a very natural valuation, the set of identities of $\mathfrak{M}$ corresponds to the set of theorems of Jankov's logic, so that the theory of mass problems provides a constructive interpretation of this logic.[4] However, Jankov's logic is structurally complete (*Prucnal 1976*). Thus one sees that the so-called weak law of excluded middle preserves the context insensitivity of CPC despite, standing in the place of full *lem*, allowing for a procedural semantics.

4 For the details of this interpretation, see *Terwijn 2004*.

In my graduate student years, several of my friends and I were thinking about bounded arithmetic because of its connections with complexity theory and because the special difficulty of representing within these theories their own consistency statements shed much light on the fine details of arithmetization. We had a running gag, which is that formal theories like PA are awfully weak, because with them one can't draw very many distinctions. The implicit punchline, of course, is that the bulk of the distinctions one can draw in theories of bounded arithmetic are among statements that are in fact equivalent. Lacking the resources to spot these equivalences is no strength!

Something perfectly analogous happens in the case of substructural logics. There are theorems of CPC that are unprovable in IPC, but not vice versa, so the latter logic is strictly weaker. Moreover, CPC proves all sorts of implications and equivalences that IPC misses. But if we stop believing for a moment, as the discipline of logic demands we do, and ask about the fine structure of interdependencies among the formulas of propositional logic, IPC delivers vastly more information. Consider just propositional functions of a single variable p. In CPC there are exactly four equivalence classes of such formulas: those inter-derivable with p, $\neg p$, $p \vee \neg p$, and $p \wedge \neg p$. In IPC the equivalence classes of these same formulas exhibit a complicated pattern of implications, forming the infinite Rieger-Nishimura lattice.

One thought one may have is that IPC should be considered an expansion of CPC: every classical tautology can be discovered with IPC via the negative translation of *Gödel 1933* & Gentzen, so with IPC one gets all the classical tautologies and a whole lot more. (Gödel at times suggested something like this attitude.) But, of course, neither the negative translation nor the very idea of a classical tautology arises within intuitionistic logic. The thought that I encourage instead is this: The logician is loath to choose between classical and intuitionistic logic because the phenomena of greatest interest are the relationships between these logical systems. Who would have guessed that the rejection of a single logical principle would generate so much complexity—an r.e. set of admissible rules with no finite basis, an infinite lattice of inter-derivability classes?

The intuitionist and the classicist have very fine systems. Perhaps with them one gains some purchase on the norms of right reasoning or the modal structure of reality. The logician claims no such insight but observes that one can hold fixed the rules of the logical particles and, by merely tweaking the calculus between single conclusion and multi conclusion, watch structural completeness come in and out of view. The same switch, he or she knows, dresses the logical connectives up in a constructive, context-sensitive interpretation in one position and divests them of this interpretation in the other. These connections between sequent calculus, constructive proof transformation, structural completeness, and *lem* are fixtures from our logical knowledge store, but they cannot seriously be thought of as a network of consequences in some allegedly correct logic.

3 Logic imposed and logic discovered

If I have conveyed my attitude successfully, then I will have inspired the following objection:

> You speak about an unwillingness to embrace any one or select few logical systems because of an interest in understanding all such systems and how their various properties relate to one another. But by making logical systems into objects of investigation, you inhabit an ambient space in which you conduct this investigation. It is legitimate to ask which logic is appropriate in this space. What is your "metalogic?"

I shall explain that this objection rests on various preconceptions that I do not share. I hope the explanation functions to aid the reader in seeing these as misconceptions. If it does, then logical nihilism will be understood.

If we agree that as logicians we are interested, not in factual truth, but in the relationships among phenomena and ideas, then the point of view we must hasten to adopt should be the one that assists us in detecting and understanding these relationships. Which relationships? Presumably, it has been suggested, those that accurately pick out grounds and consequences, those that answer the question "What rests on what?" But why stop here? What sort of purpose is served by simultaneously disregarding factual states of affairs and pledging allegiance to factual relations of ground and consequence? I have an intuitive sense of what a fact is; I have no such sense of ontological grounding. Nor have I seen any reason to expect that the study of logic can foster such a sense in me.

More appropriate seems to be a disregard for privileged relationships similar to our disregard for truth. Suppose that we are interested in detecting and understanding whatever relationships we can find. Then we might wish not to be wedded to any point of view. We might, instead, try on a few hats until some interesting patterns appear where before there seemed to have been only disorder. We might find that one hat helps time and again, but we will be well-advised not to forget that we are wearing it. For if we never take it off, then we risk forever overlooking logical relationships of considerable interest. Worse, we risk coming to think of the relationships we can detect as "in the world," "preconditions of thought," or some such thing.

Allow me to illustrate this point with an example. At least since Aristotle it was expected that great complexity could be uncovered by a proper analysis of the quantifiers that occur in natural language. Notably, quantifiers allow us to reason in succinct strokes about infinite collections. About a century ago, David Hilbert had the idea of analyzing quantifiers with consistency proofs, and he devised an intricate calculus with "transfinite axioms" which allows proof figures involving quantifiers to be transformed (in principle) into figures without them. Hilbert's idea was that reasoning about this transformation (which by its nature requires being attentive to constructibility) would expose the quantifiers as innocuous parts of our mathematical language and also make perspicuous their complexity.[5]

Hilbert conducted this investigation in the shadow of a great ideological quarrel about the validity of various logical and mathematical principles. In one quarter were dour skeptics who distrusted not only *lem* but other forms of infinitary reasoning. Chief among them was Kronecker, against whom Hilbert (*1922*, 199–201) railed because he "despised ... everything that did not seem to him to be an integer." Less famously, but perhaps more importantly, he faulted Kronecker also because "it was far from his practice to think further" about what he *did* accept, "about the integer itself." In a similar vein, Hilbert observed that Poincâre "was from the start convinced of the impossibility of a proof of the axioms of arithmetic" because of his belief that mathematical "induction is a property of the mind." Thus Hilbert viewed these figures as short-sighted, not only in their rejection of mathematical techniques that he wished to defend, but also in their belief that the manifest validity of a principle precludes any hope of our analyzing it. Both attitudes, he cautioned, "block the path to analysis."

After only a decade of partial successes, it was discovered that the sort of consistency proofs Hilbert envisioned are not available. Specifically, Gödel (*1931*) demonstrated that no proof of the consistency of a reasonably strong and consistent mathematical system could be carried out within that same system. Typically it is the recursion needed to verify that the proof transformation algorithm halts that cannot be so represented. This situation raises the question whether proving with such principles that a system is consistent is not *obscurum per obscurius*.

In *1936* Gentzen made these circumstances much more precise by providing a perspicuous proof of the consistency of PA. Gentzen's proof is carried out in the relatively weak theory PRA together with the relatively strong principle of transfinite induction up to the ordinal

5 For details, see section 2 of *Franks 2015*.

$$\epsilon_0 = \sup_{n<\infty} \omega_1^{\omega_2^{\cdot^{\cdot^{\cdot^{\omega_n}}}}}.$$

Together with Gödel's result, this proof demonstrates that transfinite induction through ϵ_0 is unprovable in PA. All of this is well known. Most people familiar with the history of logic are aware that Gentzen proved also that transfinite induction to any ordinal beneath ϵ_0, any ordinal $\omega_1^{\omega_2^{\cdot^{\cdot^{\cdot^{\omega_n}}}}}$ for $n \in \mathbb{N}$, is in fact provable in PA. But Gentzen pressed even further. One can consider fragments of PA defined by restricting the induction scheme to formulas with a maximum quantifier complexity (call these the theory's class of inductive formulas). Gentzen showed in *1943* that the height of the least ordinal sufficient for a proof of the consistency of such a fragment corresponds with the quantifier complexity of that theory's class of inductive formulas, and that transfinite induction to any smaller ordinal is provable in the fragment. So the number of quantifiers over which mathematical induction is permitted equals the number of exponentials needed to express the ordinal that measures the theory's consistency strength. One quantifier equals one exponential. Thus the theory of constructive proof transformations has turned up a precise mathematical analysis of the complexity of natural language quantifiers, a remarkable realization of Hilbert's original ambition.

Why has so little attention been given to this result? The discussion of Gentzen's work has been dominated by debate about whether or not the proof of PA's consistency can really shore up our confidence in this theory. To anyone who has witnessed a talk about ordinal analysis sabotaged by this debate, the scene will be familiar: someone reminds us that the proof uses a principle that extends the resources of PA. Someone else defends the principle despite this fact and points out that in every other way the proof is extremely elementary compared to the full strength of PA. Because the two theories PA and PRA+ti_{ϵ_0}[6] are in this basic way incomparable, the jury is out as to the gains made by reducing the consistency of one to that of the other. From the point of view of logic, however, this is all a distraction from what Gentzen actually achieved: he showed that the question of the consistency even of elementary theories can be formulated as a precise problem, and he showed that the solution to this problem requires new perspectives and techniques and carries with it unexpected insights about logical complexity. If philosophers did not harbor skepticism about PA, then they would likely not be interested in Gentzen's result one way or the other. Their disinterest in the analogous results about fragments of PA is just evidence that they harbor

6 The theory of primitive recursive arithmetic extended with a principle of transfinite induction to ϵ_0.

no skepticism about these theories. We recognize Gentzen's analysis of first-order quantifiers as one of the deepest results in the history of logic as soon as, and no sooner than, we stop believing.

I now wish to respond more directly to the objection that opened this section. Of course it is true that in the study of logical systems one must engage in reasoning of some sort or another. This reasoning can possibly be described by one or a few select logical systems. But why should anyone assume that this amount of reasoning is anything more than ways of thinking that have become habitual for us because of their proven utility? Further, why should anyone assume that there is any commonality among the principles of inference we deploy at this level over and above the fact that we do so deploy them?

To expand on the first of these points, it may be helpful to draw an analogy between rudimentary logic and set theory. Often it is thought that decisions about which principles should govern the mathematical theory of sets should be made by appealing to our intuitions about the set concept and even about the cumulative hierarchy of sets. Doubtless such appeals have figured centrally in the development of set theory. But the history of the subject suggests that a complete inversion of this dynamic has also been at play. Kanamori (*2012*, 1) explains:

> [L]ike other fields of mathematics, [set theory's] vitality and progress have depended on a steadily growing core of mathematical proofs and methods, problems and results. ... from the beginning set theory actually developed through a progression of *mathematical* moves, whatever and sometimes in spite of what has been claimed on its behalf.

Whereas one can today find endless phenomenological and metaphysical justifications of, for example, the replacement axiom, Kanamori contends that set theory in fact evolved primarily by absorption of successful techniques, like transfinite induction, devised to answer mathematical questions. "It was von Neumann's formal incorporation of this method into set theory, as necessitated by his proofs, that brought in Replacement" (33). In similar fashion, the power set existence assumption, which originally had many detractors, was not finally embraced in the wake of any argument or philosophical insight. It merely happened that "iterated cardinal exponentiation figured prominently" in Kurepa's proofs in infinitary combinatorics, "so that shedding deeper concerns the power set operation became further domesticated ..." (46). The upshot? "Set theory is a particular case of a field of mathematics in which seminal proofs and pivotal problems actually shaped the basic concepts and forged axiomatizations, these transmuting the very notion of set" (1).

The point is not that we have no reliable intuitions about the set concept, nor even that they should play no role in the development of formal set theory. The point is that those intuitions have evolved partly in response to our need to make sense of routinely counter-intuitive scientific discoveries. If our intuitions have been at least partially, perhaps largely, shaped by developments in logic, then the fact that we appeal to them on occasion in order to refine our definitions and techniques or in order to choose new axioms seems to lose its significance. Rather than develop a theory of sets that unpacks the fundamental truths we intuit, we have developed an intuition about sets that makes sense of mathematically interesting relations we have discovered.

I need not argue that rudimentary logic has taken shape in a similar fashion first because others have been persuasive on this same point and second because it does not really matter for my purposes whether, in the end, this is true. I am content simply to reveal the picture that holds us captive when we begin to think about logic. For the objection we solicited was that if our modern science of logic thrives on an unprejudiced consideration of the full gamut of properties exhibited by logical systems and the relations among them, so that issues of correctness do not arise, then it has only smuggled those issues in through the back door, in the metalogic that makes the science possible. But it is a preconception that science is made possible by ahistorical norms of right reasoning. Once one considers the possibility that logic may be studied with patterns of thought adapted to what we learn along the way, it becomes hard to understand what special status the rudimentary principles we find ourselves reflexively appealing to are supposed to have.

This brings us to the second of the points above, the idea that the principles that have found their way into our basic toolkit must presumably have some features in common that led them there. Even if these principles have been adopted over the course of time, the idea goes, there must be some reason for their being adopted instead of other principles. Perhaps this reason can be repackaged as an explanation of their being the true logical principles.

In response to this suggestion, I wish only to expose the presupposition driving it. Whoever said that there must be some property of logical validity that some principles of inference enjoy and others do not? If we knew in advance of all evidence that some such property attains, then it might be reasonable to look for it in whatever classes of inference rules we happen to find collected together. But we have no such foreknowledge and, in fact, the evidence suggests that the arrangement of our toolkit is a highly contingent matter. Are we not better off shedding this vestigial belief that among all the intricate and interesting consequence relations out there, some have a special normative status? Can we not get by with the understanding that principles of inference with a rather wide range of applicability differ from those suited only to specific inference tasks *only* in having a wider range of applicability? Had this been the understanding that our culture inherited, would anything we have learned from studying logic lead us to question it?

At the end of Plato's *Phaedrus*, Socrates explains that prior to investigating the essence of a thing, it is important to devise an extensionally adequate definition of that thing so that we will be in agreement about what we are investigating. This attitude seems right to me, and it seems to me that the familiar debates about, for example, where logic leaves off and mathematics begins violate this principle unabashedly. Suppose it were clear to everyone that some but not all patterns of reasoning are inescapable and furthermore that it were easy to tell which these are. Then we would have good reason to label this "logic" in distinction to patterns of reasoning that we all recognize as the province of some special science or particular application. It would be reasonable to wonder what accounts for the privileged role that these principles play in our lives. As things stand, however, many of us seem instead to assume that there simply *must be* patterns of reasoning that differ in kind from others. Typically, our minds are already made up about the psychological or metaphysical circumstances that underwrite this difference. This is what Wittgenstein stressed with his observation that "the crystalline purity of logic was not the result of an investigation," that instead "it was a requirement" (§107). Driven by this assumption, we thrash about looking for some extensional definition that we can hang our ready-made distinction on. These definitions are simply unconvincing on their own. They can satisfy only people who cannot tolerate the thought that there is no line to be drawn.

When Gentzen began his study of logic, he parted ways with his predecessors[7] by not first defining logical validity and then seeking out logical principles that accord with that definition. He simply observed that some patterns of reasoning abound. Consider that to define "predicate logic" he said only that it "comprises the types of inference that are continually used in all parts of mathematics" (*1934–35*, 68). This homely definition set Gentzen on a task of empirically tallying the techniques used in mathematical proofs, ignoring those that are unique to geometry, arithmetic, and other specific branches of mathematics.[8] Of course such a survey is by no means guaranteed to be exhaustively executable, and it was Gentzen's good fortune that his subject matter happens to exhibit few instances.

One should not, however, write off his success as purely a matter of luck. Gentzen devised an ingenious argument to the effect that his tally was in fact exhaustive. This involved constructing an innovative type of logical calculus that is

---

7 There are historical precedents for Gentzen's attitude: Aristotle, Condillac, Mill.

8 This is explicit in *Gentzen 1934–35* and even more vividly depicted in section 4 of *Gentzen 1936*.

at once formal and patterned on the informal reasoning recorded in mathematical proofs. This scheme enabled him to do more than construct a serial tally of proof techniques, because the inference types identifiable with it are extremely few and systematically arranged so that one can be sure that none have been overlooked. After all, if there are any inference types that went unnoticed, then for that very reason they fail to meet the criterion of ubiquity in mathematical practice that Gentzen imposed.

This empirical "completeness proof" bears little semblance to familiar conceptions of logical completeness and is interesting for this reason. I mention it now only to draw attention to the fact that while Gentzen's definition of predicate logic does pick out a well-defined body of inferences,[9] he did not concoct the definition in the service of a preconceived notion of logical validity. He did not, for example, first stipulate a semantic notion of logical consequence based on his own intuitions and then ask whether his calculi adequately capture this notion. Gentzen simply proposed that the intuitions guiding mathematicians in their research would be worth isolating and studying, and he therefore modeled a logical calculus on the inferences mathematicians actually make.[10]

Of course mathematicians also deploy proof techniques that are less universal, and the only observable difference between these and the ones that meet Gentzen's criterion of ubiquity is their relative infrequency. Mathematicians do not report a feeling that arithmetical reasoning is less valid or valid in some other way than general reasoning, and even if they did, we should be inclined to ignore these reports if they did not reflect in mathematical practice. For these reasons Gentzen could never bring himself to describe the distinction between inference rules that appear in the predicate calculus and those that belong specifically to arithmetic as a distinction between the logical and the non-logical. He only thought that he had designated *a* logical system, one that by design encodes some of the inferences he was bound to make when reasoning about it but whose logical interest derives solely from what that reasoning brings to light.

Contrast this with one of the more famous attempts to demarcate the logical: Quine's defense of first-order quantification theory. Second-order quantification, branching quantifiers, higher set-theory, and such can each be dissociated from logic for failing to have sound and complete proof systems, for violating the compactness and basic cardinality theorems, and other niceties. There is even a *1969* theorem, due to Lindström, to the effect that any logic stronger than first-

9 Actually Gentzen vacillated over the inclusion of the principle of mathematical induction, ultimately deciding against it.

10 For details of this conception of completeness, see section 3 of *Franks 2010*.

order quantification theory will fail to exhibit either compactness or the downward Löwenheim-Skolem theorem. Second-order logic, Quine concluded, is just mathematics "in sheep's clothing" because by using second order quantifiers one is already committed to non-trivial cardinality claims (*1986*, 66).

How true will these remarks ring to someone who doesn't know in advance that they are expected to distinguish logical and mathematical reasoning? Quine's consolation is telling: "We can still condone the more extravagant reaches of set theory," he writes, "as a study merely of logical relations among hypotheses" (*1991*, 243). I should have thought that this accolade, especially in light of the intricate sorts of logical relations that set-theoretical principles bear to one another and that set theory bears to other systems of hypotheses, would be used rather to enshrine a discipline squarely within the province of logic. For if we never suspected that among the plenitude of logical relations are a privileged few that capture the true inter-dependencies of propositions, what else would we mean by "logic" than just the sort of study Quine described?

As to the properties that characterize first-order quantification theory, it should now go without saying that from our perspective Lindström's theorem, far from declaring certain formal investigations extra-logical, *exemplifies* logic. So too do results of Henkin (*1949*) and others to the effect that second-order quantification theory and first-order axiomatic set theory each are complete with respect to validity over non-standard models. For a final example, I can think of none better than the recent result of Fan Yang that Väänänen's system of dependence logic (with branching quantifiers), extended with intuitionistic implication (the context sensitive and constructive operator that results from the denial of *lem*), is equivalent to full second-order quantification theory (*Yang 2013*).

Wittgenstein (§108) expected his readers to recoil from the suggestion that we shed our preconceived ideas by "turning our whole examination around." Rather than impose our intuitions about logic on our investigations by asking which principles are truly logical, let us first ask if a close look at the various inference principles we are familiar with suggests that some stand apart from others. If some do, then let us determine what it is that sets them apart. But the question he puts in our mouths—"But in that case doesn't logic altogether disappear?"—suggests that we know deep down that our empirical investigation is bound to come up empty. Various criteria will allow us to demarcate different systems of inference rules to study, but when none of these indicate more than a formal or happenstance association we will find ourselves hard pressed to explain why any one of them demarcates "the logical."

I am more optimistic than Wittgenstein. The conclusion that I expect my reader to draw from the absence of any clear demarcation of the logical is not that there is no such thing as logic. Let us agree instead that no one part of what logicians study, contingent and evolving as this subject matter is, should be idolized at the expense of everything else. Logic outstrips our preconceptions both in its range and in its depth.

4 Conclusion

Traditional debates about the scope and nature of logic do not do justice to the details of its maturation. In asking whether certain inferential practices are properly logical or more aptly viewed as part of the special sciences, for example, we ignore how modern logic has been shaped by developments in extra-logical culture. Similarly, questions about whether logic principally traces the structure of discursive thought or the structure of an impersonal world presuppose a logical subject matter unaffected by shifts in human interest and knowledge.

I mean, by saying this, not just to suggest that the principles of rudimentary logic are contingent, not different in kind from principles that we use only some of the time or very rarely and only for specific tasks. I do urge this attitude. But the caution against mistaking our default, multi-purpose habits of reasoning for something monumental is only preparation for a second, more valuable reaction. One should warm up to the trend of identifying logic with the specialized scientific study of the relationships among various systems and their properties. This is, after all, how logicians use the word. Our preference to ignore questions about a logic's correctness stems not only from an interest in exploring the properties of possible logical systems in full generality but also from an appreciation, fostered by the study of logic, that no one such system can have all the properties that might be useful and interesting.

In closing, let me re-emphasize that the idea of a true logic, one that traces the actual inter-dependencies among propositions, is unscathed by all I have said. Part of the difficulty in questioning that idea is that it is a moving target: argue against it, you feel it again in that very argument; close the door, it will try the window. But this very circumstance only underlines the fact that the idea is a presupposition, nothing that emerges from any discovery made in the study of logic. For the same reason that we can marshal no evidence against it, we see that if we can manage to forget it our future discoveries will not reveal to us that we have erred.

This realization, coupled with the observation that a fixation on the true logical relationships "out there" hinders the advancement of logic, certainly recommends nihilism on practical grounds. The question that remains is whether we are capable of sustaining a point of view with no direct argumentative support.

The proper antidote to our reflexive tendencies will surely extend an analysis of modern logic and include a rehearsal of the subject's history. I cannot offer that here.[11] I can only mention that logic as a discipline has evolved often in defiance of preconceived notions of what the true logical relations are. Logic has been repeatedly reconceived, not as a fallout from our better acquaintance with its allegedly eternal nature, but in response to the changing social space in which we reason. There is reason neither to expect nor to hope that logic will not be continually reconceived. Such reconceptions have been and likely will again be fundamental, so that what makes the moniker "logic" apt across these diverse conceptions is not an invariable essence.

In these pages I have indicated instead logic's modern contours, highlighting the fact that the deepest observations logic has to offer come with no ties to preconceptions about its essence. The richness of logic comes into view only when we stop looking for such an essence and focus instead on the accumulation of applications and conceptual changes that have made current logical investigations possible. The study of logic might be the best practical antidote to the view of it that we have inherited.

In his *Logic* of 1780, Condillac wrote: "People would like to have had philosophers presiding over the formation of languages, believing that languages would have been better made. It would, then, have required other philosophers than the ones we know" (237–8). Our interest in a better made language is an interest in a language that traces a pre-existing logical structure. Like Condillac, Wittgenstein warned that presupposing such a structure fosters dismissive attitudes about the languages we have: "When we believe that we must find that order, must find the ideal, in our actual language, we become dissatisfied with what are ordinarily called 'propositions,' 'words,' 'signs' " (§105). When we stop believing for a moment, as the discipline of logic demands we do, the structures we find immanent in our several, actual languages command our interest more than anything we could have devised in the service of our ideal.

Bibliography

[1] Chagrov A. V. 1992. "A decidable modal logic with the undecidable admissibility problem for inference rules." *Algebra and Logic* **31**: 53–55.

[2] Condillac, E. 1780. *La logique (Logic)*, translated by W. R. Albury. New York: Abaris. 1980.

11 For the details of the evolution of one central concept—that of logical completeness—in the past two centuries, see *Franks 2014*.

[3] Franks, C. 2010, "Cut as consequence," *History and Philosophy of Logic* **31**(4): 349–79.

[4] Franks, C. 2014, "Logical completeness, form, and content: an archaeology." *Interpreting Gödel: critical essays*, J. Kennedy (ed.). New York: Cambridge.

[5] Franks, C. 2015, "David Hilbert's contributions to logical theory." *The History of Logic*, A. P. Malpass (ed.) New York: Continuum.

[6] Gentzen, G. 1934–35, "Untersuchungen über das logische Schliessen." Gentzen's doctoral thesis at the University of Göttingen, translated as "Investigations into logical deduction" in *Szabo 1969*: 68–131.

[7] Gentzen, G. 1936, "Die Widerspruchsfreiheit der reinen Zahlentheorie," *Mathematische Annalen* **112**. Translated as "The consistency of elementary number theory" in *Szabo 1969*: 132–213.

[8] Gentzen, G. 1938, "Neue Fassung des Widerspruchefreiheitsbeweises für die reine Zahlentheorie," *Forschungen zur logik und zur Grundlegung der exackten Wissenschaften*, New Series **4**. Translated as "New version of the consistency proof for elementary number theory," in *Szabo 1969*: 252–86.

[9] Gentzen, G. 1943, "Beweisbarkeit und Unbeweisbarkeit von Anfangsfällen der transfiniten Induktion in der reinen Zahlentheorie." Gentzen's Habilitationschrift at the University of Göttingen, translated as "Provability and nonprovability of restricted transfinite induction in elementary number theory," in *Szabo 1969*: 287–308.

[10] Gödel, K. 1931, "Über formal unentscheidbare Sätze der Principia Mathematica und verwandter Systeme, I," *Monatshefte für Mathematik und Physik* **38**: 173–98.

[11] Gödel, K. 1932, "Zum intuitionistischen Aussagenkalkül," *Anzeiger der Akademie der Wissenschaftischen in Wien* **69**: 65–66.

[12] Gödel, K. 1933, "Zur intuitionistischen Arithmetik und Zahlentheorie," *Ergebnisse eines mathematischen Kolloquiums* **4**: 34–38.

[13] Harrop R. 1960, "Concerning formulas of the types $A \to B \vee C$, $A \to (Ex)B(x)$ in intuitionistic formal systems," *Journal of Symbolic Logic* **25**: 27–32.

[14] Henkin, L. 1949, "The completeness of the first-order functional calculus," *Journal of Symbolic Logic* **14**: 159–166.

[15] Hilbert, D. 1922, "Neubergründung der Mathematik. Erste Mitteilung," translated by William Ewald as "The new grounding of mathematics: first report" in P. Mancosu (ed). *From Brouwer to Hilbert: the debate on the foundations of mathematics in the 1920s*. New York: Oxford, 1998: 198–214.

[16] Kanamori, A. 2012, "Set theory from Cantor to Cohen," *Handbook of the history of logic, vol. 6: sets and extensions in the twentieth century*. D. M. Gabbay, A. Kanamori, and J. Woods (eds.). Amsterdam: Elsevier.

[17] Kleene, S. C. 1945, "On the interpretation of intuitionistic number theory," *Journal of Symbolic Logic* **10**: 109–124.

[18] Lindström, P. 1969, "On extensions of elementary logic," *Theoria* **35**: 1–11.

[19] Myhill, J. 1973, "Some properties of Intuitionistic Zermelo-Fraenkel set theory," *Proceedings of the 1971 Cambridge summer school in mathematical logic* (*Lecture Notes in Mathematics* **337**): 206–231.

[20] Plato. "Phaedrus," trans. by A. Nehamas and P. Woodruff. *Plato: Complete Works* (1997), ed. by J. M. Cooper, Indianapolis: Hackett.

[21] Prucnal, T. 1976, "Structural completeness of Medvedev's propositional calculus," *Reports on Mathematical Logic* **6**: 103–105.

[22] Quine, W. V. O. 1986. *Philosophy of logic*. Cambridge: Harvard.

[23] Quine, W. V. O. 1991, "Immanence and validity," *Dialectica* **45**, reprinted with emendations in W. V. O. Quine, *Selected logic papers*. Cambridge: Harvard, 1995: 242–250.
[24] Rybakov, V. 1997. *Admissibility of logical inference rules*. Amsterdam: Elsevier.
[25] Szabo, M. E. 1969. *The collected papers of Gerhard Gentzen*, London: North Holland.
[26] Terwijn, S. A. 2004, "Intuitionistic logic and computation," *Vriendenboek ofwel Liber Amicorum ter gelegenheid van het afscheid van Dick de Jongh*, Institute for Logic, Language, and Computation. Amsterdam.
[27] Tokarz, M. 1973, "Connections between some notions of completeness of structural propositional calculi," *Studia Logica* **32**(1): 77–91.
[28] Visser, A. 1999, "Rules and arithmetics," *Notre Dame Journal of Formal Logic* **40**: 116–140.
[29] Wittgenstein, L. 1953. *Philosophical investigations*, Second Edition, translated by G. E. M. Anscombe. Oxford: Blackwell. 2001.
[30] Yang, F. 2013, "Expressing second-order sentences in intuitionistic dependence logic," *Studia Logica* **101**(2): 323–342.

Pietro Galliani

The Doxastic Interpretation of Team Semantics

Abstract: We advance a *doxastic interpretation* for many of the logical connectives considered in Dependence Logic and in its extensions, and we argue that Team Semantics is a natural framework for reasoning about beliefs and belief updates.

Keywords: team semantics, dependence logic, belief representation

MSC: 03B60, 03B80

Pietro Galliani: University of Helsinki, Finland, email: pgallian@gmail.com

1 Introduction

Dependence Logic [24] is an extension of First Order Logic which adds to its language *dependence atoms* of the form $=(\vec{t}_1, \vec{t}_2)$, where $\vec{t}_1$ and $\vec{t}_2$ are tuples of terms[1], with the intended interpretation of "the value of $\vec{t}_2$ is a function of the value of $\vec{t}_1$." It is a first-order *logic of imperfect information*, like IF Logic [14, 13, 20] or Branching Quantifier Logic [12]; but rather than adding new possible patterns of dependence or independence between *quantifiers*, as these logics do, Dependence Logic isolates the notion of dependence away from the one of quantification and permits the examination of patterns of dependence and independence between *variables* or, more in general, between tuples of terms.

This different outlook makes Dependence Logic a most suitable framework for the formal study, in a first-order setting, of functional dependence itself; and, furthermore, this logic is readily adaptable to the analysis of other, non-functional notions of dependence or independence [11, 6, 8].

Like other logics of imperfect information, Dependence Logic admits both a Game Theoretic Semantics, an imperfect information variant of the one for First Order Logic, and a *Team Semantics*, a compositional semantics which is a natural adaptation of Hodges' Trump Semantics [15]. One striking peculiarity of the current state of the art of the research in Dependence Logic and its extensions is a

1 Sometimes only atoms of the form $= (\vec{t}, t')$ are taken as primitive, where t' is a single term. $=(\vec{t}_1, \vec{t}_2)$ is then equivalent to $\bigwedge_{t' \in \vec{t}_2} =(\vec{t}, t')$.

willingness to take Team Semantics – and not Game Theoretic Semantics, as for the case of much IF Logic research – as the fundamental semantic framework; and this different approach is at the root of many recent technical developments in the field, such as, for example, the characterizations of team class definability of [17], [16] and [8], the hierarchy results of [4], and the study of notions of generalized quantification of [6] and [5].

This paper is a detailed account of a *doxastic* interpretation for Team Semantics, according to which formulas are to be interpreted as assertions about beliefs and belief updates. This is not a novel idea: as a matter of fact, it is already implicit in the equivalence proof between Trump Semantics and Game Theoretic Semantics of [15]. However, the consequences of this insight are far from fully explored: until now, doxastic concerns have played very little role in the development of extensions of Dependence Logic, and the doxastic meanings of known connectives have been left largely unexamined.[2]

In what follows, we will gradually develop a formal system – in essence, a notational variant of Jouko Väänänen's Team Logic [25] – and show that many of the atoms, connectives and operators of Team Semantics arise naturally from concerns about beliefs and belief updates. It is the hope of the author that by doing this, we may succeed in drawing attention to this interesting avenue of research.

2 Belief Models

Let M be a first order model with at least two elements in its domain, let $V \subseteq \mathrm{Var}$ be a set of *state variables*, and let us consider the set of all first-order *assignments* over $\mathrm{Dom}(M)$ with domain V.

With respect to Tarski's semantics for first order logic, such an assignment s represents a possible *state of things*: in other words, once the model M is fixed, the truth value of a first order formula in an assignment depends on the elements of the model that the assignment associates to all the free variables of the formula, *and on nothing else*.

Even disregarding First Order Logic, first order assignments are very natural objects for representing states of things. For example, let us suppose that our model's domain contains all the participants to a given contest, and that our states

2 Even such important connectives as the Dependence Logic disjunction, which – after [25] – we will write $\phi \otimes \psi$, or the *intuitionistic implication* $\phi \rightarrow \psi$ of [1] do not have, to the knowledge of the author, any known doxastic interpretation. Later in this work, we will provide such interpretations: as we will see, these connectives can be understood in terms of *belief updates*.

represent the possible outcomes of the contest – and, in particular, the players who obtained the first three positions in the final ranking.

Such an outcome can be represented, in an obvious way, as an assignment s over M with domain $\{w_1, w_2, w_3\}$: here, $s(w_1)$ would be the identity of the winner, and so on.

For this particular example, of course, we would need to add a constraint requiring that no one can be placed in two different positions in the final ranking: this can be represented easily enough as the first order axiom

$$\phi := \neg\exists x((w_1 = x \wedge w_2 = x) \vee (w_2 = x \wedge w_3 = x) \vee (w_1 = x \wedge w_3 = x)) \qquad (1)$$

as a condition that must hold for *all* possible states of things s, in the sense that s is an acceptable outcome if and only if $M \models_s \phi$.

Once we have added this axiom, there is not much left to do: as long as the domain of the model is the set of all participants to the contest, any assignment which satisfies the above formula represents a possible contest outcome.

A special case of this which is of no small interest is when the domain of the model M consists of only two elements 0 and 1, or "False" and "True": then an assignment is easily seen to be equivalent to a propositional truth valuation, or, if one prefers, to the specification to a *possible world* in the sense of Kripke's Semantics for Modal Logic.

Now, let us return to our example, and let us consider an agent A who has some – not necessarily true, nor complete – belief about who will reach the first three places of our tournament. How can we represent this belief?

There are many possible choices here: for example, we could consider a probability distribution over states, or a possibility distribution [28], or even a Dempster-Shafer distribution [23, 3].

But let us limit ourselves to a very simple idea, and consider the set $X_A \subseteq (V \mapsto \text{Dom}(M))$ of all possible states of things (assignments) which our agent believes to be possible. This idea of representing beliefs as "sets of possible states" is fairly common in knowledge representation theory, and – even though other approaches, such as the ones described above, are certainly more sophisticated – it is a reasonable starting point.

Furthermore, this approach plays on the analogy between our framework and modal logic: indeed, it is easy to see that, at least for the case of Boolean models, such a belief set is exactly a set of *possible worlds* which an agent can see from the "actual" world. An important difference between our framework and modal logic, however, is that in our case the agents can reason only about *outcomes*, and not about their beliefs or about the beliefs of other agents.

Now, what can we do with beliefs? To begin with, we can describe their properties in some suitable logical formalism; and, as Section 3. will show, many primitive formulas considered in logics of imperfect information have a very natural interpretation in these terms.

But we can also *update* beliefs. In general, a (unary) *update operation* will be a function O from belief sets, or, to use the terminology in common use for logics of imperfect information, from *teams*, to sets of belief sets. We will not require these updates to be deterministic; and we will write $O(X) \mapsto Y$ as a shorthand for $Y \in O(X)$, that is, for the statement that Y is a possible outcome of updating X according to the rule O.

Binary update operators are defined analogously, as functions $\diamond$ mapping each pair of teams X and Y to a set $X \diamond Y$ of possible resulting teams; and again, we will write $X \diamond Y \mapsto Z$ for $Z \in (X \diamond Y)$, that is, for stating that the belief set Z is a possible outcome of updating X with Y according to the rule $\diamond$.

Ternary or n-ary operators can also be defined in the same way, but we will not need to consider any of them in the present work.

Once we have belief operators and a language for describing properties of belief sets we can ask a number of new questions, such as

1. Can a certain belief set be seen as the result of a certain update being applied between belief sets satisfying certain properties?
2. If we update a belief set under a certain rule, and the other belief sets used for the update (if any) satisfy certain properties, can we guarantee that the resulting belief set will satisfy certain other properties?

This kind of question is of clear practical importance in Artificial Intelligence: if some intelligent system's belief state respects a condition ϕ and our system interacts with some other system whose belief state respects another condition ψ, can we guarantee that the resulting belief states will respect some further condition θ?

The whole discipline of *belief revision* [10], for example, can be understood as a special case of this, as its fundamental problem consists in the study and comparison of the ways of updating a knowledge base K when a new statement ϕ is learned which is contradictory to it.

Our framework, in itself, is vastly more general – and, of course, vastly more computationally expensive – than any system of belief revision; but on the other hand, the update operations that we will discuss here are all much simpler than those considered in Belief Revision. What, in the opinion of the author, logics of imperfect information can provide to that field is a very general logical framework for defining update operations and reasoning about their properties.

But enough chatter. Beginning with the next section, we will define primitives and operators for a very general logic of imperfect information, containing most of the connectives which have been studied so far in the context of logics of imperfect information, and we will discuss their doxastic interpretations.

3 Atoms and First Order Formulas

In this section, we will gradually develop a logical formalism – basically, a fragment of Team Logic – and use it to state properties of belief sets.

Now, what can be said about the beliefs of an agent A, or, to be more precise, about his belief set X_A?

To begin with, we can ask, given a first order[3] formula ϕ, whether our agent A *believes* that ϕ holds. This justifies the following semantic rule:

DI-bel If ϕ is first order, $M \models_X B(\phi)$ if and only if $M \models_s \phi$ for all $s \in X$

where the expression $M \models_s \phi$ means that the assignment s satisfies ϕ in M according to the usual Tarski semantics.

As an example, let us consider again the scenario described in the previous section and the formula ϕ of Equation (1). Then "$M \models_{X_A} B(\phi)$" is a sanity condition for our agent, corresponding to the statement according to which he believes that no player will get two distinct positions in the final rankings.

As another, perhaps quite unnecessary, example, suppose that our agent A believes that the winner of the contest will be female; then, for all $s \in X_A$ we will have that $M \models_s \text{Female}(w_1)$, and hence that $M \models_{X_A} B(\text{Female}(w_1))$.

In most logics of imperfect information, one would just write $M \models_X \phi$ for what we would write here as $M \models_X B(\phi)$. Furthermore, the above condition would be given just for first-order literals, and we would rely on the connectives of our logic in order to build expressions equivalent to $B(\phi)$ for complex first-order formulas ϕ: indeed, as mentioned in ([24], Proposition 3.30), if ϕ is first order then $M \models_X \phi$ if and only if $M \models_s \phi$ for all $s \in X$.

Here, however, we will write the B operator explicitly, for three different reasons. First of all, our objective is emphatically *not* to develop a terse formalism in which to express everything that can be expressed in a logic of imperfect information: instead, we want to illustrate a possible interpretation of logics of imperfect information, and hence it will be useful to examine many doxastically significant

3 We could also use here any extension or variant of First Order Logic which admits a Tarski-style semantics, such as Transitive Closure Logic or First Order Logic augmented with the Härtig quantifier.

conditions and operators. Furthermore, it is vital for our purposes to distinguish between the first-order level of our language, which allows us to summarize the properties of all the assignments of the team *taken individually*, the more sophisticated kinds of atoms that we will describe later in this section, and the update connectives that we will introduce in Sections 4-7. The fact that first-order formulas can be decomposed in terms of first-order literals and update operators will then be, from this point of view, an interesting *theorem*, not something built in our definitions. Finally, and perhaps more practically, we want to be able to express another kind of "first-order" assertion in our language, and we need to distinguish it from belief statements as those just considered.

This new kind of first-order assertion is a *possibility* assertion, which corresponds to our agent believing some first order condition to be *possibly the case* in the "true" assignment. We can introduce this kind of assertion as follows:

DI-pos: If ϕ is first order, $M \models_X P(\phi)$ if and only if $M \models_s \phi$ for some $s \in X$

Dependence Logic and IF-Logic, the two most studied logics of imperfect information, are downwards closed[4] and hence incapable of expressing this sort of statement. The most widely known formalism capable of that is Team Logic [25], where $P(\phi)$ corresponds to $\sim \neg\phi$, where $\neg$ is the dual negation and $\sim$ is the contradictory one.

From the point of view of the present work, possibility statements are very natural: for example, using them we can express that our agent A considers it *possible* that the winner will be female, that is, that $M \models_{X_A} P(\text{Female}(w_1))$.

We could also give and justify along similar lines further "first-order" conditions over belief sets: for example, we could state that an expression of the form $\text{Most}(\phi)$ holds in a team X if and only if most of the assignments in X satisfy ϕ. But let us now move to conditions which cannot be verified by examining the truth value of a first order formula in all assignments of our belief set.

In the example which we are considering, what else could our agent A believe that we cannot express already? Well, to begin with, our agent could believe that he knows the identity of the winner.

It is easy enough to assert, using what we already have, that the agent believes that the winner will be a_0 for some $a_0 \in \text{Dom}(M)$: indeed, this is precisely the condition corresponding to $M \models_{X_A} B(w_1 = a_0)$. But what we are asking now is different: we want a formula that specifies that the agent believes that *someone* will certainly be the winner, but does not specify who this person is.

4 A logic of imperfect information is said to be *downwards closed* if for all models M, teams X and Y and expressions ϕ, $M \models_X \phi$ and $Y \subseteq X$ implies $M \models_Y \phi$.

An existential quantifier will not do the trick: writing $M \models_{X_A} B(\exists x(w_1 = x))$ corresponds only to asserting that our agent believes that someone will win the contest, not that he believes this about someone in particular![5] For example, if Tom, Bob and Jack are three participants to the tournament, it is easy to see that the belief set

$X_A =$		w_1	w_2	w_3
	s_0	Tom	Bob	Jack
	s_1	Bob	Tom	Jack

corresponding to the belief state in which our agent A knows that Jack will get third place, but is unsure about who between Tom and Bob will get the second place and who the first one, satisfies the formula $B(\exists x(w_1 = x))$ but it does not respect the condition we are talking about.

In fact, *no* expression of the form $B(\phi)$ or $P(\phi)$ will allow us to express our intended condition: indeed, those of the former sort are *flat* in the sense of [24], and hence hold in a team if and only if they hold in all singleton subteams, and those of the latter one are *upwards closed*, in the sense that if $M \models_X P(\phi)$ and $X \subseteq Y$ then $M \models_Y P(\phi)$.

What we seem to need is some way of saying that the value of w_1 is *the same* for all assignments in our team. This is precisely the semantics for *constancy atoms* of Dependence Logic.[6]

DI-con: For all terms t, $M \models_X =\!(t)$ if and only if, for all $s, s' \in X$, $t\langle s\rangle = t\langle s'\rangle$.

Given this definition, $=\!(w_1)$ characterizes precisely the condition of our example; and, more in general, it is easy to see that $=\!(t)$ is satisfied by a team X_A if and only if the corresponding agent A believes that he knows the value of t.

What if our agent instead believes that he would be able to guess the name of the winner from the name of the second placed participant? This corresponds nicely to the *dependence atom* $=\!(w_2, w_1)$, where the rule for dependence atoms is

DI-dep: $M \models_X =\!(\vec{t}_1, \vec{t}_2)$ if and only if any two $s, s' \in X$ which assign the same value to $\vec{t}_1$ also assign the same value to $\vec{t}_2$.

What else can we say about our agent A's beliefs? For example, he might think that everybody who has a chance to make it to first place has also a chance to make it

5 In brief, we are discussing here the *de dicto*/*de re* distinction.

6 Another reasonable approach could be to define an "external" existential quantifier $\exists$, and model the intended condition as $\exists x B(x = w_1)$. Later in this section, we will briefly explore this idea; but we can anticipate that this external existential quantifier is precisely the $\exists^1$ operator of [17].

to second place: then, for all $s \in X_A$ there exists a $s' \in X_A$ such that $s(w_1) = s'(w_2)$. This is represented by the *inclusion atom* $w_1 \subseteq w_2$ from [8], where

DI-inc: For all tuples of terms $\vec{t}_1$ and $\vec{t}_2$, of the same length, $M \models_X \vec{t}_1 \subseteq \vec{t}_2$ if and only if for every $s \in X$ there exists a $s' \in X$ such that $\vec{t}_1\langle s\rangle = \vec{t}_2\langle s'\rangle$.

The meaning of the expression $w_1 \subseteq w_2$ is of course different from that of $B(w_1 = w_2)$, which would instead state that the agent believes that the first and second placed players will be the same.

As an aside, possibility operators can also be constructed in terms of inclusion atoms: indeed, if the free variables of ϕ are $x_1 \dots x_n$ then it is easy to see that $P(\phi)$ can be written as $\exists w_1 \dots w_n(\phi[w_1 \dots w_n/x_1 \dots x_n] \wedge (w_1 \dots w_n \subseteq x_1 \dots x_n))$.

Again, what else? Well, our agent could also think that no one who has some chance to take first place has also some chance to take third place – only first or second. This is represented by the *exclusion atom* $w_1 \mid w_3$, also from [8], where

DI-exc: For all tuples of terms $\vec{t}_1$ and $\vec{t}_2$ of the same length, $M \models_X \vec{t}_1 \mid \vec{t}_2$ if and only if for all $s, s' \in X$, $\vec{t}_1\langle s\rangle \neq \vec{t}_2\langle s'\rangle$.

Of course, this is different, and stronger, than $B(w_1 \neq w_3)$, which would only state that the agent believes that the winner and the third placed player will not be the same.

For the last pair of atoms that we will describe here, we need to modify slightly our example. Let us suppose that w_1, w_2 and w_3 represent the winners of three *different* tournaments in three successive years, so that the same player could conceivably win more than one of them. Then a possible situation might be that, in the opinion of the agent, learning the winner of the first year would not tell him anything about who the winner of the third year that he does not know already – or, in other words, that the set of all possible third year winners is the same for each fixed first year winner. This corresponds to the *independence atoms* $w_1 \perp w_3$, where

DI-ind$_C$: Let $\vec{t}_1$ and $\vec{t}_2$ be two tuples of terms, not necessarily of the same length. Then $M \models_X \vec{t}_1 \perp \vec{t}_2$ if and only if for all $s, s' \in X$ there exists a $s'' \in X$ with $\vec{t}_1\langle s''\rangle = \vec{t}_1\langle s\rangle$ and $\vec{t}_2\langle s''\rangle = \vec{t}_2\langle s'\rangle$.

As pointed out in [11], $t \perp t$ is logically equivalent to $=(t)$. The reason for this is clear: in our interpretation, $t \perp t$ means that our agent thinks that he would not learn anything new about t by being told the value of t, and this is possible only if he believes that he knows it already.

Finally, our agent may think that, for the purpose of learning who will be the winner in the third year, learning who won in the first and second year and knowing who won in the second year alone makes no difference. Perhaps, according to this agent, learning who won in the second year *would* give him valuable information, and so would learning who won in the first year; but once he learned who won in the second year, learning who won in the first year too would be quite irrelevant.

As an example of this situation, let us consider the following belief set:

$X_A =$

	w_1	w_2	w_3
s_0	Bob	Tom	Tom
s_1	Tom	Bob	Bob
s_2	Tom	Bob	Jack
s_3	Jack	Bob	Bob
s_4	Jack	Bob	Jack

Here our agent believes that if Tom won the second year then he will win the third year too, and that if Bob won the second year then either him or Jack will win the third year. So, learning who won the second year would allow him to infer something about who will win the third year. Also, he believes that if Tom won the first year then one of Bob or Jack will win the third year, that if Jack won the first year then one of Bob or Jack will win the third year and that if Bob won the first year then Tom will certainly win the third year. So, learning who won the first year would allow him to infer something about who will win the third year. However, suppose that Tom is told who won the second year. Then learning who won in the first year too would tell him nothing new about who will win the third year: if the winner of the second year is Tom, then he is also the winner of the third year, and if the winner of the second year is Bob then, no matter who won the first year, both Bob and Jack are possible winners for the third year competition.

This is modeled by the following, more general *independence atom*, also from [11]:

DI-ind: Let $\vec{t}_1$, $\vec{t}_2$ and $\vec{t}_3$ be two tuples of terms, not necessarily of the same length. Then $M \models_X \vec{t}_2 \perp_{\vec{t}_1} \vec{t}_3$ if and only if for all $s, s' \in X$ with $\vec{t}_1\langle s\rangle = \vec{t}_1\langle s'\rangle$ there exists a $s'' \in X$ with $\vec{t}_1\vec{t}_2\langle s''\rangle = \vec{t}_1\vec{t}_2\langle s\rangle$ and $\vec{t}_1\vec{t}_3\langle s''\rangle = \vec{t}_1\vec{t}_3\langle s'\rangle$.

In particular, it is not difficult to see that the above team satisfies $w_1 \perp_{w_2} w_3$, as required.[7]

It might be of course possible to consider other notions of dependence or independence, and justify them along similar lines. And indeed, in the opinion of the author at least, one of the main future research directions in the field of logics

7 As observed in [11], we then have that $=(x, y)$ is equivalent to $y \perp_x y$, and so on. This can be justified along the same lines in which it was justified that $=(x)$ is equivalent to $x \perp x$.

of imperfect information will be the search for new, doxastically significant notions of atom and the study of the relationships between the corresponding logics of imperfect information.

However, we already have more than enough basic material for our purposes.

Let us now add the classical *conjunction* and *disjunction* to our language:

DI-or: $M \models_X \psi \vee \theta$ if and only if $M \models_X \phi$ or $M \models_X \theta$.

DI-and: $M \models_X \psi \wedge \theta$ if and only if $M \models_X \psi$ and $M \models_X \theta$.

There is not much to say here about these two connectives, as they are simply a way to join basic belief constraints into more complex ones.

Can we also add the classical quantifiers at this point? Nicely enough, this has been already done by Kontinen and Väänänen in [17] by defining the $\exists^1$ and $\forall^1$ quantifiers. These quantifiers have properties which are quite different from the ones of the "usual" existential and universal quantifiers for logics of imperfect information, which will be interpreted in Section 7 in terms of update operations. Here we will simply write $\exists$ and $\forall$ for the $\exists^1$ and $\forall^1$ of [17]. Their truth conditions are

DI-exists: $M \models_X \exists x \psi(x)$ if and only if there exists a $m \in \mathrm{Dom}(M)$ such that $M \models_X \psi(m)$;

DI-forall: $M \models_X \forall x \psi(x)$ if and only if for all $m \in \mathrm{Dom}(M)$, $M \models_X \psi(m)$

where $\psi(m)$ stands for the formula obtained by substituting m (or, to be more precise, a new constant whose interpretation is the element m) for x in $\psi(x)$.

These new connectives actually allow us to do without our many non first-order atoms: for example, it is not difficult to see at this point that $=(x, y)$ is equivalent to $\forall u \exists v B(u \neq x \vee v = y)$, $x \subseteq y$ is equivalent to $\forall u(B(u \neq x) \vee P(u = y))$, $x \mid y$ is equivalent to $\forall u(B(u \neq x) \vee B(u \neq y))$ and $y \perp_x z$ is equivalent to $\forall u_1 u_2 u_3(B(u_1 \neq x \wedge u_2 \neq y) \vee B(u_1 \neq x \wedge u_3 \neq z) \vee B(u_1 = x \wedge u_2 = y \wedge u_3 = z))$.

Examining logics of imperfect information in terms of these quantifiers, rather than in terms of some classes of non first-order atom, may actually be a promising – and, at the moment, largely unexplored – avenue of research; but in this work, we thought it better to begin by describing the atoms that have been studied so far, and only show their translations in term of basic quantifiers in a second time.

Another connective that we are missing is a negation. We can certainly negate a first-order formula inside a belief or possibility statement; but can we also consider an "external" negation, to go with our external quantifiers and our classical conjunction and disjunctions?

It turns out that we can, of course, and that such an operator is precisely the *contradictory negation* $\sim \phi$ of Team Semantics. Here we call it simply $\neg\phi$, as there is no "other" negation with which it may be confused. Its semantics is precisely the one that we would expect:

DI-not: $M \models_X \neg\psi$ if and only if $M \not\models_X \psi$.

Given such an operator, we can as usual rewrite $\phi \wedge \psi$ as $\neg(\neg\phi \vee \neg\psi)$ and $\forall x\psi$ as $\neg\exists x\neg\psi$.

Furthermore, as in the case of the "diamond" and "box" operators of Modal Logic, we can remove one of them from our list of primitives: for example, we could keep only $P(\phi)$, and define $B(\phi)$ as $\neg P(\neg\phi)$. Here, however, it must be noted that the roles of the internal and external negations are quite different: the former states that something is not true with respect to the whole team, while the negation of $\neg\phi$ states that ϕ is not true *in one specific assignment*. This is rather reminiscent, although not entirely identical, to the distinction between *contradictory negation* and *dual negation* of Team Logic.[8]

Given such a theory T in the language developed so far and a suitable model M, one can consider the set $\mathrm{Bel}_M(T)$ of all belief sets which satisfy T in M: formally, we can define

$$\mathrm{Bel}_M(T) := \{X : M \models_X T\}.$$

Given two theories T_1 and T_2 and a suitable model M, we write $T_1 \models_M T_2$ if $\mathrm{Bel}_M(T_1) \subseteq \mathrm{Bel}_M(T_2)$, and $T_1 \models T_2$ if $T_1 \models_M T_2$ for all suitable models M.

The significance of these expressions is clear: $T_1 \models_M T_2$ means that whenever our belief set over the model M satisfies all the conditions of T_1, it also describes all those described in T_2, and $T_1 \models T_2$ means that this is the case even if we do not know the underlying model M.

4 Belief Updates

Consider again our agent A with his belief set X_A, and suppose that he interacts with another agent B with a different belief set X_B. Then, our first agent could update his beliefs in many different ways, according to the degree up to which he trusts his own beliefs, to the degree up to which he trust the other agent's beliefs and on a number of other possible factors.

This can be represented by defining update operations $X_A \diamond X_B$ from pairs of teams to *sets* of teams; and, as stated in Section 2., we will write $X_A \diamond X_B \mapsto Y$

8 In particular, the Team Logic expression $\sim \phi$ corresponds to our $\neg B(\phi)$, while the expression $\neg\phi$ corresponds to our $B(\neg\phi)$.

for $Y \in (X_A \diamond X_B)$, that is, for the assertion according to which Y is a possible outcome of a $\diamond$-interaction between two agents whose beliefs are represented by X_A and X_B respectively.

There are many possible choices of update operations of this sort. Here we will consider four of them which seem, at least at first sight, to be relatively reasonable choices:

Confident update $X_A \oplus X_B$**:** A trusts his beliefs concerning the true assignment s_0, but learns and trusts in the same way the beliefs of B too. Therefore, $X_A \oplus X_B \mapsto Y$ if and only if $Y = X_A \cap X_B$.

Credulous update $X_A \otimes X_B$**:** A is willing to entertain the possibility that he is wrong and the true state is one that B believes possible and he does not. Or B may be wrong and he may be right, he does not know. Hence, $X_A \otimes X_B \mapsto Y$ if and only if $Y = X_A \cup X_B$.

Skeptical update $X_A \ominus X_B$**:** A *might* trust B's beliefs, but only if B appears to know more than A – that is, if B does not consider possible anything that A considers impossible. Otherwise, A refuses to perform the update. Therefore, $X_A \ominus X_B \mapsto Y$ if and only if $X_B \subseteq X_A$ and $Y = X_B$.

Openminded update $X_A \odot X_B$**:** A might trust B's beliefs, but only if B appears to know *less* than A – that is, if he does not consider impossible anything that A considers possible. Otherwise, A refuses to perform the update. Therefore, $X_A \odot X_B \mapsto Y$ if and only if $X_A \subseteq X_B$ and $Y = X_B$.

Note that the skeptical update and the openminded one can *fail*, that is, may not lead to any possible outcome. This is a feature, not a bug: an update operation does not need to be specified for all possible belief states, and does not need to be deterministic either – none of the update operations considered here can lead to more than one possible outcome, but nothing prevents in principle the definition of such update operators.

Furthermore, these update operators satisfy the three following properties:

Idempotence: $X \diamond X \mapsto Y$ if and only if $Y = X$;

Associativity: If $X_1 \diamond X_2 \mapsto Y$ and $Y \diamond X_3 \mapsto Z$, then there exists a W such that $X_2 \diamond X_3 \mapsto W$ and $X_1 \diamond W \mapsto Z$;

Monotonicity: If $X \diamond Y \mapsto Z$ and $Z \diamond W \mapsto X$ then $X = Z$.

The interpretation of these properties, and the reason why they may be reasonable properties to require for an update operator, should be clear. An update $X \diamond Y$ represents an interaction between two agents whose beliefs correspond to X and Y: therefore, idempotence states that whenever the two agents have the exact same beliefs, the interaction does not modify these beliefs, associativity states that, in group interactions, the order of the individual interactions is irrelevant, and monotonicity means that an agent who changed idea cannot "return back" to his previous beliefs through another interaction of the same kind.

Let us verify that the update operators that we defined satisfy these properties. For the confident and credulous updates, this is obvious; hence, we will verify the case of the skeptical update, as the one of the openminded update is completely analogous.

Idempotence: Since $X \subseteq X$, it follows at once that $X \ominus X \mapsto Y$ if and only if $X = Y$.

Associativity: Suppose that $X_1 \ominus X_2 \mapsto Y$ and $Y \ominus X_3 \mapsto Z$. Then, by the definition of the skeptical update, we have that $X_1 \supseteq X_2 = Y \supseteq X_3 = Z$. But then we have that $X_2 \ominus X_3 \mapsto X_3$, and that $X_1 \ominus X_3 \mapsto X_3 = Z$, as required.

Monotonicity: If $X \ominus Y \mapsto Z$, then $Z = Y \subseteq X$. Furthermore, if $Z \ominus W \mapsto X$ then $X = W \subseteq Z$. Hence, $Z \subseteq X \subseteq Z$, and therefore $X = Z$.

Of course, these are only a possible selection of update properties: it may well be the case that in the future interesting update operators which do not respect them will be found, or that other, more important conditions will be explored.

After this parenthesis, let us reconsider the language of Section 3. As we saw, this language is already capable of describing a number of properties of belief states. But now we have something new, that is, some ways of *updating* beliefs. This allows us to ask some very natural questions: for example, if A's belief state X_A satisfies ϕ, and if B's belief state X_B satisfies ψ, what can we say about a belief state corresponding to some update of X_A and X_B?

To answer this, we need to add some way of talking about updates to our language. A possibility is to consider, for each update $\diamond$, a connective $\phi \diamond \psi$ such that $M \models_X \phi \diamond \psi$ if and only if the belief set X can be seen as the result of a $\diamond$-update between a belief set satisfying ϕ and a belief set satisfying ψ.

In other words, the semantics of $\phi \diamond \psi$ will be

DI-$\diamond$: $M \models_X \phi \diamond \psi$ if and only if there exist teams Y and Z such that $Y \diamond Z \mapsto X$, $M \models_Y \phi$ and $M \models_Z \psi$.

This gives us at once the following operations:

DI-$\oplus$: $M \models_X \phi \oplus \psi$ if and only if $X = Y \cap Z$ for some Y and Z such that $M \models_Y \phi$ and $M \models_Z \psi$;

DI-$\otimes$: $M \models_X \phi \otimes \psi$ if and only if $X = Y \cup Z$ for some Y and Z such that $M \models_Y \phi$ and $M \models_Z \psi$;

DI-$\ominus$: $M \models_X \phi \ominus \psi$ if and only if $M \models_X \psi$ and there exists a $Y \supseteq X$ such that $M \models_Y \phi$.

DI-$\odot$: $M \models_X \phi \odot \psi$ if and only if $M \models_X \psi$ and there exists a $Y \subseteq X$ such that $M \models_X \phi$.

The credulous update connective is exactly the tensor connective of Team Logic, or the disjunction of Dependence Logic. The other ones, to the knowledge of the author, have not been studied in depth yet, but they do hold some interest; it is worth noting, in particular, that for downwards closed logics (such as Dependence Logic or Intuitionistic Dependence Logic) $\phi \oplus \psi$ and $\phi \ominus \psi$ are equivalent and correspond to the classical conjunction $\phi \wedge \psi$.

The intended interpretation of these connectives is better understood by considering expressions of the form $\phi \diamond \psi \models \theta$. According to what we just discussed, such an expression corresponds to the statement that

> *Any possible outcome of $\diamond$-update between two belief states satisfying ϕ and ψ respectively will satisfy θ.*

The significance of such a statement, and its relevance for the kind of framework that we are presenting, is then clear.

An expression of the form $\phi \models \psi \diamond \theta$ is perhaps a little more difficult to read, but its meaning is still intuitive enough: such an entailment holds if and only if any belief state which satisfies ϕ can be thought of as the result of a $\diamond$-update between a belief state such that ψ and a belief state such that θ.

This allows us to make sense of some properties of logics of imperfect information. Here we will describe only three examples relative to the credulous update:

1. It is easy to see that $(\phi \otimes \psi) \otimes \theta \models \phi \otimes (\psi \otimes \theta)$ for all ϕ, ψ and θ. This just means that the credulous update is *associative*: if a belief state can be the result of an agent, whose belief state satisfies ϕ, performing a credulous update with some agent whose belief set satisfies ψ and *then* with some other one whose belief set satisfies θ , then it can also be the result of an agent, whose belief state satisfies ϕ, performing a credulous update with some agent whose belief state *satisfied* ψ before *he* performed a credulous update with some agent whose belief state satisfied θ.
2. The credulous update is not idempotent: in general, $\phi \otimes \phi \not\models \phi$. This can be verified easily by letting ϕ be the constancy atom $=(x)$: if the value of the variable x is constant in Y and in Z, indeed, it does not necessarily follow that it is constant in $Y \cup Z$.
   The reason for this is clear: if an agent who believes that he knows the value of x performs a credulous update with another agent who also believes that he knows the value of x, and the two agents disagree on this value, then our first agent will become unsure about who, between him and the other agent, was in the right about x.
3. If ϕ, ψ and θ are downwards closed formulas – for example, if they are expressible in Intuitionistic Dependence Logic or in Exclusion Logic – then the

following "distributivity property", first pointed out by Ville Nurmi, holds:

$$(\phi \otimes \psi) \wedge (\phi \otimes \theta) \models \phi \otimes \phi \otimes (\psi \wedge \theta).$$

According to what we just discussed, this entailment can be read as follows:

If a team X can be seen as the result of a credulous update between a team such that ϕ and one such that ψ, and also as the result of a credulous update between a team such that ϕ and one such that θ, then it is can also be the result of a credulous update between two teams such that ϕ and one such that ψ and θ.

This is a nontrivial – and, at least in the opinion of the author, rather interesting – property concerning belief updates and their properties.

As an aside, the last property fails if we consider non-downwards closed formulas: for example, consider the team

$$X = \begin{array}{c|ccc} & x & y & z \\ \hline s_0 & 0 & 1 & 1 \\ s_1 & 1 & 0 & 0 \end{array}$$

in any model M with at least two elements. Then $M \models_X z = 1 \otimes x \subseteq y$: indeed, for $Y = \emptyset$ and $Z = X$ we have that $M \models_Y x = 1$, $M \models_Z x \subseteq y$, and $X = Y \cup Z$. Furthermore, $M \models_X z = 1 \otimes z = 0$, as can be easily verified by splitting X into the two subteams $\{s_0\}$ and $\{s_1\}$.

However, $M \not\models_X z = 1 \otimes z = 1 \otimes (x \subseteq y \wedge z = 0)$: indeed, otherwise we could split X into three subteams X_1, X_2 and X_3 such that $M \models_{X_1} z = 1$, $M \models_{X_2} z = 1$ and $M \models_{X_3} x \subseteq y \wedge z = 0$. Now, $s_1(z) = 0 \neq 1$, and therefore s_1 would necessarily be in X_3; but then, since $M \models_{X_3} x \subseteq y$, there should be another assignment $s \in X_3$ with $s(y) = s_1(x) = 1$. The only such assignment is s_0; but $s_0(z) = 1$, and therefore it would not be the case that $M \models_{X_3} z = 0$. This contradicts our hypothesis.

As this example shows, different fragments of our language may have different properties when it comes to the entailment relation. This may be worth exploring further in the future.

5 Adjoints

In the previous section we considered a few update operators, and for each one of them we defined a connective expressing that a given team X can be seen as the result of an update between some teams Y and Z satisfying certain properties. This increased substantially the expressive power of our formalism, and, in fact, by now our language contains the propositional fragment of most logics of imperfect information.

However, this is not all that we can do with these update operations. In particular, it may be useful to be able to make conjectures about what *would* happen if we updated the team in a certain way. In particular, any update operator $\diamond$ induces a corresponding implication $\stackrel{\diamond}{\rightarrow}$ between sets of belief sets, defined as

DI-$\stackrel{\diamond}{\rightarrow}$: $M \models_X \phi \stackrel{\diamond}{\rightarrow} \psi$ if and only if for all Y s.t. $M \models_Y \phi$ and for all Z such that $X \diamond Y \mapsto Z, M \models_Z \psi$.

The intended interpretation of these new connectives is clear: $\phi \stackrel{\diamond}{\rightarrow} \psi$ corresponds to the statement asserting that *any* $\diamond$-update between the beliefs X of our agent and the beliefs Y of *any* other agent such that $M \models_Y \phi$ will *always* result in a belief Z such that ψ.

This easily implies that

$$\phi \diamond \psi \models_M \theta \Leftrightarrow \phi \models_M \psi \stackrel{\diamond}{\rightarrow} \theta$$

for all ϕ, ψ and θ and for all models M. In other words, the operator $\stackrel{\diamond}{\rightarrow}$ is the *right adjoint* of the operator $\diamond$.

This notion of *adjointness* is precisely the one studied in [1], in which it was one of the motivations given for the definitions of the intuitionistic and linear implications.

Hence, it should come to no surprise that now our framework will allow us to recover both these implications, plus two new ones. Indeed, by instantiating our definition of $\stackrel{\diamond}{\rightarrow}$ with the update operators of the previous section we obtain the following connectives:

Confident implication: $M \models_X \phi \stackrel{\oplus}{\rightarrow} \psi$ if and only if for all Y such that $M \models_Y \phi$ it holds that $M \models_{X \cap Y} \psi$;

Credulous implication: $M \models_X \phi \stackrel{\otimes}{\rightarrow} \psi$ if and only if for all Y such that $M \models_Y \phi$ it holds that $M \models_{X \cup Y} \psi$;

Skeptical implication: $M \models_X \phi \stackrel{\ominus}{\rightarrow} \psi$ if and only if for all $Y \subseteq X$ such that $M \models_Y \psi$ it holds that $M \models_Y \theta$;

Openminded implication: $M \models_X \phi \stackrel{\odot}{\rightarrow} \psi$ if and only if for all $Y \supseteq X$ such that $M \models_Y \psi$ it holds that $M \models_Y \theta$.

Skeptical implication and credulous implication are precisely the intuitionistic and linear implications of [1]. Moreover, it is not difficult to see that whenever the antecedent satisfies the downwards closure property, the confident and the skeptical implications are equivalent (as would the credulous and the openminded ones in the case of an upwards closed antecedent). This, in particular, implies that for downwards closed logics (such as, for example, Intuitionistic Dependence Logic) these two forms of implication are interchangeable.

However, in general the skeptical and the confident implications are not equivalent. For example, consider the team

$$X = \begin{array}{c|cc} & x & y \\ \hline s_0 & 0 & 0 \\ s_1 & 1 & 1 \end{array}$$

in a model M with two elements 0 and 1. Then $M \models_X x \perp y \overset{\ominus}{\rightarrow} {=}(x)$: indeed, the only subteams of X in which x is independent of y are $\{s_0\}$ and $\{s_1\}$, and in these subteams x is clearly constant.

But $M \not\models_X x \perp y \overset{\oplus}{\rightarrow} {=}(x)$: indeed, for

$$Y = \begin{array}{c|cc} & x & y \\ \hline s_0 & 0 & 0 \\ s_1 & 1 & 1 \\ s_2 & 0 & 1 \\ s_3 & 1 & 0 \end{array}$$

we have that $M \models_Y x \perp y$, but that in $X \cap Y = X$ the value of x is not constant.

So now we have a new class of formulas which describe beliefs in terms of how they *would* change if they interacted with other beliefs; and this is, of course, of potential significance for a number of practical applications. More in general, it seems that the problem of deciding, given two formulas ϕ and ψ of our language (or of a fragment thereof) and a fixed model M, whether $\phi \models_M \psi$, is of no small relevance for the field of knowledge updating; and that the same may also be said for the problem of whether ϕ entails ψ in all models.

The interpretation just discussed also clarifies the fact, already pointed out in [1], that $=(x, y)$ is logically equivalent to $=(x) \overset{\ominus}{\rightarrow} =(y)$ (or equivalently, since constancy atoms are downwards closed, to $=(x) \overset{\oplus}{\rightarrow} =(y)$), and that more in general dependence atoms can be decomposed in terms of constancy atoms and intuitionistic implication: indeed, a belief set X satisfies $=(x) \overset{\oplus}{\rightarrow} =(y)$ if and only if the corresponding agent, by trusting the beliefs Y of some agent who believes he knows the value of x, will reach a new belief state $X \cap Y$ in which he believes to know the value of y too. This corresponds precisely to the doxastic interpretation of $=(x, y)$.

6 Minimal updates

Let $\diamond$ be an update operator satisfying the *idempotence*, *associativity* and *monotonicity* conditions described in Section 4. Then $\diamond$ defines a *partial order* over belief sets as follows:

$$X \leq^{\diamond} Y \Leftrightarrow \exists X' \text{ s.t. } X \diamond X' = Y.$$

Indeed, by idempotence we have that $X \leq^{\diamond} X$; by the associativity of the operator, we have that the $\leq^{\diamond}$ relation is transitive; and by the monotonicity of the operator, we have that if $X \leq^{\diamond} Y$ and $Y \leq^{\diamond} X$ then $X = Y$.

The interpretation of the $\leq^{\diamond}$ operator in our framework is the following: $X \leq^{\diamond} Y$ if and only if an agent, whose belief set is X, *may* reach the belief state Y through a sequence of $\diamond$-updates.

Different update operators, of course, may generate the same partial order. In particular, for the operators that we considered we have that

$$X \leq^{\oplus} Y \Leftrightarrow X \leq^{\ominus} Y \Leftrightarrow Y \subseteq X$$

and

$$X \leq^{\otimes} Y \Leftrightarrow X \leq^{\odot} Y \Leftrightarrow X \subseteq Y.$$

In other words, a belief state Y can be reached from a state X through a confident or a skeptical update if and only if Y represents a *stricter* belief than X does, and it can be reached through a credulous or an openminded statement if and only if it represents a *looser* belief than X does.

By the way, this allows us to give an alternative definition of the skeptical and openminded updates in terms of the confident and credulous ones as follows:
Skeptical update, v2: $X \ominus Y \mapsto Z$ if and only if $X \leq^{\oplus} Y$ and $X \oplus Y \mapsto Z$;
Openminded update, v2: $X \odot Y \mapsto Z$ if and only if $X \leq^{\otimes} Y$ and $X \otimes Y \mapsto Z$.

This seems to be an instance of a more general phenomenon: given an update operation $\diamond$ satisfying our three conditions, we can always generate a new operation $\diamond'$ as

$$X \diamond' Y \mapsto Z \Leftrightarrow X \leq^{\diamond} Y \text{ and } X \diamond Y = Z.$$

These new update operations $\diamond'$, in other words, are defined precisely as the older operations $\diamond$, except that now our agent – who believes that X – is willing to perform an update with Y if and only if Y itself is a belief state that he *could* possibly reach through a $\diamond$-update.

Now, let $\mathcal{V}$ be any family of belief sets, let X be a belief, and let us define $X \diamond \mathcal{V}$ as $\{Z : \exists Y \in \mathcal{V} \text{ s.t. } X \diamond Y \mapsto Z\}$. As usual, we will write $X \diamond \mathcal{V} \mapsto Z$ for $Z \in (X \diamond \mathcal{V})$: in other words, with $X \diamond \mathcal{V} \mapsto Z$ we mean that Z is a possible outcome of updating X with some $Y \in \mathcal{V}$.

Suppose now that our agent can choose which $Y \in \mathcal{V}$ to pick to update his beliefs, and also select the resulting Z if more than one exists: which strategy could he use?

A reasonable choice might be that our agent will attempt to make a $\diamond$-*minimal* update, that is, one that does not commit him any more than necessary: in particular, if he can reach both Z_1 and Z_2, and he could reach Z_2 from Z_1 through *another* $\diamond$-update, then he should pick Z_1 over Z_2. This can be defined formally as the notion of *minimal update*:

$X \,\square\, \mathcal{V} \mapsto Z$ if and only if there is a Y such that $X \diamond Y \mapsto Z$ and Z is $\leq^{\diamond}$-minimal in $X \diamond \mathcal{V}$.

Here, stating that Z is $\leq^{\diamond}$-minimal in $X \diamond \mathcal{V}$ means simply that there exists no $Z' \in X \diamond \mathcal{V}$ with $Z' <^{\diamond} Z$.

Substituting $\diamond$ with the four update operators considered so far, we get the following updates:

Minimal confident update: $X \boxplus \mathcal{V} \mapsto Z$ if and only if there exists a $Y \in \mathcal{V}$ such that $X \cap Y = Z$, and if for all $Z' \supsetneq Z$ and all $Y' \in \mathcal{V}$ it holds that $X \cap Y' \neq Z'$;

Minimal credulous update $X \boxtimes \mathcal{V} \mapsto Z$ if and only if there exists a $Y \in \mathcal{V}$ such that $X \cup Y = Z$, and for all $Z' \subsetneq Z$ and all $Y' \in \mathcal{V}$ it holds that $X \cup Y' \neq Z'$;

Minimal skeptical update: $X \boxminus \mathcal{V} \mapsto Z$ if and only $Z \subseteq X$, $Z \in \mathcal{V}$ and for all Z' with $Z \subsetneq Z' \subseteq X$ it holds that $Z' \notin \mathcal{V}$;

Minimal openminded update: $X \boxdot \mathcal{V} \mapsto Z$ if and only if $X \subseteq Z$, $Z \in \mathcal{V}$, and for all Z' with $X \subseteq Z' \subsetneq Z$ it holds that $Z' \notin \mathcal{V}$.

As before, we can at this point define connectives $\phi \,\square\, \psi$ for describing that a team X is a possible result of a minimal update of this kind, and connectives $\phi \overset{\square}{\longrightarrow} \psi$ for describing that whenever we perform a minimal update between X and a the family of teams satisfying ϕ, the result will satisfy ψ.

The formal definitions would then be

DI-$\square$: $M \models_X \phi \,\square\, \psi$ if and only if there exists a Y such that $M \models_Y \phi$ and $Y \,\square\, \mathrm{Bel}_M(\psi) \mapsto X$;

DI-$\overset{\square}{\longrightarrow}$: $M \models_X \phi \overset{\square}{\longrightarrow} \psi$ if and only if whenever $X \,\square\, \mathrm{Bel}_M(\phi) \mapsto Z$ it holds that $M \models_Z \psi$.

Here, as usual, $\mathrm{Bel}_M(\psi)$ represents the family of all teams which satisfy ψ in M.

Here we will not give the instantiations of these connectives for the four updates described above, nor will we discuss their properties.

All that we will point out is that the minimal skeptical implication connective $\phi \overset{\boxminus}{\longrightarrow} \psi$ is precisely the *maximal implication* $\phi \hookrightarrow \psi$ mentioned in [16] and defined as

DI-maximp: $M \models_X \phi \hookrightarrow \psi$ if and only if for all $Y \subseteq X$ such that $M \models_Y \phi$ and $M \not\models_Z \phi$ for all Z with $Y \subsetneq Z \subseteq X$, $M \models_Y \psi$;

Thus, even this connective can be interpreted in this framework. This notion of minimal update appears to be rather natural, and it probably deserves further study; however, for the moment we will content ourselves with having defined it and shown how to recover the $\hookrightarrow$ implication through it.

As an aside, this implication allows us to decompose independence atoms: for example, it is not difficult to see that $y \perp_x z$ is equivalent to $=(x) \hookrightarrow y \perp z$.

7 Quantifiers

So far, our operators have treated assignments as if they were point-like possible worlds. This is not the case, of course: for example, our agent may be confident about the values of certain variables, but not about the ones of others. Furthermore, we have no way so far of adding or removing variables to the domains of our teams. In this section, we will attempt to remedy this.

Let us begin with a *forgetting* operator ρx, where x is a variable, with has the effect of removing the variable x from the domain of our team: more precisely, for all belief states X we define $\rho x(X)$ as $X_{\setminus x}$, that is, as the team containing the restrictions of all assignments in X to $\mathrm{Dom}(X)\backslash\{x\}$. In the case that x is not in the domain of X to begin with, this operator has no effect.

The doxastic meaning of this operator is the one suggested by its value: after performing the update (ρx), our agents forgets everything about the value of the variable x, and even the fact that this variable exists to begin with! This is not the same as our agent simply professing ignorance about the value of x – this would be another operator, that we will examine later – as here we are *really* erasing the variable from our domain.

As in the case of the binary update operators considered in the previous sections, this forgetting operator corresponds to *two* distinct connectives, which can be formally defined as

DI-forgotten: $M \models_X (\rho x)\phi$ if and only if there exists a team X' such that $X = (\rho x)X' = X'_{\setminus x}$ and $M \models_{X'} \phi$.

DI-forgetting: $M \models_X (\eta x)\phi$ if and only if $M \models_{X'} \phi$, where $X' = (\rho x)X = X_{\setminus x}$;

In other words, $(\eta x)\phi$ holds in a team X if, starting from X and forgetting the values of the variable x, we obtain a belief state which satisfies ϕ, and $(\rho x)\phi$ holds in a team X if this team can be obtained by starting from a team X' which satisfies ϕ and forgetting the value of x. As always, these two operators are adjoints, that is,

$$(\rho x)\phi \models \psi \Leftrightarrow \phi \models (\eta x)\psi.$$

A combination of these two operators which is of particular importance is the *disbelieving operator* $Dx\phi = (\eta x)(\rho x)\phi$. The intuition here is that $M \models_X Dx\phi$ if, apart from the value of the variable x, the team X could correspond to a belief set which satisfies ϕ. This is evident by the corresponding semantic rule: indeed, by combining the rules for the forgetting and remembering operators, one can see that $M \models_X Dx\phi$ if and only if there exists a X' such that $X_{\backslash x} = X'_{\backslash x}$ and such that $M \models_{X'} \phi$.

The condition $X_{\backslash x} = X'_{\backslash x}$ is easily seen to be equivalent to the existence of a function $H : X \to \mathcal{P}(\text{Dom}(X))\backslash\{\emptyset\}$ such that $X' = X[H/x] = \{s[m/x] : s \in X, m \in H(s)\}$. Therefore, the $Dx\phi$ operator corresponds precisely to the existential quantifier rule.[9]

Just like all connectives, $Dx\phi$ can also be built directly from some belief update operator. This is our first true case of a *non-deterministic* belief update: more precisely, we can define it as

$$(Dx)X \mapsto Y \text{ if and only if } X_{\backslash x} = Y_{\backslash x}.$$

The significance of this operator in our framework should be easy to see: in brief, we have that $(Dx)X \mapsto Y$ if and only if an agent who starts from the belief X and, disbelieving his previous opinions about the possible values of the variable x, changes them *and nothing else*, can possibly reach the belief state represented by Y.

The quantifier $(Dx)\phi$ is then the unary equivalent of the $\phi \diamond \psi$ connectives considered in the previous sections: in brief, $M \models_X (Dx)\phi$ if and only if there exists a team Y with $M \models_Y \phi$ and $(Dx)Y \mapsto X$.

Of course, we also get another quantifier $(Rx)\phi$, which is satisfied by a belief state X if and only if for all Y such that $(Dx)X \mapsto Y$ we have that $M \models_Y \phi$: in other words, $M \models_X (Rx)\phi$ corresponds that our agent, whose belief state is X, is confident that ϕ would hold even if his beliefs about the value of x were wrong. Hence, we may perhaps call it the *regardless* quantifier. Thus, we have obtained a new pair of unary connectives:

DI-disbelief: $M \models_X (Dx)\phi$ if and only if there exists a Y such that $Y_{\backslash x} = X_{\backslash x}$ and $M \models_Y \phi$;

DI-regardless: $M \models_X (Rx)\phi$ if and only if for all teams Y with $Y_{\backslash x} = X_{\backslash x}$ it holds that $M \models_Y \phi$.

9 To be more precise, it corresponds to the *lax* variant of the existential quantifier discussed in [8], in which more than one possible value for x may be picked for each assignment.

The "regardless" operator is a lax version of the universal quantifier $\sim \exists x \sim \ldots$ of Team Semantics; and, as always, these two operators are adjoints, that is,

$$(Dx)\phi \models \psi \Leftrightarrow \phi \models (Rx)\psi.$$

Finally, let us consider the following scenario: our agent's belief is represented by the team X, which satisfies some property ϕ, but now the agent decides that he does not trust at all his own opinion about the value of some variable x. Then the new belief state is given by

$$¡x\phi = X[M/x] = \{s[m/x] : s \in X, m \in \mathrm{Dom}(M)\}$$

that is, the new belief state of our agent is the same as the old one, except that now our agent knows nothing at all about x. This, once again, represents a situation in which our agent doubts the validity of his beliefs about x; but where the disbelief operator Dx corresponds to the agent revising these beliefs in some arbitrary, nondeterministic way, this new operator $¡x$ has the agent taking an agnostic position about the possible values of x and moving to a belief state in which he knows nothing about it.

As in all previous cases, this allows us to develop two new connectives, that we will call the *doubted* and the *doubting* quantifiers:
DI-doubted: $M \models_X (¡x)\phi$ if and only if $X = Y[M/x]$ for some Y such that $M \models_Y \phi$.
DI-doubting: $M \models_X (!x)\phi$ if and only if $M \models_{X[M/x]} \phi$;

The $!x$ connective is exactly the $!x$ operator in Team Logic, or the one written as $\forall x$ in Dependence Logic or in many other logics of imperfect information. The other one is, to the knowledge of the author, new, but its interpretation is clear: $M \models_X (¡x)\phi$ if the belief X can be seen as the result of taking a belief state Y which satisfies ϕ, and doubting its guess about it. As always, we have that

$$¡x\phi \models \psi \Leftrightarrow \phi \models !x\psi.$$

8 Conclusion

As we saw, virtually all of the connectives and operators of Dependence Logic and its extensions admit natural interpretations in terms of descriptions of beliefs and/or belief update operations. This suggests that we can extend our formalism along at least two different lines, by adding more *dependency notions* for the specification of more properties of beliefs or by adding further forms of belief update. It will be interesting to examine how these two aspects will interact.

We mention some possible avenues for further research:

1. Modal Logics are, *de facto*, the standard logical formalisms for reasoning about beliefs; and in particular, *dynamic modal logics* [2, 21, 27] add to the basic modal framework the capability for reasoning about belief updates. It is high time for a more formal examination of the relationship between these approaches and Team Semantics, perhaps along the lines of a study of the doxastic content of Modal Dependence Logic [26] and its extensions. The results promise to be greatly rewarding for both the Dependence Logic and the Dynamic Modal Logic research communities.
2. Likewise, Belief Revision concepts and methodologies can (and, in the opinion of the author, *should*) be adapted to Team Semantics. We only dealt with very simple notions of belief update so far; but there is in principle no reason why the more complex ones which have been studied within the Belief Revision research area could not be introduced in Team Semantics. Our notion of *update minimality*, in particular, appears to have some connections with the *distance semantics for belief revision* of [18].
3. If teams are to represent belief states as set of possible assignments, and if team semantics can be thought of as a formalism for reasoning about these states and their evolution, it follows that it should be possible to develop analogous logical formalisms for finer-grained forms of belief representation. Just to mention one possibility, it may be very worthwhile to consider *probabilistic distributions* over assignments, that is, *probabilistic teams* in the sense of [7]. This might bring about interesting connections to the *Equilibrium Semantics* of [22, 9], but from a radically different perspective: rather than searching for the values of the strategic equilibria of the semantic games corresponding to sentences, as is the case for Equilibrium Semantics, we would associate formulas to conditions over probability distributions over assignments.

As commented by a reviewer, many of the ideas and results of the work of this paper does not seem to depend intrinsically from the definition of teams as sets of assignments. A – potentially very fruitful – avenue of further research may consist in abstracting from this definition and study belief representation through *algebraizations* of Team Semantics, based for example on Mann's work on IF Logic [19]: this would increase further the level of generality and abstraction of our framework.

We leave a more in-depth analysis of these ideas to future works. Here we limited ourselves to a discussion of the doxastic interpretation of the *current* state of the art in Team Semantics research; and it is the hope of the author that this presentation highlighted how this interpretation holds much promise for the further development of this fascinating family of logics.

Acknowledgement: We thank Samson Abramsky, Alexandru Baltag. Johan van Benthem and Jouko Väänänen for many insightful comments. We also thank an anonymous reviewer for their useful suggestions and corrections.

This work was supported by the European Science Foundation Eurocores programme LogICCC [FP002 - Logic for Interaction (LINT)], by the Väisäla Foundation and by grant 264917 of the Academy of Finland.

Bibliography

[1] Samson Abramsky and Jouko Väänänen. From IF to BI. *Synthese*, 167:207–230, 2009. 10.1007/s11229-008-9415-6.

[2] Alexandru Baltag, Lawrence S. Moss, and Slawomir Solecki. The logic of public announcements, common knowledge, and private suspicions. In *Proceedings of the 7th conference on Theoretical aspects of rationality and knowledge*, TARK '98, pages 43–56, San Francisco, CA, USA, 1998. Morgan Kaufmann Publishers Inc.

[3] Arthur Dempster. Upper and lower probabilities induced by a multivalued mapping. In Roland Yager and Liping Liu, editors, *Classic Works of the Dempster-Shafer Theory of Belief Functions*, volume 219 of *Studies in Fuzziness and Soft Computing*, pages 57–72. Springer Berlin / Heidelberg, 2008.

[4] Arnaud Durand and Juha Kontinen. Hierarchies in dependence logic. *CoRR*, abs/1105.3324, 2011.

[5] F. Engström and J. Kontinen. Characterizing quantifier extensions of dependence logic. arXiv:1202.5247, 2012.

[6] Fredrik Engström. Generalized quantifiers in dependence logic. *Journal of Logic, Language and Information*, 21(3):299–324, 2012.

[7] Pietro Galliani. Game Values and Equilibria for Undetermined Sentences of Dependence Logic. MSc Thesis. ILLC Publications, MoL–2008–08, 2008.

[8] Pietro Galliani. Inclusion and exclusion dependencies in team semantics: On some logics of imperfect information. *Annals of Pure and Applied Logic*, 163(1):68 – 84, 2012.

[9] Pietro Galliani and Allen L. Mann. Lottery semantics. In Juha Kontinen and Jouko Väänänen, editors, *Proceedings of Dependence and Independence in Logic*, pages 118–132. ESSLLI 2010, 2010.

[10] Peter Gardenfors, editor. *Belief Revision*. Cambridge University Press, New York, NY, USA, 1992.

[11] Erich Grädel and Jouko Väänänen. Dependence and independence. *Studia Logica*, 101(2):399–410, 2013.

[12] Leon Henkin. Some Remarks on Infinitely Long Formulas. In *Infinitistic Methods. Proc. Symposium on Foundations of Mathematics*, pages 167–183. Pergamon Press, 1961.

[13] Jaakko Hintikka. *The Principles of Mathematics Revisited*. Cambridge University Press, 1996.

[14] Jaakko Hintikka and Gabriel Sandu. Informational independence as a semantic phenomenon. In J.E Fenstad, I.T Frolov, and R. Hilpinen, editors, *Logic, methodology and philosophy of science*, pages 571–589. Elsevier, 1989.

[15] Wilfrid Hodges. Compositional Semantics for a Language of Imperfect Information. *Journal of the Interest Group in Pure and Applied Logics*, 5 (4):539–563, 1997.

[16] Juha Kontinen and Ville Nurmi. Team logic and second-order logic. In Hiroakira Ono, Makoto Kanazawa, and Ruy de Queiroz, editors, *Logic, Language, Information and Computation*, volume 5514 of *Lecture Notes in Computer Science*, pages 230–241. Springer Berlin / Heidelberg, 2009.

[17] Juha Kontinen and Jouko Väänänen. On definability in dependence logic. *Journal of Logic, Language and Information*, 3(18):317–332, 2009.

[18] Daniel Lehmann, Menachem Magidor, and Karl Schlechta. Distance semantics for belief revision. *The Journal of Symbolic Logic*, 66(1):pp. 295–317, 2001.

[19] Allen L Mann. Independence-friendly cylindric set algebras. *Logic Journal of IGPL*, 17(6):719–754, 2009.

[20] Allen L. Mann, Gabriel Sandu, and Merlijn Sevenster. *Independence-Friendly Logic: A Game-Theoretic Approach*. Cambridge University Press, 2011.

[21] Jan Plaza. Logics of public communications. *Synthese*, 158:165–179, 2007. 10.1007/s11229-007-9168-7.

[22] Merlijn Sevenster and Gabriel Sandu. Equilibrium semantics of languages of imperfect information. *Annals of Pure and Applied Logic*, 161(5):618–631, 2010. The Third workshop on Games for Logic and Programming Languages (GaLoP), Galop 2008.

[23] Glenn Shafer. *A Mathematical Theory of Evidence*. Princeton University Press, 1976.

[24] Jouko Väänänen. *Dependence Logic*. Cambridge University Press, 2007.

[25] Jouko Väänänen. Team Logic. In J. van Benthem, D. Gabbay, and B. Löwe, editors, *Interactive Logic. Selected Papers from the 7th Augustus de Morgan Workshop*, pages 281–302. Amsterdam University Press, 2007.

[26] Jouko Väänänen. Modal Dependence Logic. In Krzysztof R. Apt and Robert van Rooij, editors, *New Perspectives on Games and Interaction*. Amsterdam University Press, Amsterdam, 2008.

[27] Johan Van Benthem. Dynamic logic for belief revision. *Journal of Applied NonClassical Logics*, 17(2):129–155, 2007.

[28] Lotfi A. Zadeh. Fuzzy sets as a basis for a theory of possibility. *Fuzzy Sets and Systems*, 1(1):3 – 28, 1978.

Lauri Hella* and Jouko Väänänen

The Size of a Formula as a Measure of Complexity

***Corresponding Author: Lauri Hella:** School of Information Sciences, Mathematics, University of Tampere
Jouko Väänänen: Department of Mathematics and Statistics, University of Helsinki and Institute for Logic, Language and Computation, University of Amsterdam

1 Introduction

We propose a refinement of the usual Ehrenfeucht-Fraïssé game. The new game will help us make finer distinctions than the traditional one. In particular, it can be used to measure not only quantifier rank but also lengths of conjunctions and disjunctions needed for expressing a given property. Our game is similar to the game in [1] and in [5].

The most common measure of complexity of a first order sentence is its quantifier rank, and the method of Ehrenfeucht-Fraïssé games can be successfully used to prove lower bound results for this measure of complexity. However, the number of non-equivalent first order formulas of quantifier rank n is an exponential tower of height n with a polynomial depending on the vocabulary on the top. Thus we are not very close to knowing the formula if we merely know its quantifier rank. In this paper we measure complexity in terms of the number of symbols in the formula. The advantage of our measure is that there are only $2^{P(n)}$ non-equivalent formulas with n symbols, where $P(x)$ is a polynomial depending on the vocabulary.

Let us consider the question of the complexity of deciding whether a binary string

$$s_1 s_2 \dots s_n = 10110010...01 \tag{1}$$

has a certain pattern. One approach to this is circuit complexity. For example, it is proved in [3] that constant depth circuits cannot decide the *parity* of (1), i.e. the question whether (1) has an even number of ones. Another approach is computational complexity, where the parity of (1) is easily decided in linear time. In so called descriptive complexity the question is posed, what is the logical complexity of the simplest formula in a given logic that expresses the property of (1) in question, for example parity. Concerning this type of problems, it was proved in

[2] that the question whether a set R of $n+1$-sequences

$$(a_1, \ldots, a_{n+1}) \tag{2}$$

from a finite set $\{1, \ldots, m\}$ has an even cardinality cannot be expressed in existential second order logic, where the second order variables range over $\leq n$-ary relations, that is, there is no sentence ϕ of such existential second order logic with the property that the structure $(\{1, \ldots, m\}, R)$ satisfies ϕ if and only if R has even cardinality.

Let us return to (1). A natural setup of using logic to study the question of complexity of finding patterns in (1) is to use the propositional logic with propositional symbols

$$p_1, p_2, \ldots, p_n \tag{3}$$

with the meaning

$$p_i \text{ is true if and only if } s_i = 1. \tag{4}$$

Propositional logic gives rise to a variety of complexity measures. We focus here on the simplest of them, namely the number of occurrences of propositional symbols in a formula. Up to a constant factor this is the same as the number of symbols, that is, the length of the formula.

The parity of (1) is naturally expressed by the propositional formula

$$\bigvee \{\bigwedge_{i=1}^{n} p_i^A : A \subseteq \{1, \ldots, n\}, |A| \text{ even}\}, \tag{5}$$

where

$$p_i^A = \begin{cases} p_i & \text{if } i \in A \\ \neg p_i & \text{if } i \notin A, \end{cases} \tag{6}$$

which gives the upper bound $\frac{1}{2}n2^n$ to the size of the smallest formula expressing parity. However, there is a shorter formula ψ defined as follows: $\psi = \phi_{1,n}$, where for $1 \leq i < j \leq n$ and $k = \lfloor \frac{i+j}{2} \rfloor$

$$\phi_{i,j} \quad = \quad (\phi_{i,k} \wedge \phi_{k+1,j}) \vee (\neg\phi_{i,k} \wedge \neg\phi_{k+1,j})$$

and

$$\phi_{i,i} = \neg p_i.$$

This gives (up to a constant factor) the upper bound n^2 for the shortest propositional formula expressing parity[1]. Krapchenko (see [6, p 258]) has proved in 1971

1 More precisely, a straightforward induction shows that the size of ψ is at most $(n+1)^2$, and if n is a power of 2, then the size of ψ is exactly n^2.

that this is optimal. We prove below the same result using our refinement of the Ehrenfeucht-Fraïssé game.

For a slightly different complexity question, suppose we have n unary predicates:

$$P_1(x), \ldots, P_n(x). \tag{7}$$

We ask what is the length of the shortest formula in first order logic with the unary predicates (7) that expresses the property that every Boolean combination of the predicates is non-empty. This property can be written as the existential sentence:

$$\bigwedge\{\exists x \bigwedge_{i=1}^{n} P_i^A(x) : A \subseteq \{1, \ldots, n\}\}, \tag{8}$$

where, as above in (6), $P_i^A(x) = P_i(x)$ if $i \in A$, and $P_i^A(x) = \neg P_i(x)$ if $i \notin A$.

This gives the upper bound $(n+1)2^n$ to the size of the smallest sentence expressing the given property (we define the size of a formula to be the total number of atomic subformulas and quantifiers occurring in it). In Section 5, we show that this is optimal for existential sentences. If we give up existential sentences and allow a quantifier alternation, we have a smaller sentence ϕ defined as follows:

$$\forall x \exists y \bigwedge_{i=1}^{n} (P_i(x) \leftrightarrow P_{i+1}(y)) \wedge$$
$$\forall x \exists y (\bigwedge_{i=2}^{n} (P_i(x) \leftrightarrow P_i(y)) \wedge (P_1(x) \leftrightarrow \neg P_1(y))),$$

where addition $i+1$ is to be calculated modulo n. Since we use $\neg$, $\wedge$ and $\vee$ as basic connectives, and consider $\leftrightarrow$ only as an abbreviation, the size of ϕ is $8n+4$. Thus, ϕ is really optimal (up to a constant factor) because we cannot have such a sentence without mentioning each P_i at least once.

In our final application, we consider the minimal size of a first-order sentence expressing that the length of a linear order is at least n. It is well known that this can be expressed by a sentence with quantifier rank $\lceil \log n \rceil + 1$. Such a sentence ϕ_n can be obtained as follows: define first recursively formulas $\theta_k(x, y)$ by letting $\theta_2(x, y) := x < y$ and $\theta_k(x, y) := \exists z(\theta_l(x, z) \wedge \theta_m(z, y))$, where $l = \lfloor k/2 \rfloor + 1$ and $m = \lceil k/2 \rceil$, for $k > 2$. Thus, $\theta_k(x, y)$ says that there are at least k elements in the interval $[x, y]$. Then ϕ_n can be defined as $\exists x \exists y \theta_n(x, y)$.

Clearly the size of ϕ_n is linear in n. However, with a clever trick of recycling quantified variables it is possible to define a variant ϕ_n' of ϕ_n that has size $c\lceil \log n \rceil$, where c is a small constant. Moreover, ϕ_n' can be chosen to be in the 4-variable logic FO_4 (see [4], Section 5). On the other hand, Grohe and Schweikardt prove in [4] that the minimal size of an equivalent sentence in FO_3 is at least $\frac{1}{2}\sqrt{n}$. Their proof is based on a method which is related to the game we study in this paper, but it is not formulated in a game theoretic form.

2 A Game for Propositional Logic

We introduce now a game for measuring how long a propositional formula has to be to express a given property of binary strings

$$s = s_1 \dots s_n, \tag{1}$$

where $s_i \in \{0, 1\}$. We denote strings by s, r, t etc. Let n be the fixed length of the strings considered. We shall discuss nonempty sets S of such strings and call them *string properties*. For example, S could be the set of strings (1) where the cardinality of the set $\{i \in \{1, \dots, n\} : s_i = 1\}$ is even. Another example is the set of strings (1) with more ones than zeros. In an extreme case S can be a singleton, e.g. the singleton constant one string $\{111 \dots 1\}$.

Propositional formulas are built up from propositional symbols (3) by means of $\neg$, $\wedge$ and $\vee$. We use Greek letters ϕ, ψ etc to denote propositional formulas. The concept $s \models \phi$ of a string s satisfying a propositional formula ϕ is defined in the usual way: $s \models p_i$ if $s_i = 1$; $s \models \neg\phi$ if $s \not\models \phi$; $s \models \phi \wedge \psi$ if $s \models \phi$ and $s \models \psi$; $s \models \phi \vee \psi$ if $s \models \phi$ or $s \models \psi$.

Definition 2.1. Suppose ϕ is a propositional formula and S a string property. We say that ϕ *defines* S if for all strings s

$$s \in S \iff s \models \phi.$$

We define the *size* $\mathrm{w}(\phi)$ of the propositional formula ϕ as follows:

$$\begin{aligned} \mathrm{w}(p_i) &= 1 \\ \mathrm{w}(\neg\phi) &= \mathrm{w}(\phi) \\ \mathrm{w}(\phi\vee\psi) &= \mathrm{w}(\phi) + \mathrm{w}(\psi) \\ \mathrm{w}(\phi\wedge\psi) &= \mathrm{w}(\phi) + \mathrm{w}(\psi). \end{aligned}$$

Note that $\mathrm{w}(\phi)$ is just the number of occurrences of propositional symbols in ϕ. If ϕ is in *negation normal form*[2], then, up to a constant factor, this is the same as the number of symbols in ϕ.

2 A formula is in negation normal form, if all negations occurring in it are in front of proposition symbols.

Note that while there are, up to logical equivalence, 2^{2^n} propositional formulas over the propositional symbols $p_1, \ldots, p_n$ altogether, there are for each m, up to logical equivalence, only at most $2^m \cdot (n+2)^{2m}$ propositional formulas ϕ with $\mathrm{w}(\phi) \leq m$.

A natural question now is:

Question: Given a string property S, what is the size of the smallest propositional formula that defines S.

We shall define a game for the study of this question. In defining the game we follow mostly [1] and to a lesser degree [5]. This game resembles the usual Ehrenfeucht-Fraïssé game for the first order logic, but also the Ehrenfeucht-Fraïssé game for the independence friendly logic, presented in [5].

In the usual Ehrenfeucht-Fraïssé game we have two structures $\mathfrak{A}$ and $\mathfrak{B}$ and the game is about their similarity. Player II maintains that the structures are very similar. Player I maintains that there is a difference. During the game player I picks elements from either one of the models and player II tries to find similar elements in the other model.

In the new game we do not have just two models but instead two classes of models. Player II maintains that whatever features the models in one class have all in common, some member of the other class also has. The game to be defined is able to detect very minute differences in models and is in that sense finer than the usual Ehrenfeucht game. In particular, this game can detect differences in the size of conjunctions and disjunctions needed for separating the classes of models.

Definition 2.2. Let S and R be string properties and w a positive integer. The game $\mathrm{EF}_w(S, R)$ has two players. The number w is called the *rank* of the game. In the beginning the position is (w, S, R). Suppose the position after m moves is (w_m, S_m, R_m). There are the following two possibilities for the continuation of the game:

Left splitting move: Player I first chooses numbers u and v such that $1 \leq u, v < w$ and $u + v = w$. Then player I represents S_m as a union $C \cup D$. Now the game continues from the position (u, C, R_m) or from the position (v, D, R_m), and player II can choose which.

Right splitting move: Player I first chooses numbers u and v such that $1 \leq u, v < w$ and $u + v = w$. Then Player I represents R_m as a union $C \cup D$. Now the game continues from the position (u, S_m, C) or from the position (v, S_m, D), and player II can choose which.

The game ends in a position (w_m, S_m, R_m) and player I wins if there is a propositional symbol p_i such that either $s \models p_i$ for all $s \in S_m$ and $r \not\models p_i$ for all $r \in R_m$, or

$s \not\models p_i$ for all $s \in S_m$ and $r \models p_i$ for all $r \in R_m$. Player II wins the game if they reach a position (w_m, S_m, R_m) such that $w_m = 1$ and player I does not win in this position.

This is a game of perfect information and the concept of winning strategy is defined as usual. Since the rank w_m decreases in each move, the game always ends in a finite number of moves. Hence, by the Gale-Stewart theorem, the game is determined: exactly one of the players has a winning strategy in $\mathrm{EF}_w(S, R)$.

Note that if the players have reached a position (w_m, S_m, R_m) in the game $\mathrm{EF}_w(S, R)$, then the continuation of the game from that position onwards can be seen as a play of the game $\mathrm{EF}_{w_m}(S_m, R_m)$. Thus, player I (player II) has a winning strategy in the game $\mathrm{EF}_w(S, R)$ *from position* (w_m, S_m, R_m) onwards if an only if player I (player II, respectively) has a winning strategy in the game $\mathrm{EF}_{w_m}(S_m, R_m)$.

We say that a propositional formula ϕ *separates* string properties S and R, in symbols $(S, R) \models \phi$, if $s \models \phi$ for all $s \in S$ and $r \not\models \phi$ for all $r \in R$.

Theorem 2.3. *Suppose* (S, R) *is a pair of string properties, and let* w *be a positive integer. Then the following conditions are equivalent:*

$(1)_w$ *Player I has a winning strategy in the game* $\mathrm{EF}_w(S, R)$.

$(2)_w$ *There is a propositional formula* ϕ *of size* $\leq w$ *such that* $(S, R) \models \phi$.

Proof. We prove the equivalence of $(1)_w$ and $(2)_w$ by induction on w. Consider first the case $w = 1$. By Definition 2.2, there are no moves in the game $\mathrm{EF}_1(S, R)$, and player I wins if and only if there is a proposition symbol p_i such that either $(S, R) \models p_i$, or $(S, R) \models \neg p_i$. Since $\mathrm{w}(p_i) = \mathrm{w}(\neg p_i) = 1$, we have $(1)_1 \implies (2)_1$. On the other hand, if $\mathrm{w}(\phi) \leq 1$, then ϕ is, up to logical equivalence, either a proposition symbol, or the negation of a proposition symbol. Thus, $(2)_1$ implies $(1)_1$.

Let us then consider the case $w > 1$, and assume $(1)_v \iff (2)_v$ for all $v < w$ as an induction hypothesis. To prove $(1)_w \iff (2)_w$, assume first that player I has a winning strategy in the game $\mathrm{EF}_w(S, R)$. We have the following two cases according to the type of the first move in the winning strategy of player I.

Case 1: Player I makes a left splitting move by choosing u, v, C and D such that $1 \leq u, v < w$, $w = u+v$ and $S = C \cup D$. Since this move is given by a winning strategy, player I has a winning strategy in both of the games $\mathrm{EF}_u(C, R)$ and $\mathrm{EF}_v(D, R)$. By induction hypothesis, there are formulas ψ and θ such that $\mathrm{w}(\psi) \leq u$, $\mathrm{w}(\theta) \leq v$, $(C, R) \models \psi$ and $(D, R) \models \theta$. Thus, $s \models \psi$ for all $s \in C$ and $s \models \theta$ for all $s \in D$, whence $s \models \psi \vee \theta$ for all $s \in S$. On the other hand, for all $r \in R$, $r \not\models \psi$ and $r \not\models \theta$, whence consequently $r \not\models \psi \vee \theta$. It follows that $(S, R) \models \psi \vee \theta$. As $\mathrm{w}(\psi \vee \theta) = \mathrm{w}(\psi) + \mathrm{w}(\theta) \leq u + v = w$, $(2)_w$ is true.

Case 2: Player I makes a right splitting move by choosing u, v, C and D such that $1 \leq u, v < w$, $w = u+v$ and $R = C \cup D$. Since this move is given by a winning strategy, player I has a winning strategy in both of the games $\mathrm{EF}_u(S, C)$ and $\mathrm{EF}_v(S, D)$. By induction hypothesis there are formulas ψ and θ such that $\mathrm{w}(\psi) \leq u$, $\mathrm{w}(\theta) \leq v$, $(S, C) \models \psi$ and $(S, D) \models \theta$. Thus, for all $s \in S$, $s \models \psi$ and $s \models \theta$, whence $s \models \psi \wedge \theta$. On the other hand, $r \not\models \psi$ for all $r \in C$ and $r \not\models \theta$ for all $r \in D$, whence $r \not\models \psi \wedge \theta$ for all $r \in R$. It follows that $(S, R) \models \psi \wedge \theta$. As $\mathrm{w}(\psi \wedge \theta) = \mathrm{w}(\psi) + \mathrm{w}(\theta) \leq u + v = w$, $(2)_w$ is true.

To prove the converse implication $(2)_w \implies (1)_w$, assume that there is a formula ϕ of size $\leq w$ such that $(S, R) \models \phi$. We show that then player I has a winning strategy in the game $\mathrm{EF}_w(S, R)$. We assume without loss of generality that ϕ is in negation normal form.[3] The strategy depends on ϕ as follows.

Case 1: ϕ is a proposition symbol or the negation of a proposition symbol. Then by the definition of the game EF_w, player I wins without making any moves.

Case 2: ϕ is $\psi \vee \theta$. Let C be the set of strings $s \in S$ such that $s \models \psi$, and let D be the set of strings $s \in S$ such that $s \models \theta$. Since $s \models \phi$ for every $s \in S$, we have $S = C \cup D$. Moreover, since $r \not\models \phi$ for every $r \in R$, we have $(C, R) \models \psi$ and $(D, R) \models \theta$. Finally, as $\mathrm{w}(\phi) \leq w$, there are u and v such that $w = u + v$, $\mathrm{w}(\psi) \leq u$ and $\mathrm{w}(\theta) \leq v$. Note that $u, v < w$, and so, by induction hypothesis, player I has a winning strategy in the games $\mathrm{EF}_u(C, R)$ and $\mathrm{EF}_v(D, R)$. Thus, using $w = u + v$, $S = C \cup D$ as his first move and these winning strategies in the rest of the game, player I is guaranteed to win the game $\mathrm{EF}_w(S, R)$.

Case 3: ϕ is $\psi \wedge \theta$. Let C be the set of strings $r \in R$ such that $r \not\models \psi$, and let D be the set of strings $r \in R$ such that $s \not\models \theta$. Since $r \not\models \phi$ for every $r \in R$, we have $R = C \cup D$. Moreover, since $s \models \phi$ for every $s \in S$, we have $(S, C) \models \psi$ and $(S, D) \models \theta$. Finally, as $\mathrm{w}(\phi) \leq w$, there are u and v such that $w = u + v$, $\mathrm{w}(\psi) \leq u$ and $\mathrm{w}(\theta) \leq v$. Note that $u, v < w$, and so, by induction hypothesis, player I has a winning strategy in the games $\mathrm{EF}_u(S, C)$ and $\mathrm{EF}_v(S, D)$. Thus, using $w = u + v$, $R = C \cup D$ in his first move, and these winning strategies in the rest of the game, player I is guaranteed to win the game $\mathrm{EF}_w(S, R)$.

Q.E.D.

Corollary 2.4. *Suppose K is a string property. Then the following conditions are equivalent:*
(1) *K is definable by a propositional formula ϕ of size $\leq w$.*
(2) *Player I has a winning strategy in the game $\mathrm{EF}_w(S, R)$ for all S and R such that $S \subseteq K$ and $R \cap K = \emptyset$.*

3 A straightforward induction shows that for any formula ϕ there is an equivalent formula ϕ' in negation normal form such that $\mathrm{w}(\phi') = \mathrm{w}(\phi)$.

Proof. Suppose K is defined by ϕ of size $\leq w$. Let S and R be given so that $S \subseteq K$ and $R \cap K = \emptyset$. Thus every model in S satisfies ϕ and no model in R does, whence $(S, R) \models \phi$. By Theorem 2.3, player I has a winning strategy in $\mathrm{EF}_w(S, R)$. For the converse, suppose player I has a winning strategy in $\mathrm{EF}_w(S, R)$ for all (S, R) with $S \subseteq K$ and $R \cap K$. This holds in particular if S is K and R is the complement of K. Thus for this choice of S and R we have $(S, R) \models \phi$ for some ϕ of size $\leq w$. Thus K is defined by ϕ. **Q.E.D.**

We get the following method for showing that a string property K is *not* definable by a propositional formula of size $\leq w$: We find classes S and R such that

1. $S \subseteq K$.
2. $R \cap K = \emptyset$.
3. Player II has a winning strategy in $\mathrm{EF}_w(S, R)$.

3 The Propositional Complexity of Parity

We shall now prove that the parity of a binary string of n bits cannot be expressed with a propositional formula smaller than n^2. The result has been proved in [6] with a method which is very similar to ours, even if it does not explicitly use games. We present the proof as an example of the use of our game.

Suppose S and R are disjoint sets of binary strings of length n. Let

$$E = \{(f, g) \in S \times R : |\{i : f_i \neq g_i\}| = 1\}.$$

The *density* of the pair (S, R) is defined to be the pair

$$D(S, R) = (s, r),$$

where

$$s = \frac{|E|}{|S|}, \text{ and } r = \frac{|E|}{|R|}.$$

Lemma 3.1. *If $s > 1$ or $r > 1$, then there is no proposition symbol p_i such that $(S, R) \models p_i$ or $(S, R) \models \neg p_i$.*

Proof. Suppose $s > 1$ and $i \in \{1, \ldots, n\}$. Then there are $f \in S$ and $g, h \in R$ such that $(f, g), (f, h) \in E$. Clearly this means that $f_i = g_i$ or $f_i = h_i$, whence $f \models p_i \iff g \models p_i$ or $f \models p_i \iff h \models p_i$. Thus, if all strings in S satisfy p_i $(\neg p_i)$, then there is a string in R satisfying p_i ($\neg p_i$, respectively). It follows that neither p_i, nor $\neg p_i$ separates the sets S and R.

In the case $r > 1$ there are $f, g \in S$ and $h \in R$ such that $(f, h), (g, h) \in E$. Then we have $h_i = f_i$ or $h_i = g_i$, whence $h \models p_i \iff f \models p_i$ or $h \models p_i \iff g \models p_i$. As above, we conclude that neither p_i, nor $\neg p_i$ separates the sets S and R. **Q.E.D.**

Lemma 3.2. *(a) Suppose* $D(S,R) = (s,r)$ *and* $S = S_0 \cup S_1$*, where* $S_0 \cap S_1 = \emptyset$*. Let* $D(S_0,R) = (s_0,r_0)$ *and* $D(S_1,R) = (s_1,r_1)$*. Then* $s_0r_0 + s_1r_1 \geq sr$*.*

(b) Suppose $D(S,R) = (s,r)$ *and* $R = R_0 \cup R_1$*, where* $R_0 \cap R_1 = \emptyset$*. Let* $D(S,R_0) = (s_0,r_0)$ *and* $D(S,R_1) = (s_1,r_1)$*. Then* $s_0r_0 + s_1r_1 \geq sr$*.*

Proof. We prove (a); the proof of (b) is similar. Clearly $r_0 + r_1 = r$ and $\frac{r_0}{s_0} + \frac{r_1}{s_1} = \frac{r}{s}$. Since the harmonic mean is never greater than the arithmetic mean[4], we have

$$\frac{s_0r_0 + s_1r_1}{r} = \frac{s_0r_0 + s_1r_1}{r_0 + r_1} \geq \frac{r_0 + r_1}{\frac{r_0}{s_0} + \frac{r_1}{s_1}} = \frac{r}{\frac{r}{s}} = s,$$

and the claim follows. **Q.E.D.**

Lemma 3.3. *Suppose* $D(S,R) = (s,r)$ *and* $w < sr$*. Then player II has a winning strategy in* $\mathrm{EF}_w(S,R)$*.*

Proof. The proof is by induction on w. In the case $w = 1$, the assumption $w < sr$ implies that either $s > 1$ or $r > 1$. By Lemma 3.1, there is no proposition symbol p_i such that $(S,R) \models p_i$ or $(S,R) \models \neg p_i$. Since the game $\mathrm{EF}_1(S,R)$ ends in its initial position $(1,S,R)$, this means that player II automatically wins this game.

Assume then that $w > 1$. Using Lemma 3.1 again, we see that player I does not win the game $\mathrm{EF}_w(S,R)$ without making moves. Suppose then that player I makes a left splitting move $w = u + v$ and $S = C \cup D$. Let $S_0 \subseteq C$ and $S_1 \subseteq D$ be sets such that $S = S_0 \cup S_1$ and $S_0 \cap S_1 = \emptyset$. Let $D(S_0,R) = (s_0,r_0)$ and $D(S_1,R) = (s_1,r_1)$. Then by Lemma 3.2(a), $s_0r_0 + s_1r_1 \geq sr$. Thus $s_0r_0 > u$ or $s_1r_1 > v$, for otherwise

$$sr \leq s_0r_0 + s_1r_1 \leq u + v = w,$$

contrary to the assumption. If $s_0r_0 > u$, then by induction hypothesis, player II has a winning strategy in the game $\mathrm{EF}_u(S_0,R)$, whence, a fortiori, he has a winning strategy in $\mathrm{EF}_u(C,R)$. Otherwise, $s_1r_1 > v$, and by the same argument, player II has a winning strategy in the game $\mathrm{EF}_v(D,R)$. Thus, in any case, player II can make his move in such a way that he is guaranteed to win.

The case of a right splitting move $w = u + v$ and $R = C \cup D$ is proved in the same way by using Lemma 3.2(b). □

We are now ready to prove the promised lower bound for parity of binary strings.

Corollary 3.4 ([6]). *If* ϕ *is a propositional formula expressing the parity of strings* $s \in \{0,1\}^n$*, then the size of* ϕ *is at least* n^2*.*

---

4 We are indebted to Kerkko Luosto for pointing this out.

Proof. Let S be the set of all strings $f \in \{0,1\}^n$ such that $|\{i : f_i = 1\}|$ is even, and let R be the complement of S. Thus, $|S| = |R| = 2^{n-1}$. Furthemore, for each string $f \in S$, there are n different strings $g \in \{0,1\}^n$ such that $|\{i : f_i \neq g_i\}| = 1$, and all these strings are in R. Thus, we have $|E| = 2^{n-1}n$, and consequently $D(S,R) = (n,n)$. By Lemma 3.3, player II has a winning strategy in the game $\mathrm{EF}_w(S,R)$ whenever $w < n^2$, and we conclude that S is not definable by any formula of size less than n^2. □

As noted in the introduction, the parity of strings in $\{0,1\}^n$ can be expressed by a formula of size at most $(n+1)^2$. Thus, the lower bound n^2 cannot be essentially improved.

4 A Game for Predicate Logic

We shall next define a game that can be used for measuring the size of a first-order sentence needed for expressing properties of models. For the sake of simplicity, we will only consider models with relational vocabulary.

We need to fix some notation first. The universe of a model $\mathfrak{A}$ is denoted by A, of $\mathfrak{B}$ by B, etc. We use x_j, $j \in \mathbb{N}$, to denote variables. A variable assignment for a model $\mathfrak{A}$ is a finite partial mapping $\alpha : \mathbb{N} \to A$. The finite domain of α is denoted by $\mathrm{dom}(\alpha)$. If ϕ is a formula, then $(\mathfrak{A},\alpha) \models \phi$ means that the assignment α satisfies the formula ϕ in the model $\mathfrak{A}$. For this to be meaningful the domain of the assignment α has to include all the j for which the variable x_j is free in ϕ.

We shall discuss classes $\mathcal{A}$ of structures $(\mathfrak{A},\alpha)$, where $\mathfrak{A}$ is a model and α is an assignment. We assume that whenever $(\mathfrak{A},\alpha), (\mathfrak{B},\beta) \in \mathcal{A}$, then $\mathfrak{A}$ and $\mathfrak{B}$ have the same vocabulary, and α and β have the same domain, which we denote by $\mathrm{dom}(\mathcal{A})$. If α is an assignment on $\mathfrak{A}$, $a \in A$ and $j \in \mathbb{N}$, then $\alpha(a/j)$ is the assignment that maps j to a and agrees with α otherwise. If F is a choice function on $\mathcal{A}$, that is, F is a function defined on $\mathcal{A}$ such that $F((\mathfrak{A},\alpha)) \in A$ for all $(\mathfrak{A},\alpha) \in \mathcal{A}$, then $\mathcal{A}(F/j)$ is defined as $\{(\mathfrak{A},\alpha(F((\mathfrak{A},\alpha))/j) : (\mathfrak{A},\alpha) \in \mathcal{A}\}$. Finally, $\mathcal{A}(\star/j) = \{(\mathfrak{A},\alpha(a/j)) : (\mathfrak{A},\alpha) \in \mathcal{A}, a \in A\}$.

Let $\mathcal{A}$ and $\mathcal{B}$ be classes of structures of a fixed relational vocabulary. Assume further that $\mathrm{dom}(\mathcal{A}) = \mathrm{dom}(\mathcal{B})$, and ϕ is a formula such that $j \in \mathrm{dom}(\mathcal{A})$ for all variables x_j which are free in ϕ. As in the case of propositional logic, we say that ϕ *separates* the classes $\mathcal{A}$ and $\mathcal{B}$, $(\mathcal{A},\mathcal{B}) \models \phi$, if $(\mathfrak{A},\alpha) \models \phi$ for all $(\mathfrak{A},\alpha) \in \mathcal{A}$ and $(\mathfrak{B},\beta) \not\models \phi$ for all $(\mathfrak{B},\beta) \in \mathcal{B}$.

Definition 4.1. Let $\mathcal{A}$ and $\mathcal{B}$ be classes of structures of the same relational vocabulary with $\mathrm{dom}(\mathcal{A}) = \mathrm{dom}(\mathcal{B})$, and let w be a positive integer. The game $\mathrm{EF}_w(\mathcal{A},\mathcal{B})$ has two players. The number w is called the *rank* of the game. In the beginning the position is $(w, \mathcal{A}, \mathcal{B})$. Suppose the position after m moves is $(w_m, \mathcal{A}_m, \mathcal{B}_m)$, where $\mathrm{dom}(\mathcal{A}_m) = \mathrm{dom}(\mathcal{B}_m)$. There are the following four possibilities for the continuation of the game:

Left splitting move: Player I first chooses numbers u and v such that $1 \leq u, v < w$ and $u + v = w_m$. Then Player I represents $\mathcal{A}_m$ as a union $\mathcal{C} \cup \mathcal{D}$. Now the game continues from the position $(u, \mathcal{C}, \mathcal{B}_m)$ or from the position $(v, \mathcal{D}, \mathcal{B}_m)$, and player II can choose which.

Right splitting move: Player I first chooses numbers u and v such that $1 \leq u, v < w$ and $u + v = w_m$. Then Player I represents $\mathcal{B}_m$ as a union $\mathcal{C} \cup \mathcal{D}$. Now the game continues from the position $(u, \mathcal{A}_m, \mathcal{C})$ or from the position $(v, \mathcal{A}_m, \mathcal{D})$, and player II can choose which.

Left supplementing move: Player I chooses a natural number j and a choice function F for $\mathcal{A}_m$. Then the game continues from the position

$$(w_m - 1, \mathcal{A}_m(F/j), \mathcal{B}_m(\star/j)).$$

Right supplementing move: Player I chooses a natural number j and a choice function F for $\mathcal{B}_m$. Then the game continues from the position

$$(w_m - 1, \mathcal{A}_m(\star/j), \mathcal{B}_m(F/j)).$$

The game ends in a position $(w_m, \mathcal{A}_m, \mathcal{B}_m)$ and player I wins if there is an atomic or a negated atomic formula ϕ such that $(\mathcal{A}_m, \mathcal{B}_m) \models \phi$. Player II wins the game if they reach a position (w_m, S_m, R_m) such that $w_m = 1$ and player I does not win in this position.

This is a game of perfect information and the concept of winning strategy is defined as usual. The game is determined by the Gale-Stewart theorem.

We define the *size* $\mathrm{w}(\phi)$ of the formula ϕ of predicate logic as follows:

$$\begin{aligned}
\mathrm{w}(\phi) &= 1 \text{ for atomic } \phi \\
\mathrm{w}(\neg\phi) &= \mathrm{w}(\phi) \\
\mathrm{w}(\phi \vee \psi) &= \mathrm{w}(\phi) + \mathrm{w}(\psi) \\
\mathrm{w}(\phi \wedge \psi) &= \mathrm{w}(\phi) + \mathrm{w}(\psi) \\
\mathrm{w}(\exists x_j \phi) &= \mathrm{w}(\phi) + 1 \\
\mathrm{w}(\forall x_j \phi) &= \mathrm{w}(\phi) + 1
\end{aligned}$$

Note that if τ is a finite vocabulary, then there are for each w, up to logical equivalence, only finitely many τ-formulas of size $\leq w$.

Theorem 4.2. *Suppose $(\mathcal{A}, \mathcal{B})$ is a pair of classes of structures of the same vocabulary, and let w be a positive integer. Then the following conditions are equivalent:*

$(1)_w$ *Player I has a winning strategy in the game* $\mathrm{EF}_w(\mathcal{A}, \mathcal{B})$.
$(2)_w$ *There is a formula* ϕ *of predicate logic of size* $\leq w$ *such that* $(\mathcal{A}, \mathcal{B}) \models \phi$.

Proof. We prove the equivalence of $(1)_w$ and $(2)_w$ by induction on w. Consider first the case $w = 1$. By Definition 4.1, there are no moves in the game $\mathrm{EF}_1(\mathcal{A}, \mathcal{B})$, and player I wins if and only if there is an atomic formula ϕ such that either $(\mathcal{A}, \mathcal{B}) \models \phi$ or $(\mathcal{A}, \mathcal{B}) \models \neg\phi$. Since $\mathrm{w}(\phi) = \mathrm{w}(\neg\phi) = 1$, we have $(1)_1 \implies (2)_1$. On the other hand, if $\mathrm{w}(\phi) = 1$, then ϕ is, up to logical equivalence, either an atomic formula, or the negation of an atomic formula. Thus, $(2)_1$ implies $(1)_1$.

Let us then consider the case $w > 1$, and assume $(1)_v \iff (2)_v$ for all $v < w$ as an induction hypothesis. To prove $(1)_w \iff (2)_w$, assume first that player I has a winning strategy in the game $\mathrm{EF}_w(\mathcal{A}, \mathcal{B})$. We have the following four cases according to the type of the first move in the winning strategy of player I.

Case 1: Player I makes a left splitting move by choosing u, v, $\mathcal{C}$ and $\mathcal{D}$ such that $1 \leq u, v < w$, $w = u + v$ and $\mathcal{A} = \mathcal{C} \cup \mathcal{D}$. Since this move is given by a winning strategy, player I has a winning strategy in both of the games $\mathrm{EF}_u(\mathcal{C}, \mathcal{B})$ and $\mathrm{EF}_v(\mathcal{D}, \mathcal{B})$. By induction hypothesis, there are formulas ψ and θ such that $\mathrm{w}(\psi) \leq u$, $\mathrm{w}(\theta) \leq v$, $(\mathcal{C}, \mathcal{B}) \models \psi$ and $(\mathcal{D}, \mathcal{B}) \models \theta$. Thus, $(\mathfrak{A}, \alpha) \models \psi$ for all $(\mathfrak{A}, \alpha) \in \mathcal{C}$ and $(\mathfrak{A}, \alpha) \models \theta$ for all $(\mathfrak{A}, \alpha) \in \mathcal{D}$, whence $(\mathfrak{A}, \alpha) \models \psi \vee \theta$ for all $(\mathfrak{A}, \alpha) \in \mathcal{A}$. On the other hand, for all $(\mathfrak{B}, \beta) \in \mathcal{B}$, $(\mathfrak{B}, \beta) \not\models \psi$ and $(\mathfrak{B}, \beta) \not\models \theta$, whence consequently $(\mathfrak{B}, \beta) \not\models \psi \vee \theta$. It follows that $(\mathcal{A}, \mathcal{B}) \models \psi \vee \theta$. As $\mathrm{w}(\psi \vee \theta) = \mathrm{w}(\psi) + \mathrm{w}(\theta) \leq u + v = w$, $(2)_w$ is true.

Case 2: Player I makes a right splitting move by choosing u, v, $\mathcal{C}$ and $\mathcal{D}$ such that $1 \leq u, v < w$, $w = u + v$ and $\mathcal{B} = \mathcal{C} \cup \mathcal{D}$. Since this move is given by a winning strategy, player I has a winning strategy in both of the games $\mathrm{EF}_u(\mathcal{A}, \mathcal{C})$ and $\mathrm{EF}_v(\mathcal{A}, \mathcal{D})$. By induction hypothesis there are formulas ψ and θ such that $\mathrm{w}(\psi) \leq u$, $\mathrm{w}(\theta) \leq v$, $(\mathcal{A}, \mathcal{C}) \models \psi$ and $(\mathcal{A}, \mathcal{D}) \models \theta$. Thus, for all $(\mathfrak{A}, \alpha) \in \mathcal{A}$, $(\mathfrak{A}, \alpha) \models \psi$ and $(\mathfrak{A}, \alpha) \models \theta$, whence $(\mathfrak{A}, \alpha) \models \psi \wedge \theta$. On the other hand, $(\mathfrak{B}, \beta) \not\models \psi$ for all $(\mathfrak{B}, \beta) \in \mathcal{C}$ and $(\mathfrak{B}, \beta) \not\models \theta$ for all $(\mathfrak{B}, \beta) \in \mathcal{D}$, whence $(\mathfrak{B}, \beta) \not\models \psi \wedge \theta$ for all $(\mathfrak{B}, \beta) \in \mathcal{B}$. It follows that $(\mathcal{A}, \mathcal{B}) \models \psi \wedge \theta$. As $\mathrm{w}(\psi \wedge \theta) = \mathrm{w}(\psi) + \mathrm{w}(\theta) \leq u + v = w$, $(2)_w$ is true.

Case 3: Player I makes a left supplementing move by choosing a natural number j and a choice function F for $\mathcal{A}$. The next position in the game is then $(w - 1, \mathcal{A}(F/j), \mathcal{B}(\star/j))$. Since this move is given by a winning strategy, player I has a winning strategy in the game $\mathrm{EF}_{w-1}(\mathcal{A}(F/j), \mathcal{B}(\star/j))$. By induction hypothesis there is a formula ψ such that $\mathrm{w}(\psi) \leq w - 1$ and $(\mathcal{A}(F/j), \mathcal{B}(\star/j)) \models \psi$. Let ϕ be the formula $\exists x_j \psi$. Then $\mathrm{w}(\phi) = \mathrm{w}(\psi) + 1 \leq w$, and it suffices to show that $(\mathcal{A}, \mathcal{B}) \models \phi$. Note first that for all $(\mathfrak{A}, \alpha) \in \mathcal{A}$, $(\mathfrak{A}, \alpha(a/j)) \models \psi$, where $a = F((\mathfrak{A}, \alpha))$. Thus we have $(\mathfrak{A}, \alpha) \models \phi$ for all $(\mathfrak{A}, \alpha) \in \mathcal{A}$. On the other hand, for all $(\mathfrak{B}, \beta) \in \mathcal{B}$ and all $b \in B$, we have $(\mathfrak{B}, \beta(b/j)) \not\models \psi$, whence $(\mathfrak{B}, \beta) \not\models \phi$.

Case 4: Player I makes a right supplementing move by choosing a natural number j and a choice function F for $\mathcal{B}$. The next position in the game is then $(w-1, \mathcal{A}(\star/j), \mathcal{B}(F/j))$. Since this move is given by a winning strategy, player I has a winning strategy in the game $\mathrm{EF}_{w-1}(\mathcal{A}(\star/j), \mathcal{B}(F/j))$. By induction hypothesis there is a formula ψ such that $\mathrm{w}(\psi) \le w-1$ and $(\mathcal{A}(\star/j), \mathcal{B}(F/j)) \models \psi$. Let ϕ be the formula $\forall x_j \psi$. Then $\mathrm{w}(\phi) = \mathrm{w}(\psi) + 1 \le w$, and it suffices to show that $(\mathcal{A}, \mathcal{B}) \models \phi$. Note first that for all $(\mathfrak{A}, \alpha) \in \mathcal{A}$ and all $a \in A$, we have $(\mathfrak{A}, \alpha(a/j)) \models \psi$. Thus we have $(\mathfrak{A}, \alpha) \models \phi$ for all $(\mathfrak{A}, \alpha) \in \mathcal{A}$. On the other hand, for all $(\mathfrak{B}, \beta) \in \mathcal{B}$, $(\mathfrak{B}, \beta(b/j)) \not\models \psi$, where $b = F((\mathfrak{B}, \beta))$. Thus we have $(\mathfrak{B}, \beta) \not\models \phi$ for all $(\mathfrak{B}, \beta) \in \mathcal{B}$.

To prove the converse implication $(2)_w \implies (1)_w$, assume that there is a formula ϕ of size $\le w$ such that $(\mathcal{A}, \mathcal{B}) \models \phi$. We show that then player I has a winning strategy in the game $\mathrm{EF}_w(\mathcal{A}, \mathcal{B})$. As in the case of propositional logic, we can assume without loss of generality that ϕ is in negation normal form. The strategy of player I depends on ϕ as follows.

Case 1: ϕ is an atomic formula or the negation of an atomic formula. Then by Definition 4.1, player I wins the game $\mathrm{EF}_w(\mathcal{A}, \mathcal{B})$ without making any moves.

Case 2: ϕ is $\psi \vee \theta$. Let $\mathcal{C}$ be the class of structures $(\mathfrak{A}, \alpha) \in \mathcal{A}$ such that $(\mathfrak{A}, \alpha) \models \psi$, and let $\mathcal{D}$ be the class of structures $(\mathfrak{A}, \alpha) \in \mathcal{A}$ such that $(\mathfrak{A}, \alpha) \models \theta$. Since $(\mathfrak{A}, \alpha) \models \phi$ for every $(\mathfrak{A}, \alpha) \in \mathcal{A}$, we have $\mathcal{A} = \mathcal{C} \cup \mathcal{D}$. Moreover, since $(\mathfrak{B}, \beta) \not\models \phi$ for every $(\mathfrak{B}, \beta) \in \mathcal{B}$, we have $(\mathcal{C}, \mathcal{B}) \models \psi$ and $(\mathcal{D}, \mathcal{B}) \models \theta$. Finally, as $\mathrm{w}(\phi) \le w$, there are u and v such that $w = u + v$, $\mathrm{w}(\psi) \le u$ and $\mathrm{w}(\theta) \le v$. Note that $u, v < w$, and so, by induction hypothesis, player I has a winning strategy in the games $\mathrm{EF}_u(\mathcal{C}, \mathcal{B})$ and $\mathrm{EF}_v(\mathcal{D}, \mathcal{B})$. Thus, using $w = u + v$, $\mathcal{A} = \mathcal{C} \cup \mathcal{D}$ as his first move and these winning strategies in the rest of the game, player I is guaranteed to win the game $\mathrm{EF}_w(\mathcal{A}, \mathcal{B})$.

Case 3: ϕ is $\psi \wedge \theta$. Let $\mathcal{C}$ be the class of structures $(\mathfrak{B}, \beta) \in \mathcal{B}$ such that $(\mathfrak{B}, \beta) \not\models \psi$, and let $\mathcal{D}$ be the class of structures $(\mathfrak{B}, \beta) \in \mathcal{B}$ such that $(\mathfrak{B}, \beta) \not\models \theta$. Since $(\mathfrak{B}, \beta) \not\models \phi$ for every $(\mathfrak{B}, \beta) \in \mathcal{B}$, we have $\mathcal{B} = \mathcal{C} \cup \mathcal{D}$. Moreover, since $(\mathfrak{A}, \alpha) \models \phi$ for every $(\mathfrak{A}, \alpha) \in \mathcal{A}$, we have $(\mathcal{A}, \mathcal{C}) \models \psi$ and $(\mathcal{A}, \mathcal{D}) \models \theta$. Finally, as $\mathrm{w}(\phi) \le w$, there are u and v such that $w = u + v$, $\mathrm{w}(\psi) \le u$ and $\mathrm{w}(\theta) \le v$. Note that $u, v < w$, and so, by induction hypothesis, player I has a winning strategy in the games $\mathrm{EF}_u(\mathcal{A}, \mathcal{C})$ and $\mathrm{EF}_v(\mathcal{A}, \mathcal{D})$. Thus, using $w = u + v$, $\mathcal{A} = \mathcal{C} \cup \mathcal{D}$ in his first move, and these winning strategies in the rest of the game, player I is guaranteed to win the game $\mathrm{EF}_w(\mathcal{A}, \mathcal{B})$.

Case 4: ϕ is $\exists x_j \psi$. Since $(\mathfrak{A}, \alpha) \models \phi$ for every $(\mathfrak{A}, \alpha) \in \mathcal{A}$, there is a choice function F for $\mathcal{A}$ such that $(\mathfrak{A}, \alpha(F((\mathfrak{A}, \alpha))/j)) \models \psi$ for all $(\mathfrak{A}, \alpha) \in \mathcal{A}$. Thus, $(\mathfrak{A}, \alpha^*) \models \psi$ for every $(\mathfrak{A}, \alpha^*) \in \mathcal{A}(F/j)$. On the other hand, for all $(\mathfrak{B}, \beta) \in \mathcal{B}$ we have $(\mathfrak{B}, \beta) \not\models \phi$, whence $(\mathfrak{B}, \beta(b/j)) \not\models \psi$ for all $b \in B$. In other words, $(\mathfrak{B}, \beta^*) \not\models \psi$ for all $(\mathfrak{B}, \beta^*) \in$

$\mathcal{B}(\star/j)$. Thus we conclude that $(\mathcal{A}(F/j), \mathcal{B}(\star/j)) \models \psi$. Note that $\mathrm{w}(\psi) = \mathrm{w}(\phi) - 1 \leq w-1$, and so, by induction hypothesis, player I has a winning strategy in the game $\mathrm{EF}_{w-1}(\mathcal{A}(F/j), \mathcal{B}(\star/j))$. Thus, starting with the left supplementing move j and F, and using this winning strategy in the rest of the game, player I is guaranteed to win the game $\mathrm{EF}_w(\mathcal{A}, \mathcal{B})$.

Case 5: ϕ is $\forall x_j \psi$. Since $(\mathfrak{B}, \beta) \not\models \phi$ for every $(\mathfrak{B}, \beta) \in \mathcal{B}$, there is a choice function F for $\mathcal{B}$ such that $(\mathfrak{B}, \beta(F((\mathfrak{B}, \beta))/j)) \not\models \psi$ for all $(\mathfrak{B}, \beta) \in \mathcal{B}$. Thus, $(\mathfrak{B}, \beta^*) \not\models \psi$ for every $(\mathfrak{B}, \beta^*) \in \mathcal{B}(F/j)$. On the other hand, for all $(\mathfrak{A}, \alpha) \in \mathcal{A}$ we have $(\mathfrak{A}, \alpha) \models \phi$, whence $(\mathfrak{A}, \alpha(a/j)) \models \psi$ for all $a \in A$. In other words, $(\mathfrak{A}, \alpha^*) \models \psi$ for all $(\mathfrak{A}, \alpha^*) \in \mathcal{A}(\star/j)$. Thus we conclude that $(\mathcal{A}(\star/j), \mathcal{B}(F/j)) \models \psi$. Note that $\mathrm{w}(\psi) = \mathrm{w}(\phi) - 1 \leq w-1$, and so, by induction hypothesis, player I has a winning strategy in the game $\mathrm{EF}_{w-1}(\mathcal{A}(\star/j), \mathcal{B}(F/j))$. Thus, starting with the right supplementing move j and F, and using this winning strategy in the rest of the game, player I is guaranteed to win the game $\mathrm{EF}_w(\mathcal{A}, \mathcal{B})$. **Q.E.D.**

Corollary 4.3. *Suppose $\mathcal{K}$ is a class of models of the same vocabulary. Then the following conditions are equivalent:*

(1) *$\mathcal{K}$ is the class of models of a first order sentence ϕ of size $\leq w$.*

(2) *Player I has a winning strategy in the game $\mathrm{EF}_w(\mathcal{A}, \mathcal{B})$ for all $\mathcal{A}$ and $\mathcal{B}$ such that $\mathcal{A} \subseteq \mathcal{K}$ and $\mathcal{B} \cap \mathcal{K} = \emptyset$.*

Proof. Suppose $\mathcal{K}$ is the class of models of ϕ and $\mathrm{w}(\phi) \leq w$. Let $\mathcal{A}$ and $\mathcal{B}$ be given so that $\mathcal{A} \subseteq \mathcal{K}$ and $\mathcal{B} \cap \mathcal{K} = \emptyset$. Thus every model in $\mathcal{A}$ satisfies ϕ and no model in $\mathcal{B}$ does, whence $(\mathcal{A}, \mathcal{B}) \models \phi$. By Theorem 4.2, player I has a winning strategy in $\mathrm{EF}_w(\mathcal{A}, \mathcal{B})$. For the converse, suppose player I has a winning strategy in $\mathrm{EF}_w(\mathcal{A}, \mathcal{B})$ for all $(\mathcal{A}, \mathcal{B})$ with $\mathcal{A} \subseteq \mathcal{K}$ and $\mathcal{B} \cap \mathcal{K}$. This holds in particular if $\mathcal{A} = \mathcal{K}$ and $\mathcal{B}$ is the complement of $\mathcal{K}$. Thus for this choice of $\mathcal{A}$ and $\mathcal{B}$ we have $(\mathcal{A}, \mathcal{B}) \models \phi$ for some ϕ of size $\leq w$. Thus $\mathcal{K}$ is the class of all models of ϕ. **Q.E.D.**

We get the following method for showing that a model class $\mathcal{K}$ is *not* definable by a first order sentence of size $\leq w$: We find classes $\mathcal{A}$ and $\mathcal{B}$ such that

1. $\mathcal{A} \subseteq \mathcal{K}$.
2. $\mathcal{B} \cap \mathcal{K} = \emptyset$.
3. Player II has a winning strategy in $\mathrm{EF}_w(\mathcal{A}, \mathcal{B})$.

A Game for Existential Formulas

The game EF_w can be used for solving questions of the type

> Given a property $\mathcal{P}$ of models, what is the size of the smallest first order formula that defines $\mathcal{P}$?

But in some applications we are interested in definability by some restricted type of of formulas, rather than arbitrary first order formulas. For example, if a property $\mathcal{P}$ of models is known to be definable by an existential formula, it is natural to ask, what is the size of the smallest existential formula defining $\mathcal{P}$. We will now define a variant of the game EF_w that can be used in studying this question.

Here we say that a first order formula is *existential* if it is in negation normal form, and it does not contain any universal quantifiers. In other words, existential formulas are built from atomic formulas and negations of atomic formulas by using the connectives $\vee$, $\wedge$ and the quantifier $\exists$.

Intuitively, in the definition of the game EF_w, left and right splitting moves correspond to the connectives $\vee$ and $\wedge$, while left and right supplementing moves correspond to the quantifiers $\exists$ and $\forall$. Thus, we obtain a game for existential formulas simply by dropping right supplementing moves.

Definition 4.4. Let $\mathcal{A}$ and $\mathcal{B}$ be classes of structures of the same relational vocabulary with $\mathrm{dom}(\mathcal{A}) = \mathrm{dom}(\mathcal{B})$, and let w be a positive integer. The existential game $\mathrm{EF}_w^{\exists}(\mathcal{A}, \mathcal{B})$ has the same rules as $\mathrm{EF}_w(\mathcal{A}, \mathcal{B})$, except that player I is not allowed to make right supplementing moves.

Theorem 4.5. *Suppose $(\mathcal{A}, \mathcal{B})$ is a pair of classes of structures of the same vocabulary, and let w be a positive integer. Then the following conditions are equivalent:*
$(1)_w$ *Player I has a winning strategy in the game $\mathrm{EF}_w^{\exists}(\mathcal{A}, \mathcal{B})$.*
$(2)_w$ *There is an existential formula ϕ of size $\leq w$ such that $(\mathcal{A}, \mathcal{B}) \models \phi$.*

Proof. The claim is proved by a straightforward adaptation of the proof of Lemma 4.2. In the proof of $(1)_w \implies (2)_w$ it suffices to note that omitting Case 4, the formula ϕ separating the classes $\mathcal{A}$ and $\mathcal{B}$ will always be existential. Similarly, in the proof of the opposite implication, Case 5 never occurs as the separating formula ϕ is existential, whence the winning strategy of player I does not use right supplementing moves. **Q.E.D.**

As in the case of the full first order game EF_w, we get again a method for showing that a model class $\mathcal{K}$ is *not* definable by an existential formula of size $\leq w$: We find classes $\mathcal{A}$ and $\mathcal{B}$ such that

1. $\mathcal{A} \subseteq \mathcal{K}$.
2. $\mathcal{B} \cap \mathcal{K} = \emptyset$.

3. Player II has a winning strategy in $\mathrm{EF}_w^\exists(\mathcal{A}, \mathcal{B})$.

5 The Existential Complexity of Non-Emptiness of Boolean Combinations

We observed in the introduction that non-emptiness of all Boolean combinations of n unary predicates can be defined with a sentence of size $8n + 4$ if a quantifier alternation is allowed. If only existential sentences are allowed, this can be expressed with a sentence of size $(n + 1)2^n$. We use now our game for existential formulas to show that $(n + 1)2^n$ is the best possible value.

For each binary string $s \in \{0, 1\}^n$, let $\mathfrak{B}_s$ be the $\{P_1, \ldots, P_n\}$-structure such that for each $r \in \{0, 1\}^n$, the Boolean combination of $P_1^{\mathfrak{B}_s}, \ldots, P_n^{\mathfrak{B}_s}$ corresponding to r contains exactly two elements, b_r and c_r, except that the combination corresponding to s is empty. Furthermore, let $\mathfrak{A}$ be the structure in which each Boolean combination contains exactly one element; let a_r be the element in the Boolean combination corresponding to $r \in \{0, 1\}^n$. Let $\mathcal{A}_0$ be the class $\{(\mathfrak{A}, \emptyset)\}$, and let $\mathcal{B}_0$ be the class $\{(\mathfrak{B}_s, \emptyset) : s \in \{0, 1\}^n\}$. We will show that player II has a winning strategy in the existential game $\mathrm{EF}_w^\exists(\mathcal{A}_0, \mathcal{B}_0)$ for all $w < (n + 1)2^n$.

Since the game is existential, in any position $(u, \mathcal{A}, \mathcal{B})$, the set $\mathcal{A}$ contains only one structure $(\mathfrak{A}, \alpha)$, where α is the variable assignment that results from the moves of player I. We say that a structure $(\mathfrak{B}_s, \beta)$ is *flawless* (with respect to α), if $\mathrm{dom}(\beta) = \mathrm{dom}(\alpha)$, and for all $j \in \mathrm{dom}(\alpha)$ and all $r \in \{0, 1\}^n$, we have

$$\alpha(j) = a_r \iff \beta(j) = b_r. \tag{1}$$

Note that if $a_s \in \mathrm{ran}(\alpha)$, then there is no β such that $(\mathfrak{B}_s, \beta)$ is flawless. On the other hand, if $a_s \notin \mathrm{ran}(\alpha)$, then there is a unique β such that $(\mathfrak{B}_s, \beta)$ is flawless; we denote this β by $\beta_{s,\alpha}$.

Furthermore, we say that a structure $(\mathfrak{B}_s, \beta)$ is *good enough* (with respect to α), if it is not flawless, but $\mathrm{dom}(\beta) = \mathrm{dom}(\alpha)$, and (1) holds for all $j \in \mathrm{dom}(\alpha)$ and all $r \neq s$, and there is a string $t \in \{0, 1\}^n$ with $|\{i : s_i \neq t_i\}| = 1$ such that

$$\alpha(j) = a_s \iff \beta(j) = c_t \tag{2}$$

for all $j \in \mathrm{dom}(\alpha)$. Thus, if $a_s \in \mathrm{ran}(\alpha)$, then for each $t \in \{0, 1\}^n$ with $|\{i : s_i \neq t_i\}| = 1$ there is a unique β such that $(\mathfrak{B}_s, \beta)$ is good enough; we denote this β by $\beta_{s,t,\alpha}$.

For the rest of this section, $\mathcal{B}$ will always denote a set of structures of the form $(\mathfrak{B}_s, \beta)$, and $\mathcal{A}$ will denote a singleton set $\{(\mathfrak{A}, \alpha)\}$. The *measure* of the set $\mathcal{B}$ is defined to be

$$M(\mathcal{B}) = (n + 1) \cdot f(\mathcal{B}) + g(\mathcal{B}),$$

where $f(\mathcal{B})$ is the number of flawless structures in $\mathcal{B}$ and $g(\mathcal{B})$ is the number of good enough structures in $\mathcal{B}$.

Lemma 5.1. *If $M(\mathcal{B}) > 1$, then there is no atomic formula ϕ such that $(\mathcal{A}, \mathcal{B}) \models \phi$ or $(\mathcal{A}, \mathcal{B}) \models \neg\phi$.*

Proof. If $M(\mathcal{B}) > 1$, then either there is a flawless structure in $\mathcal{B}$, or there are at least two good enough structures in $\mathcal{B}$. If $(\mathfrak{B}_s, \beta) \in \mathcal{B}$ is flawless, then by condition (1), $(\mathfrak{A}, \alpha)$ and $(\mathfrak{B}_s, \beta)$ satisfy the same atomic formulas, whence no atomic formula separates $\mathcal{A}$ and $\mathcal{B}$.

Assume then, that $(\mathfrak{B}_s, \beta_{s,t,\alpha})$ and $(\mathfrak{B}_{s'}, \beta_{s',t',\alpha})$ are two distinct good enough structures in $\mathcal{B}$. Let ϕ be an atomic formula. If ϕ is an identity $x_j = x_k$ (with $j, k \in \mathrm{dom}(\alpha)$), then it follows easily from (1) and (2) that $(\mathfrak{A}, \alpha) \models \phi \iff (\mathfrak{B}_s, \beta_{s,t,\alpha}) \models \phi$. Thus, ϕ does not separate $\mathcal{A}$ and $\mathcal{B}$.

Consider then the case $\phi = P_l(x_j)$, where $j \in \mathrm{dom}(\alpha)$. Let $\alpha(j) = a_r$. If $r \neq s$, then $\beta_{s,t,\alpha}(j) = b_r$, and we have

$$(\mathfrak{A}, \alpha) \models \phi \iff r_l = 1 \iff (\mathfrak{B}_s, \beta_{s,t,\alpha}) \models \phi.$$

Similarly, if $r \neq s'$, then $(\mathfrak{A}, \alpha) \models \phi \iff (\mathfrak{B}_{s'}, \beta_{s',t',\alpha}) \models \phi$. Assume finally, that $r = s = s'$. Since $(\mathfrak{B}_s, \beta_{s,t,\alpha}) \neq (\mathfrak{B}_{s'}, \beta_{s',t',\alpha})$, we have $t \neq t'$. Moreover, since $|\{i : r_i \neq t_i\}| = |\{i : r_i \neq t'_i\}| = 1$, either $r_l = t_l$ or $r_l = t'_l$. Thus, it is not possible that ϕ separates $(\mathfrak{A}, \alpha)$ from both $(\mathfrak{B}_s, \beta_{s,t,\alpha})$ and $(\mathfrak{B}_{s'}, \beta_{s',t',\alpha})$. We conclude that in all cases $(\mathcal{A}, \mathcal{B}) \not\models \phi$ and $(\mathcal{A}, \mathcal{B}) \not\models \neg\phi$. **Q.E.D.**

Lemma 5.2. *(a) If $\mathcal{B} = \mathcal{C} \cup \mathcal{D}$, then $M(\mathcal{C}) + M(\mathcal{D}) \geq M(\mathcal{B})$.*
(b) If $\mathcal{A}' = \mathcal{A}(F/j)$ and $\mathcal{B}' = \mathcal{B}(\star/j)$, then $M(\mathcal{B}') \geq M(\mathcal{B}) - 1$.

Proof. (a) If $\mathcal{B} = \mathcal{C} \cup \mathcal{D}$, then obviously $f(\mathcal{C}) + f(\mathcal{D}) \geq f(\mathcal{B})$, and $g(\mathcal{C}) + g(\mathcal{D}) \geq g(\mathcal{B})$. Hence we have $M(\mathcal{C}) + M(\mathcal{D}) = (n+1)(f(\mathcal{C}) + f(\mathcal{D})) + (g(\mathcal{C}) + g(\mathcal{D})) \geq M(\mathcal{B})$.

(b) Let $F((\mathfrak{A}, \alpha)) = a_r$. Thus, $\mathcal{A}' = \{(\mathfrak{A}, \alpha')\}$, where $\alpha' = \alpha(a_r/j)$. Observe first that if $(\mathfrak{B}_s, \beta_{s,\alpha})$ is a flawless structure, and $r \neq s$, then $(\mathfrak{B}_s, \beta_{s,\alpha}(b_r/j)) = (\mathfrak{B}_s, \beta_{s,\alpha'})$ is also flawless, and clearly $(\mathfrak{B}, \beta_{s,\alpha}) \in \mathcal{B} \iff (\mathfrak{B}, \beta_{s,\alpha'}) \in \mathcal{B}'$.

Assume then that $(\mathfrak{B}_s, \beta_{s,t,\alpha})$ is a good enough structure. If $r \neq s$, then $(\mathfrak{B}_s, \beta_{s,t,\alpha}(b_r/j)) = (\mathfrak{B}_s, \beta_{s,t,\alpha'})$ is also good enough. On the other hand, if $r = s$, then $(\mathfrak{B}_s, \beta_{s,t,\alpha}(c_t/j)) = (\mathfrak{B}_s, \beta_{s,t,\alpha'})$ is good enough. In both cases, $(\mathfrak{B}, \beta_{s,t,\alpha}) \in \mathcal{B} \iff (\mathfrak{B}, \beta_{s,t,\alpha'}) \in \mathcal{B}'$.

Thus, we see that if $\mathcal{B}$ does not contain a flawless structure of the form $(\mathfrak{B}_r, \beta_{r,\alpha})$, then $f(\mathcal{B}') = f(\mathcal{B})$ and $g(\mathcal{B}') = g(\mathcal{B})$, whence the claim is true.

Assume finally, that there is a flawless structure $(\mathfrak{B}_r, \beta_{r,\alpha})$ in $\mathcal{B}$. Since $\alpha'(j) = a_r$, no structure $(\mathfrak{B}_r, \beta)$ is flawless with respect to α'. On the other hand, for each $t \in \{0,1\}^n$ with $|\{i : r_i \neq t_i\}| = 1$, there is a new good enough structure $(\mathfrak{B}_r, \beta_{r,\alpha}(c_t/j)) =$

$(\mathfrak{B}_r, \beta_{r,t,\alpha'})$ in $\mathcal{B}'$. Thus, in this case $f(\mathcal{B}') = f(\mathcal{B}) - 1$ and $g(\mathcal{B}') = g(\mathcal{B}) + n$, whence $M(\mathcal{B}') = (n+1)(f(\mathcal{B}) - 1) + g(\mathcal{B}) + n = M(\mathcal{B}) - 1$. **Q.E.D.**

Lemma 5.3. *If $w < M(\mathcal{B})$, then player II has a winning strategy in $\mathrm{EF}^{\exists}_{w}(\mathcal{A}, \mathcal{B})$.*

Proof. We prove the claim by induction on w. Consider first the case $w = 1$. By the definition of the game $\mathrm{EF}^{\exists}_{1}(\mathcal{A}, \mathcal{B})$, there are no moves, and player I wins only if there is an atomic formula ϕ such that $(\mathcal{A}, \mathcal{B}) \models \phi$ or $(\mathcal{A}, \mathcal{B}) \models \neg\phi$. Since $M(\mathcal{B}) > w = 1$, by Lemma 5.1, there is no such ϕ.

Assume then that $w > 1$, and the claim is true for all $u < w$. Since $M(\mathcal{B}) > w \geq 1$, by Lemma 5.1 again, player I does not win the game without making moves. Consider then the options of player I for his first move. Making a left splitting move $\mathcal{A} = \mathcal{C} \cup \mathcal{D}$ is not possible, since $\mathcal{A}$ is a singleton $\{(\mathfrak{A}, \alpha)\}$. Suppose then that player I makes a right splitting move $w = u + v$ and $\mathcal{B} = \mathcal{C} \cup \mathcal{D}$. Then by Lemma 5.2(a), $M(\mathcal{C}) + M(\mathcal{D}) \geq M(\mathcal{B})$, and since $w < M(\mathcal{B})$, either $u < M(\mathcal{C})$, or $v < M(\mathcal{D})$. If $u < M(\mathcal{C})$, then by induction hypothesis, player II has a winning strategy in the game $\mathrm{EF}^{\exists}_{u}(\mathcal{A}, \mathcal{C})$. Similarly, if $v < M(\mathcal{D})$, then player II has a winning strategy in the game $\mathrm{EF}^{\exists}_{v}(\mathcal{A}, \mathcal{D})$. Thus, by choosing the appropriate position $(u, \mathcal{A}, \mathcal{C})$ or $(v, \mathcal{A}, \mathcal{D})$, player II is guaranteed to win the game $\mathrm{EF}^{\exists}_{w}(\mathcal{A}, \mathcal{B})$.

Suppose then that player I starts with a left supplementing move j and F, where F is a choice function for $\mathcal{A}$. The next position in the game is then $(w - 1, \mathcal{A}', \mathcal{B}')$, where $\mathcal{A}' = \mathcal{A}(F/j)$ and $\mathcal{B}' = \mathcal{B}(\star/j))$. By Lemma 5.2(b) and our assumption $w < M(\mathcal{B})$, we have $w - 1 < M(\mathcal{B}) - 1 \leq M(\mathcal{B}')$, whence by induction hypothesis, player II has a winning strategy in the continuation of the game $\mathrm{EF}^{\exists}_{w}(\mathcal{A}, \mathcal{B})$ from position $(w - 1, \mathcal{A}', \mathcal{B}')$ onwards. **Q.E.D.**

Consider now the classes $\mathcal{A}_0$ and $\mathcal{B}_0$ defined in the beginning of this section. Since the variable assignment in the only structure in $\mathcal{A}_0$ is empty, all the 2^n structures $(\mathfrak{B}_s, \emptyset)$ in $\mathcal{B}_0$ are flawless. Thus, $M(\mathcal{B}_0) = (n+1)f(\mathcal{B}_0) + g(\mathcal{B}_0) = (n+1)2^n$, and by Lemma 5.3, player II has a winning strategy in the game $\mathrm{EF}^{\exists}_{w}(\mathcal{A}_0, \mathcal{B}_0)$ whenever $w < (n+1)2^n$. As all Boolean combinations of the predicates $P_1, \ldots, P_n$ are non-empty in $\mathfrak{A}$, but each structure in $\mathcal{B}_0$ has an empty Boolean combination, we get the desired lower bound result:

Corollary 5.4. *If ϕ is an existential first order sentence expressing the property that all Boolean combinations of n unary predicates are non-empty, then the size of ϕ is at least $(n+1)2^n$.* **Q.E.D.**

6 The Existential Complexity of the Length of Linear Order

As we noted in the introduction, for each n there is first order sentence ϕ_n of logarithmic size with respect to n which expresses the property that the length of a linear order is at least n. However, the sentence ϕ_n has an unbounded number of quantifier alternations. In this section we show that $2n - 1$ is the minimum size of an existential sentence expressing this property.

To prove the upper bound, define the following sequence of existential formulas:

$\psi_2 := \exists x_1 \exists x_2 (x_1 < x_2)$, and

$\psi_{n+1} := \exists x_{n+1}(\psi_n \wedge x_n < x_{n+1})$ for all $n \geq 2$.

Clearly ψ_n is true in a linear order if and only if its length is at least n, and an easy induction shows that $\mathrm{w}(\psi_n) = 2n - 1$.

To prove the lower bound, we will use again the existential game $\mathrm{EF}^{\exists}_{w}$. Let $\mathcal{A}_0 = \{(\mathfrak{A}, \emptyset)\}$, where $\mathfrak{A}$ is a linear order of length n, and let $\mathcal{B}_0 = \{(\mathfrak{B}, \emptyset)\}$, where $\mathcal{B}$ is a linear order of length $n - 1$. Our aim is to show that player II has a winning strategy in the game $\mathrm{EF}^{\exists}_{w}(\mathcal{A}_0, \mathcal{B}_0)$ for all $w < 2n - 1$.

Consider a position $(u, \mathcal{A}, \mathcal{B})$ in the game $\mathrm{EF}^{\exists}_{w}(\mathcal{A}_0, \mathcal{B}_0)$. Since the game is existential, $\mathcal{A}$ consists of a single structure $(\mathfrak{A}, \alpha)$. Let $a_1 <^{\mathfrak{A}} \cdots <^{\mathfrak{A}} a_l$ be the elements in $\mathrm{ran}(\alpha)$, and let a_0 and a_{l+1} be the least and the largest element in $\mathfrak{A}$, respectively. We say that a variable assignment β in $\mathfrak{B}$ is *acceptable* (with respect to α) if $\mathrm{dom}(\beta) = \mathrm{dom}(\alpha)$, there are elements $b_1 \leq^{\mathfrak{B}} \cdots \leq^{\mathfrak{B}} b_l$ such that $\mathrm{ran}(\beta) = \{b_1, \ldots, b_l\}$ and for all $i \in \{1, \ldots, l\}$ and all $j \in \mathrm{dom}(\alpha)$

$$\alpha(j) = a_i \iff \beta(j) = b_i. \tag{1}$$

In other words, β is acceptable if and only if the mapping $\alpha(j) \mapsto \beta(j)$, $j \in \mathrm{dom}(\alpha)$, preserves the relation $\leq$.

Furthermore, we say that β is *nice* (with respect to α), if it is acceptable, and in addition $|\{i \leq l : d(a_i, a_{i+1}) \neq d(b_i, b_{i+1})\}| = 1$, where $d(x, y)$ is the distance between x and y in the given linear order, and b_0 and b_{l+1} are the least and the largest element in $\mathfrak{B}$, respectively. The *distance from defect* of β is defined by $\delta(\beta) = d(b_i, b_{i+1})$, where $i \leq l$ is the unique index such that $d(a_i, a_{i+1}) \neq d(b_i, b_{i+1})$; we denote this index by $i(\beta)$. Note that $d(b_i, b_{i+1}) = d(a_i, a_{i+1}) - 1$ for $i = i(\beta)$. Note also, that for each $i \leq l$ there is exactly one nice β such that $i(\beta) = i$.

For the rest of this section, we assume that $\mathcal{A}$ is a singleton set $\{(\mathfrak{A}, \alpha)\}$ and $\mathcal{B}$ is a set of structures of the form $(\mathfrak{B}, \beta)$. The *niceness measure* of $\mathcal{B}$ is defined to be

$$N(\mathcal{B}) = \sum_{\beta \in \mathcal{N}} (2\delta(\beta) + 1),$$

where $\mathcal{N}$ is the set of all nice variable assignments β such that $(\mathfrak{B}, \beta) \in \mathcal{B}$.

Lemma 6.1. *If* $N(\mathcal{B}) > 1$, *then there is no atomic formula* ϕ *such that* $(\mathcal{A}, \mathcal{B}) \models \phi$ *or* $(\mathcal{A}, \mathcal{B}) \models \neg\phi$.

Proof. If $N(\mathcal{B}) > 1$, then either there is a nice assignment $\beta \in \mathcal{N}$ such that $\delta(\beta) \geq 1$, or there are two distinct nice assignments $\beta, \beta' \in \mathcal{N}$. Assume first that β is a nice assignment in $\mathcal{N}$, and $\delta(\beta) \geq 1$. Then there are elements $a_1, \ldots, a_l, b_1, \ldots, b_l$ such that $\mathrm{ran}(\alpha) = \{a_1, \ldots, a_l\}$, $\mathrm{ran}(\beta) = \{b_1, \ldots, b_l\}$, $a_1 <^{\mathfrak{A}} \cdots <^{\mathfrak{A}} a_l$ and $b_1 \leq^{\mathfrak{B}} \cdots \leq^{\mathfrak{B}} b_l$. Since β is nice, $d(b_i, b_{i+1}) = \delta(\beta) \geq 1$ for $i = i(\beta)$, and $d(b_i, b_{i+1}) = d(a_i, a_{i+1}) \geq 1$ for all other $i \leq l$. Thus, we have $b_1 <^{\mathfrak{B}} \cdots <^{\mathfrak{B}} b_l$. It follows easily from condition (1) that $(\mathfrak{A}, \alpha)$ and $(\mathfrak{B}, \beta)$ satisfy the same atomic formulas, whence no atomic formula separates $\mathcal{A}$ and $\mathcal{B}$.

Assume then that $\beta, \beta' \in \mathcal{N}$, and $\beta \neq \beta'$. As noted above, this means that $i(\beta) \neq i(\beta')$. Let $a_1, \ldots, a_l$ be as above, and let $\mathrm{ran}(\beta) = \{b_1, \ldots, b_l\}$ and $\mathrm{ran}(\beta') = \{b'_1, \ldots, b'_l\}$ with $b_1 \leq^{\mathfrak{B}} \cdots \leq^{\mathfrak{B}} b_l$ and $b'_1 \leq^{\mathfrak{B}} \cdots \leq^{\mathfrak{B}} b'_l$. As above, we see that $b_i <^{\mathfrak{B}} b_{i+1}$ for all $i \leq l$ except $i = i(\beta)$, and similarly $b'_i <^{\mathfrak{B}} b'_{i+1}$ for all $i \leq l$ except $i = i(\beta')$. Moreover, since $i(\beta) \neq i(\beta')$, for all $i, j \leq l$, we have either $(a_i <^{\mathfrak{A}} a_j \iff b_i <^{\mathfrak{B}} b_j)$ or $(a_i <^{\mathfrak{A}} a_j \iff b'_i <^{\mathfrak{B}} b'_j)$. Using condition (1), it is now easy to see that no atomic formula separates the sets $\mathcal{A}$ and $\mathcal{B}$. **Q.E.D.**

Lemma 6.2. *(a) If* $\mathcal{B} = \mathcal{C} \cup \mathcal{D}$, *then* $N(\mathcal{C}) + N(\mathcal{D}) \geq N(\mathcal{B})$.
(b) If $\mathcal{A}' = \mathcal{A}(F/j)$ *and* $\mathcal{B}' = \mathcal{B}(\star/j)$, *then* $N(\mathcal{B}') \geq N(\mathcal{B}) - 1$.

Proof. (a) Assume that $\mathcal{B} = \mathcal{C} \cup \mathcal{D}$. Let $\mathcal{N}$ be the set of nice assignments β such that $(\mathfrak{B}, \beta) \in \mathcal{B}$. Furthermore, let $\mathcal{O} = \{\beta \in \mathcal{N} : (\mathfrak{B}, \beta) \in \mathcal{C}\}$ and $\mathcal{P} = \{\beta \in \mathcal{N} : (\mathfrak{B}, \beta) \in \mathcal{D}\}$. Then we have

$$\begin{aligned} N(\mathcal{B}) &= \sum_{\beta \in \mathcal{N}} (2\delta(\beta) + 1) \\ &\leq \sum_{\beta \in \mathcal{O}} (2\delta(\beta) + 1) + \sum_{\beta \in \mathcal{P}} (2\delta(\beta) + 1) = N(\mathcal{C}) + N(\mathcal{D}). \end{aligned}$$

(b) Let $\mathcal{N}$ ($\mathcal{N}'$) be the set of nice assignments β such that $(\mathfrak{B}, \beta) \in \mathcal{B}$ ($(\mathfrak{B}, \beta) \in \mathcal{B}'$, respectively). Let $\mathrm{ran}(\alpha) = \{a_1, \ldots, a_l\}$ with $a_1 <^{\mathfrak{A}} \cdots <^{\mathfrak{A}} a_l$, where α is the assignment such that $\mathcal{A} = \{(\mathfrak{A}, \alpha)\}$. As usual, we denote by a_0 and a_{l+1} the least and the largest element in $\mathfrak{A}$, respectively. Furthermore, we denote $F((\mathfrak{A}, \alpha))$ by c.

Assume first that $c = \alpha(k)$ for some $k \in \mathrm{dom}(\alpha)$. It is easy to see that for any assignment β,

$$\beta \in \mathcal{N} \iff \beta(d/j) \in \mathcal{N}',$$

where $d = \beta(k)$, and moreover $\delta(\beta(d/j)) = \delta(\beta)$. Note also that if $\beta \neq \beta'$, then $\beta(d/j) \neq \beta'(d'/j)$. Thus, in this case we have $N(\mathcal{B}') = N(\mathcal{B})$.

Assume next that $c = a_0$ or $c = a_{l+1}$. Then as above, we see that $\beta \in \mathcal{N} \iff \beta(d/j) \in \mathcal{N}'$, where d is the least or the largest element in $\mathcal{B}$, respectively, and $\delta(\beta(d/j)) = \delta(\beta)$. Thus, also in this case we conclude that $N(\mathcal{B}') = N(\mathcal{B})$.

Assume finally, that $c \notin \{a_0, \ldots, a_{l+1}\}$. Then there is an index $h \leq l$ such that $a_h <^{\mathfrak{A}} c <^{\mathfrak{A}} a_{h+1}$. Let β be a nice assignment in $\mathcal{N}$, and let $\mathrm{ran}(\beta) = \{b_1, \ldots, b_l\}$ with $b_1 \leq^{\mathfrak{B}} \cdots \leq^{\mathfrak{B}} b_l$. As noted earlier, there is exactly one nice assignment β_h such that $i(\beta_h) = h$. If $\beta \neq \beta_h$, then $d(b_h, b_{h+1}) = d(a_h, a_{h+1})$, whence there is an element d such that $d(b_h, d) = d(a_h, c)$ and $d(d, b_{h+1}) = d(c, a_{h+1})$. This means that $\delta(\beta(d/j)) = \delta(\beta)$, and clearly $\beta(d/j)$ is in $\mathcal{N}'$.

On the other hand, if $\beta = \beta_h$, then $d(b_h, b_{h+1}) = d(a_h, a_{h+1}) - 1$, and there are elements d and e such that $d(b_h, d) = d(a_h, c) - 1$, $d(d, b_{h+1}) = d(c, a_{h+1})$, $d(b_h, e) = d(a_h, c)$ and $d(e, b_{h+1}) = d(c, a_{h+1}) - 1$. Let $\beta' = \beta_h(d/j)$ and $\beta'' = \beta_h(e/j)$. Then $\beta', \beta'' \in \mathcal{N}'$, and we have

$$\delta(\beta') + \delta(\beta'') = d(b_h, d) + d(e, b_{h+1}) = d(a_h, a_{h+1}) - 2 = \delta(\beta_h) - 1,$$

whence

$$(2\delta(\beta') + 1) + (2\delta(\beta'') + 1) = 2(\delta(\beta_h) - 1) + 2 = (2\delta(\beta_h) + 1) - 1.$$

Thus, if β_h is in $\mathcal{N}$, we get

$$\begin{aligned} N(\mathcal{B}') &= \sum_{\beta' \in \mathcal{N}'} (2\delta(\beta') + 1) \\ &= \Big(\sum_{\beta \in \mathcal{N} \setminus \{\beta_h\}} (2\delta(\beta) + 1)\Big) + ((2\delta(\beta_h) + 1) - 1) = N(\mathcal{B}) - 1. \end{aligned}$$

On the other hand, if β_h is not in $\mathcal{N}$, we have $N(\mathcal{B}') = N(\mathcal{B})$. **Q.E.D.**

Lemma 6.3. *If $w < N(\mathcal{B})$, then player II has a winning strategy in* $\mathrm{EF}_w^{\exists}(\mathcal{A}, \mathcal{B})$.

Proof. The proof is *verbatim* the same as for Lemma 5.3; just replace $M(\mathcal{B})$ by $N(\mathcal{B})$. **Q.E.D.**

The proof of the lower bound result is also similar to that in the previous section. Consider the sets $\mathcal{A}_0 = \{(\mathfrak{A}, \emptyset)\}$ and $\mathcal{B}_0 = \{(\mathfrak{B}, \emptyset)\}$. Clearly $\delta(\emptyset) = n - 1$, and so $N(\mathcal{B}_0) = 2(n-1) + 1 = 2n - 1$. Thus, by Lemma 6.3, player II has a winning strategy in the game $\mathrm{EF}_w^{\exists}(\mathcal{A}_0, \mathcal{B}_0)$ for all $w < 2n - 1$. Since the length of the linear order $\mathfrak{A}$ is n, while the length of $\mathfrak{B}$ is less than n, we have proved

Corollary 6.4. *If ϕ is an existential first order sentence expressing the property that the length of a linear order is at least n, then the size of ϕ is at least $2n-1$.* **Q.E.D.**

Acknowledgement: The first author was partially supported by the EUROCORES LogICCC LINT programme, and the second author by grant 251557 of the Academy of Finland and the EUROCORES LogICCC LINT programme.

Bibliography

[1] Micah Adler and Neil Immerman. An $n!$ lower bound on formula size. *ACM Trans. Comput. Log.*, 4(3):296–314, 2003.

[2] M. Ajtai. Σ_1^1-formulae on finite structures. *Ann. Pure Appl. Logic*, 24(1):1–48, 1983.

[3] Merrick Furst, James B. Saxe, and Michael Sipser. Parity, circuits, and the polynomial-time hierarchy. *Math. Systems Theory*, 17(1):13–27, 1984.

[4] Martin Grohe and Nicole Schweikardt. The succinctness of first-order logic on linear orders. *Log. Methods Comput. Sci.*, 1(1):1:6, 25, 2005.

[5] Jouko Väänänen. On the semantics of informational independence. *Log. J. IGPL*, 10(3):339–352, 2002.

[6] Ingo Wegener. *The complexity of Boolean functions*. Wiley-Teubner Series in Computer Science. John Wiley & Sons Ltd., Chichester, 1987.

Wilfrid Hodges

Notes on the History of Scope

Abstract: Recent developments in logic and linguistics, not least Jouko Väänänen's Dependence Logic [39], have thrown our ideas of scope into the melting pot. In this essay I offer a historical perspective on where those ideas of scope came from, and what people did before they had such a notion.

Keywords: scope, Frege, Hilbert, Skolem function, Avicenna

MSC: 01-02, 03B65, 01A30

Wilfrid Hodges: Herons Brook, Sticklepath, Okehampton, Devon EX20 2PY, England, e-mail: wilfrid.hodges@btinternet.com

1 Early resolutions of scope ambiguities

Here is an early example of a scope problem. It comes from Plato's dialogue *Euthydemus*, in which two ham logicians, Euthydemus and Dionysodorus, ply their trade by implicating their audience in amazing numbers of grotesque logical errors. Euthydemus is proving that unless you know everything, you know nothing at all:

> If you don't know something, you are not knowing [sc. something]. (Plato [32] 293c6.) (1)

As often happens, the original text needs cosmetic surgery before we can analyse it: in Plato's Greek the first 'something' is cliticised onto the 'if' and the second is elided altogether. Probably the best approach is to restore the elided 'something' in the second clause:

> You are not knowing something. (2)

Euthydemus is relying on a structural ambiguity which makes the meaning of (2) uncertain between 'There is something that you don't know' and 'You are not informed about anything at all'.

Plato let the fallacy pass without comment. But Aristotle in his *Sophistical Refutations* 4 (Aristotle [3] p. 280) proposed a way dealing with cases like this one; we can call it the method of combination. It reappeared in the Scholastics under the name of *compositio et divisio*. None of Aristotle's examples own examples work in English – or in Arabic, to judge by Ibn Sīnā's appreciative comment ('I can't stand examples like these, that depend on misuse of language' (Ibn Sīnā [25] 14.10)). But the English of Plato's example (2) will serve to illustrate.

The method of combination is to observe that the sentence in question has two readings, and in one of these but not the other, a certain group of words combines to form a constituent phrase of the sentence. For example in (2) the words 'not knowing' are a constituent in the first reading but not in the second. In this example we could equally well say that 'knowing something' is a constituent in the second reading but not in the first. Some modern expositors make the distinction with hyphens or parentheses:

not-knowing something; not (knowing something). (3)

Without the benefit of hyphens or parentheses, William of Sherwood in the thirteenth century developed a temporal metaphor for expressing the same thing. He would say that in the first reading in (3) the words 'not' and 'knowing' are put together first, and then 'something' arrives later (*posterius advenit*); see [37] pp. 28, 37,102f and the editor's footnotes on these pages. But more often the Scholastics distinguished the two readings as the 'combined' reading and the 'divided' one, apparently without any principled rule for deciding which name goes with which reading.

The Latin Terminists of the thirteenth century developed a second method for resolving scope ambiguities. It dealt specifically with quantifiers; it rested on the idea that universal quantifiers can be expanded as logical products, and existential quantifiers as logical sums. See for example Buridan's treatment of the sentence 'Some donkey every man sees', [4] pp. 871, 874 (from his *Sophismata* iii). I give the gist, using modern mathematical notation. Suppose we quantify universally over a set of two objects a_1, a_2 and existentially over a set of two objects b_1, b_2. Then a sentence $\phi(\Pi_i a_i, \Sigma_j b_j)$ is ambiguous between

$$(\phi(a_1, b_1) + \phi(a_1, b_2)) \times (\phi(a_2, b_1) + \phi(a_2, b_2)) \tag{4}$$

and

$$(\phi(a_1, b_1) \times \phi(a_2, b_1)) + (\phi(a_1, b_2) \times \phi(a_2, b_2)). \tag{5}$$

For a modern mathematician these are the expansions of $\Pi_i \Sigma_j \phi(a_i, b_j)$ and $\Sigma_j \Pi_i \phi(a_i, b_j)$ respectively, a view that appears in C. S. Peirce [31] and his successors in the algebra of logic tradition. This view allows us to avoid the notion of quantifier scope and rely instead on standard mathematical conventions for writing sums and products, provided that we are willing to reduce all formulas to prenex normal form with the quantifiers listed at the front.

Peirce tells us how he came to think of universal and existential quantifiers as like products and sums. An unpublished note [30] reveals what was going through his mind as he was assembling his version of first-order logic:

> The first system of relationship which logic studies is that of an indefinite collection of units. It may be represented by the schema
>
> | | | | | | | | |
>
> These constitute the universe of discourse. Various names or conventional signs, for which letters may be used, are attached to these in various ways. The study of this schema gives rise to the Boolian calculus. The logic of relatives studies a collection of units arranged in an n-dimensional block, thus:
>
> . . . . .
> . . . . .
> . . . . .
> . . . . .
>
> Distinguishing the different dimensions by the letters i, j, k, etc. we may write
>
> | | |
> |---|---|
> | $\Pi_i a_i$ | for every i has the mark a. |
> | $\Sigma_i a_i$ | for some i has the mark a. |
> | $\Pi_i \Pi_j \ell$ | Every i is in the relation ℓ to every j. |
> | $\Sigma_i \Pi_j \ell$ | Some i is in the relation ℓ to all j's. |
>
> ... The logic of relatives, so understood, coincides with Professor Mitchell's multidimensional logic ...

There is no reference here to the Terminists; perhaps Peirce wasn't aware of the parallel. Instead Peirce refers to work of his student Oscar Mitchell.

Mitchell [29] proposed an extension of Aristotle's logic, to include what he called 'propositions of two dimensions'. These propositions say that things U have properties F at times V, so that they express a relation between certain things and certain times. By speaking of 'two dimensions', Mitchell implies that the relation is a subset of the cartesian product of the universe of things and the universe of time, or in other words, a set of ordered pairs (a,b) where a is a thing and b is a time. I don't see this implication spelled out in Mitchell's paper, but it's clear from the diagram in Peirce's note above that Peirce understood Mitchell's relations as subsets of a cartesian product. The next step was to interpret quantifiers as sums or products over the first or second dimension in the product, as Peirce proceeds to do in our quotation.

The use of ordered pairs was a standard feature in Aristotelian logic long before Mitchell and Peirce. Already in the second century AD, Galen [13] xvi–xviii, p. 49ff had called attention to sentences in Euclid and elsewhere, which have (in modern terminology) two or even three prenex universal quantifiers. Presumably in response to Galen, Alexander of Aphrodisias [1] 344.9ff pointed out that the string of quantifiers can be reduced to a single quantifier by quantifying over pairs or triples; I will refer to this below as 'Alexander's trick'. Alexander's trick passed into the common stock of Aristotelian logic, and we find it used by both Ibn Sīnā and John Stuart Mill. Of course when both quantifiers are universal there is no need to compare their scopes. The key new element in Mitchell's theory was that he allowed universal quantification in one dimension and existential in the other, and it seems to have been this that led to Peirce's realisation that the quantifiers can be manipulated like mathematical products and sums in different dimensions. We will see in section 8 below that Alexander's trick probably helped Ibn Sīnā in his understanding of multiple quantification, which was different from Peirce's.

2 Possible anticipations of scope

All the writers above were in some sense dealing with scope problems. But in an important sense none of them had a notion of scope. Broadly speaking, scope is a relation which holds between an element of a sentence and one or more constituents of the sentence; the constituents are said to form the 'scope' of the element, and a common metaphor is that the element has some kind of 'command' or 'domination' within its scope.

As far as I know, no ancient texts hint at any notion of scope in this sense. Benson Mates [27] p. 30 thought he detected such a notion in a passage of Sextus Empiricus:

> "Indefinite", according to [the Stoics], are those [propositions] in which some indefinite part is dominant (*kurieúei*), as for example "Someone is sitting". ([36] II.97, p. 286)

(1)

As Mates explains it, the word 'Someone' has a 'scope' and the sentence 'lies in' this scope. But the usage of the word *kurieúei* makes it more likely that for Sextus Empiricus or his source the 'dominating' element meant the subject noun, or perhaps the first word in the sentence.

The Middle Ages provides a more plausible example. The fourteenth-century logician Walter Burley in [5] p. 26f (trans. [6] p. 109) notes that from a sentence of the form

$\phi(\text{some } X)$

(2)

we can sometimes infer that either $\phi(a)$ or $\phi(b)$ or ..., where $a, b, \ldots$ list all Xs. He adds that we destroy this property if we add a negative word immediately before 'some X', but not if we add a negative word either after 'some X' or some distance before it. He is telling us that the scope of a negation is the clause or phrase immediately following it. In fact he seems to have a word *dominium* for scope:

negatio non habet dominium supra praecedens.

(3)

Sadly this word *dominium* seems to be a flash in the pan; I've seen no evidence that it appears anywhere else in this sense. Note two other features of Burley's account. First, he defines the scope of a negative element in terms of its place in the *linear ordering* of the sentence. Second, his account of scope more or less works in Latin, but it's wildly wrong for some other languages, including his own native language of Middle English, as illustrated in [18] p. 71.

The Middle Ages also witnessed the development of the syntactic notion of government, called *regimen* in Latin and *ᶜamal* in Arabic. To say that an element X in a sentence 'governs' an element Y means, roughly speaking, that X determines the inflection of Y. One can ask whether X has a 'government scope' in the sense that if X governs Y then Y has to lie within a certain part of the sentence. Contemporary syntacticians certainly do ask this, but I don't know of any medieval thoughts along these lines.

3 Frege's introduction of quantifier scope

In 1879, in §11 of his *Begriffsschrift*, Gottlob Frege [10] presides over the birth and baptism of the notion of the scope of a variable. He calls it *Gebiet*. Using German letters for variables, he states that a universally quantified variable in a formula needs to be introduced with a quantifier symbol, so that the quantifier symbol

> *delimits the scope that the generality indicated by the letter covers. The German letter retains a fixed meaning only within its own scope.* (Frege's italics. ... *sie grenzt das Gebiet ab, auf welches sich die durch den Buchstaben bezeichnete Allgemeinheit bezieht. Nur innerhalb seines Gebietes hält der deutsche Buchstabe seine Bedeutung fest; ...*)

(1)

He explains that when two universally quantified variables have overlapping scopes within a formula, the quantifier symbols are needed in order to determine the scopes. (We ask – determine how?) He also introduces a convention that a Latin letter can also be used as a universally quantified variable, but in this case the scope is always

> *the content of the entire judgment.* (Frege's italics. ... *den Inhalt des ganzes Urtheils ...*)
>
> (2)

Frege certainly borrowed his use of Latin letters from common mathematical practice. Perhaps some mathematicians already used the word *Gebiet* in a sense that suggested its use to Frege; I have no information about this. But even without comparing with earlier texts, Frege's account in *Begriffsschrift* has the hallmarks of a first attempt. For example we are not told until (2) that the scope of a variable is a *content* rather than a part of a formula. Frege's explanations of content (more fully 'conceptual content', *begriffliche Inhalt*, [10] §3) are pretty minimal, and they certainly don't extend to explaining what it is for a symbol to have a meaning within a content. Nor has he told us what is the meaning (*Bedeutung*) of a variable letter; he surely can't mean what he calls *Bedeutung* in his later essays. Nor has he told us how the quantifier symbol delimits the scope. There is also a technical hitch: in the course of a proof we might find ourselves forming a compound from two formulas which universally quantify the same German letter. This should cause no problems of logic, because in the compound the scopes will be disjoint. But Frege's terminology will break down at this point, because he speaks of the scope of the *letter*, and in this case the same letter will have two different scopes associated with different quantifier symbols. We want to say rather that the scope belongs to certain *occurrences* of the variable; but then the scope is needed precisely in order to say which occurrences, so there is a threat of circularity!

In an essay that was written at most two years later than *Begriffsschrift*, but not published in his lifetime, Frege repairs one of the faults above by spellling out how the quantifier delimits a scope. His explanation is specific to the language of *Begriffsschrift*, in which formulas are trees. It amounts to saying that an occurrence of a quantifier marks out a unique subformula, namely the subtree immediately below it in the tree; the intended scope is then the content of this subformula ([11] p. 21 bottom). At first sight this was what he intended in *Begriffsschrift*; but there is a discrepancy. In *Begriffsschrift* he said

> Replacing a German letter everywhere in its scope by some other one is, of course, permitted. ([10] §11)
>
> (3)

That implies that the quantifier itself (the quantifier symbol plus the variable) lies inside the scope of the variable; but the clarification in [11] stipulates that the quantifier is *not* in the scope of the variable. We will come back to this point.

In *Grundgesetze der Arithmetik I* (1893) §8, Frege revised his definition of scope as follows:

> Now we call what follows a concavity accompanying a *German letter* ... the *scope* of the German letter which stands above the concavity. (Wir nennen nun das auf eine Höhlung mit einem *deutschen Buchstaben* Folgende ... das *Gebiet* des über der Höhlung stehenden deutschen Buchstaben.)
> (4)

This makes one major improvement on Frege's earlier definition: he now defines the scope as the part of the formula which is delimited by the quantifier, spelling out what counts as delimiting. In the earlier definition he had defined scope in terms of its function; the relevant part of the formula appeared only in the explanatory remarks. The new definition agrees with that in [11], and disagrees with [10], by excluding the quantifier itself from the scope. (The '...' represents an extra clause which spells out the meaning of universal quantification. It is redundant as a part of the definition, but it does confirm that the scope is now a syntactic object and that it excludes the quantifier.)

With this new definition Frege clears up all the problems I mentioned with the earlier definition, except that he still confuses the technicalities by assigning the scope to the variable letter. It was Hilbert and Ackermann who corrected this by making the scope belong to the *quantifier (occurrence)* rather than to the variable symbol:

> Further, to each universal or existential quantifier occurring in a formula, there belongs a definite constituent part of the formula to which it relates. We shall call this part the *scope* of the symbol in question. (... so gehört zu jedem in einer Formel vorkommenden All- oder Existential-zeichen ein bestimmter Bestandteil der Formel, auf den es sich bereicht. Diesen wollen wir den "*Wirkungsbereich*" des betreffenden Zeichens nennen. [16] p. 52)
> (5)

(See page 145 of [15] for an earlier formulation in Hilbert's 1917/18 lectures on 'Principien der Mathematik'.) Hilbert and Ackermann give examples to show how the quantifier scope is determined; these examples leave it unclear whether a quantifier lies within its own scope.

With hindsight it's surprising how little impact the notion of scope made before Hilbert and Ackermann. The second edition of *Principia Mathematica* [40] briefly mentions Frege's notion on page xxv (and includes the quantifier itself in the scope), but I'm not sure that the notion is ever used; Appendix ⋆8 on bound variables manages to avoid it in the style of Peirce, by always reducing to prenex normal form. Even Hilbert and Ackermann introduce the notion of scope *after* defining free and bound variables, and as a result their definition of free and bound variables is incoherent ([16] p. 52 top).

Russell's introduction of the notion of the scope of a definite description on page 70 of the first edition of *Principia* in 1913 deserves note.

4 Grammars 1: Stemmata

Both Frege and Hilbert gave their definitions of scope for formal languages of their own devising. If we go backwards in time from Frege, or forwards from Hilbert, we find logicians and linguists studying what we can broadly call scope phenomena, but in natural languages. Probably any notion of scope needs to call on some assumptions about grammatical structure, because it involves delimiting constituents of sentences. To make sense of the literature, we should distinguish between two kinds of grammar: dependency grammar and constituency grammar.

In dependency grammar a sentence is analysed by displaying which word occurrences 'depend on' which. The dependences in a sentence are usually shown as a tree branching downwards, where the nodes are identified with the occurrences of words in the sentence, and the successors of a word are the words that depend on it. I write an arrow from a dependent word to the word that it depends on. It's probably more usual to write the arrows in the other direction; but writing the arrows as I do suggests a picture that we meet in some writers. Namely, the sentence is constructed by starting with the top node and successively *attaching* successor nodes to the nodes that they depend on, so that the tree is built from the top downwards. This need not be more than a metaphor.

The tree diagram showing the analysis of a sentence is called a 'stemma' (the name is from Lucien Tesnière). We can almost read off the sentence from the stemma, roughly speaking by reading the words from left to right. The successors of a node n will run from left to right, but the stemma also needs to show the position of n among its successors. In more formal presentations of dependency grammar, the grammatical sentences of a language are characterised by the fact that they are exactly the strings that can be read off from the stemmata of the grammar. We won't pursue this idea, but we will need the assumption that each sentence of the language has a stemma, which is unique unless the sentence is structurally ambiguous. Each node determines a phrase (in fact a constituent of the sentence), which consists of that node and all nodes that can be reached by going downwards from it. The node is called the ‘head’ of the phrase that it determines.

The artificial languages of logic tend to have dependency grammars – counting the formulas as sentences and the symbols as words. This includes both Frege’s *Begriffsschrift* and Hilbert’s formal languages. For example a typical formula of propositional logic, in Hilbert style, might be written

$$(\neg(b \rightarrow \neg a) \rightarrow a) \tag{1}$$

A stemma for this formula might be written as follows (where for simplicity I ignore the parentheses and the information about placing a node relative to its successors):

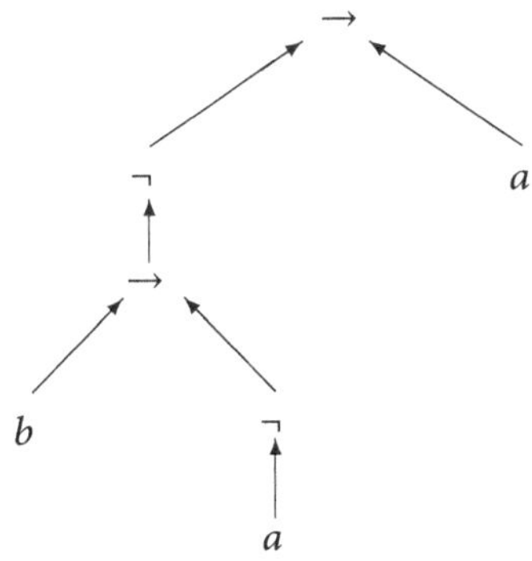

(2)

Here the lowest ¬ determines the phrase $\neg a$, then the → immediately above it determines the phrase $(b \rightarrow \neg a)$, and so on upwards. The general pattern is that an n-ary function symbol has n immediate successors, each of which determines a subformula or term which is an argument of the symbol.

Here for comparison is Frege’s notation for the same formula in *Begriffsschrift*:

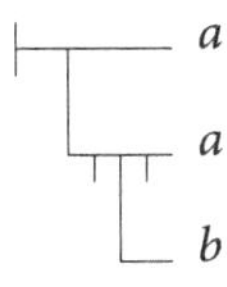

(3)

The formula is in fact its own stemma, though the tree is written roughly NW/SE rather than downwards. In Frege’s notation → becomes a branch point and ¬ becomes a short line protruding downwards.

Now in *Begriffsschrift* a universal quantifier is a function symbol of arity 1, so any occurrence of it in a stemma has a single successor node. Frege’s definition (4) says that the scope of this occurrence is the subformula determined by this successor node. This is the only reasonable general procedure for using symbols to determine phrases in a dependency grammar. Exactly the same formalism allows us to talk of the 'scope' of any occurrence of a function symbol of arity 1; in particular each occurrence of the negation symbol has a scope which is a subformula. I leave it to the reader to check how this notion of scope carries over to formulas which are strings in the Hilbert style.

5 Grammars 2: Phrase markers

By contrast a constituency grammar analyses a sentence by showing which are the grammatical constituents of the sentence, including the separate words. The analysis is written as a tree diagram branching downwards, but now the nodes (except for the terminal nodes at the bottom ends of branches) are in 1–1 correspondence not with the words but with the constituents; a line going up from a non-terminal node *m* to a node *n* indicates that the constituent for *m* is a subconstituent of the constituent for *n*. The convention is to label each non-terminal node with a symbol for the grammatical category of the corresponding constituent. The bottom non-terminal nodes are the categories of single words, and the words themselves are added as terminal nodes below these non-terminal nodes. Tree diagrams of this kind are called 'phrase markers'.

The artificial languages of logic can be given constituency grammars, though this is unusual in the logical literature. Then the phrase marker for the formula $(\neg(b \to \neg a) \to a)$ will be the following diagram or some close variant of it. Here the categories are F(ormula), U(nary truth-function) and B(inary truth-function), and again we ignore parentheses:

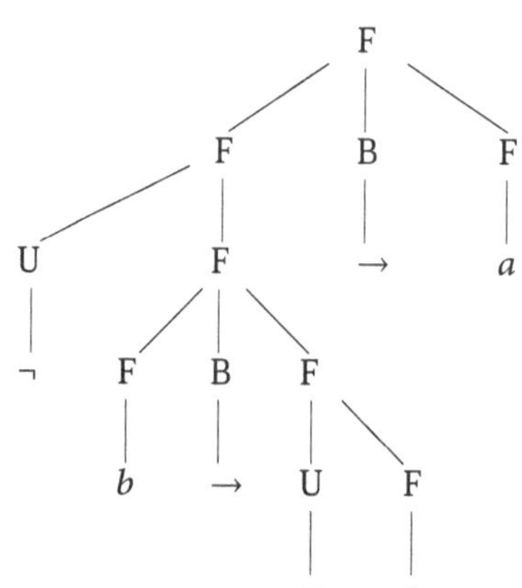

(1)

There are seven F's, corresponding to the seven subformulas of the formula.

In the mid 20th century stemmata fell almost entirely out of use in linguistics, and to a great extent phrase markers took their place. Phrase markers tend to be more elaborate than stemmata, but they beg fewer questions. For example the only constituents of a sentence that can be read off from the stemma of the sentence are those with a head word. So dependency grammar has become a minority interest. I. A. Mel'čuk is one respected linguist who maintains 'dependency syntax' [28]; I recommend his book for the light it throws on the mind set of earlier logicians and linguists who used stemmata.

Linguists find themselves examining various phenomena that involve one word w in a sentence acting on some part c of the sentence. For example

- w is a noun and c is a clause within which w governs the inflections of any verbs.
- w is a negative word and c is the part of the sentence where, because of w, we can have negative polarity phrases like 'at all'.
- w is a name and c is the part of the sentence within which every pronoun coreferential with w has to be reflexive.

So we can reasonably ask for a way of delimiting the relevant kind of scope of w, or

> defining the syntactic domain of a given node — roughly, the portion of the tree consisting of those nodes which a given node bears the structural relation to.

(2)

as Tanya Reinhart [34] p. 13 puts it. Here we meet the problem that in phrase markers, unlike stemmata, there is no canonical way for a node to delimit a constituent. So one has to search around; Reinhart goes on to describe and dismiss some early attempts. Interesting for us is her view that as an empirical fact, left-right relations in the tree 'play no role in restricting the semantic interpretation of sentences' ([34] p. 18). (So much for Burley.)

Edward Klima [26] proposed an answer, for the special case of the scope of negative words. He wanted a notion of 'constituent X is in the scope of negation Y'. With the terminology slightly adjusted, his definition reads:

> A node is *in construction with* another node if the former is below the first branching node that is above the latter. ([26] p. 297)

(3)

I don't know if Klima meant 'another' literally. Assuming not, the words that are in construction with a node n are precisely those in the smallest constituent that contains both n and at least one word not in n.

Reinhart [33] and Hodges [17] independently proposed Klima's definition as a general recipe for scope relations, Reinhart within linguistics and Hodges within logic. Reinhart reversed the direction of the relation and gave the result the name 'c-command', where 'c' is for constituent. In her formulation of [34] p. 18,

> Node A c-commands node B iff the branching node most immediately dominating A also dominates B
>
> (4)

(where 'dominates' means 'is strictly above in the tree ordering'). The definition in Hodges [17] p. 78 reads:

> If S is some sentence with a phrase-marker supplied, and P is a part of S consisting of one or more words, but not the whole of S, then we define the *scope* of P to be the smallest constituent of S which contains both P and something else besides.
>
> (5)

Reinhart credits Klima. I regret that I didn't; like Reinhart I had read [26], but I had forgotten his definition until I read Reinhart more recently.

There are some differences between Reinhart's version and mine. First, like Klima she defines a relation between an element and a *set* of constituents; I define a relation between an element and a *single* constituent. Reinhart's constituents are my constituent together with all its subconstituents; my constituent is the maximal of her constituents. So her notion and mine are interdefinable, and in fact Chomsky spells out one direction in [7]. On page 60 he defines the 'domain' of an element as 'the minimal phrase containing it'; then on page 162 he stipulates that an element c-commands the elements in its domain (adding a further condition that I discuss below). It's not clear to me whether this difference between the two definitions is more than a matter of style. One could write out reasons why the first definition is more helpful for linguistics and the second more helpful for logic, but in cold print the reasons look thin.

A second difference is in extra conditions added. We saw that already in Frege there is some equivocation about whether an element lies within its own scope. I deliberately chose a definition that does put the element within its scope, partly because I find it helpful for defining free and bound occurrences of variables, and partly because general definitions should be simple. But there are applications of scope where it seems natural to exclude the element from its scope: for example the scope of a negation. I know less about the situation in linguistics, but it's clear that linguists have similar discussions about excluding certain items from the scope. One common requirement – for example at [7] p. 162 – is that the c-command of an element E excludes all elements contained in E.

In fact Reinhart [33] (according to Reinhart [34] p. 27 Note 1) added a condition that A doesn't c-command B if A and B are distinct elements of the same branch (i.e. if one dominates the other). Reinhart states [34] p. 19 that Klima added the same condition. There is no such condition in my copy of [26], but Klima or other linguists could have added it in discussions of his paper.

6 Ibn Sīnā opens up the field of logic

There must certainly be many more things to say about the development of scope in the twentieth century. But I turn back now to a previously unnoticed corner in the history of scope. During the first half of the 11th century the logician Ibn Sīnā, known in the West as Avicenna, was working in Persia and writing mainly in Arabic. He made what as far as we know was the first attempt to deal systematically with sentences containing mixtures of universal and existential quantifiers. Dealing with these, he ran into some problems that today we understand in terms of scope. But he had no notion of scope. Also he knew that he was missing something, and he cast around for ways of filling the gap with the tools that he did have.

Those tools came from Aristotelian Logic, though not all from Aristotle himself. Ibn Sīnā was well informed — probably much better than we are today — about the work of Roman Empire scholars such as Galen, Alexander of Aphrodisias, Porphyry, Themistius, Philoponus. He could also take inspiration from the writings of earlier Arabic logicians, chiefly Al-Fārābī. Today only historians are familiar with these names. But above all Ibn Sīnā was an independent thinker who took nothing for granted, even from Aristotle. So Ibn Sīnā presents us with a rare spectacle: a highly intelligent logician trying to make sense of scope problems without using the notion of scope, and only partly succeeding.

Two lines of thought brought Ibn Sīnā to consider sentences with more than one quantifier. The first was something that he called the 'three-way division' (*taṯlīṯ fil qisma*, [23] 41.5); it consisted of three ways of reading a sentence 'Every *B* is an *A*' as a modally necessary proposition. The first reading — the equivalent of today's 'true in all possible worlds' — was that the sentence should be true at all times. This would account for the truths of theology and mathematics; for example we read a statement that God is alive as meaning

In all times and situations, God is alive. (1)

By contrast the sentence

Everything that walks moves. (2)

can't be true of all its subjects at all times, because people who walk stop walking, and later they die. Accordingly the second reading is

Everything that walks moves for so long as it is walking. (3)

Arabic logicians later than Ibn Sīnā called this the 'descriptional' (*waṣfī*) reading of the sentence. The third reading takes (2) as necessary in the sense that

Everything that walks moves for so long as it moves. (4)

Ibn Sīnā claims that this three-way division goes back to Aristotle himself.

In fact it seems likely that the division is Ibn Sīnā's own take on the three meanings of 'necessary' discussed by Aristotle's pupil Theophrastus in his lost work *Prior Analytics*. The third reading agrees with Theophrastus' third meaning as reported by Alexander of Aphrodisias *Pr. Anal.* 32a19, [9] 100B. Elsewhere Theophrastus is reported to have endorsed both the other two readings, the first by Philoponus *Pr. Anal.* 124.13, [9] 106D, and the second by Alexander *Pr. Anal.* 25a29, [9] 100A. At *Pr. Anal.* 30b37, [9] 100D, Alexander seems to be claiming to find anticipations of Theophrastus' classification of meanings of 'necessary' already in Aristotle's *De Interpretatione* and *Prior Analytics* i. See [8] Chapter Two for Theophrastus' views on necessity and their treatment by Alexander.

As far as we know (though the written record is sadly incomplete), before Ibn Sīnā the three readings were regarded as senses in which a sentence 'Every *B* is an *A*' can be counted as necessary. But Ibn Sīnā took them as three sentence-forms in their own right, with a subject quantifier 'Every' and a second quantifier which is implicit and ranges over time. One of Ibn Sīnā's most characteristic teachings was that we nearly always mean more than we say, and it falls to the logician to make explicit the elements of meaning that we leave unspoken or unwritten. So for Ibn Sīnā, the three-way division describes three types of two-quantifier sentence. But in all three cases the two quantifiers are both universal, so there is not yet a problem about comparing scopes.

The second line of thought that led Ibn Sīnā to multiple-quantifier sentences was his conviction that if logic is the tool of the sciences, then logicians should be prepared to look at what scientists actually say — not just the forms of the sentences that they use, but how an intelligent and informed reader would understand them. This led him to a much wider collection of sentence forms than he found in the Aristotelian literature, even for sentences of the form 'Every *B* is an

A':

It's unhelpful that the definitions given [by Aristotle] always interpret the content in [the style of (1) or (3)]. One ought to look for an account of all this that covers the broad range. ([23] 41.11f)

(5)

When we start to look at that broad range, as Ibn Sīnā himself does in [23] i.3, we quickly find that the time quantifier is sometimes understood as existential:

Everybody who travels from Rayy to Baghdad reaches Kermanshah. ([23] 22.12)

(6)

The normal understanding of this sentence would be that for everyone travelling from Rayy to Baghdad *there is a time during the journey* at which that person is in Kermanshah. But that reading won't work for

Everything that breathes in breathes out. ([23] 23.5)

(7)

(Ibn Sīnā reads 'breathes out' as 'breathes out sometimes', an existential quantification. This may not be a very convincing reading, but it's probably the best we can do with just existential and universal quantifiers.) And so on and so on.

7 Negations and quantifiers

Aristotelian logic taught that for every proposition ϕ there is a contradictory negation $\bar{\phi}$ which is true when ϕ is false and false when ϕ is true; also $\bar{\bar{\phi}} = \phi$. One of the tasks of a logician is to find, for every sentence form, an appropriate form for its contradictory negation. So Ibn Sīnā set to work in [23] i.5 to find contradictory negations for some of his new sentence forms. (There is a provisional commented translation of [23] i.5 on my website at `http://wilfridhodges.co.uk`.)

It's important that Ibn Sīnā's example sentences are all in Arabic, not in an artificial language and not even in a regimented logical fragment of Arabic. Ibn Sīnā was well aware that Aristotelian logicians had a tendency to adopt a language out of line with normal scholarly usage, and he was very much against this tendency:

The effect is that when we say a sentence ϕ we don't at all mean by ϕ what ϕ ought to mean. Instead of that, when we say ϕ, we say it meaning something that the artificial convention says it should mean. And you learn that 'this is a stone' and other drudgery. ([23] 45.9f)

(1)

I think the point of the remark about stones is that any artificial language that we adopt is likely to enshrine those features of language that we are already aware of; so if we work with a monadic logic for classifying objects, we will adopt conventions that tend to hide from us the relational complexities of real discourse. A little earlier he has complained that the standard Aristotelian conventions make it hard to handle some logical features of language 'intuitively' (*ᶜalā baṣīratin*, [23] 41.15).

Ibn Sīnā's broader beliefs about scientific discovery feed through into his logical practice. We learn by familiarising ourselves with many examples of the material under discussion, and noting what we can see directly about them. With familiarity we come to see the regularities, and these lead us to better definitions of the nature of the things we are studying. Then knowing the definitions, we can not only deduce the observable facts; we can also see why they are facts. Applying these ideas to the logic of contradictory negation, Ibn Sīnā calls on intuition to give us contradictory negations for sentences one at a time. Practice with these should teach us the rules, first for particular sentence forms, and then for sentences in general. The general rules should reflect the reasons for the inference relations, and these lie in the meanings of quantifiers among other words.

This view of the task governs the contents of *Qiyās* i.5, except that Ibn Sīnā reverses the order. Probably he does this to highlight the main points that he wants to make. He begins with the meanings of quantifiers, and he explains why he has difficulties in using these meanings to give general rules. Since he is unable to resolve the difficulties, he finishes with what he can do, namely to give intuitive solutions for a range of particular sentences.

Ibn Sīnā's procedure for finding the contradictory negation of a sentence is to put the Arabic negative verb *laysa* in front of the sentence, and then work the negation inwards. The following example summarises [23] 46.10–47.1. Take the sentence

Every *B* is an *A* for a limited time. (2)

We get a contradictory negation by adding *laysa* at the beginning, in effect

It is not the case that every *B* is an *A* for a limited time. (3)

Ibn Sīnā invites us to think of situations in which (3) is true. For example it's true in both of the following situations:

Always, no *B* is an *A*. (4)

Every *B* is always an *A*.

(5)

Do these and (2) exhaust all possibilities? Clearly not: it could be that

Some *B* is never an *A*.

(6)

But this reminds us that another situation where (3) is true is that

Some *B* is permanently an *A*.

(7)

Inspection shows that if (3) is true then at least one of (6) and (7) must be true. (In fact Ibn Sīnā overlooks the case where there are no *B*s, which he elsewhere explains should make (3) true and (6), (7) false; but let this pass.) So a contradictory negation for (2) can be written as

Either some *B* is permanently an *A*, or some *B* is never an *A*.

(8)

where the disjunction is inclusive.

We notice at once that Ibn Sīnā's procedure is highly informal. He seems to be inviting us to ignore the syntactic form of the sentences and ask instead 'What kind of situations would I have in mind if I said this?'. In fact it seems almost perverse that he doesn't start by noting that in (2) the leading quantifier is the 'Every' and not the temporal quantifier. Noting that, one can reason at once that 'Not every' is equivalent to 'Some not', as Ibn Sīnā had pointed out in his *ᶜIbāra*, [22] 93.1f (with the rider that the two forms come apart when they quantify over an empty set). Does Ibn Sīnā simply not notice that the quantifier 'Every' has wider scope than the temporal quantifier?

The situation is more confusing than one might think. The verb *laysa* can't always be read as 'It is not the case that'. Ibn Sīnā himself says, for example at [22] 54.6f, that the Arabic *laysa kull* (literally 'not every') can be paraphrased by *laysa baᶜḍ* (literally 'not some'). I am not sure what is happening here. There are some cross-linguistic facts that may be relevant, for example that in English 'Not every swan is white' is grammatically acceptable but 'Not some swan is white' is not. It may also be relevant that the verb *laysa*, which inflects with a personal suffix, picks up this suffix from the subject of the sentence which it introduces ([2] p. 114ff). Arabic negations are fearsomely complicated, and Ibn Sīnā doesn't hide the complications.

8 The semantics of quantifiers

At least Ibn Sīnā's informal procedure gets him plausible candidates for contradictory negations. But when he tries to reason them out in terms of the meanings of the quantifiers, he runs into difficulties:

> When we say 'Every B is an A' and we want to take into account the time in the sentence 'Not every B is an A', since this is one of [the things that have to satisfy] a condition in order to have a contradictory negation, this makes difficulties for us. ([23] 386f.)

(1)

How so?

When Ibn Sīnā intends to describe the meaning of a sentence ϕ, sometimes he gives necessary and sufficient conditions for the sentence to be true. But more often he aims to explain, in the same language as ϕ itself, what is in the mind of someone who asserts ϕ. In a long discussion at [22] 21.10–23.10 he compares two sentences:

> He is walking.

(2)

> Someone is walking.

(3)

The states of the world in which someone could use either of these two sentences to make a true statement are the same. Also the two sentences agree in not telling the listener who it is that is walking. But the difference between the two, Ibn Sīnā explains, is that in (2) the speaker has to have some individual in mind for 'he' (otherwise there is no truth value), whereas this is not needed in (3). So the primary difference between (2) and (3) is that (2) demands more knowledge on the part of the speaker. The extra knowledge required of the speaker in (2) is the same knowledge that the speaker could have expressed by saying

> Zayd is walking.

(4)

where Zayd is some individual. For Ibn Sīnā the meaning of the name 'Zayd' is an idea, a thinkable content, that identifies Zayd; so it is the same idea that the speaker of (2) needed to have in mind. Note that passing from (3) to (4) eliminates the existential quantifier.

In *Qiyās* i.5 Ibn Sīnā approaches two-quantifier sentences like 'Every human breathes' and 'Every horse sleeps'. As above, he reads these as short for

> For every human there is a time at which he breathes. (5)

and likewise with the horses. His instinct is to follow the same route that he took with (3) above, and ask what kind of knowledge would enable a person who stated one of these sentences to remove the implied existential quantifier.

The question ties in closely with Ibn Sīnā's theory of science. For example (and here I summarise a discussion in *Burhān*, [24] i.9) one typical piece of medical knowledge is that

> If a person takes a preparation of scammony, usually his bile is purged. (6)

Here there are a universal quantification over people and a 'most' quantification over times or circumstances. (As he remarks at [23] 161.13, medical knowledge tends to be about what usually happens.) Medical research must aim, he says, to remove the 'usually' by discovering the exact circumstances in which scammony has this effect. The result would be to replace (6) by some statement of the form

> If any person takes scammony under such-and-such circumstances, then his bile is purged. (7)

Here we have only universal quantifiers. Another of his favourite examples is the astronomical knowledge which allows us to spell out the circumstances in which a satellite of a planet gets eclipsed.

Ibn Sīnā notes at once that if the subject of the sentence is an individual, we are back in the position he discussed in *ᶜIbāra*:

> We should note that when there is a single [subject individual], the time of the predication can be made specific. ([23] 39.8) (8)

(Here and below, a 'time' for Ibn Sīnā is a period of time, not an instant.) But with a universally quantified subject we can't hope to remove the existential time quantifier by naming a particular time, because 'each individual has a different time' ([23] 40.1f). Instead, we would have to look for an assignment of times to individuals, or (in modern language that Ibn Sīnā didn't have) a function from individuals to times. Thus a person who asserts 'Every human breathes' has in mind that there is such an assignment.

We should compare Ibn Sīnā's position here with some other items in the logical literature. First, the assignment of times to individuals can be thought of as a subset of the two-dimensional universe of pairs (person, time). Since Ibn Sīnā regularly calls on such ordered pairs, I would be amazed if this was not how he pictured the assignment. The parallel with Alexander's trick is striking: Alexander handles two universal quantifiers by universally quantifying over the whole array of ordered pairs; Ibn Sīnā handles a universal and an existential quantifier by quantifying universally over a *subset* of the array. This comparison goes beyond what Ibn Sīnā himself says, since in this context he doesn't mention Alexander's trick. But given Ibn Sīnā's general approach, there is a clear pathway from Alexander's trick to Ibn Sīnā's function semantics. We saw earlier how there was a natural, though different, pathway from Alexander's trick to Peirce's dimensional product and sum semantics for multiple quantifiers.

Second, Ibn Sīnā is using a version of what today we call Skolem function semantics for quantifiers. Logicians know this semantics from Leon Henkin's use of it for partially ordered quantifiers, [14] Section C. (That work of Henkin is at least indirectly an ancestor of Jouko Väänänen's Dependence Logic [39].) Similar semantics for sentences like Ibn Sīnā's are now a familiar part of the landscape in formal semantics of natural languages; Mark Steedman's recent book *Taking Scope* [38] develops this approach in detail.

Ibn Sīnā evidently hopes that removing the existential quantifier will make his sentences easier to handle, and in particular it will make it easier to find their contradictory negations. With that hope he takes a sentence like (5), he skolemises the sentence, and then he puts 'It is not the case that' in front of it. We know from his procedure discussed in section 7 above that he hopes to push the negation down into the sentence. But instead he grinds to a sudden halt, believing that something has gone seriously wrong.

> We land ourselves in the following difficulties when we introduce this point. Suppose we say 'Every *B* is an *A*' and the times are different because each individual has a different time. Then if we say 'Not (*laysa*) every *B* is an *A*', it's not possible for us to make this sentence point to the time which is specific to each individual. ([23] 39.15–40.3)

(9)

He then spends several pages trying to find a way out by using different definable functions from individuals to times.

The modern parallels show that Ibn Sīnā hasn't made any major error yet, so there must be a way out of his impasse. In fact there are two. The first is to note that when an existential quantifier is inside the scope of a negation, then pushing the negation past it will turn it into a universal quantifier, and vice versa. So if we put 'It is not the case that' in front of 'Every *B* is an *A*', we should aim to skolemise the 'every *B*' rather than the 'an *A*'. Taking this route, Ibn Sīnā would have to abandon his strategy of skolemising before applying the negation. The second route, preferred by Steedman [38] p. 66f rule 2(a), is to note that putting a quantifier 'there is a function ...' inside the scope of a negation makes it behave like a quantifier 'for every function ...'; so Ibn Sīnā can stick with his assignment of times to individuals, provided he makes it an indeterminate one (*ğayr mu^c ayyan*) and quantifies universally over it – in modern terms he would need the notion of a universal function quantifier.

Either route would have tested Ibn Sīnā's logical abilities to the limit. But more than that, either route would have required him to have some sense that adding 'It is not the case that' at the front of a sentence places the whole sentence into the scope of the negation. It would have required some awareness of what a negation does to items within its scope, and in particular of how it switches positive occurrences to negative and vice versa. For some years now I have been searching through Ibn Sīnā's logical writings to find any sign of recognition of the difference between positive and negative occurrences, and so far I have drawn a complete blank. The monotonicity arguments beloved of the Scholastics [35] are completely missing in Ibn Sīnā. He never mentions the dictum de omni et nullo as a principle behind syllogisms. He has no notion of distributed and undistributed. The closest that he comes is to recognise the difference between sentences that affirm and sentences that deny; but in [19] I assemble evidence that the notion he is working with here is along the lines of the notion of sentence negation studied by some modern syntacticians, and not a logical notion at all.

(Added in August 2014:) In *Qiyās* vi.3 Ibn Sīnā [23] starts to develop an extension of syllogisms where we are not restricted to the traditional 'Every', 'No', 'Some', 'Not all' sentences, and within the new sentence frames we are allowed to take negative terms 'not-*A*' etc. (He does this within the framework of propositional logic, but this is irrelevant for present purposes.) In particular he discusses which premise-pairs give rise to a syllogistic conclusion within this extended framework. As it happens, there is a very simple criterion that answers this question: a premise-pair gives rise to a syllogistic conclusion if and only if the middle term occurs positively in one premise and negatively in the other. We find that Ibn Sīnā mistakenly claims that several premise-pairs don't have syllogistic conclusions when in fact they do. This includes some cases that he discusses at length. It seems inconceivable that he could have made these mistakes if he had known the simple criterion – and indeed he never mentions the criterion. It seems odd too that he never noticed the criterion, until we place it alongside the other evidence that he never developed any intuition of the difference between a positive and a negative occurrence.

9 Obstacles to scope

So we have to reckon that Ibn Sīnā had no concept of the scope of a negation; in fact I think he had no concept of any kind of scope. We are none of us good at creating concepts *ex nihilo*. But I can point to one obstacle that Ibn Sīnā would have had to overcome if he had tried to adopt a notion of scope.

The obstacle is of Ibn Sīnā's own making, and it is the product of insight rather than ignorance. He knows that the inference relations between sentences are determined by the meanings of the sentences; translation respects entailment. But also he is bilingual Arabic-Persian, and he knows that translation from one language to another can make a radical difference to the form of a sentence. Translation can reverse the order of two words; it can split one word into two or more, and it can contract a phrase into a single word. So the logician needs to be able to look behind the sentences and detect the structure of their meanings.

Now for Ibn Sīnā, compound meanings are constructed by attaching meanings to other meanings. The result is a kind of dependency grammar of meanings. Ibn Sīnā is reasonably consistent about what attaches to what – though for a coherent account I have added to his picture the assumption that subject and predicate terms attach to the copula. 'The subject and the predicate don't have to come in one particular order by nature' ([22] 31.3f), so they need to be distinguished by their type of attachment, shown by the labels *s* and *p*. For example the meaning structure of the sentence 'Every horse breathes' will be

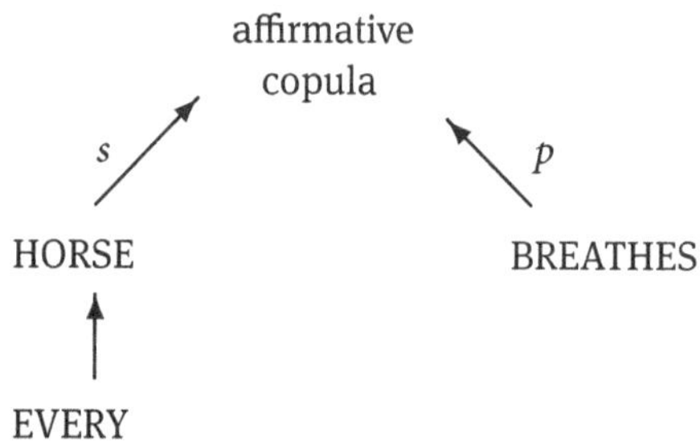

(1)

If we want to refine this representation to show the intended meaning that every horse breathes at some time, we first have to add an indeterminate reference to the time (Ibn Sīnā calls this addition a 'relation', *iḍāfa*), and then presumably we need to quantify the indeterminate. Probably the picture should be

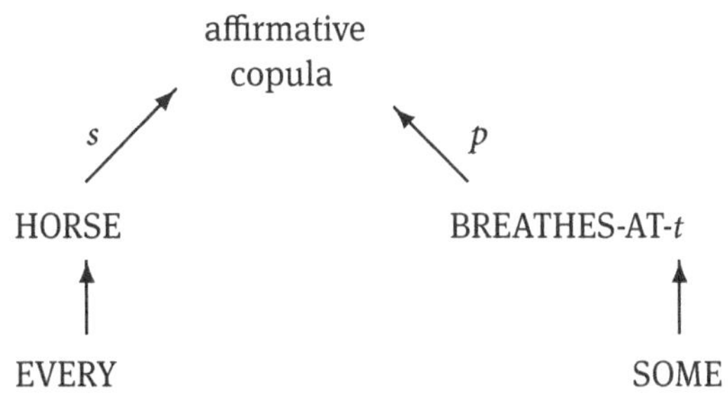

(2)

Now consider the problem of defining the scopes of the two quantifiers in terms of this diagram. As we saw earlier, there is only one natural way in a stemma to indicate that one element determines a phrase: namely where the phrase is attached to the element. But here nothing is attached to either of the quantifiers. Within this structure there is no way for the notion of comparing scopes to get off the ground.

The moral is that even if Ibn Sīnā had had a notion of quantifier scope – and there is no indication that he did have one – he wouldn't have been able to incorporate it within his semantic theory. The position for scopes of negations is similar, though the details are different. For Ibn Sīnā, a negation attaches to what it negates. So (in an example that he takes from Aristotle, *Qiyās* 492.17f) we can attach NOT to WHITE in the compound meaning WHITE STICK and get NOT-WHITE STICK; but where should we attach NOT to get NOT A WHITE STICK? To quote Mel'čuk [28] p. 28:

> ... there do exist some constructions which do not lend themselves to a satisfactory description in terms of [dependency trees] alone. These are, for instance, modification structures in which an element (X) modifying the head of a phrase can contrast semantically with the same element modifying the whole phrase ...

(3)

Ibn Sīnā doesn't say how he solves the problem. My guess is that he does think of attaching NOT to the compound phrase as a whole, but he hasn't thought through the consequences of this departure from a strict dependency grammar.

One should add that even logicians who thought of propositions as linearly ordered objects haven't always called on the linear ordering to resolve scope ambiguities. For example Mitchell [29] p. 87 distinguishes what we would write as

$\forall u \exists v F(u, v)$ and $\exists v \forall u F(u, v)$ as

> Every part of U, during some part of V, is F;
> Every part of U, during the same part of V, is F.

(4)

Ibn Sīnā uses 'same' in a similar way to resolve the scope ambiguity ([23] 40.3).

(Added in August 2014:) Two papers written more recently articulate points above in more detail. [20] gives further information on how Ibn Sīnā structures the meanings of sentences. It notes in particular that he sometimes gives his time quantification a wider scope than the other quantifier by attaching the time quantifier to the other one. [21] discusses in more detail how Ibn Sīnā handles sentences with two quantifiers, and makes further points about the relationships to Galen and Mitchell.

Bibliography

[1] Alexander of Aphrodisias, *In Aristotelis Analyticorum Priorum Librum I Commentarium*, ed. M. Wallies, Reimer, Berlin 1883.

[2] Joseph E. Aoun, Elabbas Benmamoun and Lina Choueiri, *The Syntax of Arabic*, Cambridge University Press, Cambridge 2010.

[3] Aristotle, *The Complete Works of Aristotle, Volume One*, ed. Jonathan Barnes, Princeton University Press, Princeton NJ 1984.

[4] John Buridan, *Summulae de Dialectica*, trans. Gyula Klima, Yale University Press, New Haven 2001.

[5] Walter Burleigh, *De Puritate Artis Logicae Tractatus Longior*, ed. P. Boehner, Franciscan Institute, St Bonaventure NY 1955.

[6] Walter Burley, *On the Purity of the Art of Logic: The Shorter and the Longer Treatises, trans. P. V. Spade*, Yale University Press, New Haven CT 2000.

[7] Noam Chomsky, *Knowledge of Language: Its Nature, Origin, and Use*, Praeger, New York 1986.

[8] Kevin L. Flannery, *Ways into the Logic of Alexander of Aphrodisias*, Brill, Leiden 1995.

[9] William W. Fortenbaugh et al., eds., *Theophrastus of Eresus: Sources for his Life, Writings, Thought and Influence* Part One, Brill, Leiden 1992.

[10] Gottlob Frege, *Begriffsschrift*, Nebert, Halle 1879; trans. Stefan Bauer-Mengelberg in *From Frege to Gödel*, ed. Jan van Heijenoort, Harvard University Press, Cambridge MA 1967, pp. 5–82.

[11] Gottlob Frege, 'Booles rechnende Logik und die Begriffsschrift', in *Nachgelassene Schriften* ed. H. Hermes et al., Meiner, Hamburg 1969, pp. 9–52.

[12] Gottlob Frege, *Grundgesetze der Arithmetik I*, H. Pohle, Jena 1893.

[13] Galen, *Institutio Logica*, ed. and trans. John Spangler Kieffer, Johns Hopkins Press, Baltimore 1964.

[14] L. Henkin, 'Some remarks on infinitely long formulas', *Infinitistic Methods*, Pergamon Press, Oxford and Pańswowe Wydawnictwo Naukowe, Warsaw 1961, pp. 167–183.

[15] David Hilbert, *David Hilbert's Lectures on the Foundations of Arithmetic and Logic 1917–1933*, ed. W. Ewald et al., Springer, Berlin 2013.
[16] D. Hilbert and W. Ackermann, *Grundzüge der Theoretischen Logik*, Springer, Berlin 1928; trans. of second edition (1937) by L. M. Hammond et al., *Principles of Mathematical Logic*, Chelsea, New York 1950.
[17] Wilfrid Hodges, *Logic*, Penguin, Harmondsworth 1977.
[18] Wilfrid Hodges, 'Detecting the logical content: Burley's "Purity of Logic" ', in *We Will Show Them: Essays in Honour of Dov Gabbay II*, ed. S. Artemov et al., College Publications, London 2005, pp. 69–115.
[19] Wilfrid Hodges, 'Affirmative and negative in Ibn Sīnā', in *Insolubles and Consequences: Essays in honour of Stephen Read*, ed. Catarina Dutilh Novaes and Ole Thomassen Hjortland, College Publications, London 2012, pp. 119-134.
[20] Wilfrid Hodges, 'Ibn Sīnā, Frege and the grammar of meanings', *Al-Mukhatabat* 5 (2013) 29–60.
[21] Wilfrid Hodges, 'The move from one to two quantifiers', in *The Road to Universal Logic*, ed. Arthur Buchsbaum and Arnold Koslow, Springer 2014.
[22] Ibn Sīnā, *Al-ᶜIbāra*, ed. M. El-Khodeiri et al., Cairo 1970.
[23] Ibn Sīnā, *Al-Qiyās*, ed. S. Zayed et al., Cairo 1964.
[24] Ibn Sīnā, *Al-Burhān*, ed. A. Afīfī et al., Cairo 1956.
[25] Ibn Sīnā, *Al-Safsaṭa*, ed. A. F. Ehwani et al., Cairo 1958.
[26] Edward Klima, 'Negation in English', in Jerry A. Fodor and Jerrold J. Katz eds., *The Structure of Language*, Prentice-Hall, Englewood Cliffs NJ 1964, 246–323.
[27] Benson Mates, *Stoic Logic*, University of California Press, Berkeley CA 1961.
[28] Igor A. Mel'čuk, *Dependency Syntax: Theory and Practice*, SUNY Press, Albany NY 1988.
[29] Oscar Mitchell, 'On a new algebra of logic', in C. S. Peirce et al., *Studies in Logic*, Little, Brown and co., Boston MA 1883, pp. 72–106.
[30] C. S. Peirce, Untitled fragment on the Algebra of Logic 1884, in *Writings of Charles S. Peirce, Volume 5, 1884–1886*, ed. C. J. W. Kloesel, Indiana University Press, Bloomington IN 1993, pp. 109f.
[31] C. S. Peirce, 'On the algebra of logic', *American Journal of Mathematics* 7 (1885) 180–202.
[32] Plato, *Euthydemus*, in *Platonis Opera* III, ed. J. Burnet, Clarendon Press, Oxford 1903, pp. i.271–307.
[33] Tanya Reinhart, *The Syntactic Domain of Anaphora*, PhD Dissertation, MIT, Cambridge MA 1976.
[34] Tanya Reinhart, *Anaphora and Semantic Interpretation*, University of Chicago Press, Chicago IL 1983.
[35] Víctor Sánchez Valencia, 'Monotonicity in medieval logic', in *Language and Cognition 4: Yearbook (1994)*, ed. A. de Boer, H. de Hoop and H. de Swart, Rijksuniversiteit Groningen, Groningen 1994, pp. 161–174.
[36] Sextus Empiricus, *Against the Logicians*, with trans. by R. G. Bury, Loeb, Heinemann, Cambridge MA 1961.
[37] William of Sherwood, *Treatise on Syncategorematic Words*, trans. Norman Kretzmann, University of Minnesota Press, Minneapolis MN 1968.
[38] Mark Steedman, *Taking Scope: The Natural Semantics of Quantifiers*, MIT Press, Cambridge MA 2012.
[39] Jouko Väänänen, *Dependence Logic*, Cambridge University Press, Cambridge 2007.

[40] Alfred North Whitehead and Bertrand Russell, *Principia Mathematica* I, Cambridge University Press, Cambridge, 1st ed. 1913, 2nd ed. 1927.

Jan Hubička* and Jaroslav Nešetřil

Universal Structures with Forbidden Homomorphisms

Abstract: We relate the existence problem of universal objects to the properties of corresponding enriched categories (lifts or expansions). In particular, extending earlier results, we prove that for every (possibly infinite) regular set $\mathcal{F}$ of finite connected structures there exists a (countable) ω-categorical universal structure $\mathbf{U}$ for the class $\mathrm{Forb}_{\mathrm{h}}(\mathcal{F})$ (of all countable structures not containing any homomorphic image of a member of $\mathcal{F}$). We employ a technique known as homogenization: The universal object $\mathbf{U}$ is the shadow (reduct) of an ultrahomogeneous structure $\mathbf{U}'$.

We also put the results of this paper in the context of homomorphism dualities and constraint satisfaction problems. This leads to an alternative proof of the characterization of finite dualities (given by Tardif and Nešetřil) as well as of the characterization of infinite-finite dualities for classes of relational trees given by P. L. Erdős, Pálvölgyi, Tardif and Tardos.

The notion of regular families of structures is motivated by the recent characterization of infinite-finite dualities for classes of relational forests (itself related to regular languages). We show how the notion of a regular family of relational trees can be extended to regular families of relational structures. This gives a partial characterization of the existence of a (countable) ω-categorical universal object for classes $\mathrm{Forb}_{\mathrm{h}}(\mathcal{F})$.

Keywords: amalgamation, Fraïssé, universality, homomorphism duality

MSC: 05C99

***Corresponding Author: Jan Hubička, Jaroslav Nešetřil:** Computer Science Institute of Charles University (IUUK), Charles University, 118 00 Praha 1, Czech Republic, e-mail: {hubicka,nesetril}@iuuk.mff.cuni.cz

1 Introduction

We review first a few well known concepts and facts.

A *relational structure* (or simply *structure*) $\mathbf{A}$ is a pair $(A, (R^i_{\mathbf{A}} : i \in I))$, where $R^i_{\mathbf{A}} \subseteq A^{\delta_i}$ (i.e., $R^i_{\mathbf{A}}$ is a δ_i-ary relation on A). The family $(\delta_i : i \in I)$ is called the *type* Δ. The type is usually fixed and understood from the context. We consider only finite types. If the set A is finite we call $\mathbf{A}$ a *finite structure*. We consider only countable or finite structures. The class of all (countable) relational structures of type Δ will be denoted by Rel(Δ). The class Rel(Δ), $\Delta = (\delta_i; i \in I)$, is fixed throughout this paper. Unless otherwise stated all structures $\mathbf{A}, \mathbf{B}, \ldots$ belong to Rel(Δ).

A *homomorphism* $f : \mathbf{A} \to \mathbf{B} = (B, (R^i_{\mathbf{B}} : i \in I))$ is a mapping $f : A \to B$ such that $(x_1, x_2, \ldots, x_{\delta_i}) \in R^i_{\mathbf{A}}$ implies $(f(x_1), f(x_2), \ldots, f(x_{\delta_i})) \in R^i_{\mathbf{B}}$, for each $i \in I$. For given structures $\mathbf{A}$ and $\mathbf{B}$ we will denote the existence of homomorphism $f : \mathbf{A} \to \mathbf{B}$ by $\mathbf{A} \to \mathbf{B}$ and the non-existence by $\mathbf{A} \nrightarrow \mathbf{B}$. If f is one-to-one then f is called a *monomorphism*. A monomorphism f such $(x_1, x_2, \ldots, x_{\delta_i}) \in R^i_{\mathbf{A}}$ if and only if $(f(x_1), f(x_2), \ldots, f(x_{\delta_i})) \in R^i_{\mathbf{B}}$ for each $i \in I$ is called an *embedding*.

Given a family of relational structures $\mathcal{F}$, by $\mathrm{Forb_h}(\mathcal{F})$ we denote the class of all relational structures $\mathbf{A}$ for which there is no homomorphism $\mathbf{F} \to \mathbf{A}$, for any $\mathbf{F} \in \mathcal{F}$. Formally,

$$\mathrm{Forb_h}(\mathcal{F}) = \{\mathbf{A}; \forall_{\mathbf{F} \in \mathcal{F}} \mathbf{F} \nrightarrow \mathbf{A}\}.$$

Given a class $\mathcal{K}$ of countable structures, an object $\mathbf{U} \in \mathcal{K}$ is called *universal* for $\mathcal{K}$ if for every object $\mathbf{A} \in \mathcal{K}$ there exists an embedding $\mathbf{A} \to \mathbf{U}$.

For a class $\mathcal{K}$ of countable relational structures, we denote by Age($\mathcal{K}$) the class of all finite structures isomorphic to a substructure of some $\mathbf{A} \in \mathcal{K}$ and call it the *age of* $\mathcal{K}$. Similarly, for a relational structure $\mathbf{A}$, the age of $\mathbf{A}$, Age($\mathbf{A}$), is Age($\{\mathbf{A}\}$).

A structure $\mathbf{A}$ is *ultrahomogeneous* (sometimes also simply called *homogeneous*) if every isomorphism between two induced finite substructures of $\mathbf{A}$ can be extended to an automorphism of $\mathbf{A}$. A structure $\mathbf{G}$ is *generic* for the class $\mathcal{K}$ if it is universal for $\mathcal{K}$ and ultrahomogeneous.

The key property of the age of any ultrahomogeneous structure is described by the following concept. Let $\mathbf{A}, \mathbf{B}, \mathbf{C}$ be relational structures, α an embedding of $\mathbf{C}$ into $\mathbf{A}$, and β an embedding of $\mathbf{C}$ into $\mathbf{B}$. An *amalgamation of* $(\mathbf{A}, \mathbf{B}, \mathbf{C}, \alpha, \beta)$ is a triple $(\mathbf{D}, \gamma, \delta)$, where $\mathbf{D}$ is a relational structure, γ an embedding $\mathbf{A} \to \mathbf{D}$ and δ an embedding $\mathbf{B} \to \mathbf{D}$ such that $\gamma \circ \alpha = \delta \circ \beta$. Less formally, an amalgamation "glues together" the structures $\mathbf{A}$ and $\mathbf{B}$ into a single substructure of $\mathbf{D}$ such that copies of $\mathbf{C}$ coincide. See Figure 1.

Often the vertex sets of structures $\mathbf{A}$, $\mathbf{B}$ and $\mathbf{C}$ can be chosen in such a way that the embeddings α and β are identity mappings. In this case, for brevity, we shall call an amalgamation of $(\mathbf{A}, \mathbf{B}, \mathbf{C}, \alpha, \beta)$ simply an *amalgamation of* $\mathbf{A}$ *and* $\mathbf{B}$ *over* $\mathbf{C}$. Similarly, for an amalgamation $(\mathbf{D}, \gamma, \delta)$ of a given $(\mathbf{A}, \mathbf{B}, \mathbf{C}, \alpha, \beta)$ we are often interested in the structure $\mathbf{D}$ alone. In this case we shall call the structure $\mathbf{D}$ an amalgamation of $(\mathbf{A}, \mathbf{B}, \mathbf{C}, \alpha, \beta)$ (omitting the embeddings γ and δ).

We say that an amalgamation is *strong* when

$$\gamma(A) \cap \delta(B) = \gamma(\alpha(C)) = \delta(\beta(C)).$$

Less formally, a strong amalgamation glues together $\mathbf{A}$ and $\mathbf{B}$ with an overlap no greater than the copy of $\mathbf{C}$ itself. A strong amalgamation is *free* if there are no

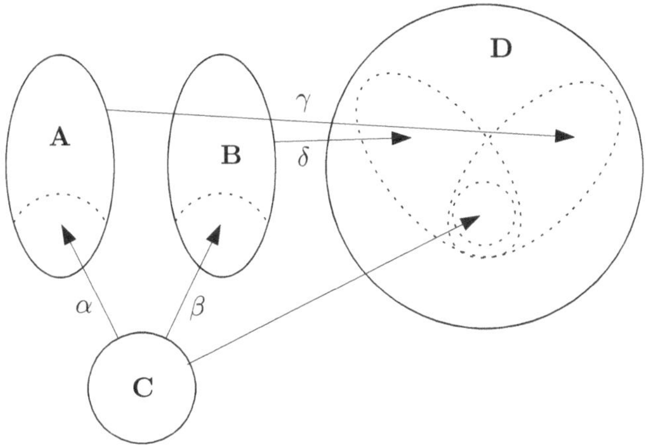

Fig. 1. Amalgamation of $(\mathbf{A}, \mathbf{B}, \mathbf{C}, \alpha, \beta)$.

relations of $\mathbf{D}$ spanning both vertices of $\gamma(A)$ and $\delta(B)$ that are not images of some relations of structure $\mathbf{A}$ or $\mathbf{B}$ via the embedding γ or δ, respectively.

A class $\mathcal{K}$ of finite relational structures is called an *amalgamation class* if the following conditions hold:

1. (*Hereditary property*) For every $\mathbf{A} \in \mathcal{K}$ and induced substructure $\mathbf{B}$ of $\mathbf{A}$ we have $\mathbf{B} \in \mathcal{K}$.
2. (*Amalgamation property*) For $\mathbf{A}, \mathbf{B}, \mathbf{C} \in \mathcal{K}$ and α an embedding of $\mathbf{C}$ into $\mathbf{A}$, β an embedding of $\mathbf{C}$ into $\mathbf{B}$, there exists $(\mathbf{D}, \gamma, \delta)$, $\mathbf{D} \in \mathcal{K}$, that is an amalgamation of $(\mathbf{A}, \mathbf{B}, \mathbf{C}, \alpha, \beta)$.
3. $\mathcal{K}$ is closed under isomorphism.
4. $\mathcal{K}$ has only countably many mutually non-isomorphic structures. (This is always the case in our setting of finite types).

The following classical result establishes the correspondence between amalgamation classes and ultrahomogeneous structures.

Theorem 1.1 (Fraïssé [11, 13]). *(a) A class $\mathcal{K}$ of finite structures is the age of a countable ultrahomogeneous structure $\mathbf{G}$ if and only if $\mathcal{K}$ is an amalgamation class.*

(b) If the conditions of (a) are satisfied then the structure $\mathbf{G}$ is unique up to isomorphism.

The ultrahomogeneous structure $\mathbf{G}$ such that $\mathrm{Age}(\mathbf{G}) = \mathcal{K}$ is called the *Fraïssé limit* of $\mathcal{K}$. We say that structure $\mathbf{A}$ is *younger* than structure $\mathbf{B}$ if $\mathrm{Age}(\mathbf{A})$ is a subset of $\mathrm{Age}(\mathbf{B})$. Every ultrahomogeneous structure $\mathbf{G}$ has the property that it is universal (and also generic) for the class $\mathcal{K}$ of all countable structures younger than $\mathbf{G}$.

A countably infinite structure is called *ω-categorical* if all countable models of its first order theory are isomorphic. We use the following characterization of ω-categorical structures given by Engeler [7], Ryll-Nardzewski [21] and Svenonius [22].

Theorem 1.2. *For a countable first order structure $\mathbf{A}$, the following conditions are equivalent:*
1. *$\mathbf{A}$ is ω-categorical.*
2. *The automorphism group of $\mathbf{A}$ has only finitely many orbits on n-tuples, for every n.*

1.0.0.5 Lifts and shadows.

Let $\Delta' = (\delta'_i; i \in I')$ be a type containing type Δ. (By this we mean $I \subseteq I'$ and $\delta'_i = \delta_i$ for $i \in I$.) Then every structure $\mathbf{X} \in \mathrm{Rel}(\Delta')$ may be viewed as a structure $\mathbf{A} = (A, (R^i_{\mathbf{A}}; i \in I)) \in \mathrm{Rel}(\Delta)$ together with some additional relations for $i \in I' \setminus I$. To make this more explicit, these additional relations will be denoted by $X^i_{\mathbf{X}}, i \in I' \setminus I$. Thus a structure $\mathbf{X} \in \mathrm{Rel}(\Delta')$ will be written as

$$\mathbf{X} = (A, (R^i_{\mathbf{A}}; i \in I), (X^i_{\mathbf{X}}; i \in I' \setminus I)),$$

and, abusing notation, more briefly as

$$\mathbf{X} = (\mathbf{A}, X^1_{\mathbf{X}}, X^2_{\mathbf{X}}, \dots, X^N_{\mathbf{X}}).$$

We call $\mathbf{X}$ a *lift* of $\mathbf{A}$ and $\mathbf{A}$ is called the *shadow* of $\mathbf{X}$. In this sense the class $\mathrm{Rel}(\Delta')$ is the class of all lifts of $\mathrm{Rel}(\Delta)$. Conversely, $\mathrm{Rel}(\Delta)$ is the class of all shadows of $\mathrm{Rel}(\Delta')$. If all extended relations are unary, the lift is called *monadic*. In the context of monadic lifts, the *color* of vertex v is the set $\{i; (v) \in X^i_{\mathbf{U}}\}$. Note that a lift is also in the model-theoretic setting called an *expansion* (as we are expanding our relational language) and a shadow a *reduct* (as we are reducing it). (Our terminology is motivated by a computer science context; see [15].) Unless stated explicitely, we shall use letters $\mathbf{A}, \mathbf{B}, \mathbf{C}, \dots$ for shadows (in $\mathrm{Rel}(\Delta)$) and letters $\mathbf{X}, \mathbf{Y}, \mathbf{Z}$ for lifts (in $\mathrm{Rel}(\Delta')$).

For a lift $\mathbf{X} = (\mathbf{A}, X^1_{\mathbf{X}}, \dots, X^N_{\mathbf{X}})$ we denote by $\mathrm{Sh}(\mathbf{X})$ the relational structure $\mathbf{A}$, i.e. its shadow. (Sh is called a *forgetful functor*.) Similarly, for a class $\mathcal{K}'$ of lifted objects, we denote by $\mathrm{Sh}(\mathcal{K}')$ the class of all shadows of structures in $\mathcal{K}'$.

1.0.0.6 Homogenization.

Many naturally defined classes $\mathcal{K}$ of relational structures contain universal structures that are ω-categorical. Because ω-categoricity can be seen as a weaker notion of ultrahomogeneity, it is natural to construct ω-categorical universal structures as shadows of ultrahomogeneous structures. Such a construction is called *homogenization*. Covington [6] provided a sufficient condition for the existence of a universal structure for a given class $\mathcal{K}$ that is a shadow of an ultrahomogeneous structure by means of amalgamation failures. This concept in fact relaxes the Fraïssé Theorem.

However, not all universal structures are constructed by means of homogenization. A necessary and sufficient condition for the existence of a universal structure for the class defined by forbidden monomorphisms from a finite family $\mathcal{F}$ of connected graphs was given by Cherlin, Shelah and Shi [3]. Here the classes are characterized by means of local finiteness of the algebraic closure operator. The techniques of [3] are motivated by proofs of the non-existence of a universal structure for a given class. The universal structure is not constructed by an explicit amalgamation argument.

1.0.0.7 Our motivation and results.

Our motivation stems from several sources. First, we seek a more streamlined and combinatorial proof of the following corollary of the aforementioned result of Cherlin, Shelah and Shi.

Theorem 1.3 ([3]). *For every finite family $\mathcal{F}$ of finite connected graphs there is an ω-categorical universal graph for the class* $\mathrm{Forb_h}(\mathcal{F})$.

We prove a stronger form of Theorem 1.3 by an explicit amalgamation argument. A similar construction can be also be obtained by characterizing the amalgamation failures and applying homogenization method. Our lifts are, however, different and more effective.

We are interested in the structure of lifts constructed for a given (possibly infinite) family $\mathcal{F}$. In special cases we relate lifts to the concept of homomorphism dualities. Motivated by a recent characterization of infinite-finite dualities by P. L. Erdős, Pálvölgyi, Tardif, Tardos [8], we introduce a notion of regular families of relational structures. These (possibly infinite) families of structures generalize regular forests, used in [8] to characterize infinite-finite dualities. In fact the regular families of trees (and forests) corresponds to well established notion of recognizable tree languages, see [5].

In Section 3 we strengthen Theorem 1.3 by proving the existence of a universal structure for $\mathrm{Forb_h}(\mathcal{F})$, where $\mathcal{F}$ is a regular family of finite connected structures.

In Section 4 we show the non-existence of an ω-categorical universal structure for $\mathrm{Forb_h}(\mathcal{F})$ for certain non-regular families $\mathcal{F}$, and give a partial characterization of such families.

Finally, in Section 5 we relate our results to homomorphism dualities and constraint satisfaction problems. We show that for the classes $\mathcal{F}$ consisting of regular relational trees the universal structure has a finite retract. This gives an alternative construction of graph duals and also an alternative proof of the characterization of homomorphism dualities.

2 Regular families of structures and $\mathcal{F}$-lifts

Let $\mathcal{F}$ be a fixed family of finite connected relational structures. For the construction of a universal structure of $\mathrm{Forb_h}(\mathcal{F})$ we use special lifts, called $\mathcal{F}$-lifts. The definition of an $\mathcal{F}$-lift is easy and resembles decomposition techniques standard in graph theory, and thus we adopt a similar terminology. First we review some elementary graph-theoretic notions, see [16, 2] for details.

For a structure $\mathbf{A} = (A, (R^i_{\mathbf{A}}, i \in I))$, the *Gaifman graph* (in combinatorics often called *2-section*) is the graph $G_{\mathbf{A}}$ with vertices A and all those edges which are a subset of a tuple of a relation of $\mathbf{A}$, i.e., $G = (A, E)$, where $x, y \in E$ if and only if $x \neq y$ and there exists a tuple $\vec{v} \in R^i_{\mathbf{A}}$, $i \in I$, such that $x, y \in \vec{v}$.

We adopt the following standard graph-theoretic notions for relational structures. We call a structure $\mathbf{A}$ *connected* if its Gaifman graph $G_{\mathbf{A}}$ is connected. For a structure $\mathbf{A}$ and subset of its vertices $B \subseteq A$, we denote by $N_{\mathbf{A}}(B)$ the *neighborhood* of the set B, that is all vertices of $A \setminus B$ connected in the Gaifman graph $G_{\mathbf{A}}$ by an edge to a vertex of B. We denote by $\mathbf{A} \setminus B$ the structure induced on $A \setminus B$ by $\mathbf{A}$. Similarly we denote by $G_{\mathbf{A}} \setminus B$ the graph created from the Gaifman graph $G_{\mathbf{A}}$ by removing the vertices in B.

A *g-cut* in $\mathbf{A}$ is a subset C of A such that the Gaifman graph $G_{\mathbf{A}}$ is disconnected by removing the set C. That is, there are vertices $u, v \in A \setminus C$ that belong to the same connected component of $G_{\mathbf{A}}$ but to different connected components of $G_{\mathbf{A}} \setminus C$. A *cut* in $\mathbf{A}$ is subset C of A such that there are vertices $u, v \in A \setminus C$ that belong to the same connected component of $\mathbf{A}$ but to different connected components of $\mathbf{A} \setminus C$.

Observe that not every cut is a g-cut. With relations of arity greater than 2, $G_{\mathbf{A} \setminus C}$ may be different from $G_{\mathbf{A}} \setminus C$.

For a g-cut C in the relational structure $\mathbf{A}$, a structure $\mathbf{A}_1$ is a *g-component* of $\mathbf{A}$ with g-cut C if $\mathbf{A}_1$ is induced by $\mathbf{A}$ on some connected component of $G_{\mathbf{A}} \setminus C$.

We will make use of the following simple observation about the neighborhood and g-components.

Observation 2.1. Let $\mathbf{A}_1$ be a g-component of $\mathbf{A}$ with g-cut C. Then the neighborhood $N_{\mathbf{A}}(A_1)$ is a subset of C. Moreover $N_{\mathbf{A}}(A_1)$ is a g-cut and $\mathbf{A}_1$ is one of the g-components of $\mathbf{A}$ with g-cut $N_{\mathbf{A}}(A_1)$.

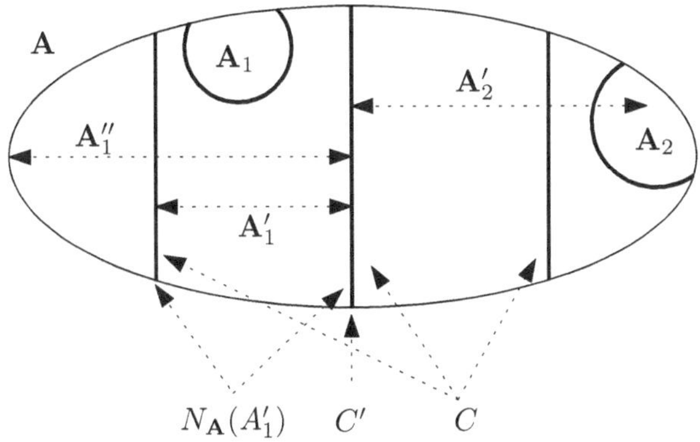

Fig. 2. Construction of a minimal g-separating g-cut C' separating $\mathbf{A}_1$ and $\mathbf{A}_2$ in $\mathbf{A}$.

Given a structure $\mathbf{A}$ with g-cut C and two (induced) substructures $\mathbf{A}_1$ and $\mathbf{A}_2$, we say that C *g-separates* $\mathbf{A}_1$ and $\mathbf{A}_2$ if there are g-components $\mathbf{A}_1' \neq \mathbf{A}_2'$ of $\mathbf{A}$ with g-cut C such that $A_1 \subseteq A_1'$ and $A_2 \subseteq A_2'$.

Definition 2.2. Let C be a g-cut in structure $\mathbf{A}$. Let $\mathbf{A}_1 \neq \mathbf{A}_2$ be two g-components of $\mathbf{A}$ with g-cut C. We call C *minimal g-separating g-cut* for $\mathbf{A}_1$ and $\mathbf{A}_2$ in $\mathbf{A}$ if $C = N_{\mathbf{A}}(A_1) = N_{\mathbf{A}}(A_2)$.

For brevity, we can omit one or both g-components when speaking about a minimal g-separating g-cut. Explicitly, we call a g-cut C *minimal g-separating* for $\mathbf{A}_1$ in $\mathbf{A}$ if there exists another structure $\mathbf{B}$ such that C is minimal g-separating for $\mathbf{A}_1$ and $\mathbf{B}$ in $\mathbf{A}$. A g-cut C is *minimal g-separating* in $\mathbf{A}$ if there exists structures $\mathbf{B}_1$ and $\mathbf{B}_2$ such that C is minimal g-separating for $\mathbf{B}_1$ and $\mathbf{B}_2$ in $\mathbf{A}$.

The name of minimal g-separating g-cut is justified by the following (probably folkloristic) proposition.

Proposition 2.3. *Let $\mathbf{A}$ be a connected relational structure, C a g-cut in $\mathbf{A}$ and $\mathbf{A}_1$ and $\mathbf{A}_2$ (induced) substructures of $\mathbf{A}$ g-separated by C. Then there exists a minimal g-separating g-cut $C' \subseteq C$ that g-separates $\mathbf{A}_1$ and $\mathbf{A}_2$ in $\mathbf{A}$. Moreover if $N_{\mathbf{A}}(A_1) \subseteq C$ (or, equivalently, $\mathbf{A}_1$ is a g-component of $\mathbf{A}$ with g-cut C), then $C' \subseteq N_{\mathbf{A}}(A_1)$.*

Proof. We will construct a series of g-cuts and g-components as depicted in Figure 2.

Denote by $\mathbf{A}'_1$ the g-component of $\mathbf{A}$ with g-cut C containing $\mathbf{A}_1$ (and thus not containing $\mathbf{A}_2$). By Observation 2.1, $N_{\mathbf{A}}(A'_1) \subseteq C$ is a g-cut that g-separates $\mathbf{A}'_1$ and $\mathbf{A}_2$ (because $\mathbf{A}'_1$ is also g-component of $\mathbf{A}$ with cut $N_{\mathbf{A}}(A'_1)$ and $\mathbf{A}'_1$ does not contain $\mathbf{A}_2$).

Now consider g-component $\mathbf{A}'_2$ of $\mathbf{A}$ with g-cut $N_{\mathbf{A}}(A'_1)$ containing $\mathbf{A}_2$. Put $C' = N_{\mathbf{A}}(A'_2)$. By Observation 2.1, $C' \subseteq N_{\mathbf{A}}(A'_1) \subseteq C$ is g-cut and $\mathbf{A}'_2$ (not containing $\mathbf{A}_1$) is one of its g-components.

Denote by $\mathbf{A}''_1$ the g-component of $\mathbf{A}$ with cut C' containing $\mathbf{A}_1$. It follows that C' g-separates $\mathbf{A}''_1$ (that contains $\mathbf{A}_1$) and $\mathbf{A}'_2$ (that contains $\mathbf{A}_2$).

To see that C' is minimal g-separating for $\mathbf{A}''_1$ and $\mathbf{A}'_2$ it remains to show that every vertex in $C' = N_{\mathbf{A}}(A'_2)$ is also in $N_{\mathbf{A}}(A''_1)$. This is true because every vertex of C' is in $N_{\mathbf{A}}(A'_1)$ and $\mathbf{A}'_1$ is substructure of $\mathbf{A}''_1$.

□

Observe that every inclusion minimal g-cut is also minimal g-separating, but not vice versa. Every minimal g-separating g-cut $C' \subset C$ that g-separates $\mathbf{A}_1$ and $\mathbf{A}_2$ is however also inclusion minimal g-cut that separates $\mathbf{A}_1$ and $\mathbf{A}_2$.

If C is a set of vertices then $\overrightarrow{C}$ will denote a tuple (of length $|C|$) of all the elements of C. Alternatively, $\overrightarrow{C}$ is an arbitrary linear ordering of C. A *rooted structure* $\mathcal{P}$ is a pair $(\mathbf{P}, \overrightarrow{R})$ where $\mathbf{P}$ is a relational structure and $\overrightarrow{R}$ is a tuple consisting of distinct vertices of $\mathbf{P}$. $\overrightarrow{R}$ is called the *root* of $\mathcal{P}$ and the size of $\overrightarrow{R}$ is the *width* of $\mathcal{P}$. We say that rooted structures $\mathcal{P}_1 = (\mathbf{P}_1, \overrightarrow{R}_1)$ and $\mathcal{P}_2 = (\mathbf{P}_2, \overrightarrow{R}_2)$ are *isomorphic* if there is a function $f : P_1 \to P_2$ that is an isomorphism of structures $\mathbf{P}_1$ and $\mathbf{P}_2$ and f restricted to $\overrightarrow{R}_1$ is a monotone bijection between $\overrightarrow{R}_1$ and $\overrightarrow{R}_2$ (we denote this $f(\overrightarrow{R}_1) = \overrightarrow{R}_2$).

The following is the principal notion of this paper:

Definition 2.4. Let $\mathbf{A}$ be a connected relational structure and R a minimal g-separating g-cut for g-component $\mathbf{C}$ in $\mathbf{A}$. A *piece* of a relational structure $\mathbf{A}$ is then a rooted structure $\mathcal{P} = (\mathbf{P}, \overrightarrow{R})$, where the tuple $\overrightarrow{R}$ consists of the vertices of the g-cut R in a (fixed) linear order and $\mathbf{P}$ is a structure induced by $\mathbf{A}$ on $C \cup R$.

Note that since $\mathbf{P}$ is the union of a g-component and its neighborhood it follows that the pieces of a connected structure are always connected structures. As an example, pieces of the Petersen graph are shown in Figure 3.

Given rooted structures $(\mathbf{P}, \overrightarrow{R})$ and $(\mathbf{P}', \overrightarrow{R}')$ such that $|R| = |R'|$, denote by $(\mathbf{P}, \overrightarrow{R}) \oplus (\mathbf{P}', \overrightarrow{R}')$ the (possibly rooted) structure created as a free amalgam of $\mathbf{P}$ and $\mathbf{P}'$ with corresponding roots being identified (in the order of $\overrightarrow{R}$ and $\overrightarrow{R}'$). Note

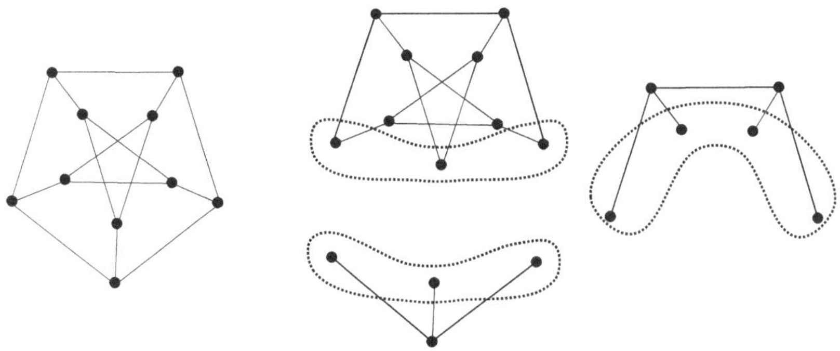

Fig. 3. Pieces of the Petersen graph up to isomorphism (and permutations of roots). G-cut of size 4 splits the graph into 3 isomorphic pieces.

that $(\mathbf{P}, \vec{R}) \oplus (\mathbf{P}', \vec{R}')$ is defined only if the rooted structure induced by $\mathbf{P}$ on $\vec{R}$ is isomorphic to the rooted structure induced by $\mathbf{P}'$ on $\vec{R}'$.

Recall that $\mathcal{F}$ is a fixed family of finite connected relational structures. A piece $\mathcal{P} = (\mathbf{P}, \vec{R})$ is *incompatible* with a rooted structure $\mathcal{A}$ if $\mathcal{P} \oplus \mathcal{A}$ is defined and there exists $\mathbf{F} \in \mathcal{F}$ that is isomorphic to $\mathcal{P} \oplus \mathcal{A}$. (In other words, there exists $\mathbf{F}'$ isomorphic to some $\mathbf{F}'' \in \mathcal{F}$, such that $\mathcal{P}$ is a piece of $\mathbf{F}'$ and $\mathcal{A}$ is the structure induced on $F' \setminus (P \setminus R)$ by $\mathbf{F}'$ rooted by $\vec{R}$.)

Assign to each piece $\mathcal{P}$ a set $\mathcal{I}_{\mathcal{P}}$ containing all rooted structures that are incompatible with $\mathcal{P}$. For two pieces $\mathcal{P}_1$ and $\mathcal{P}_2$ put $\mathcal{P}_1 \sim \mathcal{P}_2$ if and only if $\mathcal{I}_{\mathcal{P}_1} = \mathcal{I}_{\mathcal{P}_2}$. Observe that every equivalence class of $\sim$ contains pieces of the same width n. We also call n the *width* of the equivalence class of $\sim$.

Definition 2.5. A family of finite structures $\mathcal{F}$ is called *regular* if there are only finitely many equivalence classes of $\sim$ on the family of all pieces of structures in family $\mathcal{F}$.

The notion of regular family is a generalization of that of a regular family of forests, introduced in [8]. The term used in [8] was motivated by the connection to regular languages. This is explained in the following examples.

2.0.0.8 Example.

All finite families $\mathcal{F}$ of finite structures are regular. Examples of infinite families $\mathcal{F}$ include the following:

1. The family $\mathcal{F}_{\text{odd}}$ consisting of all graph cycles of odd length. All pieces of $\mathcal{F}_{\text{odd}}$ are paths rooted by initial vertex and terminal vertex. There are only

two equivalence classes of the pieces: paths of odd length and paths of even length.

2. The family $\mathcal{F}_{\text{oriented}}$ consisting of those orientations of graph cycles where all edges are oriented in the same direction. Pieces of $\mathcal{F}_{\text{oriented}}$ are oriented paths with all edges in a forward direction with roots on initial and terminal vertex. Consequently there are only two equivalence classes of pieces: paths with first root on initial vertex and second root on terminal vertex, and paths with first root on terminal vertex and second root on initial vertex.
3. Oriented paths can be described by words on alphabet $\{\leftarrow, \rightarrow\}$. It follows that every language of words on this alphabet corresponds to a family of oriented paths. It is not difficult to show that all regular languages correspond to a regular family of paths. Consequently regular families may have a rich structure; see [8].
   Consider for example the family created by words of the form $\rightarrow\rightarrow (\rightarrow\leftarrow\rightarrow)^n \rightarrow\rightarrow$, $n \geq 1$, where $(\rightarrow\leftarrow\rightarrow)^n$ stands for n repetitions of $\rightarrow\leftarrow\rightarrow$. All these paths are cores and form an antichain. Several other examples of regular families of directed graphs are discussed in [9].

We continue our construction with the following:

Definition 2.6. We denote by $E_1, \dots E_N$ the equivalence classes of $\sim$ corresponding to pieces of structures in $\mathcal{F}$. Put $I' = \{1, 2, \dots, N\}$. A relational structure $\mathbf{X} = (\mathbf{A}, (X^i_{\mathbf{X}} : i \in I'))$ is called an *$\mathcal{F}$-lift* of the relational structure $\mathbf{A}$ when the arities of the relations $X^i_{\mathbf{X}}, i \in I'$, correspond to the width of E_i.

For a relational structure $\mathbf{A}$, we define the *canonical lift*

$$L(\mathbf{A}) = (\mathbf{A}, X^1_{L(\mathbf{A})}, X^2_{L(\mathbf{A})}, \dots, X^N_{L(\mathbf{A})})$$

by putting $(v_1, v_2, \dots, v_l) \in X^i_{L(\mathbf{A})}$ if and only if there is a piece $\mathcal{P} = (\mathbf{P}, \overrightarrow{R}) \in E_i$ such that there is a homomorphism $f : \mathbf{P} \to \mathbf{A}$ with $f(\overrightarrow{R}) = (v_1, v_2, \dots, v_l)$.

2.0.0.9 Example.

As an illustration we provide an explicit description of some canonical lifts of the regular families discussed above.

1. For the family $\mathcal{F}_{\text{odd}}$ there are two new binary relations. In a canonical lift $L(\mathbf{A})$ there is $(u, v) \in X^1_{L(\mathbf{A})}$ if and only if there is a walk of odd length between vertices u and v and $(u, v) \in X^2_{L(\mathbf{A})}$ if and only if there is a walk of even length between u and v.

For $\mathbf{A} \in \mathrm{Forb_h}(\mathcal{F}_{\mathrm{odd}})$ there is no (u, v) such that $(u, v) \in X^1_{L(\mathbf{A})}$ and $(u, v) \in X^2_{L(\mathbf{A})}$. This means that odd cycles can be recognized by the existence of both a walk of even length and a walk of odd length in between a given pair of vertices.

2. For the family $\mathcal{F}_{\mathrm{oriented}}$ there are two new binary relations. In a canonical lift $L(\mathbf{A})$ there is $(u, v) \in X^1_{L(\mathbf{A})}$ if and only if there is an oriented walk from u to v and $(u, v) \in X^2_{L(\mathbf{A})}$ if and only if there is an oriented walk from v to u.
   For $\mathbf{A} \in \mathrm{Forb_h}(\mathcal{F}_{\mathrm{oriented}})$ there is no (u, v) such that both $(u, v) \in X^1_{L(\mathbf{A})}$ and $(v, u) \in X^1_{L(\mathbf{A})}$. The same holds for $X^2_{L(\mathbf{A})}$ and in fact the second relation is fully redundant in our construction and can be ignored.

3 Construction of universal structures

Theorem 3.1. *Let $\mathcal{F}$ be a regular family of finite connected relational structures (of a finite type). Then there exists an ultrahomogeneous lift $\mathbf{U}'$ with only finitely many new relations such that its shadow $\mathrm{Sh}(\mathbf{U}')$ is a universal structure for the class $\mathrm{Forb_h}(\mathcal{F})$.*

Moreover, the lift $\mathbf{U}'$ can be constructed in the following way. Denote by $\mathcal{L}$ the class of all induced substructures (sublifts) of canonical lifts $L(\mathbf{A})$, $\mathbf{A} \in \mathrm{Forb_h}(\mathcal{F})$. Then $\mathrm{Age}(\mathcal{L})$ is a amalgamation class (closed for strong amalgams whose shadows are free amalgams) and $\mathbf{U}'$ is the Fraïssé limit of $\mathrm{Age}(\mathcal{L})$.

Denote by n the maximal size of a minimal g-separating g-cut in a structure in $\mathcal{F}$ (by regularity of $\mathcal{F}$ the size of g-cuts is bounded). Then the arity of extended relations is bounded by n.

This theorem will be proved in the rest of this section. We take time for a simple Lemma.

Given a piece $\mathcal{P} = (\mathbf{P}, \vec{R})$ of structure $\mathbf{F}$, we call $\mathcal{P}' = (\mathbf{P}', \vec{R}')$ a *subpiece* of $\mathcal{P}$ if $\mathcal{P}'$ is a piece of $\mathbf{F}$, $P' \subset P$. We show that a subpiece can be freely replaced by an equivalent subpiece without changing the equivalence class of a given piece.

Lemma 3.2. *Let $\mathcal{P}_1 = (\mathbf{P}_1, \vec{R}_1)$ be a piece of structure $\mathbf{F}_1 \in \mathcal{F}$, $\mathcal{P}'_1 = (\mathbf{P}'_1, \vec{R}'_1)$ be a subpiece of $\mathcal{P}_1$ and $\mathcal{P}'_2 = (\mathbf{P}'_2, \vec{R}'_2)$ a piece such that $\mathcal{P}'_1 \sim \mathcal{P}'_2$.*

Create $\mathcal{P}_2 = (\mathbf{P}_2, \vec{R}_2)$ as a copy of $\mathcal{P}_1$ with $\mathbf{P}'_1$ replaced by $\mathbf{P}'_2$ identifying $\vec{R}'_1$ with $\vec{R}'_2$. Then $\mathcal{P}_2$ is an isomorphic copy of a piece of some $\mathbf{F}_2 \in \mathcal{F}$, and moreover $\mathcal{P}_1 \sim \mathcal{P}_2$.

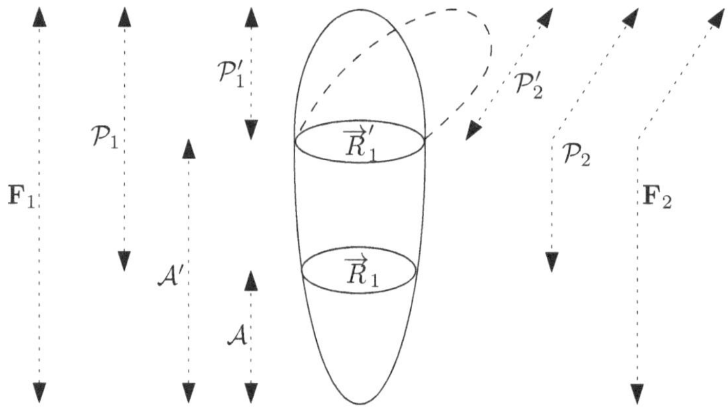

Fig. 4. Replacing subpieces.

Proof. See Figure 4. Consider some $\mathcal{A} \in \mathcal{I}_{\mathcal{P}_1}$. By definition $\mathbf{F}_1 = \mathcal{P}_1 \oplus \mathcal{A}$ is isomorphic to some structure $\mathcal{F}$. $\mathcal{P}'_1$ is also a piece of $\mathbf{F}_1$ and thus there exists a rooted structure $\mathcal{A}'$ such that $\mathcal{P}'_1 \oplus \mathcal{A}' = \mathbf{F}_1$. It follows that $\mathcal{A}' \in \mathcal{I}_{\mathcal{P}'_1}$. Because $\mathcal{P}'_2 \sim \mathcal{P}'_1$ and thus $\mathcal{I}_{\mathcal{P}'_2} = \mathcal{I}_{\mathcal{P}'_1}$, we also know that $\mathbf{F}_2 = \mathcal{P}'_2 \oplus \mathcal{A}'$ is isomorphic to some structure in $\mathcal{F}$. $\mathcal{P}_2$ is a copy of $\mathcal{P}_1$ with $\mathbf{P}'_1$ replaced by $\mathbf{P}'_2$. By construction of $\mathcal{A}'$ we have $\mathcal{P}_2 \oplus \mathcal{A} = \mathcal{P}'_2 \oplus \mathcal{A}' = \mathbf{F}_2$ and consequently $\mathcal{A} \in \mathcal{I}_{\mathcal{P}_2}$. We get $\mathcal{I}_{\mathcal{P}_1} \subseteq \mathcal{I}_{\mathcal{P}_2}$. By symmetry we also have $\mathcal{I}_{\mathcal{P}_2} \subseteq \mathcal{I}_{\mathcal{P}_1}$. □

For $\mathbf{X} \in \mathcal{L}$ we denote by $W(\mathbf{X})$ one of the structures $\mathbf{A} \in \mathrm{Forb_h}(\mathcal{F})$ such that the structure $\mathbf{X}$ is induced on X by $L(\mathbf{A})$. $W(\mathbf{X})$ is called a *witness* of the fact that $\mathbf{X}$ belongs to $\mathcal{L}$. Note that in this definition, the witness of a finite lift may be infinite structure, because $\mathrm{Forb_h}(\mathcal{F})$ contains infinite structures.

Proof of Theorem 3.1. Clearly it suffices to prove the second part of the theorem. By definition the class $\mathrm{Age}(\mathcal{L})$ is hereditary, isomorphism-closed, and has the joint embedding property. Assuming that $\mathrm{Age}(\mathcal{L})$ has the amalgamation property (with restrictions described), the rest of the theorem follows from the Fraïssé Theorem and the fact that $\mathcal{L}$ is the class of all lifts younger than the Fraïssé limit $\mathbf{U}'$ of $\mathrm{Age}(\mathcal{L})$ and thus $\mathbf{U}'$ is generic for $\mathcal{L}$.

We show the amalgamation property. Consider $\mathbf{X}, \mathbf{Y}, \mathbf{Z} \in \mathrm{Age}(\mathcal{L})$. Assume that structure $\mathbf{Z}$ is a substructure induced by both $\mathbf{X}$ and $\mathbf{Y}$ on Z and without loss of generality assume that $X \cap Y = Z$.

Put

$$\mathbf{A} = W(\mathbf{X}),$$

$$\mathbf{B} = W(\mathbf{Y}),$$

$$\mathbf{C} = \mathrm{Sh}(\mathbf{Z}).$$

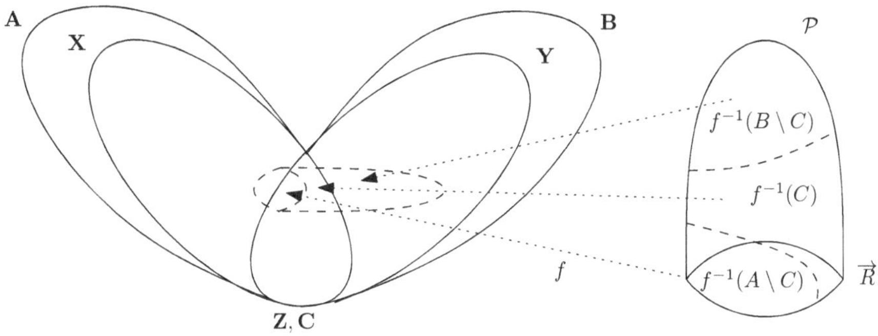

Fig. 5. Construction of an amalgamation.

Because Age($\mathcal{L}$) is closed under isomorphism, we can assume that **A** and **B** are vertex-disjoint with the exception of vertices of **C**.

Let **D** be the free amalgamation of **A** and **B** over vertices of **C**: the vertices of **D** are $A \cup B$ and there is $\vec{v} \in R^i_{\mathbf{D}}$ if and only if $\vec{v} \in R^i_{\mathbf{A}}$ or $\vec{v} \in R^i_{\mathbf{B}}$.

We claim that the structure

$$\mathbf{V} = L(\mathbf{D})$$

is a strong amalgamation of $L(\mathbf{A})$ and $L(\mathbf{B})$ over **Z** and thus also an amalgamation of **X**, **Y** over **Z**. The situation is depicted in Figure 5.

First we show that the substructure induced by **V** on A is $L(\mathbf{A})$ and that the substructure induced by **V** on B is $L(\mathbf{B})$. In the other words, no new tuples to $L(\mathbf{A})$ or $L(\mathbf{B})$ (and thus none to **X** or **Y** either) have been introduced. Assume to the contrary that there is a new tuple $(v_1, \ldots, v_t) \in X^k_{\mathbf{V}}$. By symmetry we can assume that $\{v_1, \ldots, v_t\} \subseteq A$. Explicitly, we assume that there is a piece $\mathcal{P} = (\mathbf{P}, \vec{R}) \in E_k$ and a homomorphism f from **P** to **D** such that $f(\vec{R}) = (v_1, v_2, \ldots, v_t) \notin X^k_{L(\mathbf{A})}$.

The set of vertices of **P** mapped to $L(\mathbf{A})$, $f^{-1}(A)$, is nonempty, because it contains all vertices of $\vec{R}$. The set $f^{-1}(B \setminus C)$ is nonempty f is not homomorphism from **P** to **A** (otherwise we would have $(v_1, v_2, \ldots, v_t) \in X^k_{L(\mathbf{A})}$). Because there are no tuples spanning both vertices $A \setminus C$ and vertices $B \setminus C$ in **D**, and because pieces are connected, we also have $f^{-1}(C)$ nonempty.

We will reason about the decomposition of $\mathcal{P}$ given by $f^{-1}(C)$ and create sub-pieces containing vertices of $f^{-1}(B \setminus C)$. This requires some careful analysis. The process is depicted in Figure 6. Denote by $\mathbf{F} \in \mathcal{F}$ the structure such that $\mathcal{P}$ is a piece of **F**. The vertices of $f^{-1}(C)$ form a g-cut in **F** g-separating any vertex in $f^{-1}(B \setminus C)$ from any vertex in $f^{-1}(A \setminus C)$ (if such a vertex exists) as well as any vertex of $\mathbf{F} \setminus P$.

We further strengthen our assumption on the choice of counter-example (consisting of the choice of **X**, **Y**, **A**, **B**, the piece $\mathcal{P}$ and the homomorphism f):

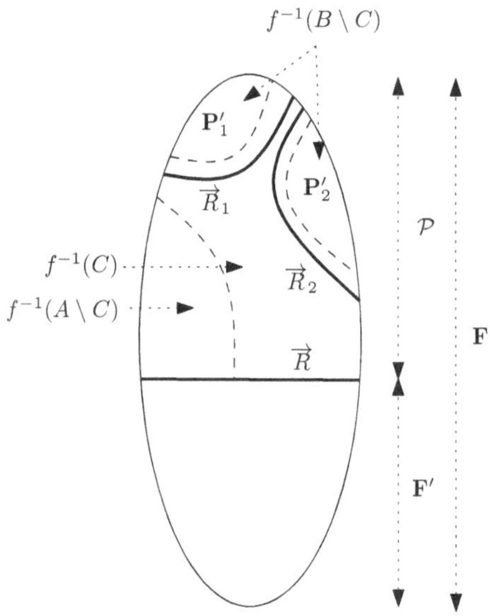

Fig. 6. The decomposition of piece $\mathcal{P}$.

(*a*) No subpiece $\mathcal{P}' = (\mathbf{P}', \overrightarrow{R}')$ of $\mathcal{P}$ is a counter-example with homomorphism f. More precisely if $f(R') \subseteq A$, then $f(\overrightarrow{R'}) \in X^i_{L(\mathbf{A})}$ for i such that $\mathcal{P}' \in E_i$. Similarly if $f(R') \subseteq B$, then $f(\overrightarrow{R'}) \in X^i_{L(\mathbf{B})}$ for i such that $\mathcal{P}' \in E_i$.

(*b*) If there is some g-component $\mathbf{P}'$ of $\mathbf{F}$ with g-cut $f^{-1}(C)$ contained in $f^{-1}(B \setminus C)$ such that $R \subseteq N_{\mathbf{P}}(P')$, then there exists a g-component $\mathbf{P}''$ of $\mathbf{F}$ with g-cut $f^{-1}(C)$ contained in $f^{-1}(A \setminus C)$ such that $R \subseteq N_{\mathbf{P}}(P'')$.

It easily follows that the existence of counter-example implies the existence of counter-example satisfying (*a*) and (*b*). (*a*) can be made to hold by considering the smallest subpiece of $\mathcal{P}$ that is still a counter-example. If (*b*) fails we can exchange $\mathbf{A}$ and $\mathbf{B}$ as well as exchange $\mathbf{X}$ and $\mathbf{Y}$. This is possible because $f^{-1}(C)$ g-separates any vertex in $f^{-1}(A \setminus C)$ and any vertex in $f^{-1}(B \setminus C)$ and thus the existence of $\mathbf{P}'$ implies $R \subseteq f^{-1}(C)$.

Denote by $\mathbf{F}'$ a substructure of $\mathbf{F}$ so that R is a minimal g-separating g-cut of $\mathbf{F}'$ and $\mathbf{P} \setminus R$ in $\mathbf{F}$. Denote by $\mathbf{P}'_1, \mathbf{P}'_2, \ldots, \mathbf{P}'_l$ the substructures induced on $\mathbf{P}$ by all connected components of $G_{\mathbf{P}} \setminus f^{-1}(A)$. We aim to find, for every $i = 1, 2, \ldots l$, a minimal g-separating g-cut $R_i \subseteq f^{-1}(C)$ of $\mathbf{F}$ that separates $\mathbf{P}'_i$ from $\mathbf{F}'$. Moreover, we will choose R_i so that $R_i \nsubseteq R$. This implies the existence of $\mathcal{P}_i = (\mathbf{P}_i, \overrightarrow{R}_i)$ that is a subpiece of $\mathcal{P}$ containing $\mathbf{P}'_i$.

We consider two cases:

1. $R \not\subseteq N_\mathbf{P}(P_i')$: Construct $R_i \subseteq N_\mathbf{P}(P_i')$ as a minimal g-separating g-cut that g-separates $\mathbf{F}'$ and $\mathbf{P}_i'$ in $\mathbf{F}$ (given by Proposition 2 for structure $\mathbf{F}$ and g-cut $f^{-1}(C)$).
   Since pieces are connected and R is a minimal g-separating g-cut in $\mathbf{F}$ for $\mathbf{P} \setminus R$ and $\mathbf{F}'$, we know that $\overrightarrow{R}_i$ must contain some vertex $v \notin R$.
2. $R \subseteq N_\mathbf{P}(P_i')$: In this case consider structure $\mathbf{P}''$ given by (b). Construct R_i as a minimal g-separating g-cut that g-separates $\mathbf{P}''$ and $\mathbf{P}_i'$ in $\mathbf{F}$.
   We show that $R_i \supset R$. Because every vertex $v \in R$ is connected in $G_\mathbf{P}$ to a vertex in P_i and a vertex in P'', we have $R_i \supseteq R$. Moreover because pieces with roots removed are g-components, R does not g-separate $\mathbf{P}''$ and $\mathbf{P}_i'$ and thus $R_i \supset R$. Since $R_i \supset R$, R_i g-separates $\mathbf{P}_i'$ and $\mathbf{F}'$ in $\mathbf{F}$.

We have constructed a family of subpieces $\mathcal{P}_1, \mathcal{P}_2, \dots, \mathcal{P}_l$ such that $\mathbf{P}_i'$ is contained in $\mathbf{P}_i \setminus \overrightarrow{R}$. It is possible that $\mathcal{P}_i$ is a subpiece of $\mathcal{P}_j$ for some $i \neq j$. Without loss of generality assume that $\mathcal{P}_1, \mathcal{P}_2, \dots \mathcal{P}_{l'}$ is the maximal subset of pieces $\mathcal{P}_1, \mathcal{P}_2, \dots, \mathcal{P}_l$ such that no piece is a subpiece of any other. Obviously $P \setminus f^{-1}(A) = \cup_{i=1,2,\dots l} P_i'$ is a subset of $\cup_{i=1,2,\dots l'} (P_i \setminus R_i)$.

Let e_i be the index of the equivalence class of $\sim$ such that $\mathcal{P}_i \in E_{e_i}$. Now we use assumption (a). All the pieces $\mathcal{P}_i$, $i = 1, \dots, l'$ are subpieces of $\mathcal{P}$. Thus we have that $f(\overrightarrow{R}_i) \in X^{e_i}_{L(\mathbf{D})} \implies f(\overrightarrow{R}_i) \in X^{e_i}_{L(\mathbf{A})}$. Thus there exists a piece $\mathcal{P}_i^A = (\mathbf{P}_i^A, \overrightarrow{R}_i^A)$, $\mathcal{P}_i^A \sim \mathcal{P}_i$, and a homomorphism f_i^A from $\mathbf{P}_i^A$ to $\mathbf{A}$ such that $f_i^A(\overrightarrow{R}_i^A) = f(\overrightarrow{R}_i)$, for every $i = 1, 2, \dots, l'$.

In this situation we want to create $\mathcal{P}^A = (\mathbf{P}^A, \overrightarrow{R}^A)$ as a copy of $\mathcal{P}$ with $\mathcal{P}_i$ replaced by $\mathcal{P}_i^A$ for every $i = 1, \dots l'$. By (repeated) application of Lemma 3.2 we will then have $\mathcal{P}^A \sim \mathcal{P}$. To make this possible, we must show that no root vertex v of $\mathcal{P}_i$ is contained in some $\mathbf{P}_j \setminus R_j$ for $1 \leq i \leq l'$, $1 \leq j \leq l'$. (Otherwise replacing $\mathcal{P}_j$ by $\mathcal{P}_j^A$ may make it impossible to replace $\mathcal{P}_i$ by $\mathcal{P}_i^A$.) Assume, to the contrary that there is such a choice of $\mathcal{P}_i$, $\mathcal{P}_j$ and v. Because $\mathcal{P}_i$ is not subpiece of $\mathcal{P}_j$, nor vice versa, there is a root v' of $\mathcal{P}_i$ that is not contained in $\mathbf{P}_j$. Because v is in $\mathbf{P}_j \setminus R_j$ that is a g-component of $\mathbf{F}$ with cut R_j and v' is not, we conclude that v and v' are g-separated by R_j in $\mathbf{F}$. This leads to the fact that $R_j \cap P_i$ is g-cut in $\mathbf{P}_i$ g-separating v and v'. This is not possible because, by our construction, $v, v' \in N_\mathbf{P}(P_i')$ and $\mathbf{P}_i'$ is a g-component of $\mathbf{P}$ with g-cut $f^{-1}(C)$. Thus v and v' are connected by a walk in $G_{\mathbf{P}_i}$ containing only vertices of $\mathbf{P}_i'$. It is not possible for vertex of $\mathbf{P}_i'$ to be in R_j, because vertices of R_j are in $f^{-1}(C)$, while vertices of $\mathbf{P}_i'$ are in $f^{-1}(B \setminus C)$.

Finally define $f^A : P^A \to A$ as follows:
1. $f^A(x) = f_i^A(x)$ when $x \in P_i^A$ for some $i = 1, 2, \dots, l'$;
2. $f^A(x) = f(x)$ otherwise.

It is easy to see that f^A is a homomorphism from $\mathbf{P}^A$ to $\mathbf{A}$ such that $f^A(\overrightarrow{R}^A) = (v_1, v_2, \dots v_t)$. This is a contradiction with $(v_1, v_2, \dots v_t) \notin X^k_{L(\mathbf{A})}$.

It remains to verify that $\mathbf{D} \in \mathrm{Forb_h}(\mathcal{F})$. We proceed analogously. Assume that f is a homomorphism from some $\mathbf{F} \in \mathcal{F}$ to $\mathbf{D}$ and further assume that the counter-example is chosen in a way so $\mathbf{F} \setminus f^{-1}(A)$ has minimal number of vertices.

Because $\mathbf{A}, \mathbf{B} \in \mathrm{Forb_h}(\mathcal{F})$, f must use vertices of $\mathbf{D} \setminus A$ and vertices of $\mathbf{D} \setminus B$ and, because $\mathbf{F}$ is connected, also vertices of $\mathbf{C}$. Analogously to the previous part, $f^{-1}(C)$ forms a g-cut in $\mathbf{F}$. Denote by $\mathbf{F}_A$ a g-component of $\mathbf{F}$ with g-cut $f^{-1}(C)$ contained in $\mathbf{F} \setminus f^{-1}(B)$ and by $\mathbf{F}_B$ a g-component of $\mathbf{F}$ with g-cut $f^{-1}(C)$ contained in $\mathbf{F} \setminus f^{-1}(A)$. Denote by R a minimal g-separating g-cut in $\mathbf{F}$ contained in $f^{-1}(C)$ that g-separate $\mathbf{F}_A$ and $\mathbf{F}_B$ (given by Proposition 2). Denote by $\mathcal{P} = (\mathbf{P}, \overrightarrow{R})$ a piece of $\mathbf{F}$ containing $\mathbf{F}_B$. We have shown that $f(\overrightarrow{R}) \in X^i_{L(\mathbf{A})}$, for i such that $\mathcal{P} \in E_i$. Consequently there is $\mathcal{P}' = (\mathbf{P}', \overrightarrow{R}')$, $\mathcal{P} \sim \mathcal{P}'$, and a homomorphism $f^A : \mathbf{P}' \to \mathbf{A}$ such that $f^A(\overrightarrow{R}') = f(\overrightarrow{R})$. Now denote by $\mathbf{F}'$ a structure created from $\mathbf{F}$ by replacing piece $\mathcal{P}$ by piece $\mathcal{P}'$ and identifying $\overrightarrow{R}$ with $\overrightarrow{R}'$. Because $\mathcal{P} \sim \mathcal{P}'$, $\mathbf{F}'$ is isomorphic to some structure in $\mathcal{F}$. Now consider the homomorphism $f' : \mathbf{F}' \to \mathbf{D}$ defined as follows:
1. $f'(x) = f^A(x)$ for $x \in P'$,
2. $f'(x) = f(x)$ otherwise.

The size of $\mathbf{F}' \setminus f'^{-1}(A)$ is strictly smaller than the size of $\mathbf{F} \setminus f^{-1}(A)$, a contradiction with the minimality of the counter-example.

This finishes the proof of the amalgamation property of $\mathrm{Age}(\mathcal{L})$: while $\mathbf{V}$ may be infinite (because witnesses may be infinite), the lift $\mathbf{V}'$ induced on vertices $X \cup Y$ by $\mathbf{V}$ is the finite amalgamation of $\mathbf{X}, \mathbf{Y}$ over $\mathbf{Z}$ and thus $\mathbf{V}' \in \mathrm{Age}(\mathcal{L})$. □

4 Non-existence of universal structures for classes $\mathrm{Forb_h}(\mathcal{F})$.

In this section we show that there exists infinite families $\mathcal{F}$ such that there is no countable ω-categorical universal structure for $\mathrm{Forb_h}(\mathcal{F})$. This is in contrast with the finite case, where the universal structures always exists.

Theorem 4.1. *Let $\mathcal{F}$ be a family of finite connected relational structures (of finite type). Assume that:*
(i) The size of all minimal g-separating g-cuts of structures in $\mathcal{F}$ is bounded by n.

(*ii*) *Let* $\mathcal{P} = (\mathbf{P}, \overrightarrow{R})$ *and* $\mathcal{P}' = (\mathbf{P}', \overrightarrow{R}')$ *be two pieces (of some structures in* $\mathcal{F}$*). Denote by* $\mathcal{A}$ *a rooted structure such that* $\mathcal{P} \oplus \mathcal{A} = \mathbf{F} \in \mathcal{F}$*. If* $\mathcal{P}' \oplus \mathcal{A}$ *is defined and* $\mathcal{P}' \oplus \mathcal{A} \notin \mathrm{Forb_h}(\mathcal{F})$*, then there is* $\mathbf{F}' \in \mathcal{F}$ *isomorphic to* $\mathcal{P}' \oplus \mathcal{A}$*.*

Then the following conditions are equivalent:

(*a*) $\mathcal{F}$ *is a regular family of connected structures.*

(*b*) *There is an ultrahomogeneous lift* $\mathbf{U}'$ *with only finitely many new relations such that its shadow* $\mathrm{Sh}(\mathbf{U}')$ *is a universal structure for the class* $\mathrm{Forb_h}(\mathcal{F})$.

(*c*) *There exists an countable* ω*-categorical universal structure for* $\mathrm{Forb_h}(\mathcal{F})$.

Proof. (*a*) $\Longrightarrow$ (*b*) follows from Theorem 3.1 for the class $\mathcal{F}$.

(*b*) $\Longrightarrow$ (*c*) is immediate. The shadow of every ultrahomogeneous structure with finitely many relations is ω-categorical.

To see that (*c*) $\Longrightarrow$ (*a*), assume to the contrary the existence of $\mathcal{F}$ satisfying (*i*) and (*ii*) which is not regular such that there is a universal structure $\mathbf{U} \in \mathrm{Forb_h}(\mathcal{F})$ which is ω-categorical.

Because the sizes of minimal g-separating g-cuts are bounded by n, we know that there is $n' \leq n$ with infinitely many pieces $\mathcal{P}_1, \mathcal{P}_2, \dots$ of width n' such that the corresponding sets $\mathcal{I}_{\mathcal{P}_1}, \mathcal{I}_{\mathcal{P}_2}, \dots$ are all different.

From Theorem 1.2 it follows that there are only finitely many orbits of n'-tuples. Denote by k the number of orbits of n' tuples. Now assign every piece $\mathcal{P}_i$ the set O_i of all orbits such that there exists a rooted embedding from $\mathcal{P}_i$ to $\mathbf{U}$ (that is a rooted homomorphism that is also an embedding of $\mathbf{P}_i$ to $\mathbf{U}$) sending the root of $\mathcal{P}_i$ to the orbit. All the sets O_i are finite of size at most k. By the pigeonhole principle there is $i \neq j$ such that $O_i = O_j$.

By our assumption (*ii*) we know that for two pieces $\mathcal{P} = (\mathbf{P}, \overrightarrow{R})$ and $\mathcal{P}' = (\mathbf{P}', \overrightarrow{R}')$ such that $\mathbf{P}, \mathbf{P}' \notin \mathrm{Forb_h}(\mathcal{F})$ we have $\mathcal{I}_{\mathcal{P}} = \mathcal{I}_{\mathcal{P}'}$ if and only if the rooted structure induced by $\mathbf{P}$ on $\overrightarrow{R}$ is identical to the rooted structure induced by $\mathbf{P}'$ on $\overrightarrow{R}'$. Consequently all those pieces belong to the same class of $\sim$ and there are only finitely many classes $\sim$ containing piece $\mathcal{P} = (\mathbf{P}, \overrightarrow{R})$ such that $\mathbf{P} \notin \mathrm{Forb_h}(\mathcal{F})$. We can thus assume that $\mathbf{P}_i, \mathbf{P}_j \in \mathrm{Forb_h}(\mathcal{F})$ and thus there is an isomorphic copy of both $\mathbf{P}_i$ and $\mathbf{P}_j$ in $\mathbf{U}$ (we where choosing i and j from infinitely many equivalence classes of $\sim$).

Now, because $\mathcal{I}_{\mathcal{P}_i} \neq \mathcal{I}_{\mathcal{P}_j}$ there is a rooted structure $\mathcal{A}$ that distinguishes $\mathcal{I}_{\mathcal{P}_i}$ from $\mathcal{I}_{\mathcal{P}_j}$. Without loss of generality assume that $\mathcal{A} \in \mathcal{I}_{\mathcal{P}_i}$. By our assumption (*ii*), $\mathcal{A} \oplus \mathcal{P}_j \notin \mathcal{F}$ implies $\mathcal{A} \oplus \mathcal{P}_j \in \mathrm{Forb_h}(\mathcal{F})$. Consequently there is an embedding from $\mathcal{A} \oplus \mathcal{P}_j$ to $\mathbf{U}$. This embedding must map the root of $\mathcal{A} \oplus \mathcal{P}_j$ to a tuple within an orbit $o \in O_j$. Since $\mathcal{A} \oplus \mathcal{P}_i \notin \mathrm{Forb_h}(\mathcal{F})$, we also have $o \notin O_i$. This is in contradiction to $O_i = O_j$. □

4.0.0.10 Example.
Consider relational structures with two binary relations (red and blue directed edges). The family consist $\mathcal{F}_{\mathrm{rb-balanced}}$ of all red-blue balanced cycles, defined as follows. Take two paths P_1 and P_2 with initial and terminal vertex specified where P_1 consists only of red edges, P_2 consists only of blue edges, and P_1, P_2 have have identical non-zero algebraic length. (The algebraic length of a path is the number of forwarding edges minus number of backwarding edges on the way from the initial vertex to the terminal vertex.) A *red-blue balanced cycle* is created as a disjoint union of P_1 and P_2 with the corresponding initial and terminal vertices identified.

The pieces of $\mathcal{F}_{\mathrm{rb-balanced}}$ consist of paths starting with a (possibly empty) blue segment followed by a (possibly empty) red segment. The equivalence class of a piece can be characterized by

1. the existence of red segment and the existence of blue segment,
2. whether the first root is in red or blue segment,
3. the algebraic length of red segment (if any) and the algebraic length of blue segment (if any).

It is not difficult to see that decision whether $\mathcal{P} \oplus \mathcal{A}$ is in $\mathrm{Forb_h}(\mathcal{F}_{\mathrm{rb-balanced}})$ depends precisely on the information above. There are infinitely many different algebraic lengths and thus also equivalence classes of $\sim$. Consequently $\mathcal{F}_{\mathrm{rb-balanced}}$ is not a regular family.

Finally we verify that $\mathcal{F}_{\mathrm{rb-balanced}}$ satisfies the assumptions of Theorem 4.1. Cycles have minimal separating cuts of size 2 giving (*i*). We now verify (*ii*). The only rooted structures $\mathcal{A}$ such that $\mathcal{P} \oplus \mathcal{A}$ is isomorphic to some red-blue balanced cycle are again pieces of red-blue balanced cycles. The algebraic lengths are preserved in homomorphic images and because the algebraic lengths are non-zero, no red-blue balanced cycle can be mapped to a path. Consequently for every piece of red-blue balanced cycle $\mathcal{P}$, $\mathcal{P}' \oplus \mathcal{A}$ is a cycle and thus $\mathcal{P}' \oplus \mathcal{A} \notin \mathrm{Forb_h}(\mathcal{F}_{\mathrm{rb-balanced}})$ imply that $\mathcal{P}' \oplus \mathcal{A}$ is a red-blue balanced cycle giving (*ii*). Consequently, by Theorem 4.1, there is no ω-categorical universal structure for $\mathrm{Forb_h}(\mathcal{F}_{\mathrm{rb-balanced}})$.

There are non-regular families $\mathcal{F}$ where universal graph for $\mathrm{Forb_h}(\mathcal{F})$ exists. Consider relational structures with one binary relation. For simplicity, assume that loops and bi-directional edges are forbidden so we consider oriented graphs. Family $\mathcal{F}_{\mathrm{balanced}}$ of all balanced orientations of graph cycles (i.e., orientations having the same number of forward and backward edges) is not a regular family. Here all pieces are all oriented paths. The equivalence class of a piece depends on the algebraic length of the path. On the other hand, however, homomorphic image of balanced 4-cycle is an oriented edge. Consequently the countably infinite graph with no edges is the universal graph for $\mathrm{Forb_h}(\mathcal{F}_{\mathrm{balanced}})$.

The family $\mathcal{F}_{\text{balanced}}$ do not satisfy requirement of *Theorem* 4.1, *(ii)*. Let $\mathcal{P}$ be a piece of balanced cycle and $\mathcal{A}$ an oriented path rooted in the endpoints. $\mathcal{P} \oplus \mathcal{A}$ is a oriented cycle and because it contains an edge we know that $\mathcal{P} \oplus \mathcal{A} \notin \mathrm{Forb_h}(\mathcal{F})$. Every oriented cycle can be constructed this way. The smallest family $\mathcal{F}'$ containing $\mathcal{F}_{\text{balanced}}$ that satisfy *(ii)* is thus the family of all oriented cycles. Such $\mathcal{F}'$ is a regular family.

Further applications of Theorem 4.1 are given in the following section.

5 Homomorphism dualities and constraint satisfaction problems

A constraint satisfaction problem (CSP) is the following decision problem:

Instance: A finite structure $\mathbf{A}$.
Question: Does there exist a homomorphism $\mathbf{A} \to \mathbf{D}$?

We denote by $\mathrm{CSP}(\mathcal{D})$ the class of all finite structures $\mathbf{A}$ with $\mathbf{A} \to \mathbf{D}$ for some $\mathbf{D} \in \mathcal{D}$.

Recall that a *homomorphism duality* (for structures of given type) is any equation

$$\mathrm{Forb_h}(\mathcal{F}) = \mathrm{CSP}(\mathcal{D}).$$

When both $\mathcal{F}$ and $\mathcal{D}$ are finite sets of finite structures, we call the pair $(\mathcal{F}, \mathcal{D})$ *a finite duality pair* [17, 18, 12]. When $\mathcal{F}$ is an infinite set of finite structures, and $\mathcal{D}$ is a finite set of finite structures, we call it an *infinite-finite duality* [8].

Dualities play a role not only in complexity problems but also in logic, model theory, the theory of partial orders and category theory. In particular, it follows from [1] and [20] that dualities coincide with those first-order definable classes which are homomorphism-closed.

For the sake of simplicity, in the following discussion we shall restrict ourselves to the case where $\mathcal{D}$ consists of a single element $\mathbf{D}$. $\mathbf{D}$ is called the *dual of* $\mathcal{F}$ (it is easy to see that $\mathbf{D}$ is up to homomorphism-equivalence uniquely determined).

The notion of universal structures and duals is related. Given a class $\mathcal{K}$ of countable structures, an object $\mathbf{U} \in \mathcal{K}$ is called *hom-universal* for $\mathcal{K}$ if for every object $\mathbf{A} \in \mathcal{K}$ there exists a homomorphism $\mathbf{A} \to \mathbf{U}$. The following is immediate from the definitions:

Proposition 5.1. *Let $\mathcal{F}$ be a family of relational structures. Structure $\mathbf{D}$ is the dual of $\mathcal{F}$ if and only if $\mathbf{D}$ is hom-universal for* $\mathrm{Forb_h}(\mathcal{F})$.

In this section we shall show how to turn the universal structure constructed in Section 3 into a finite dual. This is possible only in the special cases where a finite dual exists. First we review some results characterizing dualities.

A (relational) tree can be defined as follows (see [18]): The *incidence graph* $\mathrm{IG}(\mathbf{A})$ of relational structure $\mathbf{A}$ is the bipartite graph with parts A and $\mathrm{Block}(A)$, where

$$\mathrm{Block}(A) = \{(i, (a_1, \ldots, a_{\delta_i})) : i \in I, (a_1, \ldots, a_{\delta_i}) \in R^i_{\mathbf{A}}\},$$

and edges $[a, (i, (a_1, \ldots, a_{\delta_i}))]$ such that $a \in (a_1, \ldots, a_{\delta_i})$. (Here we write $x \in (x_1, \ldots, x_n)$ when there exists an index k such that $x = x_k$; $\mathrm{Block}(A)$ is a multigraph.) Relational structure $\mathbf{A}$ is called a *(relational) tree* when $\mathrm{IG}(\mathbf{A})$ is a graph tree (see e.g. [16]). The definition of relational trees by the incidence graph $\mathrm{IG}(\mathbf{A})$ allows us to use graph terminology for relational trees.

Theorem 5.2 ([18]). *For every finite family $\mathcal{F}$ of finite relational trees there exists a dual $\mathbf{D}$. Up to homomorphism-equivalence there are no other finite dualities with only one dual.*

Various constructions of duals of a given $\mathcal{F}$ are known [19]. More recently, infinite-finite dualities have been characterized:

Theorem 5.3 ([8]). *All regular families $\mathcal{F}$ of relational trees have a finite dual $\mathbf{D}$.*

Theorem 5.4 ([8]). *The family $\mathcal{F}$ of relational trees has a finite dual if and only if its upward closure $\mathrm{UP}(\mathcal{F})$ is regular.*

Here the *upward closure*, $\mathrm{UP}(\mathcal{F})$, is the class of all relational trees $\mathbf{T}_1$ such that there is $\mathbf{T}_2 \in \mathcal{F}$ and $\mathbf{T}_2 \to \mathbf{T}_1$.

We remark that all these characterizations extend naturally to duality pairs $(\mathcal{F}, \mathcal{D})$ where structures in the class $\mathcal{F}$ are not necessarily connected (i.e., they are relational forests). In this case however $\mathcal{D}$ generally consists of one or more structures. See [10, 8] for details.

The construction of Section 3 may be used to obtain an alternative way of constructing a dual in the proof of Theorems 5.2 and 5.3:

Corollary 5.5 (of Theorem 3.1). *Let $\mathcal{F}$ be a regular set of finite relational trees. Then there exists a class $\mathcal{L}$ of monadic lifts such that:*

(i) $\mathrm{Age}(\mathcal{L})$ *is an amalgamation class with free amalgamation;*
(ii) The Fraïssé limit of $\mathrm{Age}(\mathcal{L})$ *is an ultrahomogeneous structure* $\mathbf{U}'$ *such that* $\mathrm{Sh}(\mathbf{U}') = \mathbf{U}$ *is universal for* $\mathrm{Forb_h}(\mathcal{F})$*;*
(iii) $\mathbf{U}'$ *has a finite retract* $\mathbf{D}'$ *and consequently* $\mathrm{Sh}(\mathbf{D}') = \mathbf{D}$ *is a dual of* $\mathcal{F}$*.*

Proof. Observe that the minimal g-separating g-cuts of a relational tree all have size 1. Thus for a fixed family $\mathcal{F}$ of finite relational trees Theorem 3.1 establishes the existence of a class $\mathcal{L}$ and lift $\mathbf{U}'$ satisfying *(ii)*. Class Age($\mathcal{L}$) is closed under strong amalgams that are free in the shadow. With only unary relations added to the lift, we immediately get that Age($\mathcal{L}$) is closed under free amalgamation, too, thereby obtaining *(ii)*.

We show *(iii)*. We find finite $\mathbf{D}'$ which is a retract of $\mathbf{U}'$, $\mathrm{Sh}(\mathbf{D}') \in \mathrm{Forb_h}(\mathcal{F})$, and for which there is a homomorphism $\mathbf{A} \to \mathrm{Sh}(\mathbf{D}')$ if and only if there is a homomorphism $\mathbf{A} \to \mathrm{Sh}(\mathbf{U}')$ for every relational structure $\mathbf{A}$.

Construct $\mathbf{D}'$ from $\mathbf{U}'$ by identifying all vertices of the same color (recall that the color of vertex v is the set $\{i; (v) \in X^i_{\mathbf{U}}\}$). Denote by r the homomorphism (retraction) $\mathbf{U}' \to \mathbf{D}'$. Obviously, if $f : \mathbf{Y} \to \mathbf{U}'$ is a homomorphism then $f \circ r : \mathbf{Y} \to \mathbf{D}'$ is also a homomorphism. Thus $\mathbf{A} \to \mathrm{Sh}(\mathbf{U}')$ implies $\mathbf{A} \to \mathrm{Sh}(\mathbf{D}')$.

It remains to show that $\mathrm{Sh}(\mathbf{D}') \in \mathrm{Forb_h}(\mathcal{F})$. Suppose, to the contrary, that there is a tree $\mathbf{F} \in \mathcal{F}$ and a homomorphism $f : \mathbf{F} \to \mathrm{Sh}(\mathbf{D}')$. Let $\mathbf{X}$ be a lift created from $\mathbf{F}$ by adding an extended relation $(v) \in X^i_{\mathbf{X}}$ if and only if $(f(v)) \in X^i_{\mathbf{D}'}$, for every $i = 1, 2, \ldots N$. Obviously f is also a homomorphism $\mathbf{X} \to \mathbf{D}'$. Consider lift $\mathbf{Y}$ induced by $\mathbf{X}$ on elements of some tuple $\vec{v} \in R^j_{\mathbf{X}}$. Since $\mathbf{F}$ is a relational tree, the shadow $\mathrm{Sh}(\mathbf{Y})$ has only one tuple. Because $\mathbf{D}'$ is retract of $\mathbf{U}'$ we know that the homomorphic image of $\mathbf{Y}$ is a substructure of $\mathbf{U}'$ and thus it is in $\mathcal{L}$.

Because $\mathbf{F}$ is a relational tree, it is possible to construct a homomorphic copy of $\mathbf{X}$ by starting with the homomorphic image of $\mathbf{Y}$ in $\mathcal{L}$ and using free amalgamation (over a one-element set) to add lifts of homomorphic images of all other tuples of $\mathbf{F}$. It follows that the homomorphic image of $\mathbf{X}$ is in $\mathcal{L}$, a contradiction. □

5.0.0.11 Remark.
We stress the fact that families of trees are not the only regular families $\mathcal{F}$ of relational structures where the universal structure for $\mathrm{Forb_h}(\mathcal{F})$ can be described as a shadow of an ultrahomogeneous monadic lift $\mathbf{U}'$. For example, consider relational structures created from a relational tree by replacing tuples by an arbitrary irreducible structure (recall that a structure is *irreducible* if it has no vertex cuts). Such structures have all minimal g-separating g-cuts of size 1. One can easily construct continuum many such examples. There is however no finite retract of $\mathbf{U}'$ satisfying the statement of Corollary 5.5. (Such structures cannot be constructed from individual tuples by the aid of free amalgamation.) Of course such structures may have infinite chromatic numbers.

Note that it is also possible to construct the dual $\mathbf{D}$ of $\mathcal{F}$ without using the Fraïssé limit. Also this follows by our construction in Theorem 3.1. For every possible combination of new relations on a single vertex, create a single vertex of $\mathbf{D}$ and then keep adding tuples as long as possible so that $\mathbf{D}$ is still in $\mathcal{L}$ (in a similar way to the proof of Proposition 5.6).

We have shown that special cases of universal structures can be used to construct duals. Now we show the opposite: every dual can be turned into a universal structure by an especially simple monadic lift.

Proposition 5.6. *For a family $\mathcal{F}$ of relational structures the following two statements are equivalent:*
(i) *There is a finite dual* $\mathbf{D}$ *of* $\mathcal{F}$.
(ii) *There exists a finite family* $\mathcal{F}'$ *of monadic lifts* $\mathbf{X}$ *whose shadow* $\mathrm{Sh}(\mathbf{X})$ *has one tuple with the following property:*
Denote by $\mathcal{L}$ *the class of all lifts* $\mathbf{Y}$ *such that*
(a) $\mathbf{Y} \in \mathrm{Forb_h}(F')$, *and*
(b) *every vertex of* $\mathbf{Y}$ *is in precisely one extended relation* $X^i_{\mathbf{Y}}$.
There is a generic lift $\mathbf{U}'$ *for* $\mathcal{L}$ *and its shadow* $\mathrm{Sh}(\mathbf{U}')$ *is an* ω*-categorical universal structure for* $\mathrm{Forb_h}(\mathcal{F})$.

Loosely speaking, the class $\mathcal{L}$ is described by forbidden colors of vertices and forbidden colorings of edges.

Proof. (i) $\Longrightarrow$ (ii): Fix a dual $\mathbf{D}$ with vertices $\{1, 2, \ldots N\}$ and consider lifts with N extended unary relations. Let $\mathcal{F}'$ be the family of all structures $\mathbf{X}$ such that:
1. the vertex set of $\mathbf{X}$ is $X \subseteq \{1, 2, \ldots N\}$;
2. there is a tuple $\vec{v} \in R^j_{\mathbf{X}}$ for some $j \in I$ such that $\vec{v} \notin R^j_{\mathbf{D}}$;
3. for every $i \in X$ there is a tuple $(i) \in X^i_{\mathbf{X}}$;
4. there are no other tuples in $\mathbf{X}$ and there are no vertices in X except ones in $\vec{v}$.

By definition, $\mathrm{Age}(\mathcal{L})$ is obviously an (free) amalgamation class (all forbidden substructures are irreducible).

We show that the shadow of $\mathcal{L}$ is $\mathrm{Forb_h}(\mathcal{F})$. For every $\mathbf{A} \in \mathrm{Forb_h}(\mathcal{F})$ and homomorphism $f : \mathbf{A} \to \mathbf{D}$ construct a lift $\mathbf{X}$ by putting $(v) \in X^i_{\mathbf{X}}$ if and only if $f(v) = i$. It is easy to see that $\mathbf{X} \in \mathcal{L}$. On the other hand, for every structure $\mathbf{A}$ and lift $\mathbf{X} \in \mathcal{L}$, a homomorphism $\mathbf{A} \to \mathrm{Sh}(\mathbf{X})$ can be interpreted as an $\mathbf{D}$-coloring of $\mathbf{A}$ and thus $\mathbf{A} \in \mathrm{Forb_h}(\mathcal{F})$.

The rest of statement follows by Fraïssé theorem analogously as Theorem 3.1.

In the opposite direction assume the existence of $\mathcal{F}'$, $\mathcal{L}$ and $\mathbf{U}'$ satisfying the statement of the proposition. Construct the retract $\mathbf{D}'$ of $\mathbf{U}'$ by unifying all vertices of the same color. This gives a homomorphism (retraction) $r : \mathrm{Sh}(\mathbf{U}') \to \mathbf{D}'$. Put $\mathbf{D} = \mathrm{Sh}(\mathbf{D}')$. We show that $\mathbf{D}$ is the dual of $\mathcal{F}$.

For every $\mathbf{A} \in \mathrm{Forb_h}(\mathcal{F})$ there is an embedding $e : \mathbf{A} \to \mathrm{Sh}(\mathbf{U}')$. It follows that $\mathbf{A} \in \mathrm{CSP}(\{\mathbf{D}\})$ because $e \circ r$ is a homomorphism $\mathbf{A} \to \mathbf{D}$. To see that $\mathbf{D} \notin \mathrm{Forb_h}(\mathcal{F})$, assume for a contradiction that there is $\mathbf{F} \in \mathcal{F}$ and a homomorphism $f : \mathbf{F} \to \mathbf{D}$. Create lift $\mathbf{X}$ from $\mathbf{F}$ by adding a tuple (v) to $X^i_{\mathbf{X}}$ if and only if $(f(v)) \in X^i_{\mathbf{D}'}$. Lift $\mathbf{X}$

satisfy condition (a) of the definition of $\mathcal{L}$. For every $\mathbf{F}' \in \mathcal{F}$ a homomorphism $\mathbf{F}' \to \mathbf{X}$ implies a homomorphism $\mathbf{F}' \to \mathbf{U}$ giving (b) and thus $\mathbf{X} \in \mathcal{L}$. A contradiction with $\mathbf{U}'$ being generic for $\mathcal{L}$ and $\mathrm{Sh}(\mathbf{U}') \in \mathrm{Forb}_h(\mathcal{F})$. □

Theorem 4.1 and Proposition 5.6 imply Theorem 5.4. It is easy to see that $\mathrm{UP}(\mathcal{F})$ have the size of separating cuts o bounded by 1 and moreover it is closed for free amalgams of trees.

To show both implications of Theorem 5.2 it is necessary to show the non-existence of monadic lifts as described in Proposition 5.6 for families $\mathcal{F}$ not consisting of relational trees. This is possible with a more systematic study of the minimal arities needed in the lift for a given family $\mathcal{F}$, and by giving a more explicit description of the lifts via forbidden substructures, as shown in [14].

Acknowledgement: We would like to thank to Manuel Bodirsky for informing us about [6] and to Jan Foniok for pointing out the connection to the classification of infinite-finite dualities in [8].

We are also grateful to Andrew Goodall for remarks and comments that led to improvements in the quality of this paper.

Finally we would like to thank to the anonymous referee for correcting two principal examples in the paper and number of useful remarks.

The Computer Science Institute of Charles University (IUUK) is supported by grant ERC-CZ LL-1201 of the Czech Ministry of Education and CE-ITI P202/16/6061 of GAČR.

Bibliography

[1] A. Atserias: On digraph coloring problems and tree with duality. In: LICS'05, IEEE (2005), 106–115.

[2] B. Bolobás: Modern Graph Theory, Graduate Texts in Mathematics, vol. 184, Springer, 1998.

[3] G. L. Cherlin, S. Shelah, N. Shi: Universal Graphs with Forbidden Subgraphs and Algebraic Closure, Advances in Applied Mathematics 22 (1999), 454–491.

[4] G. L. Cherlin, N. Shi: Graphs omitting a finite set of cycles, J. Graph Theory 21 (1996), 351–355.

[5] H. Comon, M. Dauchet, R. Gilleron, C. Löding, F. Jacquemard, D. Lugiez, S. Tison, M. Tommas: Tree Automata Techniques and Applications, Available on: `http://www.grappa.univ-lille3.fr/tata`.

[6] J. Covington: Homogenizable Relational Structures, Illinois J. Mathematics, 34(4) (1990), 731–743.

[7] E. Engeler: Äquivalenz von n-Tupeln, Z. Math. Logik Grundl. Math. 5 (1959), 121–131.

[8] P. L. Erdős, D. Pálvölgyi, C. Tardif, G. Tardos: On infinite-finite tree-duality pairs of relational structures, arXiv:1207.4402v1 (submitted) (2012).
[9] P. L. Erdős, C. Tardif, G. Tardos: On infinite-finite duality pairs of directed graphs, arXiv:arXiv:1207.4402v1 (subitted) (2012).
[10] J. Foniok, J. Nešetřil, C. Tardif: Generalised dualities and maximal finite antichains in the homomorphism order of relational structures. Eur. J. Comb. 29(4): 881-899 (2008)
[11] R. Fraïssé: Théorie des relations, North Holland, 1986.
[12] P. Hell, J. Nešetřil: Graphs and homomorphisms, Oxford University Press (2004).
[13] W. Hodges: Model Theory, Cambridge University Press (1993).
[14] J. Hubička, J. Nešetřil: Homomorphism and embedding universal structures for restricted classes of graphs, arXiv:0909.4939v1 (2009).
[15] G. Kun, J. Nešetřil: NP by means of lifts and shadows, MFCS'07 LNCS 4708 (2007), 171–181.
[16] J. Matoušek, J. Nešetřil: Invitation to discrete mathematics, Oxford Univ. Press (1998), 2nd edition (2007).
[17] J. Nešetřil, A. Pultr: On classes of relations and graphs determined by subobjects and factorobjects, Discrete Math. 22 (1978), 287–300.
[18] J. Nešetřil, C. Tardif: Duality Theorems for Finite Structures (Characterizing Gaps and Good Characterizations), J. Combin. Theory B, 80 (2000), 80–97.
[19] J. Nešetřil, C. Tardif: Short answers to exponentially long questions: Extremal aspects of homomorphism duality, SIAM J. Discrete Math. 19, No. 4 (2005), 914–920.
[20] B. Rossman: Existential positive types and preservation under homomorphisms. J. ACM 55 (3) (2008), 1–53.
[21] C. Ryll-Nardzewski, One the categoricity in power $\aleph_0$, Bull. Acad. Polon. Sci. Ser. Math. 7 (1959), 545–548.
[22] N. Svenonius: $\aleph_0$-categoricity in first-order predicate calculus, Theoria 25 (1955), 82–94.

Tapani Hyttinen

Counting Measure and Forking in Finite Models

Abstract: Based on the recent work by E. Hrushovski, using the counting measure, we construct an independence notion to nice classes of finite structure and study when the independence notion is the same notion that we get from the non-forking in the ultraproduct of a representative collection of models from the class.

Keywords: forking, counting measure, finite structures

MSC2010: primary 03C45, secondary 03C13

Tapani Hyttinen: Department of Mathematics and Statistics, University of Helsinki, e-mail: tapani.hyttinen@helsinki.fi

In the paper [9], Jouko Väänänen argues that pseudo-finite models (i.e. ultraproducts of finite models) are a good framework for studying first-order logic on finite models and he demonstrates this in several ways, e.g. he describes how to use pseudo-finite models to prove various non-definability results for finite models. In the spirit of [9], in this paper we look at one way of defining an independence notion on a class of finite structures.

A lot of work has been done in which stability theory for finite structures is developed, used or which study questions that can be looked at naturally also from this point of view. E.g. the work [5] by C. Hill on effective algorithms inverting L^2-invariants, many papers by V. Koponen, see e.g. [2], my own contributions, see e.g. [8], the study of asymptotic classes of finite structure (and smoothly approximated structures), see e.g. [4], and the recent work [7] by E. Hrushovski from which we will use several ideas.

So let $\mathcal{K}$ be a collection of finite models in a countable language and for simplicity let Δ be the set of quantifier free formulas. If the class has the joint embedding property, we can find a subset $\{\mathcal{A}_i \mid i < \omega\}$ of $\mathcal{K}$ so that for all $i < \omega$, $\mathcal{A}_i$ is a substructure of $\mathcal{A}_{i+1}$ and every member of $\mathcal{K}$ embeds to some $\mathcal{A}_i$, $i < \omega$, and one can think of the sequence $(\mathcal{A}_i)_{i<\omega}$ as a monster model for the class, especially when one can choose the sequence $(\mathcal{A}_i)_{i<\omega}$ well, see [8]. Now one can ask, how to find an independence notion to $(\mathcal{A}_i)_{i<\omega}$, something like $\phi(x, a)$ is free over b for Δ-formulas $\phi(x, y)$.

If U is a non-principle ultrafilter on ω, then the non-forking in

$$\mathbf{M} = \Pi_{i<\omega}\mathcal{A}_i/U$$

gives one such notion when we think each $\mathcal{A}_i$ as a substructure of $\mathbf{M}$ in the natural way. But what does this notion tell about the structures $\mathcal{A}_i$, $i < \omega$?

Using ideas from [7], from the counting measure on the models $\mathcal{A}_i$, $i < \omega$, one gets another such notion. This notion is the main novelty of this paper. We will study various properties of it, in particular we ask when it behaves nicely and when the notion is the same as the one, we get from the non-forking in $\Pi_{i<\omega}\mathcal{A}_i/U$. We will show that they are the same in nice classes.

Everything we do is very close to what is done in the theory of asymptotic classes. The main difference in the starting points is that we do not require the existence of the dimensions, instead we work directly with the counting measures inside all (Δ-)definable subsets of $\mathcal{A}_i^n$, not just those of the form $\mathcal{A}_i^n$, see Example 2.3 below. Also we ask slightly different questions, in particular we are a lot less ambitious.

There is also a close connection to the theory of NIP theories and to Keisler measure. In particular, Proposition 4.7 in [6] is close to what we prove here. See also [3] Lemma 8.41.

In the end of this paper, many examples are given. However we mention one of them already here to make it easier for the reader to follow our definitions (which are rather tedious sometimes). Choose the models $\mathcal{A}_i$ so that $\mathcal{A}_i = (A_i, E_k)_{k<\omega}$, where A_i consists of those functions $f : \omega \to i+1$ such that for all $x > i$, $f(x) = 0$ and $f\ E_k\ g$ if $f \restriction (k+1) = g \restriction (k+1)$. And let Δ be the set of all quantifier free formulas.

1 Counting measure and forking

We let L be a countable vocabulary and we fix a strictly increasing sequence $(\mathcal{A}_i)_{i<\omega}$ of finite L-structures (i.e. L-structures with finite universe) such that for all $i < \omega$, $\mathcal{A}_i$ is a submodel of $\mathcal{A}_{i+1}$. Δ is a collection of first-order L-formulas which contains all atomic formulas and is closed under boolean combinations and replacing variables by other variables (and adding dummy variables). The main examples of Δ we have in mind are the set of all quantifier free formulas and the set of all first-order formulas. In some cases we could let Δ be also something other than a collection of first-order formulas, see the end of this section.

By a, b etc. we mean finite sequences of elements from some model and by x, y etc. we mean finite sequences of variables. $a \in \mathcal{A}_i$ means that $a \in (A_i)^n$ for suitable n. By $t_\Delta(a/A;\mathcal{B})$ we mean the complete Δ-type of a over A in a model $\mathcal{B}$ i.e. $\{\phi(x,b)|\ \phi\in\Delta,\ b\in A,\ \mathcal{B}\models\phi(a,b)\}$. By $t(a/A;\mathcal{B})$ we mean the usual complete first-order type. We let $\mathcal{A}=\cup_{i<\omega}\mathcal{A}_i$. By formulas, definability, elementary etc. mean the usual first-order concepts (not with respect to Δ). So e.g. we say that $X\subseteq\mathcal{A}_i^n$ is definable if for some first-order formula $\phi(x,y)$ and $b\in\mathcal{A}_i$, $X=\phi(\mathcal{A}_i,b)=\{a\in\mathcal{A}_i|\ \mathcal{A}_i\models\phi(a,b)\}$. The number n is called the arity of X and in this case we also say that X is definable over b. We add Δ to the notion (e.g. Δ-definable) when we use only formulas from Δ (i.e. Δ-formulas).

We fix a non-principal ultrafilter U on ω.

We write $\mathcal{A}\models_U\phi(a)$ if $\{i<\omega|\ a\in\mathcal{A}_i,\ \mathcal{A}_i\models\phi(a)\}\in U$ and by $\phi^U(\mathcal{A},a)$ we mean the set $\{b\in\mathcal{A}|\ \mathcal{A}\models_U\phi(b,a)\}$ and $\phi^U(\mathcal{A}_i,a)=\phi^U(\mathcal{A},a)\cap\mathcal{A}_i$. Also we say that $(\mathcal{A}_i)_{i<\omega}$ is a Δ-sequence if for all $i<\omega$, Δ-formula ϕ and $a\in\mathcal{A}_i$, $\mathcal{A}_i\models\phi(a)$ iff $\mathcal{A}_{i+1}\models\phi(a)$ (i.e. $\mathcal{A}_i$ is a Δ-elementary submodel of $\mathcal{A}_{i+1}$). Then if $(\mathcal{A}_i)_{i<\omega}$ is a Δ-sequence, $\phi(x,y)$ a Δ-formula and $a,b\in\mathcal{A}_i$, $\phi^U(\mathcal{A}_i,b)=\phi(\mathcal{A}_i,b)$ and $\mathcal{A}_i\models\phi(a,b)$ iff $\mathcal{A}\models_U\phi(a,b)$.

Abusing the notation, we say that $X=(X_i)_{i<\omega}$ is (Δ-) definable if there are a (Δ-) formula $\phi(x,y)$ and $a_i\in\mathcal{A}_i$ such that for all $i<\omega$, $X_i=\phi(\mathcal{A}_i,a_i)$. If for all $i<\omega$, we can choose $a_i=a\in\mathcal{A}$ (X_i is e.g. $\emptyset$ if $a\notin\mathcal{A}_i$) we say that X is definable over a and that its arity is $length(x)$. We write also $\mathcal{A}$ for the sequence $(\mathcal{A}_i)_{i<\omega}$ of models and $\mathcal{A}^n$ for $(\mathcal{A}_i^n)_{i<\omega}$. It will be clear from the context, whether we mean $(\cup_{i<\omega}\mathcal{A}_i)$ or $(\mathcal{A}_i)_{i<\omega}$. In fact, practically always, by $\mathcal{A}$ we mean the model $\cup_{i<\omega}\mathcal{A}_i$, but the meaning of some notions is not revealed from the union alone, we also need to know how the union was got (e.g. $\mathcal{A}\models_U$ above). So one can think $\mathcal{A}$ as the union which also carries the information on how it was got.

We let μ_i be the counting measure on $\mathcal{A}_i$ i.e. for $A,B\subseteq\mathcal{A}_i$, $\mu_i(A/B)=|A\cap B|/|B|$ (say $\mu_i(A/B)=1$ if $B=\emptyset$) and for all definable sequences X and Y, we write $\mu_i(X/Y)=|X_i\cap Y_i|/|Y_i|$. By $\mu(X/Y)$ we mean $lim_{i\to\infty}\mu_i(X/Y)$ if this limit exists. We define $\mu^*(X/Y)$ to be the infimum of those rationals q such that $\{i<\omega|\ \mu_i(X/Y)\leq q\}\in U$. Notice that for all Y the map $X\mapsto\mu^*(X/Y)$ is some kind of measure on the set of all definable sequences. Notice also that $\mu(X/Y)=0$ implies $\mu^*(X/Y)=0$ or more generally $\mu(X/Y)=r$ implies $\mu^*(X/Y)=r$. So, at least in many cases, $\mu^*(X/Y)$ can be calculated without knowing U; see the examples in Section 2.

Definition 1.1. (i) We say that $(\mu_i)_{i<\omega}$ is U-continuous if for every Δ-formula $\phi(x,y)$, $a\in\mathcal{A}$ and $\epsilon>0$, there is a first-order formula $\psi(y)$ such that $\mathcal{A}\models_U\psi(a)$ and for all $b\in\mathcal{A}$, if $\mathcal{A}\models_U\psi(b)$, then

$$\{i<\omega|\ |1-(|\phi(\mathcal{A}_i,a)|/|\phi(\mathcal{A}_i,b)|)|\leq\epsilon\}\in U,$$

where $0/0=1$ and $n/0=2$ for $n>0$.

(ii) We say that $(\mu_i)_{i<\omega}$ is uniformly U-continuous if for every Δ-formula $\phi(x,y)$ and $\epsilon>0$, there are $n<\omega$ and first-order formulas $\psi_j(y)$, $j\leq n$, such that $\mathcal{A}\models_U$

$\forall y \bigvee_{j\leq n} \psi_j(y)$ and for all $j \leq n$

$$\{i < \omega |\, \textit{for all}\ a, b \in \psi_j(\mathcal{A}_i),\ |1 - (|\phi(\mathcal{A}_i, a)|/|\phi(\mathcal{A}_i, b)|)| \leq \epsilon\} \in U,$$

where again $0/0 = 1$ and $n/0 = 2$ for $n > 0$.

(iii) We say that $(\mu_i)_{i<\omega}$ is U, Δ-continuous if (i) holds with the additional requirement that the formula ψ can always be chosen from Δ. Uniformly U, Δ-continuous is defined similarly.

Notice that e.g. the following class is U-continuous but not uniformly U-continuous: For each $i < \omega$, let $\mathcal{A}_i = (\bigcup_{j\leq i}\{j\} \times (j+1), E)$, where $(j, n)E(j', n')$ if $j = j'$ and let Δ be the set of all quantifier free formulas. The class is not U, Δ-continuous.

Lemma 1.2. *(i) Uniform U-continuity implies U-continuity.*

(ii) Suppose $(\mu_i)_{i<\omega}$ is U-continuous, $\epsilon > 0$ and that $X = (\phi(\mathcal{A}_i, a))_{i<\omega}$ and $Y = (\psi(\mathcal{A}_i, a))_{i<\omega}$ have the same arity and ϕ and ψ are Δ-formulas. Then there is a first-order formula $\theta(y)$ such that $\mathcal{A} \models_U \theta(a)$ and for all b, if $\mathcal{A} \models_U \theta(b)$, then, letting $X' = (\phi(\mathcal{A}_i, b))_{i<\omega}$ and $Y' = (\psi(\mathcal{A}_i, b))_{i<\omega}$,

$$|\mu^*(X/Y) - \mu^*(X'/Y')| \leq \epsilon.$$

(iii) Suppose $(\mu_i)_{i<\omega}$ is uniformly U-continuous, $\epsilon > 0$ and ϕ and ψ are Δ-formulas. Then there are first-order formulas $\theta_i(y)$, $i \leq n$, such that $\mathcal{A} \models_U \forall y \bigvee_{i\leq n} \theta_i(y)$ and for all $j \leq n$ and $a_i, b_i \in \mathcal{A}_i$, $i < \omega$, if $\{i < \omega |\, \mathcal{A}_i \models \theta_j(a_i) \wedge \theta_j(b_i)\} \in U$, then letting $X = (\phi(\mathcal{A}_i, a_i))_{i<\omega}$, $Y = (\psi(\mathcal{A}_i, a_i))_{i<\omega}$, $X' = (\phi(\mathcal{A}_i, b_i))_{i<\omega}$ and $Y' = (\psi(\mathcal{A}_i, b_i))_{i<\omega}$,

$$\{i < \omega |\, |\mu_i(X_i/Y_i) - \mu_i(X'_i/Y'_i)| \leq \epsilon\} \in U.$$

Proof. We prove (iii), (ii) is similar (notice that although the claims look different, the only difference is in the order of quantifiers) and (i) is clear. Let $\epsilon > 0$ and $\delta = min\{\epsilon, 1/2\}/4$. By the assumption, it is enough to show that if $|1 - |X_i|/|X'_i|| \leq \delta$ and $|1 - |Y_i|/|Y'_i|| \leq \delta$, then $||X_i|/|Y_i| - |X'_i|/|Y'_i|| \leq \epsilon$ (assuming that $X_i \subseteq Y_i$ and $X'_i \subseteq Y'_i$ and so $|X_i|/|Y_i| \leq 1$ and $|X'_i|/|Y'_i| \leq 1$). Now

$$(|X_i|/|X'_i|)/(|Y_i|/|Y'_i|) \leq (1 + \delta)/(1 - \delta)$$

and

$$(|Y_i|/|Y'_i|)/(|X_i|/|X'_i|) \leq (1 + \delta)/(1 - \delta)$$

and so

$$||X_i|/|Y_i| - |X'_i|/|Y'_i|| \leq |1 - (1 + \delta)/(1 - \delta)| \leq 2\delta/(1 - \delta) \leq 4\delta \leq \epsilon.$$

□

Before further studies of this measure from the point of view of independence, we notice that there is a connection between the continuity properties defined above and the elimination of quantifiers. This is a side track, but we feel that it needs to be addressed.

If $(\mathcal{A}_i)_{i<\omega}$ is a Δ-sequence, then we say that $\mathcal{A}$ is Δ-atomic if for all $a \in \mathcal{A}$ there is $\phi(x) \in t_\Delta(a/\emptyset;\mathcal{A})$ such that for all $b \in \mathcal{A}$, $\mathcal{A} \models \phi(b)$ implies $t_\Delta(b/\emptyset;\mathcal{A}) = t_\Delta(a/\emptyset;\mathcal{A})$. We say that $\mathcal{A}$ is strongly Δ-atomic if for all $a \in \Pi_{i<\omega}\mathcal{A}_i/U$ there is $\phi(x) \in t_\Delta(a/\emptyset;\Pi_{i<\omega}\mathcal{A}_i/U)$ such that for all $b \in \Pi_{i<\omega}\mathcal{A}_i/U$, $\Pi_{i<\omega}\mathcal{A}_i/U \models \phi(b)$ implies $t_\Delta(b/\emptyset;\Pi_{i<\omega}\mathcal{A}_i/U) = t_\Delta(a/\emptyset;\Pi_{i<\omega}\mathcal{A}_i/U)$.

Notice that if L is finite and relational and Δ is the set of all quantifier free formulas, then for all $n < \omega$, the number of $t_\Delta(a/\emptyset;\mathcal{A})$, $a \in \mathcal{A}^n$ is finite and also it is easy to see that the number of these types is finite iff $\mathcal{A}$ is strongly Δ-atomic (cf. Ryll-Nardzewski). However, we will not use the direction from right to left of this fact in the proof of the next lemma, since in some cases in which these measures may work (e.g. continuous logic), the direction is not true.

Lemma 1.3. *Assume that $(\mathcal{A}_i)_{i<\omega}$ is a Δ-sequence and that Δ is closed under subformulas.*

(i) Every $\mathcal{A}_i$ is a Δ-elementary submodel of $\mathcal{A}$ and $\mathcal{A}$ is a Δ-elementary submodel of $\Pi_{i<\omega}\mathcal{A}_i/U$.

(ii) Assume that $(\mu_i)_{i<\omega}$ is U,Δ-continuous and $\mathcal{A}$ is Δ-atomic. Then $\mathcal{A}$ is Δ-homogeneous i.e. if a and b have the same Δ-type then there is an automorphism f of $\mathcal{A}$ such that $f(a) = b$.

(iii) If $(\mu_i)_{i<\omega}$ is uniformly U,Δ-continuous and $\mathcal{A}$ is strongly Δ-atomic, then $\mathcal{A}$ is an elementary submodel of $\Pi_{i<\omega}\mathcal{A}_i/U$ and $\Pi_{i<\omega}\mathcal{A}_i/U$ has Δ-elimination of quantifiers, i.e, in $\Pi_{i<\omega}\mathcal{A}_i/U$, every first-order formula is equivalent with a Δ-formula.

Proof. (i) is immediate by the assumption on Δ (the same argument works that shows that every member of an elementary chain is an elementary substructure of the union of the chain).

(ii): It is clearly enough to show that if $t_\Delta(a/\emptyset;\mathcal{A}) = t_\Delta(b/\emptyset;\mathcal{A})$, then the second player wins the Ehrenfeuch-Fraıssé-game of length ω played between $(\mathcal{A},a)$ and $(\mathcal{A},b)$. For this it is enough to show that for all $c \in \mathcal{A}$ then there is $d \in \mathcal{A}$ such that $t_\Delta(ac/\emptyset;\mathcal{A}) = t_\Delta(bd/\emptyset;\mathcal{A})$. By Δ-atomicity, let $\phi(x,y)$ be the Δ-formula that isolates $t_\Delta(ac/\emptyset;\mathcal{A})$. By U,Δ-uniformicity, there is a Δ-formula $\psi(x)$ such that $\mathcal{A} \models \psi(a)$ and if $\mathcal{A} \models \psi(a')$, then $\phi(a',\mathcal{A})$ is not empty. Since $t_\Delta(a/\emptyset;\mathcal{A}) = t_\Delta(b/\emptyset;\mathcal{A})$, d can be found.

(iii): For the first claim of (iii), it is again enough to show that the second player wins the Ehrenfeuch-Fraıssé-game of length ω played between $(\mathcal{A},e)$ and $(\Pi_{i<\omega}\mathcal{A}_i/U,e)$, where $e \in \mathcal{A}$ (so we prove more than what we claimed). This is easy: She plays so that the sequences of the elements played so far have the same Δ-

type over e. We show first how she makes her second move in the case $e = \emptyset$ and on the first move $a \in \mathcal{A}$ and $b \in \Pi_{i<\omega}\mathcal{A}_i/U$ are chosen and on the second round the first player has chosen c from $\mathcal{A}$ and then we look at the case when the first player chooses from $\Pi_{i<\omega}\mathcal{A}_i/U$. Then the general case goes exactly the same way.

Let $\phi(x, y)$ be the Δ-formula that isolates

$$t_\Delta(ca/\emptyset; \mathcal{A}) = t_\Delta(ca/\emptyset; \Pi_{i<\omega}\mathcal{A}_i/U)$$

even in $\Pi_{i<\omega}\mathcal{A}_i/U$. Let $\psi_j(y)$ be as in the definition of uniform U, Δ-continuity for $\phi(x, y)$ and $\epsilon = 1/2$ be chosen so that $\mathcal{A} \models \psi_j(a)$. Then $\Pi_{i<\omega}\mathcal{A}_i/U \models \psi_j(b)$ and let $X \in U$ be such that for all $i \in X$, $\mathcal{A}_i \models \psi_j(b_i)$ (where $b = (b_i)_{i<\omega}/U$). By the choice of ψ_j, there is $Y \in U$ such that $Y \subseteq X$ and for all $i \in Y$, $|1 - |\phi(\mathcal{A}_i, a)|/|\phi(\mathcal{A}_i, b_i)|| \leq 1/2$. Clearly we may assume that Y is chosen so that $a, c \in \mathcal{A}_i$ for all $i \in Y$. But then for $i \in Y$, $\phi(\mathcal{A}_i, b_i)$ can not be empty and thus the second player can choose $d = (d_i)_{i<\omega}/U \in \Pi_{i<\omega}\mathcal{A}_i/U$ such that for all $i \in Y$, $\mathcal{A}_i \models \phi(d_i, b_i)$.

Suppose then that instead of choosing from $\mathcal{A}$, the first player has chosen $c = (c_i)_{i<\omega}/U \in \Pi_{i<\omega}\mathcal{A}_i/U$. Let $\phi(x, y)$ be the Δ-formula that isolates $t_\Delta(cb, \emptyset; \Pi_{i<\omega}\mathcal{A}_i/U)$ and $\psi(y)$ be the Δ-formula that isolates $t_\Delta(b, \emptyset; \Pi_{i<\omega}\mathcal{A}_i/U)$. Let $i < \omega$ be such that $\mathcal{A} \models \phi(c_i, b_i) \wedge \psi(b_i)$. Now by (ii), the second player can find the required element.

For the second claim of (iii), let $\phi(x)$ be an arbitrary first-order formula. By (ii) (i.e. if the sequences have the same Δ-type, they satisfy the same first-order formulas) there are Δ-formulas ψ_{kj}, $k \in I$ and $j \in J_k$, such that

$$\mathcal{A} \models \forall x(\phi(x) \leftrightarrow \bigvee_{k\in I} \bigwedge_{j\in J_k} \psi_{kj}).$$

By the proof above,

$$\Pi_{i<\omega}\mathcal{A}_i/U \models \forall x(\phi(x) \leftrightarrow \bigvee_{k\in I} \bigwedge_{j\in J_k} \psi_{kj}).$$

Since $\Pi_{i<\omega}\mathcal{A}_i/U$ is ω_1-saturated,

$$Th(\Pi_{i<\omega}\mathcal{A}_i/U) \models \forall x(\phi(x) \leftrightarrow \bigvee_{k\in I} \bigwedge_{j\in J_k} \psi_{kj}).$$

From this the claim follows by compactness. □

We extend the vocabulary L to a vocabulary L^* by adding new predicates $I^{\phi,\psi}$ for all Δ-formulas $\phi = \phi(x, y)$ and $\psi = \psi(x, y)$. We interpret these in $\mathcal{A}$ and in each $\mathcal{A}_i$ so that $(a, b) \in I^{\phi,\psi}$ if $\mu^*((\phi(\mathcal{A}_i, a))_{i<\omega}/(\psi(\mathcal{A}_i, b))_{i<\omega}) = 0$. We write $\mathcal{A}_i^*$ for what we get from $\mathcal{A}_i$ after adding the interpretations for the predicates $I^{\phi,\psi}$ and $\mathcal{A}^*$ is defined similarly. We let $\mathbf{M}^*$ be $\Pi_{i<\omega}\mathcal{A}_i^*/U$ and the embedding $a \mapsto (a_i)_{i<\omega}/U$, $a \in \mathcal{A}$, is considered as the identity map, where $a_i = a$ for all $i < \omega$.

We let $\mathbf{M}$ be the reduct of $\mathbf{M}^*$ to the vocabulary L and we recall that by Lemma 1.3 (iii), if $(\mathcal{A}_i)_{i<\omega}$ is a Δ-sequence, $\mathcal{A}$ is strongly Δ-atomic and $(\mu_i)_{i<\omega}$ is uniformly U,Δ-continuous, then $\mathcal{A}$ is an elementary submodel of $\mathbf{M}$.

Remark 1.4. For all $a, b \in \mathcal{A}$, $\mathcal{A}^* \models I^{\phi,\psi}(a,b)$ iff $\mathbf{M}^* \models I^{\phi,\psi}(a,b)$.

Proof. Immediate by the definitions. □

For the rest of this paper, we make the assumption below. This is done because we are interested in the properties of the models $\mathcal{A}_i$ that are first-order in the original similarity type (at least first-order in the sense of continuous logic). However, many of the results below hold if instead of $\mathbf{M}$ we work in $\mathbf{M}^*$ and modify definitions and assumptions suitably, see Section 2 in [7].

Assumption 1.5. From now on, we assume that $(\mu_i)_{i<\omega}$ is U-continuous.

Lemma 1.6. *(i) For all $a, b, c, d \in \mathcal{A}$, if $t(ab/\emptyset;\mathbf{M}) = t(cd/\emptyset;\mathbf{M})$, then $(a,b) \in I^{\phi,\psi}$ iff $(c,d) \in I^{\phi,\psi}$. If in addition $(\mu_i)_{i<\omega}$ is uniformly U-continuous, this holds for all $a, b, c, d \in \mathbf{M}$.*

(ii) If $\phi(x,y)$ and $\phi'(x,y)$ are equivalent in $\mathbf{M}$, i.e. $\{i < \omega \mid \mathcal{A}_i \models \forall x \forall y(\phi \leftrightarrow \phi')\} \in U$, and similarly for $\psi(x,y)$ and $\psi'(x,y)$, then the interpretations of $I^{\phi,\psi}$ and $I^{\phi',\psi'}$ are the same.

Proof. (i): Immediate by Lemma 1.2.

(ii): Immediate by the definitions. □

Suppose Y is a Δ-definable sequence over b. We write $a \in I_b^{\phi,Y}$ if $(a,b) \in I^{\phi,\psi}$, where ψ is any formula such that $(\psi(\mathcal{A}_i, b))_{i<\omega} = Y$.

Let X and Y be definable subsets of $\mathbf{M}$, say $X = \phi(\mathbf{M}, a)$ and $Y = \psi(\mathbf{M}, b)$, where $a = (a_i)_{i<\omega}/U$ and $b = (b_i)_{i<\omega}/U$. We write $\mu^*(X/Y)$ for the infimum of the rationals q such that

$$\{i < \omega \mid \mu_i(\phi(\mathcal{A}_i, a_i)/\psi(\mathcal{A}_i, b_i)) \leq q\} \in U.$$

We also write X_i for $\phi(\mathcal{A}_i, a_i)$ (and so $\mu^*(X/Y) = \mu^*((X_i)_{i<\omega}/(Y_i)_{i<\omega})$). Notice that X_i may depend on the choice of ϕ and a but not too much, only in a small set. In fact:

Lemma 1.7. *(i) The definition of μ^* does not depend on the choice of $\phi(x,a)$ and $\psi(x,b)$ nor on the choice of the representatives $(a_i)_{i<\omega}$ and $(b_i)_{i<\omega}$.*

(ii) For all X and Y Δ-definable over $a \in \mathcal{A}$ by $\phi_0(x,a)$ and $\phi_1(x,a)$, respectively, and $\epsilon > 0$, there is a first-order formula $\psi(y)$ such that $\mathcal{A} \models_U \psi(a)$ and for all $b \in \mathcal{A}$ if $\mathcal{A} \models_U \psi(b)$ and $X' = \phi_0(\mathbf{M}, b)$ and $Y' = \phi_1(\mathbf{M}, b)$, then $|\mu^*(X/Y) - \mu^*(X'/Y')| \leq \epsilon$.

(iii) For all X and Y Δ-definable over $a \in \mathbf{M}$ by $\phi_0(x, a)$ and $\phi_1(x, a)$, respectively, and $\epsilon > 0$, there is a first-order formula $\psi(y)$ such that $\mathbf{M} \models \psi(a)$ and for all $b \in \mathbf{M}$ if $\mathbf{M} \models \psi(b)$ and $X' = \phi_0(\mathbf{M}, b)$ and $Y' = \phi_1(\mathbf{M}, b)$, then $|\mu^*(X/Y) - \mu^*(X'/Y')| \leq \epsilon$.

Proof. (i): If $\mathbf{M} \models \forall x(\phi(x, a) \leftrightarrow \phi'(x, a'))$, then

$$W = \{i < \omega | \mathcal{A}_i \models \forall x(\phi(x, a_i) \leftrightarrow \phi'(x, a_i'))\} \in U.$$

Thus for all $i \in W$ and $c \in \mathcal{A}_i$, $c \in \phi(\mathcal{A}_i, a_i)$ iff $c \in \phi(\mathcal{A}_i, a_i')$ i.e. for all $i \in W$, $|\phi(\mathcal{A}_i, a_i)| = |\phi(\mathcal{A}_i, a_i')|$. And so the claim is immediate by the definitions.

(ii) and (iii) are immediate by Lemma 1.2 and the definitions. □

Notice that for all Δ-definable subsets Y of $\mathbf{M}^n$, $X \mapsto \mu^*(X/Y)$ is a kind of measure on the Δ-definable subsets of $\mathbf{M}^n$ (the domain of μ^* is not a σ-algebra) and letting I_Y be the set of all Δ-definable subsets X of $\mathbf{M}$ of measure 0 (i.e. $\mu^*(X/Y) = 0$), then I_Y is an ideal. Notice also that if Y is definable over $b \in \mathcal{A}$, then for all $a \in \mathcal{A}$, $\phi(\mathbf{M}, a) \in I_Y$ iff $a \in I_b^{\phi, Y'}$ where $Y' = (Y_i)_{i<\omega}$.

When we speak about order indiscernible sequences, we mean that they are order indiscernible in $\mathbf{M}$ (or in some elementary extension of $\mathbf{M}$) and in the sense of the first-order logic.

Definition 1.8. (i) We say that I_Y is an S1-ideal if for all Δ-formulas $\phi(x, y)$, for all b such that Y is Δ-definable over b and order-indiscernible $(a_i)_{i<\omega}$ over b, if for some $n < \omega$, $\bigcap_{i \leq n} \phi(\mathbf{M}, a_i) \in I_Y$, then $\phi(\mathbf{M}, a_0) \in I_Y$.

(ii) We say that I_Y is an S1-ideal over $(b, \mathcal{A})$ if Y is Δ-definable over $b \in \mathcal{A}$ and the following holds: for all Δ-formulas $\phi(x, y)$ and order-indiscernible $(a_i)_{i<\omega}$ over b, if for all $i < \omega$, $a_i \in \mathcal{A}$ and $\bigcap_{i<2} \phi(\mathbf{M}, a_i) \in I_Y$, then $\phi(\mathbf{M}, a_0) \in I_Y$.

Notice that assuming uniform U-continuity, the requirement in Definition 1.8 (i) above does not depend on the choice of b: By Erdös-Rado and ω_1-saturation, for any c and order-indiscernible $(a_i)_{i<\omega}$ over b, there is an order-indiscernible sequence $(b_i)_{i<\omega}$ over bc such that $t((b_i)_{i\leq n}/b; \mathbf{M}) = t((a_i)_{i\leq n}/b; \mathbf{M})$ for any $n < \omega$.

For the switch from n in (i) to 2 in (ii), notice that if $(a_i)_{i<\omega}$ is order-indiscernible over b, then so is $(c_i)_{i<\omega}$, where $c_i = (a_j | mi \leq j < m(i+1))$ for any $0 < m < \omega$.

The following Lemma and its corollary are based on ideas from [7].

Lemma 1.9. *(i) For all $Y \subseteq \mathbf{M}$ Δ-definable over $b \in \mathcal{A}$, I_Y is an S1-ideal over $(b, \mathcal{A})$.*

(ii) If $(\mu_i)_{i<\omega}$ is uniformly U-continuous, then for all Δ-definable Y, I_Y is an S1-ideal.

Proof. (i) Suppose not. Let $(a_i)_{i<\omega}$ and $\phi(x, y)$ witness that the claim is not true. Let $\epsilon = \mu^*(\phi(\mathbf{M}, a_0)/Y) > 0$, $m \geq 4/\epsilon$ be a natural number and $\delta = \epsilon/m^3$. By U-continuity (i.e. Lemma 1.7 (ii)) and the definition of μ^*, we can find $i < \omega$ such that for all $j < m$, $\mu_i(\phi(\mathcal{A}_i, a_j)/Y_i) > 3\epsilon/4$ and for all $j < k < m$, $\mu_i((\phi(\mathcal{A}_i, a_j) \cap \phi(\mathcal{A}_i, a_k))/Y_i) < \delta$. But then we have a contradiction since $\mu_i(\bigcup_{k<m} \phi(\mathcal{A}_i, a_k)/Y_i) > 1$:

Let $X = \bigcup_{i<m}(\{i\} \times |\phi(\mathcal{A}_i, a_j) \cap Y_i|)$ and let $f : X \to \mathbf{M}^{length(x)}$ be such that for all $i < m$, $f \restriction (\{i\} \times |\phi(\mathcal{A}_i, a_j) \cap Y_i|)$ is onto $\phi(\mathcal{A}_i, a_j) \cap Y_i$ and one to one. Let Z be the set of those $c \in X$ such that for some $d \in X - \{c\}$, $f(d) = f(c)$. By the choice of i and ϵ, $|X| \geq 3|Y_i|$ and by the choice of Z, $|X - Z| \leq |\bigcup_{k<m}(\phi(\mathcal{A}_i, a_k) \cap Y_i)|$, since $f \restriction (X - Z)$ is one to one. Thus it suffices to show that $|Z| \leq |Y_i|$, since then $|\bigcup_{k<m}(\phi(\mathcal{A}_i, a_k) \cap Y_i)| \geq |X| - |Z| \geq 3|Y_i| - |Y_i| = 2|Y_i|$. But

$$|Z| \leq \Sigma_{k<j<m} m|\phi(\mathcal{A}_i, a_k) \cap \phi(\mathcal{A}_i, a_j) \cap Y_i|$$

and thus $|Z| \leq m^2(m(\delta|Y_i|)) = \epsilon|Y_i| \leq |Y_i|$.

(ii) As (i). Notice that the witness for the counter assumption that I_Y is not an S1-ideal can always be chosen so that $n = 2$, see the discussion after Definition 1.8. □

Notice that in the proof above we showed a bit more than what we claim: If Y is Δ-definable over $b \in \mathcal{A}$, $\phi(x, a)$, $a \in \mathcal{A}$, is a Δ-formula and $\mu^*(\phi(\mathbf{M}, a)/Y) = \epsilon > 0$, then for any natural number $m > 4/\epsilon$, there are no $a_i \in \mathcal{A}$, $i < m$, such that $a_0 = a$, $\mu^*(\phi(\mathbf{M}, a_0) \wedge \phi(\mathbf{M}, a_1))/Y) = 0$ and for all $k < j < m$, $t(a_k a_j/b; \mathbf{M}) = t(a_0 a_1/b; \mathbf{M})$.

Recall that $\phi(x, a)$ divides over (countable) B if there is a sequence $(a_i)_{i<\omega}$ and $n < \omega$ such that $a_0 = a$, $(a_i)_{i<\omega}$ is order-indiscernible over B and

$$\mathbf{M} \models \neg\exists x \bigwedge_{i \leq n} \phi(x, a_i).$$

We say that $\phi(x, a)$ forks over B if there are formulas $\phi_i(x, a_i)$, $i < n$, such that each ϕ_i divides over B and

$$\mathbf{M} \models \forall x(\phi \to \vee_{i<n} \phi_i).$$

We write $a \downarrow_C B$, if no $\phi(x, b) \in t(a/CB; \mathbf{M})$ forks over C.

We say that $\phi(x, a)$, $a \in \mathcal{A}$, splits strongly over $b \in \mathcal{A}$ inside $\mathcal{A}$ if there is a sequence $(a_i)_{i<\omega} \subseteq \mathcal{A}$ order-indiscernible over b such that $a_0 = a$ and $\phi(x, a_0) \wedge \phi(x, a_1)$ is not realized in $\mathbf{M}$.

Corollary 1.10. *(i) If $a \in \mathcal{A}$, the Δ-formula $\phi(x, a)$ splits strongly over $b \in \mathcal{A}$ inside $\mathcal{A}$ and $Y \subseteq \mathbf{M}^{length(x)}$ is a non-empty set Δ-definable over b, then $\phi(\mathbf{M}, a) \in I_Y$.*

(ii) Suppose $(\mu_i)_{i<\omega}$ is uniformly U-continuous. If the Δ-formula $\phi(x, a)$, $a \in \mathbf{M}$, divides over $b \in \mathcal{A}$ and $Y \subseteq \mathbf{M}^{length(x)}$ is a non-empty set Δ-definable over b, then $\phi(\mathbf{M}, a) \in I_Y$.

(iii) Suppose $(\mu_i)_{i<\omega}$ is uniformly U-continuous. If Δ-formulas $\psi_i(x, a_i)$, $a_i \in \mathbf{M}$ and $i \leq n$, divide over $b \in \mathcal{A}$ and $Y \subseteq \mathbf{M}^{length(x)}$ is a non-empty set Δ-definable over b, then $\vee_{i\leq n} \psi_i(\mathbf{M}, a_i) \in I_Y$.

Proof. The proof of (i) is similar to the proof of (ii). For (ii), let a_i, $i < \omega$, and $n < \omega$ witness that $\phi(x, a)$ divides over b. By ω_1-saturation of $\mathbf{M}$, we may assume that $a_0 = a$. Then $X = \bigcap_{i \le n} \phi(\mathbf{M}, a_i) = \emptyset$ and thus $X \in I_Y$. By Lemma 1.9, $\phi(\mathbf{M}, a) = \phi(\mathbf{M}, a_0) \in I_Y$.

(iii): So suppose each $\psi_i(x, a_i)$ divides over b. Then by (ii),

$$\mu^*(\vee_{i \le n} \psi_i(\mathbf{M}, a_i)/Y) \le \Sigma_{i \le n} \mu^*(\psi_i(\mathbf{M}, a_i)/Y) = 0.$$

□

Definition 1.11. (i) We say that the Δ-formula $\phi(x, a)$ μ^*-forks over b if for all non-empty sets Y of arity $length(x)$ Δ-definable over b in $\mathbf{M}$, $\phi(\mathbf{M}, a) \in I_Y$ (which is the same as saying that $\mu^*(\phi(\mathbf{M}, a)/Y) = 0$).

(ii) Let p be a Δ-type (i.e. for some x, p is a collection of Δ-formulas $\phi(x, a)$, $a \in \mathbf{M}$). We say that p μ^*-forks over b if there are $n \in \omega$ and formulas $\phi_i(x, a_i) \in p$, $i \le n$, such that $\wedge_{i \le n} \phi_i(x, a_i)$ μ^*-forks over a.

(iii) For countable $C \subseteq \mathbf{M}$ and finite $B \subseteq \mathbf{M}$, we write $a \downarrow^*_B C$ if the type $t_\Delta(a/BC; \mathbf{M})$ does not μ^*-fork over B.

(iv) For countable $A, C \subseteq \mathbf{M}$ and finite $B \subseteq \mathbf{M}$, we write $A \downarrow^*_B C$ if for all (finite sequences) $a \in A$, $t_\Delta(a/BC; \mathbf{M})$ does not μ^*-fork over B.

Above there is a slight misuse of notation: If $A = \{a_0, ..., a_n\}$ and $a = (a_0, ..., a_n)$, then it may happen that $a \downarrow^*_B C$ but $A \not\downarrow^*_B C$ i.e. unlike in the case of non-forking, now the difference between the definition of independence for sequences and sets may be significant, see Example 2.4.

Lemma 1.12. *Let $\phi(x, y)$ be a Δ-formula and p a countable Δ-type over $\mathbf{M}$.*

(i) If $\phi(x, a)$, $a \in \mathbf{M}$, does not μ^*-fork over A and $A \subseteq B$ (both finite), then $\phi(x, a)$ does not μ^*-fork over B.

(ii) If $a, b, c, d \in \mathcal{A}$ and $t(ab/\emptyset; \mathbf{M}) = t(cd/\emptyset; \mathbf{M})$, then $\phi(x, a)$ μ^*-forks over b iff $\phi(x, c)$ μ^*-forks over d. If $(\mu_i)_{i<\omega}$ is uniformly U-continuous, then this holds for all $a, b, c, d \in \mathbf{M}$.

(iii) If $a, b \in \mathcal{A}$ and $\phi(x, a)$ does not μ^*-fork over b, then there is $\psi(y, z) \in t(ab/\emptyset; \mathbf{M})$ such that if $c, d \in \mathcal{A}$ and $\mathbf{M} \models \psi(c, d)$, then $\phi(x, c)$ does not μ^*-fork over d. If $(\mu_i)_{i<\omega}$ is uniformly U-continuous, then this holds for $\mathbf{M}$ in place of $\mathcal{A}$.

(iv) If $p \vdash \phi(x, b)$ and $\phi(x, b)$ is a Δ-formula such that it μ^*-forks over a, then p μ^*-forks over a.

(v) If $t_\Delta(a/b; \mathbf{M})$ is algebraic, then $a \downarrow^*_b c$ for all c.

(vi) If $t_\Delta(a/b; \mathbf{M})$ is not algebraic but $t_\Delta(a/bc; \mathbf{M})$ is, then $a \not\downarrow^*_b c$.

Proof. (i) follows from the fact that if Y witnesses that $\phi(x, a)$ does not μ^*-fork over A, then it is Δ-definable over A and thus also over B and thus it witnesses that $\phi(x, a)$ does not μ^*-fork over B.

(v) is clear since if $\phi(\mathbf{M}, b)$ contains n elements, then for all Δ-definable X which contain a, $\mu^*(X/\phi(M, b)) \geq 1/n > 0$.

(vi): Clearly we can find algebraic $\phi(x, bc) \in t_\Delta(a/bc : \mathbf{M})$ such that

$$\phi(x, bc) \vdash t_\Delta(a/bc; \mathbf{M}).$$

But then if Y is Δ-definable over b and $\phi(\mathbf{M}, bc) \cap Y \neq \emptyset$, Y is not algebraic. Thus $\mu^*(\phi(\mathbf{M}, bc)/Y) = 0$.

The rest are immediate by Lemma 1.7 and definitions. □

We say that $\downarrow^*$ is symmetric if for all $a, b, c \in \mathbf{M}$, $a \downarrow^*_b c$ implies $c \downarrow^*_b a$.

The proof of Lemma 1.13 (iii) below uses ideas from [7].

Lemma 1.13. *Let $a, b, c, d \in \mathbf{M}$, $\phi(x, y)$ be a Δ-formula and p a countable Δ-type over $\mathbf{M}$.*

(i) If $\phi(\mathbf{M}, a) \neq \emptyset$, then $\phi(x, a)$ does not μ^*-fork over a.

(ii) $a \downarrow^*_b cd$ implies $a \downarrow^*_{bc} d$ and $a \downarrow^*_b c$. In particular, if $\downarrow^*$ is symmetric, $a \downarrow^*_b c$ and d is a subsequence of a, then $d \downarrow^*_b c$.

(iii) If p does not μ^*-fork over b, then for all countable $A \subseteq \mathbf{M}$ there is $a \in \mathbf{M}$ such that $a \models p$ and $a \downarrow^*_b A$.

Proof. For (i), just let Y in the definition of μ^*-forking be $\phi(\mathbf{M}, a)$ and (ii) is immediate by Lemma 1.12 (i). So we prove (iii): By compactness and the fact that $\mathbf{M}$ is ω_1-saturated, it is enough to show that

$$\theta = \wedge_{i \leq n} \phi_i(x, b^i) \wedge \wedge_{i \leq n} \neg\psi_i(x, c^i)$$

is consistent, where $\phi_i(x, b^i) \in p$ and $\psi_i(x, c^i)$ are such that they μ^*-fork over b and $c^i \in A$. Let Y witness that $\wedge_{i \leq n} \phi_i(x, b^i)$ does not μ^*-fork over b i.e. Y is Δ-definable over b and $\mu^*(\wedge_{i \leq n} \phi_i(\mathbf{M}, b^i)/Y) \geq \epsilon > 0$. Now by the definition of μ^*-forking, one can see that $X \in U$, when X is the set of all $j < \omega$ such that for all $i \leq n$, $\mu_j(\psi_i(\mathcal{A}_j, c^i_j)/Y_j) < \epsilon/(2n+2)$ and $\mu_j(\wedge_{k \leq n} \phi_k(\mathcal{A}_j, b^i_j)/Y_j) > \epsilon/2$. Then we can find $a = a/U \in \mathbf{M}$ such that for all $j \in X$, $a_j \in \cap_{i \leq n} \phi_i(\mathcal{A}_i, b^i_j) - \cup_{j \leq n} \psi_i(\mathcal{A}_j, c^i_j)$. Clearly a realizes θ. □

Corollary 1.14. *Assume that $\downarrow^*$ is symmetric. Suppose $B, C \subseteq \mathbf{M}$, B finite and C countable and $A \subseteq \mathbf{M}$ is countable. If $A \downarrow^*_B C$, then for all countable $D \subseteq \mathbf{M}$, there is A' such that*

$$t_\Delta(A'/BC; \mathbf{M}) = t_\Delta(A/BC; \mathbf{M})$$

and $A \downarrow^*_B D$.

Proof. Write $A = (a_i)_{i<\omega}$ and by x, x_k etc. we mean finite sequences of variables from the set $\{v_i | \ i < \omega\}$. Let $\phi(x, b) \in t_\Delta((a_i)_{i<\omega}/BC; \mathbf{M})$ and $\psi_k(x_k, d_k)$, $k < n$ and

$d_k \in D$, be such that each $\psi_k(x_k, d_k)$ μ^*-forks over B. Again it is enough to show that $\theta = \phi(x,b) \wedge \bigwedge_{k<n} \neg\psi(x_k, a_k)$ is realized in $\mathbf{M}$. Let $m < \omega$ be such that if v_i appears in x or in some x_k, then $i < m$. By Lemma 1.13 (iii), there is $(a'_i)_{i<m} \in \mathbf{M}$ such that $t_\Delta((a'_i)_{i<m}/BC; \mathbf{M}) = t_\Delta((a_i)_{i<m}/BC)$ and $(a'_i)_{i<m} \downarrow^*_B D$. By Lemma 1.13 (ii), $(a'_i)_{i<m}$ realizes θ. □

Let $\mathbf{M}^{**}$ be some large very saturated elementary extension of $\mathbf{M}$.

Definition 1.15. Suppose $(\mu_i)_{i<\omega}$ is uniformly U-continuous.

(i) For $a, b \in \mathbf{M}$, we say that $t(a/b; \mathbf{M})$ is good if for all countable $C \subseteq \mathbf{M}$ there is a' such that $t(a'/b; \mathbf{M}) = t(a/b; \mathbf{M})$, $a' \downarrow^*_b C$ and $a' \notin C$.

(ii) For $a, b, c \in \mathbf{M}^{**}$, we write $a \downarrow^*_b c$ if for all (some), $a', b', c' \in \mathbf{M}$ with $t(a'b'c'/\emptyset; \mathbf{M}) = t(abc/\emptyset; \mathbf{M}^{**})$, $a \downarrow^*_b c$

Notice also that if $t(a/b; \mathbf{M})$ is not algebraic and $t(a/b; \mathbf{M}) \not\vdash \bigvee_{i\leq n} \psi_i(x, b_i)$ for any Δ-formulas $\psi_i(x, b_i)$ which μ^*-fork over b, then $t(a/b; \mathbf{M})$ is good.

Lemma 1.16. *Suppose $(\mu_i)_{i<\omega}$ is uniformly U-continuous and $\mathbf{M}$ is stable. Then for all $a, b, c \in \mathbf{M}$, the following holds: if $t(c/b; \mathbf{M})$ is good and $a \downarrow^*_b c$, then either $a \not\downarrow_b c$ or $c \downarrow^*_b a$. In particular, if $\mathbf{M}$ has Δ-elimination of quantifiers, $\downarrow^*$ is symmetric.*

Proof. Suppose not i.e. $a \downarrow_b c$ and $c \not\downarrow^*_b a$. Since $t(c/b; \mathbf{M})$ is good and $\mathbf{M}$ is ω_1-saturated by compactness as in the proof of Lemma 1.3 (iii), we can find $a_i, c_i \in \mathbf{M}^{**}$, $i < \kappa = \beth_{(2^\omega)^+}$ so that for all $i < \kappa$, $t(a_i c_i/b; \mathbf{M}^{**}) = t(ac/b; \mathbf{M})$ and $a_i \downarrow_{bc_i} \cup_{j<i} a_j c_j$ (and so $a_i \downarrow_b \cup_{j<i} a_j c_j$), $c_i \notin \cup_{j<i} b a_j c_j$ and $c_i \downarrow^*_b d$ for all $d \in \cup_{j<i} a_j b_j$. By Erdös-Rado, we may assume that $(a_i c_i)_{i<\omega}$ is an (infinite) indiscernible sequence over b and by ω_1-saturation of $\mathbf{M}$ that $a_i c_i \in \mathbf{M}$ for all $i < \omega$ and that $a_0 = a$ and $c_0 = c$. By the choice of c_i, $c_i \downarrow^*_b a_j$ if $i > j$. If $i < j$, then $stp(c_i/b; \mathbf{M}) = stp(c/b; \mathbf{M})$ and $stp(a_j/b; \mathbf{M}) = stp(a/b; \mathbf{M})$ and since both $a \downarrow_b c$ and $a_j \downarrow_b c_i$, $t(ac/b; \mathbf{M}) = t(a_j c_i/b; \mathbf{M})$ and so $c_i \not\downarrow^*_b a_j$. By Lemma 1.12 (ii), this is a contradiction with the stability of $\mathbf{M}$ (we have an infinite definable ordering).

For the in particular part we assume that $\mathbf{M}$ has Δ-elimination of quantifiers and that $a \downarrow^*_b c$ and show that $c \downarrow^*_b a$. We observe first that by Lemma 1.13 (i) and (iii) and Lemma 1.12 (v), if $t(c/b; \mathbf{M})$ is not algebraic, then it is good (Lemma 1.12 (v) guarantees that the μ^*-non-forking extension is not algebraic). But if $t(c/b; \mathbf{M})$ is algebraic, then by Lemma 1.12 (v), $c \downarrow^*_b a$. So we may assume that $t(c/b; \mathbf{M})$ is good. By Δ-elimination of quantifiers, Corollary 1.10 (iii) and Lemma 1.12 (iv), $a \downarrow_b c$ and so by what we showed above $c \downarrow^*_b a$. □

Theorem 1.17. *Suppose $(\mu_i)_{i<\omega}$ is uniformly U-continuous, $\mathbf{M}$ is stable and has Δ-elimination of quantifiers. Then for all $a, b, c \in \mathbf{M}$, $a \downarrow_b c$ iff $a \downarrow^*_b c$.*

Proof. By Δ-elimination of quantifiers, we have already shown the implication from right to left in Lemma 1.12 (iv) and Corollary 1.10 (iii). So we prove the other direction. By Corollary 1.14 and Lemma 1.16 together with Δ-elimination of quantifiers (i.e. symmetry), we can find a countable elementary submodel $\mathcal{B}$ of $\mathbf{M}$ such that $b \in \mathcal{B}$ and $\mathcal{B} \downarrow_b^* ac$. Then $\mathcal{B} \downarrow_b ac$ and so $a \downarrow_b \mathcal{B}c$, in particular $a \downarrow_{\mathcal{B}} c$. Also by monotonicity and symmetry (i.e. Lemma 1.16), $a \downarrow_b^* \mathcal{B}$. So by Corollary 1.14 and Δ-elimination of quantifiers, there is a' such that $t(a'/\mathcal{B};\mathbf{M}) = t(a/\mathcal{B};\mathbf{M})$ and $a' \downarrow_b^* \mathcal{B}c$. Since the implication from right to left holds $a' \downarrow_{\mathcal{B}} c$. Thus $t(a/bc;\mathbf{M}) = t(a'/bc;\mathbf{M})$. Since $a' \downarrow_b^* c$, Lemma 1.12 (ii) implies that $a \downarrow_b^* c$. □

Before the examples, we want to observe, that although we have assumed that Δ consists of first-order formulas, there are cases in which we may loosen this assumption: Suppose that the vocabulary L is finite and relational and let $k \in \omega$ be greater than the arity of any relation in L. By L^k we mean the set of all $L_{\omega_1\omega}$-sentences such that at most k variables appear in them. We let Δ be the closure of L^k under (finite) boolean combinations (so in Δ-formulas, any finite number of variables may appear). Finally suppose that $(\mathcal{A}_i)_{i<\omega}$ is such that $\mathcal{A}_i$ is L^k-elementary submodel of $\mathcal{A}_{i+1}$. Then each $\mathcal{A}_i$ is Δ-elementary submodel of $\mathbf{M} = \Pi_{i<\omega}\mathcal{A}_i/U$ and we may think of the formulas of L^k as atomic formulas. When we do this, $(\mathcal{A}_i)_{i<\omega}$ is a Δ-sequence and $\mathcal{A}$ is strongly Δ-atomic, in fact, for each n the number of $t_\Delta(a/\emptyset;\mathcal{A})$, $a \in \mathcal{A}^n$, is finite. Now it is easy to see that everything we have done above goes through also in this situation.

Also one expects that what we have done above can also be done for continuous logic (~ fuzzy logic).

Open question 1.18. If in Theorem 1.17 uniform U-continuity is replaced by just U-continuity, can one still prove that $a \downarrow_b c$ iff $a \downarrow_b^* c$, for all $a, b, c \in \mathcal{A}$.

2 Examples

In this section we give examples. The first three in a sense positive examples, at least mostly, and the last two contain counterexamples.

Example 2.1. Let F be a finite field and for all $i < \omega$, let $\mathcal{A}_i$ be an $i+1$-dimensional vector space over F chosen so that $\mathcal{A}_i$ is a subspace of $\mathcal{A}_{i+1}$. We let Δ be the set of all quantifier free formulas. Now $(\mathcal{A}_i)_{i<\omega}$ is a Δ-sequence, the number of $t_\Delta(a/\emptyset;\mathcal{A})$, $a \in \mathcal{A}^n$, is finite for all $n < \omega$ and $(\mu_i)_{i<\omega}$ is uniformly U,Δ-continuous, in fact, for Δ-formulas $\phi(x,y)$ and $\psi(x,y)$ and $a \in \mathcal{A}$ such that $\mathcal{A} \models \forall x(\phi(x,a) \rightarrow \psi(x,a))$, letting deg be the Morley degree,

$$\mu(\phi(\mathcal{A},a)/\psi(\mathcal{A},a)) = \mu^*(\phi(\mathcal{A},a)/\psi(\mathcal{A},a)) = deg(\phi(x,a))/deg(\psi(x,a))$$

if the generic element in $\phi(\mathbf{M},a)$ and the generic element in $\psi(\mathbf{M},a)$ have the same dimension i.e. the sets have the same Morley rank and otherwise

$$\mu(\phi(\mathcal{A},a)/\psi(\mathcal{A},a)) = \mu^*(\phi(\mathcal{A},a)/\psi(\mathcal{A},a)) = 0.$$

Also this value is clearly definable (and not only approximately) with a Δ-formula and since the number of possible Δ-types is finite, this can be done uniformly. Of course, $\mathbf{M}$ is stable and has Δ-elimination of quantifiers.

Example 2.2. Let p be a prime and $N_i = \Pi_{n\leq i} p_n^i$, where $(p_n)_{n<\omega}$ is the usual enumeration of primes. Let $\mathcal{A}_i$ be a field of size p^{N_i} chosen so that $\mathcal{A}_i$ is a subfield of $\mathcal{A}_{i+1}$ and Δ be the set quantifier free formulas. Then $\mathcal{A}$ is algebraically closed. Now for this class, $(\mu_i)_{i<\omega}$ is uniformly U-continuous. This can be seen essentially as in Example 2.1 using (the proof of) Main Theorem in [1]: We write $MR(\phi(x,a))$ for the Morley rank of $\phi(x,a)$, $a \in \mathcal{A}_i \subseteq \mathcal{A}$, as calculated in the theory of $\mathcal{A}$. By [1], for every Δ-formula $\phi(x,y)$ there are $n < \omega$ and for $j < n$, a formula $\psi_j(y)$ and a rational number μ_j, such that for all $a \in \mathcal{A}$ the following holds: For all $\epsilon > 0$ and for all $i < \omega$ large enough

$$|q - |\phi(\mathcal{A}_i,a)|/|\mathcal{A}_i^{MR(\phi(x,a))}|| < \epsilon$$

for some

$$q \in \{\mu_j |\ j < n\}$$

and for all $q \in \{\mu_j |\ j < n\}$ for all $\epsilon > 0$ small enough and $i < \omega$ large enough,

$$|q - |\phi(\mathcal{A}_i,a)|/|\mathcal{A}_i^{MR(\phi(x,a))}|| < \epsilon$$

iff $\mathcal{A}_i \models \bigvee\{\psi_j(y)|\ \mu_j = q\}$. Finally, for all $j < n$ and $i < \omega$, $\mathcal{A}_i \models \psi_j(a) \wedge \psi_j(b)$ implies $MR(\phi(x,a)) = MR(\phi(x,b))$. From this the claim follows immediately. Notice also that $\mathbf{M}$ is not algebraically closed and thus $\mathcal{A}$ and $\mathbf{M}$ are not elementarily equivalent. In particular, $\mathcal{A}$ is not strongly Δ-atomic (of course, the Δ-type of a non-algebraic element is not isolated). For more on this example and variations, see the theory of pseudo-finite fields.

Our next example is the one we already mentioned in the introduction.

Example 2.3. For all $i < \omega$, let $\mathcal{A}_i = (A_i, E_k)_{k<\omega}$ be such that A_i consist of functions $f : \omega \to i+1$ such that for all $x > i$, $f(x) = 0$ and $f\ E_k\ g$ if $f \restriction (k+1) = g \restriction (k+1)$. Again, let Δ be the set of quantifier free formulas. Then $(\mu_i)_{i<\omega}$ is uniformly U,Δ-continuous, $\mathbf{M}$ is stable and has Δ-elimination of quantifiers. Notice that this class is not an N-dimensional asymptotic class for any $N < \omega$.

Example 2.4. For all $i < \omega$, let $\mathcal{A}_i = (\mathcal{A}_i, E)$ be such that $\mathcal{A}_i$ consists of pairs $(k, m) \in \omega^2$ such that $k \leq i$ and if $k = 0$, then $m \leq i^2$ and otherwise $m \leq i$. $(k, m)\ E\ (k', m')$ if $k = k'$. Δ is the set of quantifier free formulas. Now the class is not U-continuous (since for any $k < \omega$, for all large enough i, in $\mathcal{A}_i$, all the elements have the same type upto quantifier rank k) and in $\mathbf{M}$, $E(x, (0, 0))$ forks over $\emptyset$ but it does not μ^*-fork over $\emptyset$.

If instead of $k \leq i$ we require that $k < 2$, we get a class which still is not U-continuous and in which $E(x, (1, 0))$ does not fork over $\emptyset$ but it μ^*-forks over $\emptyset$.

We also want to point out what happens in the latter example with the formula $\phi(x) = \neg E(y, z)$ in relation with Lemma 1.13. Now letting $a = (0, 0)$ and $b = (1, 1)$, $t_\Delta((a, b)/\{(1, 0)\}; \mathbf{M})$ does not μ^*-fork over $\emptyset$ ($\phi(\mathbf{M})$ witnesses this) while $t_\Delta((1.0)/\{a, b\}; \mathbf{M})$ does μ^*-fork over $\emptyset$ and so symmetry fails. Of course also $t_\Delta(b/\{(1, 0)\}; \mathbf{M})$ μ^*-forks over $\emptyset$ and thus also the conclusion of the in particular part of Lemma 1.13 (ii) fails.

Example 2.5. For all $i < \omega$, define N_i so that $N_0 = 3$ and $N_{i+1} = 2N_i + 1$. Let $\mathcal{A}_i = (\mathcal{A}_i, <)$, where $\mathcal{A}_i$ is the set of all integers between $-N_i$ and N_i and $<$ is the natural ordering. By replacing $\mathcal{A}_i$, $i > 0$, by isomorphic copies, we may assume that the embeddings $x \mapsto 2x$ are identical maps. Let Δ be the set of quantifier free formulas. Then $(\mu_i)_{i<\omega}$ is not U-continuous (and $\mathbf{M}$ is not stable) and $-1 < x \wedge x < 1$ ($-1, 1 \in \mathcal{A}_0$) divides over $\emptyset$ but it does not μ^*-fork over $\emptyset$. Notice that here $\mathbf{M}$ is elementarily equivalent with $\omega + \mathbb{Z} \times \mathbb{Q} + \omega^*$.

Acknowledgement: The author was partially supported by the Academy of Finland, grant 1123110.

Bibliography

[1] Z. Chatzidakis, L. van den Dries and A. Macintyre, Definable sets over finite fields, Journal für die Reine und Angewandte Mathematik, vol. 427, 1992, 107-135.

[2] M. Djordjevic, Finite variable logic, stability and finite models, Journal of Symbolic Logic, vol. 66, 2001, 837-858.

[3] G. Cherlin and E. Hrushovski, Finite Structures with Few Types, Annals of Mathematics Studies, Number 152, Princeton University Press, Princeton, 2003.

[4] R. Elwes, Asymptotic classes of finite structures, Journal of Symbolic Logic, vol. 72, 2007, 418-438.

[5] C.D. Hill, Efficiently inverting the L^2-invariant through stability theory, extended abstract in Logical Approaches to Barriers in Computing and Complexity, 2010.

[6] E. Hrushovski and A. Pillay, On NIP and invariant measures, J. Eur. Math. Soc., vol. 13, 2011, 1005-1061.

[7] E. Hrushovski, Stable group theory and approximate subgroups, J. Amer. Math. Soc., vol. 25, 2012, 189-243.
[8] T. Hyttinen, On stability in finite models, Archive for Mathematical Logic, vol. 39, 2000, 89-102.
[9] J. Väänänen, Pseudo-finite model theory, Matematica Contemporanea, vol. 24, 2003, 169-183.

Richard Kaye* and Tin Lok Wong

The Model Theory of Generic Cuts

Abstract: We survey some properties of the recently discovered notion of *generic cuts* [6, 8], and discuss the model-theoretic context of these results for theories of pairs (M, I) where I is a generic cut of a model M of Peano arithmetic.
The main new results are that such pairs possess certain model completeness properties (Section 4), and they are existentially closed in a suitable category (Section 5).

Keywords: generic cuts, models of arithmetic

MSC: 03C62, 03H15, 03C10

***Corresponding Author: Richard Kaye:** School of Mathematics, University of Birmingham, United Kingdom
Tin Lok Wong: Department of Mathematics, National University of Singapore, Singapore

1 Preliminaries

For background in model theory, see Hodges [4]. Unless otherwise stated, we follow the notation in the books by Kaye [5] and by Kossak–Schmerl [12]. We fix some notation and repeat a few more relevant definitions here.

The usual language for arithmetic $\{0, 1, +, \times, <\}$ is denoted by $\mathscr{L}_{\mathrm{A}}$. Let $\mathscr{L}_{\mathrm{Sk}}$ denote the Skolemised language for arithmetic, i.e., $\mathscr{L}_{\mathrm{Sk}}$ contains, in addition to the symbols in $\mathscr{L}_{\mathrm{A}}$, a function symbol $f_\theta(\bar{x})$ for each formula $\theta(\bar{x}, y) \in \mathscr{L}_{\mathrm{A}}$, intending to mean the least y satisfying $\theta(\bar{x}, y)$. By 'definable', we always mean 'definable with parameters'. All our types can only contain finitely many parameters. We write $\mathrm{Q}x \ldots$ for 'there are cofinally many x such that ...'. Peano arithmetic is abbreviated as PA.

A *cut* of a model of arithmetic is a nonempty proper initial segment I that has no maximum element. We write $I \subseteq_{\mathrm{e}} M$ for 'I is a cut of M'. If I is, in addition, an elementary substructure of the model, then we say that I is an *elementary cut*.

Let $\bar{c} \in M \models \mathrm{PA}$. We denote by $\mathrm{Aut}(M)$ the automorphism group of M, and by $\mathrm{Aut}(M, \bar{c})$ the pointwise stabiliser of $\bar{c}$ in $\mathrm{Aut}(M)$. If $X \subseteq M$ and f is a function with domain M, then X^f denotes the image of X under f, i.e.,

$$X^f = \{f(x) : x \in X\}.$$

Similarly, we sometimes write x^f for $f(x)$. Two cuts $I, J \subseteq_e M$ are *conjugate* over $\bar{c}$ if there is $g \in \mathrm{Aut}(M, \bar{c})$ such that $I^g = J$.

2 Generic cuts

Generic cuts were discovered by the first author during an axiomatic study on *indicators*. The motivation was to understand structures of the form (M, I), where I is a cut of a model of arithmetic M, in a model-theoretic context. We view a generic cut as representing the majority of cuts in a model. This will be made more precise at the end of this section. For now, let us first review some basic definitions related to generic cuts.

Generic cuts are constructed using a notion of forcing, i.e. a Banach–Mazur game in an appropriate topological space. The open sets are generated by certain *intervals* in a model of arithmetic. If $I \subseteq_e M \models \mathrm{PA}$ and $[a, b] \subseteq M$, then we write $I \in [a, b]$ to mean $a \in I < b$. We will regard an interval both as a set of numbers and as a set of cuts, which of these we are thinking of will be clear from context. The consistency condition for our forcing comes from a neighbourhood system.

Definition 2.1. A nonempty set $\mathcal{N}$ of intervals of the form $[x, y]$ from M is a *neighbourhood system* if the following properties hold.
(1) (Nontriviality.) For all $x \in M$ there is some $[x, y] \in \mathcal{N}$.[1]
(2) For each $B \in M$, there exists a recursive Σ_1-type $p(x, y)$ over M such that

$$\forall x, y{<}B \left([x, y] \in \mathcal{N} \Leftrightarrow M \models \bigwedge\!\!\!\bigwedge p(x, y)\right).$$

(3) $y > (x + 1)^2$ for all $[x, y] \in \mathcal{N}$.
(4) If $[x, y] \in \mathcal{N}$ and $z \in M$, then either $[x, z] \in \mathcal{N}$ or $[z, y] \in \mathcal{N}$.
(5) If $[x, y] \in \mathcal{N}$ and $[x', y'] \supseteq [x, y]$, then $[x', y'] \in \mathcal{N}$.

We will be particularly interested in neighbourhood systems that are invariant under the action of the automorphism group of the model. An important example of a neighbourhood system is the set $\mathcal{N}_{\mathrm{elt}}$ of intervals $[a, b]$ for which there is an elementary cut $I \in [a, b]$. This is automorphism-invariant and type-definable in a model of PA; if the model is recursively saturated it is in fact locally type-definable by a recursive Σ_1-type in the sense of (2) above.

1 The nontriviality clause has been added to the definition given in our previous paper [8] though it was not required for all the key results there. It implies that the corresponding set of cuts does not have any isolated points.

Neighbourhood systems arise naturally from indicators, as introduced by Paris and Kirby [9, 10].

Definition 2.2. Let $M \models \mathrm{PA}$ be nonstandard. An *indicator* on M is a function $Y\colon M^2 \to M$ that satisfies the following properties.

(1) (Nontriviality.) For all $x \in M$, there exists $y \in M$ such that $Y(x, y) > \mathbb{N}$.

(2) Y is *piecewise definable*, i.e.,

$$\{\langle x, y, Y(x, y)\rangle : x, y < B\}$$

is definable in M for every $B \in M$.

(3) $y > (x + 1)^2$ for all $x, y \in M$ with $Y(x, y) > \mathbb{N}$.

(4) $M \models \forall x, y\, (x \geqslant y \to Y(x, y) = 0)$.

(5) $M \models \forall x, y, x', y'\, (x \leqslant x' \wedge y' \leqslant y \to Y(x', y') \leqslant Y(x, y))$.

(6) If $x, y \in M$ such that $Y(x, y) > \mathbb{N}$ and $z \in [x, y]$, then either $Y(x, z) > \mathbb{N}$ or $Y(z, x) > \mathbb{N}$.

(7) $M \models \forall x, y, z\, \big(Y(x, y) \geqslant z \to \exists x', y'\, (Y(x', y) = z \wedge Y(x, y') = z)\big)$.

We will often treat an indicator as if it were definable. Formally, one has to replace the function with the code of some large enough initial part of it.

An indicator Y clearly gives a neighbourhood system

$$\mathscr{N}_Y = \{[x, y] : Y(x, y) > \mathbb{N}\}.$$

Conversely, provided the model has countable set-theoretic cofinality, we can obtain an indicator Y from any neighbourhood system which is locally defined by recursive types of bounded complexity $n \in \mathbb{N}$. The idea is that $Y(a, b)$ counts the number of formulas in $p(x, y)$ satisfied by the pair a, b, where $p(x, y)$ is as given in clause (2) of the definition of neighbourhood systems. This can be expressed within $\mathscr{L}_{\mathrm{A}}$ using a Σ_n-complete formula Sat_{Σ_n}. The countable cofinality assumption comes in when one tries to combine a family of 'partial indicators' into a single function — it is not hard to piece two 'partial indicators' together, but we do not see how to deal with infinitely many of them at one time.

Strictly speaking, the material in this paper concerns neighbourhood systems, in the sense that we are only interested in indicators up to the neighbourhood system they determine. However a single indicator function is often considerably easier to work with than a neighbourhood system with its type-definition.

Definition 2.3. Let $M \models \mathrm{PA}$ be nonstandard, and Y be an indicator on M.

- A Y-*interval* is an interval $[a, b] \subseteq M$ such that $Y(a, b) > \mathbb{N}$. Double square brackets $[\![\cdot, \cdot]\!]$ will be reserved for denoting Y-intervals.
- A Y-*cut* is a cut $I \subseteq_{\mathrm{e}} M$ that satisfies $Y(x, y) > \mathbb{N}$ for all intervals $[x, y]$ containing I.

– Y is *invariant* if its neighbourhood system $\mathcal{N}_Y$ is $\mathrm{Aut}(M)$-invariant.

Suppose we have an indicator Y on a nonstandard model $M \models \mathrm{PA}$. We think of Y-intervals as being 'large' enough to 'catch' at least one Y-cut. It is straightforward to verify that the collection of Y-intervals generates a topology on the class of all Y-cuts, and if M is countable, this topological space is homeomorphic to the Cantor set 2^ω (using the nontriviality clause in the definition). Instead of defining generic cuts in terms of forcing, we define them using this topology and the automorphisms of M.

Definition 2.4. Let $M \models \mathrm{PA}$ be nonstandard and Y be an indicator on M. A cut $I \subseteq_{\mathrm{e}} M$ is *Y-generic* if it is contained in a class of Y-cuts $\mathcal{G}$ that has the following properties.

(a) $\mathcal{G}$ is *closed under* $\mathrm{Aut}(M)$, i.e., $I^g \in \mathcal{G}$ for all $I \in \mathcal{G}$ and all $g \in \mathrm{Aut}(M)$.
(b) $\mathcal{G}$ is *dense* in the space of Y-cuts, i.e., every Y-interval contains a cut in $\mathcal{G}$.
(c) For each $I \in \mathcal{G}$ and each $\bar{c} \in M$, there is an interval $[a, b] \subseteq M$ containing I in which all cuts in $\mathcal{G}$ are conjugate over $\bar{c}$.

Theorem 2.5. *Let M be a countable arithmetically saturated model of* PA, *and Y be an invariant indicator on M.*

(1) There is a unique class $\mathcal{G}$ of Y-cuts that satisfies (a)–(c) in the definition of Y-generic cuts.
(2) The class of Y-generic cuts is the smallest comeagre set in the space of Y-cuts.

The proof of this [8] goes via the combinatorial notion of *pregeneric intervals*, which we do not want to delve into here. (Pregenericity will, nevertheless, make a brief appearance in Theorem 3.11.) Actually, generic cuts were defined in terms of these intervals in our previous paper [8].

Notice part (1) of this theorem implies that around any generic cut, there is an interval in which all generic cuts are conjugate. This is the key property of generic cuts that we will use over and over again.

The name 'generic cuts' comes from part (2) of this theorem. It says that generic cuts satisfy all the properties that are possessed by almost all cuts, when 'almost all' means 'comeagre'.

The previous paper, the definition of generic cuts, and Theorem 2.5 concern neighbourhood systems, indicators and generic cuts in a single model of PA. Some neighbourhood systems, such as $\mathcal{N}_{\mathrm{elt}}$, and their corresponding indicators, make sense across a range of models of PA. When they do we will be interested in their model-theoretic properties, and this is the main focus of this paper. We will get to this in the final section.

3 Saturation

We work with a fixed countable arithmetically saturated model $M \models \mathrm{PA}$ and an invariant indicator Y on M throughout this section.

Before we show how saturated generic cuts are, we need to describe the language involved more explicitly.

Definition 3.1. Define $\mathscr{L}_{\mathrm{Sk}}^{\mathrm{cut}}$ to be the language obtained from $\mathscr{L}_{\mathrm{Sk}}$ by adding one new unary predicate symbol $\mathbb{I}$, which is intended to be interpreted as a cut. So with some abuse of notation, we sometimes write $t \notin \mathbb{I}$ as $t > \mathbb{I}$. $\mathrm{PA}^{\mathrm{cut}}$ is the $\mathscr{L}_{\mathrm{Sk}}^{\mathrm{cut}}$-theory that consists of the axioms of PA in $\mathscr{L}_{\mathrm{Sk}}$, the definitions for the Skolem functions, and an axiom saying $\mathbb{I}$ is a cut.

We divide the $\mathscr{L}_{\mathrm{Sk}}^{\mathrm{cut}}$-formulas into levels as in usual model theory.

Definition 3.2. The formula classes $\forall_n$ and $\exists_n$ in the language $\mathscr{L}_{\mathrm{Sk}}^{\mathrm{cut}}$ will be referred to as Π_n^{cut} and Σ_n^{cut} respectively, for all $n \in \mathbb{N}$.

For example, since $\mathscr{L}_{\mathrm{Sk}}$ is Skolemised, all formulas in $\mathscr{L}_{\mathrm{Sk}}$ are Π_1^{cut}. The theory $\mathrm{PA}^{\mathrm{cut}}$ contains both Π_1^{cut}- and Σ_1^{cut}-formulas. So it is at most Π_2^{cut}. This implies that models of $\mathrm{PA}^{\mathrm{cut}}$ are closed under unions of chains. Using the preservation theorems, it is a straightforward exercise to show that $\mathrm{PA}^{\mathrm{cut}}$ is neither Π_1^{cut} nor Σ_1^{cut}.

Most of the other concepts we will see are either Σ_1^{cut} or Π_1^{cut}. For instance, if $\chi(x, y, \bar{z})$ is an $\mathscr{L}_{\mathrm{A}}$-formula, then 'there is an interval $[x, y]$ containing $\mathbb{I}$ that satisfies $\chi(x, y, \bar{z})$' is Σ_1^{cut}. This is more than just a random example.

Lemma 3.3. *Every Σ_1^{cut}-formula is equivalent over $\mathrm{PA}^{\mathrm{cut}}$ to a formula of the form*

$$\exists x{\in}\mathbb{I}\ \exists y{>}\mathbb{I}\ \chi(x, y, \bar{z}),$$

where $\chi \in \mathscr{L}_{\mathrm{A}}$.

Proof. Note that every Σ_1^{cut}-formula is equivalent to a finite disjunction of formulas of the form

$$\exists \bar{v} \Big(\bigwedge_{i<m} t_i(\bar{v}, \bar{z}) \in \mathbb{I} \wedge \bigwedge_{j<n} s_j(\bar{v}, \bar{z}) > \mathbb{I} \wedge \eta(\bar{v}, \bar{z}) \Big),$$

where the t_i's and s_j's are Skolem functions and η is an $\mathscr{L}_{\mathrm{A}}$-formula, and this displayed formula is equivalent over $\mathrm{PA}^{\mathrm{cut}}$ to

$$\exists x{\in}\mathbb{I}\ \exists y{>}\mathbb{I}\ \exists \bar{v}\ \big(x = \max_{i<m}(t_i(\bar{v}, \bar{z})) \wedge y = \min_{j<n}(s_j(\bar{v}, \bar{z})) \wedge \eta(\bar{v}, \bar{z})\big). \qquad \square$$

Let us come back to the saturation properties of generic cuts. We start with a notion defined in Kirby's thesis [9].

Definition 3.4. The Y-*index* of a cut $I \subseteq_e M$ is defined to be

$$\{n \in M : (M, I) \models \forall x{\in}\mathbb{I}\ \forall y{>}\mathbb{I}\ Y(x, y) > n\}.$$

The index of a cut is clearly an initial segment of the model.

Proposition 3.5. *If I is a Y-generic cut in M, then the Y-index of I is $\mathbb{N}$.*

Proof. Every natural number is in the Y-index of I because I is a Y-cut. Take any nonstandard $v \in M$. Recall that Y is piecewise definable. Pick any $B \in M \setminus I$, and let $\hat{Y} \in M$ be the code of $\{\langle x, y, Y(x, y)\rangle : x, y < B\}$. Using the genericity of I, let $[a, b]$ be an interval containing I in which all Y-generic cuts are conjugate over $\langle \hat{Y}, v\rangle$. Choosing $b = B - 1$ if necessary, we may assume $b < B$. Let $[u, v] \subseteq [a, b]$ such that $\mathbb{N} < Y(u, v) < v$, and J be a Y-generic cut in $[u, v]$. By the choice of $[a, b]$, we know that I and J are conjugate over $\langle \hat{Y}, v\rangle$. Since

$$(M, J) \models \exists x{\in}\mathbb{I}\ \exists y{>}\mathbb{I}\ \hat{Y}(x, y) \leqslant v,$$

the same formula is true in (M, I) too. So v is not in the Y-index of I. □

Remark 3.6. There is a notion that is similar to the index, called the *cofinality*, of a cut. It turns out that the cofinality of a generic cut depends on the indicator chosen. See Theorem 4.13 in Kaye [6] for the details.

This proposition implies a non-saturation property of generic cuts. Recall the following definition: if $\mathscr{L}$ is a recursive language and Γ is a class of $\mathscr{L}$-formulas, then an $\mathscr{L}$-structure $\mathfrak{M}$ is *Γ-recursively saturated* if and only if all recursive types that just consist of formulas in Γ are realised in $\mathfrak{M}$.

Corollary 3.7. *If I is a Y-generic cut in M, then (M, I) is not Π_1^{cut}-recursively saturated.*

Proof. Consider the type

$$p(v) = \{v > n : n \in \mathbb{N}\} \cup \{\forall x{\in}\mathbb{I}\ \forall y{>}\mathbb{I}\ \hat{Y}(x, y) > v\},$$

where $\hat{Y}$ is a code for some suitable initial part of Y. □

This has consequences on the non-definability of generic cuts, in the sense that no first-order $\mathscr{L}_{\mathrm{Sk}}^{\mathrm{cut}}$-theory captures only generic cuts.

Corollary 3.8. *Let T be a consistent $\mathscr{L}_{\mathrm{Sk}}^{\mathrm{cut}}$-theory extending $\mathrm{PA}^{\mathrm{cut}}$. Then there is a model $(K, J) \models T$ such that J is not Y-generic for all indicators Y on K.*

Proof. Take $(K, J) \models T$ to be countable and recursively saturated. □

In particular no first-order $\mathscr{L}_{\mathrm{Sk}}^{\mathrm{cut}}$-theory T captures the notion of 'elementary generic', i.e. generic for $\mathscr{N}_{\mathrm{elt}}$.

We will show that the failure of Π_1^{cut}-recursive saturation is best possible for generic cuts. The proof uses a lemma that is worth stating on its own.

Lemma 3.9. *Let $[\![a,b]\!]$ be a Y-interval and $p(v)$ be a recursive set of Σ_1^{cut}-formulas that involves only finitely many parameters from M. Then the following are equivalent.*

(1) *There exists a Y-interval $[\![r,s]\!] \subseteq [\![a,b]\!]$ such that for all Y-cuts I in $[\![r,s]\!]$ the set $p(v)$ is realised in (M,I).*
(2) *There is a Y-cut $I \in [\![a,b]\!]$ such that $p(v)$ is realised in (M,I).*
(3) *There is a Y-cut $I \in [\![a,b]\!]$ such that $p(v)$ is finitely satisfied in (M,I).*
(4) *For each finite $p'(v) \subseteq p(v)$, there is a Y-cut $I \in [\![a,b]\!]$ such that $(M,I) \models \exists v \bigwedge p'(v)$.*

Proof. It is clear that $(i) \Rightarrow (i+1)$ for $i = 1, 2$, and 3. So it suffices to show (4) $\Rightarrow$ (1).

By Lemma 3.3, we may assume

$$p(v) = \{\exists \bar{x} \in \mathbb{I}\ \exists \bar{y} > \mathbb{I}\ \theta_i(v,\bar{x},\bar{y},\bar{c}) : i \in \mathbb{N}\},$$

where $\bar{c} \in M$ and $\theta_0, \theta_1, \ldots \in \mathscr{L}_{\mathrm{A}}$. Consider the recursive set

$$\begin{aligned} q(v,r,s) = {} & \{Y(r,s) > n : n \in \mathbb{N}\} \cup \{a \leqslant r \wedge s \leqslant b\} \\ & \cup \{\exists \bar{x} < r\ \exists \bar{y} > s\ \theta_i(v,\bar{x},\bar{y},\bar{c}) : i \in \mathbb{N}\} \end{aligned}$$

of $\mathscr{L}_{\mathrm{A}}$-formulas. If this is realised in M, then (1) is true. So by recursive saturation, it suffices to show $q(v,r,s)$ is finitely satisfied in M. We can prove

$$M \models \exists v\ \exists [r,s] \subseteq [\![a,b]\!] \left(Y(r,s) > n \wedge \bigwedge_{i<n} \exists \bar{x} < r\ \exists \bar{y} > s\ \theta_i(v,\bar{x},\bar{y},\bar{c}) \right)$$

for every $n \in \mathbb{N}$ from the hypothesis that (4) holds. □

Proposition 3.10. *If I is a Y-generic cut in M, then (M,I) is Σ_1^{cut}-recursively saturated.*

Proof. Let $p(v)$ be a Σ_1^{cut}-recursive type that is finitely satisfied in (M,I), and $\bar{c} \in M$ be the parameters that appear in $p(v)$. Using genericity, pick an interval $[a,b] \subseteq M$ containing I in which all Y-generic cuts are conjugate over $\bar{c}$. By the previous lemma, we can find a Y-interval $[\![r,s]\!] \subseteq [\![a,b]\!]$ in which all Y-cuts J make $p(v)$ realised in (M,J). Inside such an interval, there is a Y-generic cut that is conjugate to I over $\bar{c}$. So $p(v)$ is realised in (M,I) too. □

Strangely, generic cuts seem to possess more than what Σ_1^{cut}-recursive saturation can offer. For example, the following apparently needs Π_1^{cut}-recursive saturation. The extra bit comes from the strength of $\mathbb{N}$.

Theorem 3.11. *Let I be a Y-generic cut in M. Then for every $\bar{c} \in M$, there exists an interval $[a, b]$ containing I such that for every $\mathscr{L}_\mathrm{A}$-formula $\chi(x, y, \bar{z})$,*

$$(M, I) \models \exists x{\in}\mathbb{I}\ \exists y{>}\mathbb{I}\ \chi(x, y, \bar{c}) \to \exists x{<}a\ \exists y{>}b\ \chi(x, y, \bar{c}).$$

Proof. Let $[u, v]$ be an interval around I in which all Y-generic cuts are conjugate over $\bar{c}$. Using recursive saturation, find $f \in M$ coding a function $\mathbb{N} \to M$ such that

$$f(\chi) = (\max n)(\exists [r, s]{\subseteq}[u, v]\ (Y(r, s) \geqslant n \wedge \exists x{<}r\ \exists y{>}s\ \chi(x, y, \bar{c}))),$$

for all $\mathscr{L}_\mathrm{A}$-formulas $\chi \in \mathbb{N}$. Let $d \in M \setminus \mathbb{N}$ such that

$$f(\chi) > \mathbb{N} \Leftrightarrow f(\chi) > d,$$

for all $\chi \in \mathbb{N}$. Such d exists because $\mathbb{N}$ is strong in M. We claim that for every $\chi \in \mathbb{N}$, we have $f(\chi) > d$ if and only if

$$(M, I) \models \exists x{\in}\mathbb{I}\ \exists y{>}\mathbb{I}\ \chi(x, y, \bar{c}).$$

This suffices because we can then write our type as

$$p(a, b) = \{a \in \mathbb{I} \wedge \mathbb{I} < b\} \cup \{f(\chi) > d \to \exists x{<}a\ \exists y{>}b\ \chi(x, y, \bar{c}) : \chi \in \mathscr{L}_\mathrm{A}\},$$

which is recursive and Σ_1^{cut}.

Let $\chi \in \mathbb{N}$. If $f(\chi) \leqslant d$, then $f(\chi) \in \mathbb{N}$, and so

$$(M, I) \models \forall x{\in}\mathbb{I}\ \forall y{>}\mathbb{I}\ \neg\chi(x, y, \bar{c})$$

since I is a Y-cut. Conversely, suppose $f(\chi) > d$. Let $[r, s] \subseteq [u, v]$ such that $Y(r, s) = f(\chi) > \mathbb{N}$ and $M \models \exists x{<}r\ \exists y{>}s\ \chi(x, y, \bar{c})$. By our choice of $[u, v]$, the cut I is conjugate over $\bar{c}$ to some Y-generic cut $J \in [r, s]$. For any such J, we have

$$(M, J) \models \exists x{\in}\mathbb{I}\ \exists y{>}\mathbb{I}\ \chi(x, y, \bar{c}).$$

Therefore, this formula is also true in (M, I). □

We present two applications of this theorem, the first of which is easy.

Corollary 3.12. *Let I be a Y-generic cut in M. Then for all $\bar{c} \in M$, there exists $a \in I$ such that for every $\mathscr{L}_\mathrm{A}$-formula $\eta(v, \bar{z})$,*

$$(M, I) \models \exists x{\in}\mathbb{I}\ \forall v{\in}\mathbb{I}\ (v > x \to \eta(v, \bar{c})) \to \exists x{<}a\ \forall v{\in}\mathbb{I}\ (v > x \to \eta(v, \bar{c})).$$

Proof. Let $\chi(x, y, \bar{c})$ be $\forall v{\in}[x, y]\ \eta(v, \bar{c})$. □

The second one seems slightly more tricky. Recall first the following.

Definition 3.13. Let $\mathfrak{M}$ be a structure in some first-order language $\mathscr{L}$, and $\bar{c}, r \in \mathfrak{M}$. The *existential type* of r over $\bar{c}$, denoted by $\mathrm{etp}_{\mathfrak{M}}(r/\bar{c})$, is defined to be

$$\{\phi(w,\bar{c}) \in \exists_1 : \mathfrak{M} \models \phi(r,\bar{c})\}.$$

As is commonly done in models of PA, we regard each element of M as an M-finite set here. See our previous paper [7] for details. Our consequence of Theorem 3.11 now follows.

Corollary 3.14. *Let I be a Y-generic cut in M. Then the existential type of any tuple $\bar{c}$ in (M, I) is* coded, *i.e., there exists $t \in M$ such that for all $\phi(\bar{w}) \in \Sigma_1^{\mathrm{cut}}$,*

$$(M, I) \models \ulcorner\phi\urcorner \in t \leftrightarrow \phi(\bar{c}).$$

Proof. By Lemma 3.3, it suffices to consider only Σ_1^{cut}-formulas of the form

$$\exists x{\in}\mathbb{I}\ \exists y{>}\mathbb{I}\ \chi(x, y, \bar{z}),$$

where $\chi \in \mathscr{L}_{\mathrm{A}}$. By Theorem 3.11, we can find an interval $[a, b]$ containing I such that

$$(M, I) \models \exists x{\in}\mathbb{I}\ \exists y{>}\mathbb{I}\ \chi(x, y, \bar{c}) \leftrightarrow \exists x{<}a\ \exists y{>}b\ \chi(x, y, \bar{c}),$$

for every $\mathscr{L}_{\mathrm{A}}$-formula $\chi(x, y, \bar{z})$. So recursive saturation of M implies that

$$p(t) = \{\ulcorner\exists x{\in}\mathbb{I}\ \exists y{>}\mathbb{I}\ \chi(x, y, \bar{z})\urcorner \in t \leftrightarrow \exists x{\in}\mathbb{I}\ \exists y{>}\mathbb{I}\ \chi(x, y, \bar{c}) : \chi \in \mathscr{L}_{\mathrm{A}}\}$$

is realised in M. □

4 Model completeness

Throughout this section, we work with a fixed countable arithmetically saturated model $M \models \mathrm{PA}$, an invariant indicator Y on M, and a Y-generic cut I in M.

In our previous paper [8], we established a weak quantifier elimination result. We reformulate this theorem here in more model-theoretic terms, and list some consequences related to model completeness.

Theorem 4.1. *Let $\bar{c} \in M$ and J be a Y-generic cut in M. If $\mathrm{etp}_{(M,I)}(\bar{c}) = \mathrm{etp}_{(M,J)}(\bar{c})$, then $(M, I, \bar{c}) \cong (M, J, \bar{c})$.*

Proof sketch. Without loss, assume $I < J$. Using genericity, pick intervals $[a, b]$ and $[u, v]$ containing I and J respectively in which all Y-generic cuts are conjugate over $\bar{c}$. Let $x \in I$ that is above a. We claim that there is $h \in \mathrm{Aut}(M, \bar{c})$ such that $x^h \in [u, v]$, which will give us what we want. By recursive saturation, it suffices to show that the $\mathscr{L}_A$-type of x over $\bar{c}$ is finitely satisfied in $[u, v]$. Let $\theta(x, \bar{w}) \in \mathscr{L}_A$ such that $M \models \theta(x, \bar{c})$. Then $(M, I) \models \mathrm{Q}y{\in}\mathbb{I}\ \theta(y, \bar{c})$ by (the proof of) Corollary 3.12. Notice $\mathrm{Q}y{\in}\mathbb{I}\ \theta(y, \bar{w})$ is Π_1^{cut}, and $\neg\mathrm{Q}y{\in}\mathbb{I}\ \theta(y, \bar{w}) \notin \mathrm{etp}_{(M,I)}(\bar{c})$. Thus $\neg\mathrm{Q}y{\in}\mathbb{I}\ \theta(y, \bar{w}) \notin \mathrm{etp}_{(M,J)}(\bar{c})$, which means $(M, J) \models \mathrm{Q}y{\in}\mathbb{I}\ \theta(y, \bar{c})$. In particular, there is $y \in J$ above u such that $M \models \theta(y, \bar{c})$. □

Corollary 4.2. *For any $\bar{c}, r, s \in M$, if $\mathrm{etp}_{(M,I)}(r/\bar{c}) = \mathrm{etp}_{(M,I)}(s/\bar{c})$, then $(M, I, \bar{c}, r) \cong (M, I, \bar{c}, s)$.*

Proof. Suppose $\mathrm{etp}_{(M,I)}(r/\bar{c}) = \mathrm{etp}_{(M,I)}(s/\bar{c})$. Using recursive saturation, let $g \in \mathrm{Aut}(M, \bar{c})$ such that $s^g = r$. Setting $J = I^g$ gives $(M, I, \bar{c}, s) \cong (M, J, \bar{c}, r)$, and so by hypothesis,

$$\mathrm{etp}_{(M,I)}(\bar{c}, r) = \mathrm{etp}_{(M,I)}(\bar{c}, s) = \mathrm{etp}_{(M,J)}(\bar{c}, r).$$

The previous theorem then does the rest. □

Definable elements in models of $\mathrm{PA}^{\mathrm{cut}}$ were studied in Kossak–Bamber [11]. Corollary 4.2 gives us more information about these definable elements in the case of generic cuts.

Corollary 4.3. *All $\mathscr{L}_{\mathrm{Sk}}^{\mathrm{cut}}$-definable elements of (M, I) are Σ_1^{cut} definable with the same parameters.*

Proof. Let $a, \bar{c} \in M$ and $\theta(w, \bar{z}) \in \mathscr{L}_{\mathrm{Sk}}^{\mathrm{cut}}$ such that a is the unique element w that satisfies $\theta(w, \bar{c})$ in (M, I). Consider $p(w) = \mathrm{etp}_{(M,I)}(a/\bar{c})$. If $a' \in M$ that satisfies all of $p(w)$ in (M, I), then $(M, I, \bar{c}, a) \cong (M, I, \bar{c}, a')$ by Corollary 4.2, and so $a = a'$ because they both satisfy $\theta(w, \bar{c})$. It follows that

$$q(w) = p(w) \cup \{w \neq a\}$$

is not realised in (M, I). The set $q(w)$ can be rewritten as a recursive set of Σ_1^{cut}-formulas using Corollary 3.14. So by Σ_1^{cut}-recursive saturation, this set is not finitely satisfied in (M, I). Therefore, there is $\phi(w, \bar{c}) \in p(w)$ such that

$$(M, I) \models \forall w\ (\phi(w, \bar{c}) \rightarrow w = a),$$

as required. □

Remark 4.4. Typical Σ_1^{cut}-definable elements are

$$(\max w \in \mathbb{I})(\theta(w, \bar{c})) \quad \text{and} \quad (\min w > \mathbb{I})(\theta(w, \bar{c})),$$

where $\theta(w, \bar{c})$ is in $\mathscr{L}_A$.

Question 4.5. Can one make the map $\theta(w, \bar{z}) \mapsto \phi(w, \bar{z})$, which is given by the proof above, independent of $\bar{c}$?

Recall that a definition of model completeness says: a theory T is *model complete* if and only if every $\forall_1$-formula is equivalent modulo T to a $\exists_1$-formula.

Corollary 4.6. *For every Π_1^{cut}-formula $\theta(w, \bar{z})$ and every $\bar{c} \in M$, there exists a set $\Phi(w, \bar{z})$ of Σ_1^{cut}-formulas such that*

$$(M, I) \models \forall w \left(\theta(w, \bar{c}) \leftrightarrow \bigvee\!\!\!\bigvee \Phi(w, \bar{c})\right).$$

Proof. Let $A = \{w \in M : M \models \theta(w, \bar{c})\}$. For the moment, work with a fixed $a \in A$. Set $p_a(w) = \mathrm{etp}_{(M,I)}(a/\bar{c})$. By Corollary 4.2,

$$(M, I) \models \forall w \left(\bigwedge\!\!\!\bigwedge p_a(w) \to \theta(w, \bar{c})\right).$$

Therefore, the set $p_a(w) \cup \{\neg\theta(w, \bar{c})\}$ is not realised in (M, I). As in the previous proof, we use Σ_1^{cut}-recursive saturation to pick $\phi_a(w, \bar{c}) \in p_a(w)$ such that

$$(M, I) \models \forall w\, (\phi_a(w, \bar{c}) \to \theta(w, \bar{c})).$$

For every $a \in A$, we can find such a formula $\phi_a(w, \bar{c})$. It can be verified that $\Phi(w, \bar{z}) = \{\phi_a(w, \bar{z}) : a \in A\}$ does what we want. □

Question 4.7. Is the corollary above true for all $\mathscr{L}_{\mathrm{Sk}}^{\mathrm{cut}}$-formulas $\theta(w, \bar{z})$?

In Corollary 4.6 above, we cannot guarantee that the set $\Phi(w, \bar{z})$ is finite. This can be proved using Corollary 3.7 and Proposition 3.10. Therefore, $\mathrm{Th}(M, I)$ is not model complete.

5 Existential closure

In this section, we show that generic cuts are existentially closed in a suitable category. Existentially closed models of arithmetic were studied by Goldrei, Macintyre, Simmons [2, 13], Hirschfeld, Wheeler [3], and others in the 1970s. A more recent reference is Adamowicz–Bigorajska [1].

Recall that an *existentially closed* model of a theory T is a model $\mathfrak{M} \models T$ such that whenever $\mathfrak{K} \models T$ that extends $\mathfrak{M}$, if $\sigma(\bar{z}) \in \exists_1$ and $\bar{c} \in \mathfrak{M}$, then $\mathfrak{K} \models \sigma(\bar{c})$ implies $\mathfrak{M} \models \sigma(\bar{c})$. This property requires us to consider indicators across several models.

Definition 5.1. Let $\mathbf{K}$ be a class of models of PA. An *(invariant) indicator* over $\mathbf{K}$ is a sequence $(Y(x, y) = n)_{n \in \mathbb{N}}$ of $\mathscr{L}_{\mathrm{A}}$-formulas with a recursive definition that can be extended to $(Y(x, y) = n)_{n \in M}$ to obtain an (invariant) indicator on M for every $M \in \mathbf{K}$. For simplicity, we will often write an indicator over $\mathbf{K}$ as Y instead of

$(Y(x, y) = n)_{n\in\mathbb{N}}$. For an indicator Y over some class of models of PA, define $\mathrm{PA}_Y^{\mathrm{cut}} = \mathrm{PA}^{\mathrm{cut}} \cup \{\forall x{\in}\mathbb{I}\ \forall y{>}\mathbb{I}\ Y(x, y) \geqslant n : n \in \mathbb{N}\}$, where $Y(x, y) \geqslant n$ means $\bigwedge\!\!\!\bigwedge_{i<n} Y(x, y) \neq i$.

Note that $\mathrm{PA}_Y^{\mathrm{cut}}$ is Π_2^{cut}.

Almost all indicators that appeared in the literature are definable, and their definitions make sense over all models of PA. This allows one to prove independence results for PA using indicators. See Chapter 14 in Kaye [5] for an exposition on this line.

There is a uniform way of defining an indicator Y_{elt} for the neighbourhood system $\mathcal{N}_{\mathrm{elt}}$ in the sense above over the class of all recursively saturated models of PA. For this Y the theory $\mathrm{PA}_Y^{\mathrm{cut}}$ is the theory of models of PA with a distinguished proper elementary cut.

We need one technical lemma before the theorem.

Lemma 5.2. *Fix a countable arithmetically saturated model $M \models \mathrm{PA}$, an invariant indicator Y on M, and a Y-generic cut I in M. If Θ is an $\mathcal{L}_{\mathrm{A}}$-definable function under which I is closed, then there is $n \in \mathbb{N}$ such that for all sufficiently large $x \in I$, we have $Y(x, \Theta(x)) < n$.*

Proof. Find an interval $[a, b]$ around I in which all Y-generic cuts are conjugate over the parameters $\bar{c}$ needed to define Θ. Suppose I is closed under Θ, but for all $n \in \mathbb{N}$, there are cofinally many $x \in I$ such that $Y(x, \Theta(x)) \geqslant n$. Then

$$M \models \exists x{\in}[a, b]\ (\Theta(x) \in [a, b] \wedge Y(x, \Theta(x)) \geqslant n),$$

for all $n \in \mathbb{N}$. Using recursive saturation, find nonstandard $v \in M$ and $x \in [a, b]$ such that

$$M \models \Theta(x) \in [a, b] \wedge Y(x, \Theta(x)) \geqslant v.$$

Note $[x, \Theta(x)]$ is a Y-interval. So it contains a Y-generic cut, say J. However J is not closed under Θ, and hence I cannot be conjugate to J over $\bar{c}$. This contradicts our choice of $[a, b]$. □

Theorem 5.3. *If M is a countable arithmetically saturated model of PA, and I is a Y-generic cut in M, then (M, I) is an existentially closed model of $\mathrm{PA}_Y^{\mathrm{cut}}$.*

Proof. Let (K, J) be an extension of (M, I) that satisfies $\mathrm{PA}_Y^{\mathrm{cut}}$. Notice we must have $M \prec K$ since $\mathcal{L}_{\mathrm{Sk}}$ is Skolemised. Suppose $(K, J) \models \sigma(c)$, where $c \in M$ and $\sigma(z) \in \Sigma_1^{\mathrm{cut}}$. By Lemma 3.3, we may assume $c > I$ and $\sigma(z)$ is of the form

$$\exists x\ \exists y\ (\theta(x, y, z) \wedge x \in \mathbb{I} \wedge y \notin \mathbb{I}),$$

where $\theta \in \mathscr{L}_{\mathrm{A}}$. Define

$$\Theta(x) = \begin{cases} (\max y)(\theta(x, y, c)), & \text{if it exists;} \\ c, & \text{if } \forall u\ \exists y{>}u\ \theta(x, y, c); \\ 0, & \text{if } \neg\exists y\ \theta(x, y, c). \end{cases}$$

The fact that $(K, J) \models \sigma(c)$ implies that J is not closed under Θ.

Suppose first that we have $a \in I$ such that $\Theta(x) > J$ for some $x < a$ in J. Then

$$(K, J) \models (\min b)(\forall x{<}a\ (\Theta(x) \leqslant b)) > \mathbb{I}.$$

This transfers to (M, I), and so $(M, I) \models \sigma(c)$.

Suppose next that such an a cannot be found. Then I is closed under Θ. By the previous lemma, there is $n \in \mathbb{N}$ such that for all sufficiently large $x \in I$,

$$(M, I) \models Y(x, \Theta(x)) < n.$$

By overspill there is $[a, b] \subseteq M$ containing I such that

$$M \models \forall x \in [a, b]\ Y(x, \Theta(x)) < n.$$

This transfers to (K, J) as $M \prec K$. However, such a statement cannot be true in (K, J) because J is a Y-cut under which Θ is not closed. □

It follows immediately that if I is a generic cut in a model $M \models \mathrm{PA}$, then the existential type of any $c \in (M, I)$ is *maximal* over $\mathrm{PA}_Y^{\mathrm{cut}}$, i.e., for all Σ_1^{cut}-formulas $\sigma(z)$, either $\mathrm{PA}_Y^{\mathrm{cut}} + \mathrm{etp}_{(M,I)}(c) \vdash \sigma(c)$ or $\mathrm{PA}_Y^{\mathrm{cut}} + \mathrm{etp}_{(M,I)}(c) \vdash \neg\sigma(c)$.

Question 5.4. If I is a cut of a countable arithmetically saturated model $M \models \mathrm{PA}$ such that for all $\bar{c} \in M$ the existential type $\mathrm{etp}_{(M,I)}(\bar{c})$ is maximal over $\mathrm{PA}_Y^{\mathrm{cut}}$, then is the cut I always Y-generic for some indicator Y on M?

A strengthening of existential closure is *existential universality*, which has been around in the literature for some time [3]. Recall that an *existentially universal* model of a theory T is a model $\mathfrak{M} \models T$ such that whenever $\mathfrak{K} \models T$ that extends $\mathfrak{M}$, if $p(v, \bar{z})$ is a set of $\exists_1$-formulas and $\bar{c} \in \mathfrak{M}$, then $\mathfrak{K} \models \exists v \bigwedge p(v, \bar{c})$ implies $\mathfrak{M} \models \exists v \bigwedge p(v, \bar{c})$. No countable nonstandard model $M \models \mathrm{PA}$ is existentially universal: just consider

$$p(v) = \{i \in v : i \in S\} \cup \{i \notin v : i \in \mathbb{N} \setminus S\},$$

where $S \subseteq \mathbb{N}$ that is not in $\mathrm{SSy}(M)$. For the same reason, no countable model of $\mathrm{PA}^{\mathrm{cut}}$ is existentially universal. To avoid this, we relax the definition a bit.

Definition 5.5. Let T be a theory in a recursive language $\mathscr{L}$. Then a *recursively existentially universal* model of T is a model $\mathfrak{M} \models T$ such that whenever $\mathfrak{K} \models T$

that extends $\mathfrak{M}$, if $p(v, \bar{z})$ is a recursive set of $\exists_1$-formulas and $\bar{c} \in \mathfrak{M}$, then $\mathfrak{K} \models \exists v \bigwedge p(v, \bar{c})$ implies $\mathfrak{M} \models \exists v \bigwedge p(v, \bar{c})$.

Corollary 5.6. *Fix an invariant indicator Y over some class of models of* PA. *If M is a countable arithmetically saturated model of* PA, *and I is a Y-generic cut in M, then (M, I) is a recursively existentially universal model of* $\mathrm{PA}_Y^{\mathrm{cut}}$.

Proof. Take a tuple $\bar{c} \in M$ and a recursive set of Σ_1^{cut}-formulas $p(v, \bar{z})$ that is realised in some extension of (M, I) satisfying $\mathrm{PA}_Y^{\mathrm{cut}}$. Then $p(v, \bar{c})$ is finitely satisfied in (M, I) by existential closure. So Σ_1^{cut}-recursive saturation guarantees that $p(v, \bar{c})$ is realised in (M, I). □

It is standard model-theoretic fact that all embeddings between existentially closed models are $\exists_2$-elementary. This applies to generic cuts too. The previous corollary suggests that generic cuts may actually satisfy something stronger. Recall that all embeddings between existentially universal models are elementary [2, 3]. We only have a partial result towards this.

Corollary 5.7. *Let Y be an invariant indicator over a class* **K** *of models of* PA, *and let I, J be Y-generic cuts in countable arithmetically saturated models $M, K \in$* **K** *respectively. If* $\mathrm{SSy}(M) = \mathrm{SSy}(K)$, *then every embedding $(M, I) \to (K, J)$ is elementary.*

Proof. Let $h\colon (M, I) \to (K, J)$ be an embedding, $\bar{c} \in M$ and $\bar{d} = \bar{c}^h$. Suppose $(K, J) \models \exists w\, \theta(w, \bar{d})$ where $\theta \in \mathscr{L}_{\mathrm{Sk}}^{\mathrm{cut}}$, and take $s \in K$ with $(K, J) \models \theta(s, \bar{d})$. Then $p(w, \bar{d}) = \mathrm{etp}_{(K,J)}(s/\bar{d})$ is coded in K by Corollary 3.14, and hence it is coded in M too. Also $\mathrm{etp}_{(M,I)}(\bar{c}) = \mathrm{etp}_{(K,J)}(\bar{d})$, since the inclusion $\subseteq$ is obvious and $\mathrm{etp}_{(M,I)}(\bar{c})$ is maximal by Theorem 5.3. Thus $p(w, \bar{c})$ is finitely satisfied in (M, I) and by Proposition 3.10 is realised in (M, I), by some $r \in M$ say. Let $s' = r^h \in K$. Then $\mathrm{etp}_{(K,J)}(s/\bar{d}) = \mathrm{etp}_{(K,J)}(s'/\bar{d})$ by maximality again and so $\mathrm{tp}_{(K,J)}(s/\bar{d}) = \mathrm{tp}_{(K,J)}(s'/\bar{d})$ by Corollary 4.2. Therefore $(K, J) \models \theta(s', \bar{d})$ as required. □

Question 5.8. Can the assumption $\mathrm{SSy}(M) = \mathrm{SSy}(K)$ in Corollary 5.7 be removed?

We also ask a related question that is specific to Y_{elt}-generic cuts.

Question 5.9. Let I, J be elementary generic cuts in countable arithmetically saturated models $M, K \models$ PA respectively. Is it necessarily the case that if $(M, \mathbb{N}) \equiv (K, \mathbb{N})$, then $(M, I) \equiv (K, J)$?

Acknowledgement: Part of this work is from the second author's doctoral dissertation [14], written under the supervision of the first author. While this paper was being written, the second author was supported financially by the Department of Mathematics at the National University of Singapore. He would like to thank Professor Chitat Chong for this. We also wish to thank the anonymous referee, whose comments helped improve the quality of this paper.

Bibliography

[1] Zofia Adamowicz and Teresa Bigorajska. Existentially closed structures and Gödel's second incompleteness theorem. *The Journal of Symbolic Logic*, 66(1):349–356, March 2001.
[2] Derek C. Goldrei, Angus Macintyre, and Harold Simmons. The forcing companions of number theories. *Israel Journal of Mathematics*, 14:317–337, 1973.
[3] Joram Hirschfeld and William H. Wheeler. *Forcing, Arithmetic, Division Rings*, volume 454 of *Lecture Notes in Mathematics*. Springer-Verlag, Berlin, 1975.
[4] Wilfrid Hodges. *Building Models by Games*. Dover Publications, Inc., Mineola, New York, 2006. Unabridged republication of the work originally published as volume 2 in the *London Mathematical Society Student Texts* series by Cambridge University Press, Cambridge, in 1985.
[5] Richard Kaye. *Models of Peano Arithmetic*, volume 15 of *Oxford Logic Guides*. Clarendon Press, Oxford, 1991.
[6] Richard Kaye. Generic cuts in models of arithmetic. *Mathematical Logic Quarterly*, 54(2):129–144, 2008.
[7] Richard Kaye and Tin Lok Wong. On interpretations of arithmetic and set theory. *Notre Dame Journal of Formal Logic*, 48(4):497–510, October 2007.
[8] Richard Kaye and Tin Lok Wong. Truth in generic cuts. *Annals of Pure and Applied Logic*, 161(8):987–1005, May 2010.
[9] Laurence A.S. Kirby. *Initial segments of models of arithmetic*. PhD thesis, Manchester University, July 1977.
[10] Laurence A.S. Kirby and Jeff B. Paris. Initial segments of models of Peano's axioms. In Alistair Lachlan, Marian Srebrny, and Andrzej Zarach, editors, *Set Theory and Hierarchy Theory V*, volume 619 of *Lecture Notes in Mathematics*, pages 211–226, Berlin, 1977. Springer-Verlag.
[11] Roman Kossak and Nicholas Bamber. On two questions concerning the automorphism groups of countable recursively saturated models of PA. *Archive for Mathematical Logic*, 36(1):73–79, December 1996.
[12] Roman Kossak and James H. Schmerl. *The Structure of Models of Peano Arithmetic*, volume 50 of *Oxford Logic Guides*. Clarendon Press, Oxford, 2006.
[13] Harold Simmons. Existentially closed models of basic number theory. In Robin O. Gandy and J. Martin E. Hyland, editors, *Logic Colloquium 76*, volume 87 of *Studies in Logic and the Foundations of Mathematics*, pages 325–369, Amsterdam, 1977. North-Holland Publishing Company.
[14] Tin Lok Wong. *Initial segments and end-extensions of models of arithmetic*. PhD thesis, University of Birmingham, May 2010.

Juha Kontinen

On Natural Deduction in Dependence Logic

Abstract: We study the natural deduction system of dependence logic introduced in [6]. We introduce a single rule that allows us to generalize the completeness theorem of [6] from sentences to open formulas. By using a variant of our rule for independence logic, an analogous generalization of the corresponding completeness theorem of [5] is also obtained.

Keywords: dependence logic, axiomatization, team

MSC: 03B60

Juha Kontinen: Department of Mathematics and Statistics,University of Helsinki, e-mail:juha.kontinen@helsinki.fi

1 Introduction

Dependence logic was introduced in [7]. It extends first-order logic by new atomic formulas

$$=(x_1, \ldots, x_n, y)$$

with the meaning that the values of the variables $x_1, \ldots, x_n$ functionally determine the values of y. The set of valid formulas of dependence logic has the same complexity as that of full second-order logic, and hence it is not possible to give a complete axiomatization of dependence logic. One classical way to circumvent this problem is to modify the semantics - this is the line adopted in Henkin's Completeness Theorem for second-order logic. For dependence logic this direction has been pursued in [3]. Another way to proceed is to restrict to a fragment of dependence logic. This is the line of study pursued in [6], where a natural deduction inference system was formulated for dependence logic. This deduction system was shown to be strong enough to derive all first-order consequences of dependence logic sentences.

Independence logic [4] is a variant of dependence logic in which the dependence atoms are replaced by independence atoms $\overline{x} \perp_{\overline{z}} \overline{y}$. In [5] the deduction system, and the analogous completeness theorem, of [6] was formulated for independence logic. For both dependence and independence logic, an open question remained whether the completeness result hold for open formulas also. In this article we introduce a single rule for both dependence and independence logic that allows us to generalize both completeness theorems to open formulas.

2 Preliminaries

In this section we define Dependence Logic ($\mathcal{D}$) and recall some basic results about it.

Definition 2.1 ([7]). The syntax of $\mathcal{D}$ extends the syntax of FO, defined in terms of $\vee$, $\wedge$, $\neg$, $\exists$ and $\forall$, by new atomic formulas (dependence atoms) of the form

$$=(t_1, \ldots, t_n), \tag{1}$$

where $t_1, \ldots, t_n$ are terms. For a vocabulary τ, $\mathcal{D}[\tau]$ denotes the set of τ-formulas of $\mathcal{D}$.

The meaning of the dependence atom (1) is that the value of the term t_n is functionally determined by the values of the terms $t_1, \ldots, t_{n-1}$. As singular cases we have $=()$ which we take to be universally true, and $=(t)$ meaning that the value of t is constant.

The set $\mathrm{Fr}(\phi)$ of free variables of a formula $\phi \in \mathcal{D}$ is defined as for first-order logic, except that we have the new case

$$\mathrm{Fr}(=(t_1, \ldots, t_n)) = \mathrm{Var}(t_1) \cup \cdots \cup \mathrm{Var}(t_n),$$

where $\mathrm{Var}(t_i)$ is the set of variables occurring in the term t_i. If $\mathrm{Fr}(\phi) = \emptyset$, we call ϕ a sentence.

In order to define the semantics of $\mathcal{D}$, we first need to define the concept of a *team*. Let $\mathfrak{A}$ be a model with domain A. *Assignments* of $\mathfrak{A}$ are finite mappings from variables into A. The value of a term t in an assignment s is denoted by $t^{\mathfrak{A}}\langle s\rangle$. If s is an assignment, x a variable, and $a \in A$, then $s(a/x)$ denotes the assignment (with domain $\mathrm{Dom}(s) \cup \{x\}$) which agrees with s everywhere except that it maps x to a.

Let A be a set and $\{x_1, \ldots, x_k\}$ a finite (possibly empty) set of variables. A *team* X of A with domain $\mathrm{Dom}(X) = \{x_1, \ldots, x_k\}$ is any set of assignments from the variables $\{x_1, \ldots, x_k\}$ into the set A. We denote by $rel(X)$ the k-ary relation of A corresponding to X

$$rel(X) = \{(s(x_1), \ldots, s(x_k)) : s \in X\}.$$

If X is a team of A, and $F\colon X \to A$, we use $X(F/x_n)$ to denote the (supplemented) team $\{s(F(s)/x_n) : s \in X\}$ and $X(A/x_n)$ the (duplicated) team $\{s(a/x_n) : s \in X \text{ and } a \in A\}$.

We call a formula of $\mathcal{D}$ *first-order* if it does not contain any dependence atoms. An is [6], we assume that the scope of negation is always a first-order formula.

We are now ready to define the semantics of dependence logic. In this definition $\mathfrak{A} \models_s \phi$ refers to satisfaction in first-order logic.

Definition 2.2 ([7]). Let $\mathfrak{A}$ be a model and X a team of A. The satisfaction relation $\mathfrak{A} \models_X \phi$ is defined as follows:
1. If ϕ is first-order, then $\mathfrak{A} \models_X \phi$ iff for all $s \in X$, $\mathfrak{A} \models_s \phi$.
2. $\mathfrak{A} \models_X =\!(t_1, \ldots, t_n)$ iff for all $s, s' \in X$ such that $t_1^{\mathfrak{A}}\langle s\rangle = t_1^{\mathfrak{A}}\langle s'\rangle, \ldots, t_{n-1}^{\mathfrak{A}}\langle s\rangle = t_{n-1}^{\mathfrak{A}}\langle s'\rangle$, we have $t_n^{\mathfrak{A}}\langle s\rangle = t_n^{\mathfrak{A}}\langle s'\rangle$.
3. $\mathfrak{A} \models_X \psi \wedge \phi$ iff $\mathfrak{A} \models_X \psi$ and $\mathfrak{A} \models_X \phi$.
4. $\mathfrak{A} \models_X \psi \vee \phi$ iff $X = Y \cup Z$ such that $\mathfrak{A} \models_Y \psi$ and $\mathfrak{A} \models_Z \phi$.
5. $\mathfrak{A} \models_X \exists x_n \psi$ iff $\mathfrak{A} \models_{X(F/x_n)} \psi$ for some $F\colon X \to A$.
6. $\mathfrak{A} \models_X \forall x_n \psi$ iff $\mathfrak{A} \models_{X(A/x_n)} \psi$.

Above, we assume that the domain of X contains the variables free in ϕ. Finally, a sentence ϕ is true in a model $\mathfrak{A}$, $\mathfrak{A} \models \phi$, if $\mathfrak{A} \models_{\{\emptyset\}} \phi$.

Next we define the notions of logical consequence and equivalence for formulas of dependence logic.

Definition 2.3. Let T be a set of formulas of dependence logic with only finitely many free variables. The formula ψ is a *logical consequence* of T,

$$T \models \psi,$$

if for all models $\mathfrak{A}$ and teams X, with $\mathrm{Fr}(\psi) \cup \bigcup_{\phi \in T} \mathrm{Fr}(\phi) \subseteq \mathrm{Dom}(X)$, and $\mathfrak{A} \models_X T$ we have $\mathfrak{A} \models_X \psi$. The formulas ϕ and ψ are *logically equivalent*,

$$\phi \equiv \psi,$$

if $\phi \models \psi$ and $\psi \models \phi$.

The following basic properties of dependence logic will be used in this article. Let X be a team with domain $\{x_1, \ldots, x_k\}$ and $V \subseteq \{x_1, \ldots, x_k\}$. Denote by $X \upharpoonright V$ the team $\{s \upharpoonright V : s \in X\}$ with domain V. The following lemma shows that the truth of a formula depends only on the interpretations of the variables occurring free in the formula.

Proposition 2.4. *Suppose* $V \supseteq \mathrm{Fr}(\phi)$. *Then* $\mathfrak{A} \models_X \phi$ *if and only if* $\mathfrak{A} \models_{X \upharpoonright V} \phi$.

Note that by Proposition 2.4, we may use the notation $\psi(\overline{x})$ to mean that $\mathrm{Fr}(\psi)$ is contained in the variables of the tuple $\overline{x}$ without problems. The following fact is also a very basic property of all formulas of dependence logic:

Proposition 2.5 (Downward closure). *Let* ϕ *be a formula of dependence logic,* $\mathfrak{A}$ *a model, and* $Y \subseteq X$ *teams. Then* $\mathfrak{A} \models_X \phi$ *implies* $\mathfrak{A} \models_Y \phi$.

3 Sentences versus open formulas

It is well known that the relationship between open formulas and sentences of dependence logic is more complicated than, e.g., in first-order logic. In FO, an open formula can be translated into a sentence (in an extended vocabulary) by replacing the free variables of it by fresh constant symbols. In dependence logic, this is not possible as exemplified by the following example.

Example 3.1. Let $\mathfrak{A}$ be a structure over $\tau = \{P\}$ such that $A = \{0, 1\}$ and $P^{\mathfrak{A}} = 0$. Consider the following team of A with domain $\{x_0\}$:

	x_0
s_0	0
s_1	1

(1)

It is straightforward to verify that $\mathfrak{A} \not\models_X P(x_0)$ and $\mathfrak{A} \not\models_X \neg P(x_0)$. Suppose we extend $\mathfrak{A}$ by a new constant symbol $c^{\mathfrak{A}}$. Now $c^{\mathfrak{A}} \in \{0, 1\}$ and, in either case, it holds that $\mathfrak{A} \models P(c)$ or $\mathfrak{A} \models \neg P(c)$, and hence the behaviour of the formula $P(x_0)$ cannot be captured by $P(c)$.

The problem in Example 3.1 can be fixed by representing the team X by a new relation symbol. In [7] it was shown that any formula of dependence logic can be translated to an equivalent sentence of ESO that has an extra predicate for the team (i.e., X is represented by $rel(X)$). For our purposes the following variant of this result will be useful.

Proposition 3.2 ([1]). *Let τ be a vocabulary and ϕ a $\mathcal{D}[\tau]$-formula with free variables $x_1, \ldots, x_s$. Then for all $\tau \cup \{R\}$-structures $\mathfrak{A}$ and X with* $\mathrm{Dom}(X) = \{x_1, \ldots, x_s\}$,

$$\mathfrak{A} \models_X \phi \iff (\mathfrak{A}, rel(X)) \models \forall x_1 \ldots \forall x_s(\neg R(x_1, \ldots, x_s) \vee \phi).$$

The proof of Proposition 3.2 uses the downwards closure of all formulas of dependence logic and the flatness property of formulas without dependence atoms (see Clause 1 of Definition 2.2).

3.1 Generalizing completeness for open formulas

In this section we formulate an inference rule that allow us to extend the completeness theorem of [6] to cover also open formulas. Let us recall this result:

Theorem 3.3 ([6]). *Let T be a set of sentences of dependence logic and $\phi \in$ FO. Then the following are equivalent:*

1. $T \models \phi$,
2. $T \vdash_{\mathcal{D}} \phi$.

In clause 2, the symbol $\vdash_{\mathcal{D}}$ refers to derivability with respect to the natural deduction system of dependence logic in [6]. We will not discuss the details of this inference system in this article but refer to [6] for details. The open problem addressed in this article is whether Theorem 3.3 holds if the set T is allowed to contain also open formulas. One of the crucial steps in the proof of Theorem 3.3 is the approximation of dependence logic sentences by an infinite set of first-order sentences. It seems that some modifications would be needed in the approximations if our dependence logic formulas are allowed to have free variables.

It turns out that, by adopting the following auxiliary inference rule to the deduction system $\vdash_{\mathcal{D}}$ of [6], the generalization of Theorem 3.3 can be proved. Note that Proposition 3.2 is applied in the formulation of this rule:

$$\frac{A(\overline{x}) \qquad \begin{array}{c}[\forall\overline{x}(\neg R(\overline{x}) \vee A(\overline{x}))] \\ \vdots \\ \forall\overline{x}(\neg R(\overline{x}) \vee B(\overline{x}))\end{array}}{B(\overline{x})} \tag{2}$$

where the prerequisite for applying this rule is that R does not occur in A or B, nor in any non-discharged assumption in the subderivation, except for $\forall\overline{x}(\neg R(\overline{x}) \vee A(\overline{x}))$. The soundness of Rule (2) can be shown using Proposition 3.2. We will write $T \vdash^* \psi$ to mean that ψ can be derived from T using the rules of [6] and rule (2).

Lemma 3.4. *Let $T \cup \{\psi\}$ be a set of formulas of dependence logic. If $T \vdash^* \psi$, then $T \models \psi$.*

Proof. We will show how to extend the Soundness Theorem of [6] to cover also rule (2). Assume that we have a natural deduction proof of a formula $B(\overline{x})$ from the assumptions

$$\{A_1, \ldots, A_k\}$$

with the last rule (2) applied to $A(\overline{x})$. Without loss of generality, we may assume that $\overline{x}$ contains the variables that appear free in any of these formulas. By the assumption, we have a shorter deduction of $A(\overline{x})$ from the same assumptions, and a deduction of

$$\forall\overline{x}(\neg R(\overline{x}) \vee B(\overline{x}))$$

using assumptions $\{A_{i_1}, \ldots, A_{i_l}, \forall\overline{x}(\neg R(\overline{x}) \vee A(\overline{x}))\}$, where $A_{i_j} \in \{A_1, \ldots, A_k\}$ for $1 \le j \le l$.

Let $\mathfrak{A}$ be of vocabulary τ and X (with $\overline{x}$ as domain) be such that $\mathfrak{A} \models_X A_i$, for $1 \le i \le k$. We need to show that $\mathfrak{A} \models_X B(\overline{x})$. Denote by $\mathfrak{C}$ the reduct of $\mathfrak{A}$ to the vocabulary $\tau \setminus \{R\}$.

By the induction assumption, we get that $\mathfrak{A} \models_X A(\overline{x})$, and, since R does not occur in $A(\overline{x})$, $\mathfrak{C} \models_X A(\overline{x})$. Therefore, by Proposition 3.2

$$(\mathfrak{C}, rel(X)) \models \forall\overline{x}(\neg R(\overline{x}) \vee A(\overline{x})). \tag{3}$$

Since R does not occur in A_{i_j}, we also get that

$$(\mathfrak{C}, rel(X)) \models A_{i_j}, \tag{4}$$

for $1 \leq j \leq l$. Now by (3), (4), and the induction assumption it follows that

$$(\mathfrak{C}, rel(X)) \models \forall\overline{x}(\neg R(\overline{x}) \vee B(\overline{x})).$$

Using Proposition 3.2 again, we get $\mathfrak{C} \models_X B(\overline{x})$. Now, since R does not occur in $B(\overline{x})$, $\mathfrak{A} \models_X B(\overline{x})$ follows. □

We are now ready to state the generalization of the completeness theorem of [6]. Recall that by assumption only finitely many distinct variables may appear free in the formulas of T below. We write $T(\overline{x})$ to indicate that the free variables of T are contained in $\overline{x}$.

Theorem 3.5. *Let $T(\overline{x})$ be a set of formulas of dependence logic and $\phi(\overline{x}) \in$ FO. Then the following are equivalent:*

1. $T(\overline{x}) \models \phi(\overline{x})$,
2. $T(\overline{x}) \vdash^* \phi(\overline{x})$.

Proof. By Lemma 3.4 it suffices to show the implication from 1 to 2. Note first that there has to be a finite $T' \subseteq T$ such that $T'(\overline{x}) \models \phi(\overline{x})$. If no such finite T' exists, then by the compactness theorem of $\mathcal{D}$ [7] the following set of sentences (recall Proposition 3.2)

$$\{\forall\overline{x}(\neg R(\overline{x}) \vee \theta_i(\overline{x})) : \theta_i \in T\},$$

and the sentence

$$\neg\forall\overline{x}(\neg R(\overline{x}) \vee \phi(\overline{x})) \tag{5}$$

are consistent. This would imply (again using Proposition 3.2) that there is a structure $\mathfrak{A}$ and a team X such that $\mathfrak{A} \models_X T(\overline{x})$ and $\mathfrak{A} \not\models_X \phi(\overline{x})$ contradicting assumption 1. Note that, for this argument it is essential that $\phi(\overline{x})$, and hence also (5), is first-order. In particular, the following now holds:

$$\Psi \models \forall\overline{x}(\neg R(\overline{x}) \vee \phi(\overline{x})),$$

where Ψ is the sentence

$$\forall\overline{x}(\neg R(\overline{x}) \vee \bigwedge_{\theta_i \in T'} \theta_i(\overline{x})).$$

By Theorem 3.3, we get that

$$\Psi \vdash_{\mathcal{D}} \forall \overline{x}(\neg R(\overline{x}) \vee \phi(\overline{x})). \qquad (6)$$

Now by combining the derivation (6) with one application of the rule (2) we get

$$\bigwedge_{\theta_i \in T'} \theta_i(\overline{x}) \vdash^* \phi(\overline{x}),$$

which by $T' \subseteq T$ (and the conjunction introduction rule) concludes the proof. □

3.2 The case of independence logic

In this section we prove the analogue of Theorem 3.5 for independence logic and thereby generalize the completeness result of [5].

Independence logic $\mathcal{I}$ replaces the dependence atoms of $\mathcal{D}$ by independence atoms of the form $\overline{y} \perp_{\overline{x}} \overline{z}$ with the following semantics:
$\mathfrak{A} \models_X \overline{y} \perp_{\overline{x}} \overline{z}$ iff for all $s, s' \in X$: if $s(\overline{x}) = s'(\overline{x})$, then there is $s_0 \in X$ such that $s_0(\overline{x}) = s(\overline{x})$, $s_0(\overline{y}) = s(\overline{y})$ and $s_0(\overline{z}) = s'(\overline{z})$.

Dependence atoms can be easily expressed using independence atoms implying that independence logic contains dependence logic as a sublogic. This containment is in fact proper, since, for example, the analogue of Proposition 2.5 (downwards closure) does not hold for independence logic. On the other hand, for sentences, the expressive power of both $\mathcal{D}$ and $\mathcal{I}$ coincides with that of existential second-order logic.

It worth noting that adopting the semantics of Definition 2.2 for the existential quantifier makes independence logic non-local, i.e. Proposition 2.4 does not hold for it. On the other hand, by the results of [2], locality can be restored if the co-domain of $F\colon X \to A$ (clause 5 of Definition 2.2) is changed to $\mathcal{P}(A) \setminus \{\emptyset\}$. This variant of the semantics (Lax-semantics) is also used in [5].

We will use the following proposition which is analogous to Proposition 3.2.

Proposition 3.6. *Let τ be a vocabulary and ϕ a $\mathcal{I}[\tau]$-formula with free variables $x_1, \ldots, x_s$. Then for all $\tau \cup \{R\}$-structures $\mathfrak{A}$ and X with* $\mathrm{Dom}(X) = \{x_1, \ldots, x_s\}$,

$$\mathfrak{A} \models_X \phi \iff (\mathfrak{A}, rel(X)) \models \forall \overline{x}(\neg R(x_1, \ldots, x_s) \vee (R(x_1, \ldots, x_s) \wedge \phi)). \qquad (7)$$

The author is grateful to Pietro Galliani for pointing out that, due to the lack of downward closure, the conjunct $R(x_1, \ldots, x_s)$ is needed in the formula in (7) (compare to Proposition 3.2).

We can now formulate the new inference rule for independence logic:

$$\dfrac{A(\overline{x}) \qquad \begin{array}{c}[\forall\overline{x}(\neg R(\overline{x}) \vee (R(\overline{x}) \wedge A(\overline{x})))]\\ \vdots \\ \forall\overline{x}(\neg R(\overline{x}) \vee (R(\overline{x}) \wedge B(\overline{x})))\end{array}}{B(\overline{x})} \tag{8}$$

where the prerequisite for applying this rule is that R does not occur in A or B, nor in any non-discharged assumption in the subderivation, except for $\forall\overline{x}(\neg R(\overline{x}) \vee (R(\overline{x}) \wedge A(\overline{x})))$. The soundness of rule (8) can be proved analogously to Lemma 3.4. We will write $T \vdash^{*} \psi$ to mean that ψ can be derived from T using the rules of [5] and rule (8). The generalization of the completeness theorem of [5] can be now stated.

Theorem 3.7. *Let $T(\overline{x})$ be a set of formulas of independence logic and $\phi(\overline{x}) \in$ FO. Then the following are equivalent:*
1. $T(\overline{x}) \models \phi(\overline{x})$,
2. $T(\overline{x}) \vdash^{*} \phi(\overline{x})$.

Proof. The proof is analogous to the proof of Theorem 3.5. In particular, compactness theorem, which was used in the proof of Theorem 3.5, also holds for sentences of independence logic. □

Acknowledgement: I would like to thank the anonymous referee for useful comments. This work was supported by grant 264917 of the Academy of Finland.

Bibliography

[1] Arnaud Durand and Juha Kontinen. Hierarchies in dependence logic. *ACM Transactions on Computational Logic*, 13(4):1–21, 2012.
[2] Pietro Galliani. Inclusion and exclusion dependencies in team semantics - on some logics of imperfect information. *Annals of Pure and Applied Logic*, 163(1):68–84, 2012.
[3] Pietro Galliani. General models and entailment semantics for independence logic. *Notre Dame Journal of Formal Logic*, 54(2):253–275, 2013.
[4] Erich Grädel and Jouko A. Väänänen. Dependence and independence. *Studia Logica*, 101(2):399–410, 2013.
[5] Miika Hannula. Axiomatizing first order consequences in independence logic. Manuscript, 2012.
[6] Juha Kontinen and Jouko Väänänen. Axiomatizing first-order consequences in dependence logic. *Annals of Pure and Applied Logic*, 64(11):1101 – 1117, 2013.
[7] Jouko Väänänen. *Dependence logic*, volume 70 of *London Mathematical Society Student Texts*. Cambridge University Press, Cambridge, 2007.

Steven Lindell*, Henry Towsner, and Scott Weinstein

Infinitary Methods in Finite Model Theory

Abstract: We exhibit the usefulness of infinitary methods in finite model theory.

***Corresponding Author: Steven Lindell:** Haverford College
Henry Towsner, Scott Weinstein: University of Pennsylvania

1 Introduction

This paper explores applications of the infinitary techniques of classical model theory in the context of finite model theory, and thus elaborates on themes explored by [13]. We are grateful for the opportunity to celebrate Jouko's inclusive vision of logic in this way.

2 Preliminaries

We suppose the reader is familiar with the notion of a relational structure of similarity type τ. By a class of structures, we understand a collection of relational structures of common similarity type which is closed under isomorphism. We use FO to denote the collection of sentences of first order logic.

Definition 2.1. Let K be a class of structures of similarity type τ and let Γ be a set of first-order sentences.

1. $\mathsf{Th}(K) = \{\varphi \in \mathsf{FO} \mid (\forall A \in K) A \models \varphi\}$.
2. $\mathsf{Mod}(\Gamma) = \{A \mid A \models \Gamma\}$.
3. Let A be a structure of similarity type τ. A is a *pseudo-K structure* if and only if $A \in (\mathsf{Mod}(\mathsf{Th}(K)) - K)$.

An alternative formulation of Definition 2.13 is: A is a pseudo-K structure if and only if $A \notin K$ but for every first order sentence φ, if $A \models \varphi$, then for some $B \in K$, $B \models \varphi$. Intuitively, a structure A is pseudo-K just in case it is indistinguishable from structures in K by first order sentences.

Let F_τ be the class of finite structures of similarity type τ. A structure A of similarity type τ is *pseudo-finite* if and only if A is pseudo-F_τ. In general, if there is no finite upper bound on the size of structures in $K \subseteq F_\tau$, then a structure A is pseudo-K just in case A satisfies the *limit theory of* K ($\mathrm{Th}_\infty(K)$) defined as follows.

$$\mathrm{Th}_\infty(K) = \{\varphi \in \mathrm{FO} \mid (\exists n \in \omega)(\forall B \in K)(|B| \geq n \rightarrow B \models \varphi)\}.$$

Suppose F is a class of structures, not necessarily finite, and $K \subseteq F$. We say that K is first order definable over F if and only if there is a first order sentence φ such that for all $A \in F$, $A \models \varphi$ if and only if $A \in K$.

3 An easy inexpressibility result – connectivity

In this section, we give some examples which apply the notion of pseudo-K structure to inexpressibility results in finite model theory. The source of these results is the following simple observation.

Proposition 3.1. *Let $K \subseteq F_\tau$. Suppose there is a structure A of similarity type τ which is both pseudo-K and pseudo-$(F_\tau - K)$. Then K is not first order definable over F_τ.*

The first example is due to [9].

Proposition 3.2. *Connectivity is not definable in first order logic over finite simple graphs.*

Proof: Let K be the class of connected finite simple graphs and let G be the class of all finite simple graphs. By Proposition 3.1, it suffices to exhibit a structure A which is both pseudo-K and pseudo-$(G - K)$. And for this, it suffices to show that there is a structure A which is both pseudo-C_1 and pseudo-C_2, where C_1 is the class of simple cycles and C_2 is the class of pairs of disjoint simple cycles. We claim that

3.3. *every 2-regular, acyclic simple graph is both pseudo-C_1 and pseudo-C_2.*

Let T be the first order theory which specifies that a graph is 2-regular, acyclic and simple. T can be axiomatized by the following infinite set of first order sentences.

1. $(\forall x)\neg Exx \wedge (\forall x)(\forall y)(Exy \rightarrow Eyx)$
2. $(\forall x)(\exists y)(\exists z)(y \neq z \wedge (\forall w)(Exw \leftrightarrow (w = y \vee w = z))$
3. for each $n \geq 3$ the sentence

$$\neg(\exists x_1)\ldots(\exists x_n)(\bigwedge_{1\leq i<j\leq n} x_i \neq x_j \wedge \bigwedge_{1\leq i<n} Ex_ix_{i+1} \wedge Ex_nx_1)$$

It is easy to see that every model of T consists of a disjoint union of bi-infinite simple chains. It follows at once that for every $\kappa > \aleph_0$, T is κ-categorical. Therefore, by the Łos-Vaught Test, T is complete. Note that every finite sub-theory of T has a model C which consists of a single simple cycle and a model D which consists of a disjoint pair of simple cycles. It follows at once, by the Compactness Theorem, that the same holds for any sentence φ which is a consequence of T; this establishes 3.3, thereby concluding the proof. □

In the next section, we give a "classical" proof that connectivity is not monadic-Σ^1_1, as a corollary to the Hanf threshold-locality theorem.

Other investigators have noted the effectiveness of classical model-theoretic techniques in establishing limits on the expressive power of first order logic over finite structures. Antedating the classical proof of [9], [12] showed, using ultraproducts of 2-regular finite graphs, that connectivity is not first-order definable over finite simple graphs; extending this, [13] uses ultraproducts to show that connectivity is not monadic-Σ^1_1 over finite graphs. [13] also uses ultraproducts to show that even cardinality of finite structure is not first order definable; [11] gives a beautiful proof of this result using initial segments of "non-standard models" of the first order theory of $\langle\omega, <\rangle$. Moreover, [6] used classical model theoretic techniques to derive the 0-1 law for first-order logic with respect to finite graphs with constant edge probability, from which a raft of inexpressibility results follows. Very recently, Philip Dittmann has applied ultraproducts to obtain a variety of results in finite model theory, as reported in [2].

4 Locality Theorems

Locality theorems characterize various senses in which properties of a relational structure which are definable in a particular way depend on "local information" about the structure, typically, isomorphism types of neighborhoods in the Gaifman-graph of the structure. The next definition formulates some topological notions requisite for describing locality precisely.

Definition 4.1. Let G be a graph, $a, b \in G$ and $r \geq 0$.
1. We write $\delta_G(a, b)$ for the distance between a and b in G.
2. $\mathsf{B}_r(G, a) = \{c \in G \mid \delta_G(a, c) \leq r\}$, for $r \in \omega$. $\mathsf{B}_\infty(G, a) = \bigcup_{r\in\omega} \mathsf{B}_r(G, a)$.
3. Let A be a relational structure of similarity type τ. The *Gaifman-graph* of A (denoted G_A) is the simple graph whose node set is A and whose edge relation holds between a pair of nodes a, b with $a \neq b$ if and only if for some $R \in \tau$ and tuple $\bar{c} \in R^A$ there are i, j with $c_i = a$ and $c_j = b$.
4. We say a structure A has *bounded degree* if and only if there is a $d \in \omega$ such that for every $a \in A$ the degree of a in G_A is at most d; we say A has *finite degree* if and only if for every $a \in A$, the degree of a in G_A is finite.

5. Let $r \in \omega$ or $r = \infty$, and let $\overline{a}$ be an n-tuple of elements of A.

$$A^r(\overline{a}) = \langle (A \restriction \bigcup_{1 \leq i \leq n} \mathrm{B}_r(G_A, a_i)), \overline{a} \rangle.$$

   The structure $A^r(\overline{a})$ is often called the r-neighborhood of $\overline{a}$ in A.

Note that for any finite relational signature τ, there is a first-order formula $\delta^r_\tau(x, y)$ such that for every τ-structure A and $a, b \in A$, $A \models \delta^r_\tau[a, b]$ if and only if the distance between a and b in G_A is at most r. If $\varphi(x_1, \ldots, x_k)$ is a formula, we write $\varphi^r(x_1, \ldots, x_k)$ for the formula which results from $\varphi(x_1, \ldots, x_k)$ by relativizing all its quantifiers (Qy) to the formula $\delta^r_\tau(x_1, y) \vee \ldots \vee \delta^r_\tau(x_k, y)$. We call such a formula an *r-local formula*. If $\overline{a}$ is a k-tuple of elements of A, the *local type* of $\overline{a}$ in A is the set of formulas $\varphi^r(\overline{x})$, for all $r \in \omega$, such that $A(\overline{a}) \models \varphi^r(\overline{a})$. An n-ary τ-query is an isomorphism invariant map φ from τ-structures A to n-ary relations on A; we write $\varphi[A]$ for the φ image of A.

4.1 Gaifman locality

The next definition formulates a first notion of locality.

Definition 4.2. Suppose φ is an n-ary τ-query, $n > 0$. We say that φ is *Gaifman r-local* if and only if for every τ-structure A and n-tuples $\overline{a}, \overline{b}$ of elements of A, if $A^r(\overline{a}) \cong A^r(\overline{b})$, then $\overline{a} \in \varphi[A] \iff \overline{b} \in \varphi[A]$. A query is *Gaifman local* if and only if it is Gaifman r-local for some $r \in \omega$.

Theorem 4.3. *Every first order definable query is Gaifman local.*

Proof: Suppose that $\varphi(\overline{x})$ is a first order formula defining an n-ary query on τ-structures which is not Gaifman local.

4.4. *Then for every $r \in \omega$ there is a τ-structure A and $\overline{a}, \overline{b} \in A$ such that*
1. *$\overline{a} \in \varphi[A]$ and $\overline{b} \notin \varphi[A]$, and*
2. *$A^r(\overline{a}) \cong A^r(\overline{b})$.*

Expand the signature τ with $2n$ new constants $\overline{a}$ and $\overline{b}$ and a unary function symbol f. Choose, for each $r \in \omega$, a first order formula ψ^r in the expanded signature to uniformly express the condition that f is an isomorphism from $A^r(\overline{a})$ onto $A^r(\overline{b})$. Let $T = \{\varphi(\overline{a}), \neg\varphi(\overline{b})\} \cup \{\psi^r \mid r \in \omega\}$. It follows at once from 4.41 and 2, via the compactness theorem, that T is satisfiable. Let $A' \models T$ and let A be the reduct of A' to τ. It follows at once that there is an isomorphism g mapping $A^\infty(\overline{a})$ onto $A^\infty(\overline{b})$. We extend g to an automorphism h of A thereby contradicting $A \models \varphi[\overline{a}]$ and $A \not\models \varphi[\overline{b}]$.

For all $a \in A$, if $a \notin A^{\infty}(\overline{a}) \cup A^{\infty}(\overline{b})$, we let $h(a) = a$. For all $a \in (A^{\infty}(\overline{b}) - A^{\infty}(\overline{a}))$, there is a unique $b \in A^{\infty}(\overline{a})$ and $m \leq n$ such that $b \notin \mathrm{ran}(g)$ and $g^m(b) = a$. We let $h(a) = b$. It is easy to verify that h so defined is an automorphism. □

We make use of the following lemma in our proof of Theorem 4.7.

Lemma 4.5. *Let $\overline{a}$ and $\overline{b}$ be non-empty tuples of the same local type in ω-saturated models A and B respectively. Then $A^{\infty}(\overline{a}) \equiv B^{\infty}(\overline{b})$.*

Proof: We use the Ehrenfeucht-Fraïssé game to show that $A^{\infty}(\overline{a}) \equiv B^{\infty}(\overline{b})$. We proceed by induction on the number of moves. In particular, at the k-th round of the game, we maintain the property that $\langle a_1, \ldots, a_k, \overline{a}\rangle$ and $\langle b_1, \ldots, b_k, \overline{b}\rangle$ have the same local type.

Basis: For $k = 0$, $\langle \overline{a}\rangle$ and $\langle \overline{b}\rangle$ have the same local type in A and B by hypothesis.

Induction Step: By induction hypothesis, $\langle a_1, \ldots, a_{k-1}, \overline{a}\rangle$ and $\langle b_1, \ldots, b_{k-1}, \overline{b}\rangle$ have the same local type $\Lambda(x_1, \ldots, x_{k-1}, \overline{x})$. We will show that the local type $\Lambda'(x_1, \ldots, x_k, \overline{x})$ of every extension in $A^{\infty}(\overline{a})$ is also realized in $B^{\infty}(\overline{b})$, and vice versa. Suppose without loss of generality we are given a_k in $A^{\infty}(\overline{a})$ (resp. b_k in $B^{\infty}(\overline{b})$) such that $A^{\infty}(\overline{a}) \models \Lambda'[a_1, \ldots, a_k, \overline{a}]$. We want to find b_k in $B^{\infty}(\overline{b})$ (resp. a_k in $A^{\infty}(\overline{a})$) such that $B^{\infty}(\overline{b}) \models \Lambda'[b_1, \ldots, b_k, \overline{b}]$. Take any finite collection of local formulas $\Phi \subseteq \Lambda'$. Their conjunction is equivalent to an r-local formula $\varphi^r(x_1, \ldots, x_k, \overline{x})$ by taking r to be the maximum of the radii in Φ. Let d be the distance from a_k to $\overline{a}$ and observe that $\psi(x_1, \ldots, x_{k-1}, \overline{x}) = (\exists x_k)[\delta^d(\overline{x}, x_k) \wedge \varphi^r(x_1, \ldots, x_k, \overline{x})]$ is $(d+r)$-local by the triangle inequality. So $A^{d+r}(a_1, \ldots, a_{k-1}, \overline{a}) \supseteq A^r(a_1, \ldots, a_k, \overline{a}) \models \varphi[a_1, \ldots, a_k, \overline{a}]$ implies $A^{d+r}(a_1, \ldots, a_{k-1}, \overline{a}) \models \psi[a_1, \ldots, a_{k-1}, \overline{a}]$, and hence $\psi \in \Lambda$. This means $B^{\infty}(\overline{b}) \models \psi[b_1, \ldots, b_{k-1}, \overline{b}]$, and hence there is a b_k of distance d from $\overline{b}$ such that $B^r(b_1, \ldots, b_k, \overline{b}) \models \varphi[b_1, \ldots, b_k, \overline{b}]$, and hence $B^{\infty}(\overline{b}) \models \varphi^r[b_1, \ldots, b_k, \overline{b}]$, which makes $\Lambda'(b_1, \ldots, b_{k-1}, x_k, \overline{b})$ finitely consistent in $B^{\infty}(\overline{b})$. So by ω-saturation, Λ' is realized by a single b_k such that $B^{\infty}(\overline{b}) \models \Lambda'[b_1, \ldots, b_{k-1}, b_k, \overline{b}]$. But notice that since the distance from b_k to $\overline{b}$ must be d, it is in fact in $B^{\infty}(\overline{b})$. This completes the induction.

To finish the proof, notice that if $\langle a_1, \ldots, a_k, \overline{a}\rangle$ and $\langle b_1, \ldots, b_k, \overline{b}\rangle$ have the same local type, then the radius 1 formulas in that type witness a partial isomorphism. Therefore, $A^{\infty}(\overline{a}) \equiv B^{\infty}(\overline{b})$, since we have shown we can play the game between them for any number of moves. □

The following definition and theorem are due to [8].

Definition 4.6. A *basic local sentence* is one which says there are k disjoint neighborhoods each satisfying the same local formula:

$$(\exists x_1) \ldots (\exists x_k) \bigwedge \{d(x_i, x_j) > 2r \wedge \varphi^r(x_i) \mid 1 \leq i < j \leq k\}.$$

Theorem 4.7. *Every first-order sentence is equivalent to a Boolean combination of basic local sentences.*

Proof: By Corollary 1.0.7 of [4], it suffices to show that any two models A and B which satisfy the same basic local sentences are elementarily equivalent. This is obvious if one of the models is finite, because a basic local sentence can completely describe any finite model up to isomorphism. In case A and B are infinite, we may suppose, without loss of generality, that they are ω-saturated, since every structure has an ω-saturated elementary extension. If A has k distinct isomorphic copies of some component C, choose any element a in C, and let Σ be the infinite set of formulas which says its k variables are infinitely far apart and have the same local type as a:

$$\Sigma(x_1 \dots x_k) = \{d(x_i, x_j) > 2r \wedge \varphi^r(x_i) \mid 1 \le i < j \le k; A^r(a) \models \varphi[a]; 1 \le r < \infty\}.$$

The existential closure of every finite subset of Σ is equivalent to a basic local sentence true in A, hence true in B, by hypothesis. So Σ is finitely consistent over B. But since B is saturated, B satisfies Σ, and therefore has k elements of the same local type that lie in separate components. By Lemma 4.5, these components in B are all elementarily equivalent to each other and to the corresponding components in A. Repeating this argument the other way around, from B to A, we see that each model must have either the same finite number of each component up to elementary equivalence, or an infinite number. Therefore, we can play any finite number of rounds of the Ehrenfeucht-Fraïssé game, and hence $A \equiv B$. □

4.2 Hanf locality

The next definition formulates a second notion of locality.

Definition 4.8. Suppose κ, λ, and μ are cardinals, A and B are τ-structures, and K is a class of τ-structures.

1. $\kappa \sim_\mu \lambda$ if and only if $\kappa = \lambda$ or $\kappa \ge \mu$ and $\lambda \ge \mu$.
2. Let $r \in \omega$ and $t \in \omega \cup \{\omega\}$. A is *Hanf threshold r, t-equivalent to B* (written $A \approx_{r,t} B$) if and only if for every pointed τ-structure C,
$$|\{a \in A \mid A^r(a) \cong C\}| \sim_t |\{b \in B \mid B^r(b) \cong C\}|.$$
3. A class of τ-structures J is *Hanf r, t-local on K* if and only if for all $A, B \in K$, if $A \approx_{r,t} B$, then $A \in J \iff B \in J$.
4. A class of τ-structures J is *Hanf threshold-local on K* if and only if J is Hanf r, t-local on K for some $r, t \in \omega$.

In the proof of the next theorem, we use the notion of a model pair, which is extensively used by [3] in similar spirit. Suppose A and B are τ-structures and that

$A \cap B = \emptyset$. The *model pair* $\langle A, B \rangle$ is the $\tau \cup \{P\}$ structure with universe $A \cup B$, where the extension of the unary predicate P is A and the interpretation of all τ vocabulary is the union of its interpretation of A and B. Observe that for every first order τ sentence φ, there are $\tau \cup \{P\}$ sentences φ_A and φ_B such that

$$\langle A, B \rangle \models \varphi_A \iff A \models \varphi \text{ and } \langle A, B \rangle \models \varphi_B \iff B \models \varphi.$$

The following theorem is due to [7].

Theorem 4.9. *If J is first order definable and K is a class of structures of bounded degree, then J is Hanf threshold-local on K.*

Proof: Let J be a class of τ-structures defined by a first order sentence φ, let K be a class of τ-structures of degree bounded by d, and suppose that J is not Hanf threshold-local on K. We formulate a theory T in the language of pairs of τ-structures expanded with a new binary predicate R which describes this situation. Observe that since the structures in K are of uniformly bounded degree, for each $r \in \omega$ there is a first order $\tau \cup \{P\}$ formula $\psi^r(x, y)$ such that for all $a \in A$ and $b \in B$,

$$\langle A, B \rangle \models \psi^r[a, b] \iff A^r(a) \cong B^r(b).$$

Let T be the following set of sentences.

1. $\varphi_A \wedge \neg\varphi_B$
2. the degree of A and of B are bounded by d
3. $(\forall x)(\forall y)(R(x, y) \to \psi^r(x, y))$, for every $r \in \omega$
4. $[(\forall x_1 \dots x_t)(\bigwedge_i P(x_i) \to (\exists y_1 \dots y_t)(\bigwedge_i \neg P(y_i) \wedge \bigwedge_{i<j}(x_i \neq x_j \to y_i \neq y_j) \wedge \bigwedge_i R(x_i, y_i)))] \wedge$
   $[(\forall x_1 \dots x_t)(\bigwedge_i \neg(Px_i) \to (\exists y_1 \dots y_t)(\bigwedge_i P(y_i) \wedge \bigwedge_{i<j}(x_i \neq x_j \to y_i \neq y_j) \wedge \bigwedge_i R(y_i, x_i)))]$, for every $t \in \omega$

It follows at once from our hypotheses, via the compactness and Löwenheim-Skolem theorems, that T is satisfiable in a countable model pair $\langle A, B \rangle$ expanded by an interpretation F of R which is a bijection of A onto B with the property that

4.10. *for all $a \in A$ and for all $r \in \omega$,*

$$A^r(a) \cong B^r(F(a)).$$

Since A is of bounded degree, for each $r \in \omega$, $A^r(a)$ is finite, hence there are only finitely many isomorphisms from $A^r(a)$ onto $B^r(F(a))$. Observe that the set of all isomorphisms from $A^r(a)$ onto $B^r(F(a))$ for all $r \in \omega$ forms a finitely branching tree under inclusion, hence, by the König Infinity Lemma, this tree has an infinite path which represents an isomorphism from $A^\infty(a)$ onto $B^\infty(F(a))$. Thus, F is an isomorphism preserving bijection between the *pointed* components of A and B. We say a component C of A has finite character if for some $a \in C$, there are only finitely many $b \in C$ such that $A^{\infty}(a) \cong A^{\infty}(b)$, and similarly for components of B. It is easy to see that F induces a bijection between the components of A and B of finite character. By compactness, we may take countable elementary extensions A' of A and B' of B such that every component of A' which is not of finite character occurs infinitely often in A', up to isomorphism, and similarly for B'. It follows at once that $A' \cong B'$, which contradicts $A' \models \varphi$ and $B' \not\models \varphi$. □

As a corollary to Theorem 4.9, we show that connectivity is not mon-Σ^1_1 definable over the class of finite simple graphs, a result first established by [5]; our proof is an "infinitary" version of an argument presented in [7]. (A formula of second order logic is mon-Σ^1_1 if and only if it is of the form $(\exists\overline{P})\varphi$, where $\overline{P}$ is a sequence of unary predicate variables and φ is a first order formula.)

Theorem 4.11. *Connectivity is not mon-Σ^1_1 definable over the class of finite simple graphs.*

Proof: Suppose that

4.12. *the mon-Σ^1_1 sentence $(\exists\overline{P})\varphi(\overline{P})$ defines connectivity over the class of finite simple graphs.*

Let K_1 be the class of finite $\overline{P}$-colored simple cycles which satisfy $\varphi(\overline{P})$ and let K_2 be the class of pairs of all finite $\overline{P}$-colored simple cycles (a $\overline{P}$-colored simple graph has, in addition to a loop-free undirected edge relation, interpretations of the unary predicates in $\overline{P}$ – we shorten "$\overline{P}$-colored" to "colored" for the remainder). By Theorem 4.9, it follows from 4.12 that for some r_0 and t_0, if $D \approx_{r_0,t_0} C$ for some $C \in K_1$, then $D \in K_1$, for all finite colored simple graphs D. It is easy to see that

4.13. *for every r, t, and d, there is a first-order sentence characterizing graphs of degree at most d up to $\approx_{r,t}$.*

Therefore, in order to derive a contradiction from 4.12, it suffices to show that

4.14. *for every A, if A is pseudo-K_1, then A is pseudo-K_2.*

In order to verify 4.14, suppose that A is pseudo-K_1; by the Löwenheim-Skolem Theorem, we may suppose, without loss of generality, that $|A| > (2^{\aleph_0})^+$. It is easy to see that in this case A consists of a collection of colored bi-infinite simple chains. Moreover, it follows from 4.13 and the fact that A is pseudo-K_1 that for every r, t, there is a colored simple cycle $C_{r,t}$ with $A \approx_{r,t} C_{r,t}$. To complete the verification of 4.14, it therefore suffices to show that there is a colored simple cycle D such that

4.15. $C_{r,t} \oplus D \approx_{r,t} A$.

Since $|A| > (2^{\aleph_0})^+$, it follows at once that there is a colored bi-infinite chain $B \subseteq A$ which occurs infinitely often up to isomorphism in A. It is easy to see, by the pigeon-hole principle, that there is a $2r + 1$-neighborhood N in B which occurs infinitely often up to isomorphism. Let N_1 and N_2 be disjoint copies of N occurring in B. It follows at once that every r-neighborhood occurring between them in B occurs infinitely often in A up to isomorphism. Let D be the colored cycle formed by snipping B at the midpoints of N_1 and N_2 and identifying the endpoints of the of the resulting finite simple chain. It is easy to see that this choice of D satisfies 4.15. □

5 Ultraproducts and directed reachability

In this section we present Henry Towsner's proof of the following result which was established by Miklos Ajtai and Ronald Fagin (see [1]).

Theorem 5.1. *Reachability is not mon-Σ_1^1 definable over the class of finite directed graphs.*

Proof: We are considering models of the language $\Rightarrow$ (a binary, asymmetric, irreflexive relation), $=$, and two constant symbols s, t, possibly augmented by finitely many unary predicate symbols. We write $u \Leftrightarrow v$ for $(u \Rightarrow v) \vee (v \Rightarrow u)$. Suppose that

5.2. *the mon-Σ_1^1 sentence $(\exists P_1)\dots(\exists P_n)\varphi(P_1, \dots, P_n)$ defines s, t-connectivity over the class of finite directed graphs.*

We derive a contradiction to 5.2 via the following lemmas. The use of the ultraproduct construction obviates the need for any game argument. On a finite structure $(X_n, \Rightarrow_n)$, there is a canonical choice of measure, the normalized counting measure $\mu_n(S) = \frac{|S|}{|X_n|}$. On an ultraproduct of finite structures, as in Lemma 5.3, μ denotes the Loeb-measure, which canonically lifts the normalized counting measure on the finite structures $\mathcal{X}_n$ to a measure on the σ-algebra generated by the internal subsets of their ultraproduct $\mathcal{X}$ (see [10]).

Lemma 5.3. *There exists a sequence of models $\mathcal{X}_i = (X_i, \Rightarrow_i)$ with $|X_i| \to \infty$ such that s, t are connected in each $\mathcal{X}_i$, and in any ultraproduct (by a nontrivial ultrafilter) $\mathcal{X}$,*

- *for μ-almost every u, the $\Leftrightarrow$-connected component of u is a tree (i.e. has no loops),*

- *if $A \subseteq X$ is internal with $\mu(A) > 0$ then for almost every $u \in X$, there are infinitely many $b \in A$ with $u \Leftrightarrow b$.*

Proof: The construction of the models $(\mathcal{X}_i, \Rightarrow_i)$ is essentially the one in Proposition 8.3 of [1]. Suppose that, in $\mathcal{X}$, the set L of u such that the $\Leftrightarrow$-connected component of u contains a loop has measure $\epsilon > 0$. Observe that $L = \bigcup_d L_d$ where L_d is the set of u such that there is a loop in the ball of radius d around u (where distance is $\Leftrightarrow$-distance). So if $\mu(L) > 0$ then there is some d with $\mu(L_d) > 0$. The set L_d is definable, so $L_d = \langle L_{d,i} \rangle$ where $L_{d,i}$ is the set of $u \in X_i$ such that there is a loop in the ball of radius d around u. Therefore there are infinitely many i such that $\mu_i(L_{d,i}) > 0$.

Write $\#(A, u) = |\{b \in A \mid u \Leftrightarrow b\}|$ if this is finite and $\#(A, u) = \infty$ otherwise. Let $A = \langle A_i \rangle$ be internal. Suppose that $\mu(A) > 0$ and the set of u with $\#(A, u) < \infty$ has positive measure. Then for some d, the set of u with $\#(A, u) < d$ has positive measure. Then there must be infinitely many i where the set of u with $\#(A_i, u) < \infty$ has positive measure.

So it suffices to arrange for the sequence of models $\mathcal{X}_i$ to have the following properties:
- for every d and every $\epsilon > 0$ there is an i such that for all $j \geq i$, $\mu_j(L_{d,j}) < \epsilon$, and
- for every d and every $\epsilon > 0$ there is an n such that for all $j \geq i$ and every set $A \subseteq X_j$ with $\mu_j(A) \geq \epsilon$, the set of u with $\#(A, u) < d$ has μ_j-measure $< \epsilon$.

The models $(X_i, \Rightarrow_i)$ will have the following form: $X_i = [1, i]$, $s^{X_i} = 1$, $t^{X_i} = i$, for all $a < i$, $a \Rightarrow_i a + 1$, and in addition we select a small set of pairs $a < b$ so that $b \Rightarrow_i a$. The only discretion we have is in choosing the set of "backwards" arrows.

Consider the random process where for each pair $a < b$, we decide whether $b \Rightarrow_i a$ holds randomly with probability $p(i) = \ln i/i$, with each pair determined independently. (As in [1], what really matters is that $p(i)$ grows faster than k/i for all k and slower than i^s/i for all $s > 0$.) We show that the properties we want hold with high probability, and so certainly at least one model with the desired properties exists.

We recall, for repeated use, the Chernoff bound. Suppose Y is a Bernoulli random variable—that is, $Y = \sum_{i \leq m} Y_i$ where each $Y_i \in \{0, 1\}$ is chosen randomly and independently with the probability $\mathbb{P}[Y_i = 1] = p$. Then the expected value of Y is $\mu = mp$ and for any $\gamma \geq 0$,

$$\mathbb{P}[Y > (1 + \gamma)\mu] \leq e^{-\gamma^2 \mu/(2+\gamma)}, \ \mathbb{P}[Y < (1 - \gamma)\mu] \leq e^{-\gamma^2 \mu/(2+\gamma)}.$$

Let us write $B_r(u)$ for the ball of radius r around u. Observe that $|B_1(u)| = Y + 3$ (or $Y + 2$ if u is an endpoint) where Y is a Bernoulli random variable with mean $\ln n$, and so for a given u, the probability that $|B_1(u)| \geq 4 \ln i = (1 + 3) \ln i$ is bounded by $e^{-9\ln i/5} = i^{-9/5}$. Therefore, the probability that *some* u has $|B_1(u)| \geq 4\ln i$ is bounded by $i^{-4/5}$, and in particular goes to 0 as $i \to \infty$. It follows that whenever $r \ll i$, with high probability for every u, $|B_r(u)| \leq \ln^k i$ for some constant k depending only on r.

We show that, with high probability, there are few cycles. Specifically, for fixed $r, d, \epsilon > 0$, we show that as i goes to ∞, with high probability the set of v such that $B_r(v)$ contains a loop of length $\leq d$ has density $< \epsilon$. Consider the following method for generating $\Rightarrow_i$. We first define a relation $\Rightarrow_i'$. We have $a \Rightarrow_i' a+1$ for all a, and we place the pairs $a < b$ in an arbitrary order, and successively decide in this order whether $b \Rightarrow_n' a$ will hold with probability $p(i)$, except that if $b \Rightarrow_i' a$ would create a loop of length $\leq d$, we do not include an edge. We then extend $\Rightarrow_i'$ to $\Rightarrow_i$ by deciding, for each pair we skipped in the creation of $\Rightarrow_i'$, whether $\Rightarrow_i$ should hold between this pair. This is equivalent to the original method for generating $\Rightarrow_i$ (we are making the same choices with the same probability, and the choices are independent; the only difference is the order in which we make the choices). Further, $\Rightarrow_i$ has a loop of length $\leq d$ exactly if we add an edge in the second stage. By taking i big enough, with high probability we have $|B_d(u)| \leq \ln^k i$ for all u. Therefore, for each u, we consider at most $\ln^k i$ possible edges, so there are $i\ln^k i$ possible edges added in the second stage, each with probability $\ln i/i$. The number of edges added is a Bernoulli random variable, and by the Chernoff bound again, when i is large, the probability is close to 1 that we add fewer than $4\ln^{k+1} i$ edges. Every loop in the graph contains one of these edges; for each of these edges, taking u to be an arbitrary endpoint, $|B_r(u)| \leq \ln^{k'} i$. Since $B_r(v)$ contains a loop of length $\leq d$ iff $v \in B_r(u)$ where u is one of the chosen endpoints, there are at most $4\ln^{k'(k+1)} i$ vertices v such that $B_r(v)$ contains a loop of length $\leq d$.

Now suppose we fix a set $A \subseteq X_i$ with $|A| \geq \epsilon i$ and fix some u and then generate $\Rightarrow_i$. Then $\#(A, u)$ is a binomial random variable with expected value $\epsilon \ln i$, so letting i be large enough so that $\epsilon \ln i \geq 2d$,

$$\mathbb{P}(\#(A, u) \leq d) \leq e^{-3\epsilon \ln i/8}.$$

Let $\delta = e^{-3\epsilon \ln i/8}$; observe that $\delta \to 0$ as $i \to \infty$. For any fixed set B with $|B| \geq \epsilon i$, the probability that for every $u \in B$, $\#(A, u) \leq d$ is bounded by $\delta^{|B|} \leq \delta^{\epsilon i}$. There are 2^i choices for A and 2^i possible choices for a set B, so the probability that there is any A and B with both $|A| \geq \epsilon i$, $|B| \geq \epsilon i$, and for every $u \in B$, $\#(A, u) \leq d$, is bounded by $2^i 2^i \delta^{\epsilon i} = e^{n(\ln 4 + \epsilon \ln \delta)}$. By taking i sufficiently large, and so δ sufficiently small, we may ensure that this probability approaches 0. Therefore, with high probability, for every set A with $|A| \geq \epsilon i$, $|\{u \mid \#(A, u)\}| < \epsilon i$. □

We may make the following simple observation:

Lemma 5.4. *Every* $\Leftrightarrow$*-connected component has measure* 0.

Proof: Suppose some connected component A has positive measure. Since almost every point belongs to a component with no loops, A must contain no loops. A itself is not internal, but it is a union of internal sets: pick any $u \in A$; then $A = \bigcup_r B_r(u)$. In particular, some $B_r(u)$ must have positive measure; let r be least such that $B_r(u)$ has positive measure. Since $B_r(u) \setminus B_{r-1}(u)$ is an internal set of positive measure, almost every v satisfies $\#(B_r(u) \setminus B_{r-1}(u), v) = \infty$. In particular, we may pick two points $a_0, a_1 \in B_r(u)$ such that $a_i \Leftrightarrow v$. Further, we may assume $v \notin B_{r-1}(u)$. This means a_0, a_1 each have paths to u not including v or a_{1-i}, and therefore combining these paths with v gives a loop in A, which is the desired contradiction. □

It follows from Lemma 5.2, 5.3, and Łoś's Theorem that there are internal sets $A_1, \dots, A_n$ such that the expansion $\mathcal{X} = (X, \Rightarrow, A_1, \dots, A_n)$ satisfies $\varphi(P_1, \dots, P_n)$.

Definition 5.5. If ϕ is a formula with a single free variable, we write $\mu(\phi)$ for $\mu(\{a \in X \mid \mathcal{X} \vDash \phi(a)\})$.

We say $tp(x)$ is *wide* if whenever $\phi \in tp(x)$, $\mu(\phi) > 0$. We say x is *central* if for every $\phi \in tp(x)$, $\mu(\{a \mid a < x \wedge \mathcal{X} \vDash \phi(a)\}) > 0$ and $\mu(\{a \mid x < a \wedge \mathcal{X} \vDash \phi(a)\}) > 0$.

Clearly central elements have wide type.

Lemma 5.6. *Almost every x is central.*

Proof: We show that for every ϵ, the non-central elements are contained in a set of measure $\leq \epsilon$. Let $\phi_1, \dots, \phi_n, \dots$ be an enumeration of the formulas with exactly one free variable (and, ambiguously, the subsets of $\mathcal{X}$ they define). We will define $U = \bigcup U_i$ where $\mu(U_i) \leq \epsilon 2^{-i}$. If $\mu(\phi_i) = 0$, set $U_i = \phi_i$. If $\mu(\phi_i) > 0$, define

$$\begin{aligned} U_i = &\{x \mid \mu(\{a \mid a < x \wedge \mathcal{X} \vDash \phi_i(a)\}) < \epsilon 2^{-(i+1)}\} \\ &\cup \{x \mid \mu(\{a \mid x < a \wedge \mathcal{X} \vDash \phi_i(a)\}) < \epsilon 2^{-(i+1)}\}. \end{aligned}$$

That is, U_i is the "bottom" and "top" $\epsilon 2^{-(i+1)}$ of the set defined by ϕ_i.

If x is not central then there is some ϕ_i witnessing this, and by construction, we have included x in $U_i \subseteq U$. □

Now we argue as follows. Consider those x such that x and $x + 1$ are both central, and such that the $\Leftrightarrow$-connected component of x contains no loops; this is a set of measure 1, so we may take some c with these properties.

Lemma 5.7. *There are infinitely many a with $tp(a) = tp(c + 1)$ and $c \Rightarrow a$.*

Proof: For any $\phi \in tp(c + 1)$, since $c + 1$ is central, $\mu(\{a \mid a < c + 1 \wedge \mathcal{X} \vDash \phi(a)\}) > 0$. Therefore, by choice of $\mathcal{X}$, there are infinitely many a such that $\mathcal{X} \vDash \phi(a)$, $a < c$, and $c \Leftrightarrow a$; but this implies $c \Rightarrow a$. Since this holds for each ϕ, saturation ensures that there are actually infinitely many a with $tp(a) = tp(c + 1)$ and $c \Rightarrow a$. □

Similarly,

Lemma 5.8. *There are infinitely many b with* $tp(b) = tp(c)$ *and* $c + 1 \Rightarrow b$.

Lemma 5.9. *There is a d with* $tp(d) = tp(c)$ *not in the same* $\Leftrightarrow$*-connected component as c.*

Proof: For every r, $\mu(B_r(c)) = 0$, while for each $\phi \in tp(c)$, $\mu(\phi) > 0$, so $\mu(\phi \setminus B_r(c)) > 0$ as well. Since this holds for each r and each $\phi \in tp(c)$, saturation implies that there is a d with $tp(d) = tp(c)$ and $d \notin \bigcup_r B_r(c)$. □

Let $\mathcal{X}'$ be the model identical to $\mathcal{X}$ except that $c \not\Rightarrow c + 1$. It is immediate that $\mathcal{X}'$ is an ultraproduct of structures $\mathcal{X}'_i$ almost all of which are not s, t-connected; hence, $\mathcal{X}' \not\models \varphi(P_1, \ldots, P_n)$. On the other hand, we claim $\mathcal{X}$ is isomorphic to $\mathcal{X}'$, which completes the *reductio*.

We define $\pi : \mathcal{X} \to \mathcal{X}'$ to be the identity outside the connected components of c and d.

We define π on the connected component of c so $\pi(c) = c$ but the image does not include $c+1$: let $A = \{a \mid tp(a) = tp(c+1) \wedge c \Rightarrow a\}$, and we define $\pi''A = A \setminus \{c+1\}$, which we can do since A is infinite. Since the connected component of c is a tree and the type of a point determines the type of its neighbors, we can define π on the rest of the $\Leftrightarrow$-connected component of c.

On the connected component of d, we do the reverse: let $B = \{b \mid tp(b) = tp(c + 1) \wedge d \Rightarrow b\}$, and define $\pi''B = B \cup \{c + 1\}$. □

Acknowledgement: Henry Towsner is partially supported by NSF grant DMS-1340666. We are grateful to Roman Rabinovich and Marlene Gregorek for valuable comments on an earlier draft of this paper.

Bibliography

[1] Miklos Ajtai and Ronald Fagin. Reachability is harder for directed than for undirected finite graphs. *Journal of Symbolic Logic*, 55:113–150, 1990.

[2] P. Dittmann. Ultraproducts as a tool for first-order inexpressibility in the finite and infinite. arXiv:1310.3137, 2013.

[3] K. Doets. Uniform short proofs for classical theorems. *Notre Dame Journal for Formal Logic*, 42:121–127, 2001.

[4] Heinz-Dieter Ebbinghaus and Jörg Flum. *Finite model theory*. Springer, 1995.

[5] R. Fagin. Monadic generalized spectra. *Zeitschrift für Mathematische Logik und Grundlagen der Mathematik*, 21:89–96, 1975.

[6] R. Fagin. Probabilities on finite models. *Journal of Symbolic Logic*, 41:50–58, 1976.

[7] R. Fagin, L. Stockmeyer, and M. Y. Vardi. On monadic NP vs. monadic co-NP. *Information and Computation*, 120(1):78–92, 1995.

[8] H. Gaifman. On local and nonlocal properties. In J. Stern, editor, *Logic Colloquium '81*, pages 105–135. North Holland, 1982.

[9] H. Gaifman and M. Y. Vardi. A simple proof that connectivity is not first-order definable. *Bulletin of the European Association for Theoretical Computer Science*, 26:43–45, 1985.

[10] Peter A. Loeb. Conversion from nonstandard to standard measure spaces and applications in probability theory. *Transactions of the American Mathematical Society*, 211:113–122, 1975.

[11] Martin Otto. Model theoretic methods for fragments of FO and special classes of (finite) structures. In Javier Esparza, Christian Michaux, and Charles Steinhorn, editors, *Finite and Algorithmic Model Theory*, pages 271–341. Cambridge University Press, 2011.

[12] G. Turan. On the definability properties of finite graphs. *Discrete Mathematics*, 49:291–302, 1984.

[13] J. Väänänen. Pseudo-finite model theory. *Matematica Contemporanea*, 24:169–183, 2003.

Maryanthe Malliaris* and Saharon Shelah

Saturating the Random Graph with an Independent Family of Small Range

Dedicated to Jouko Väänänen on the occasion of his 60th birthday

Abstract: The first part of this paper is an expository overview of the authors' recent work on Keisler's order, a far-reaching program of understanding basic model-theoretic structure through the lens of regular ultrapowers. We motivate the problem and explain how this work connects model theory and set theory, leading to theorems on both sides. In the second part of the paper, we prove a new theorem which shows that, in some sense, saturating ultrapowers of the random graph is much less complex than it appeared. More precisely, we prove that for a class of regular filters $\mathcal{D}$ on I, $|I| = \lambda > \aleph_0$, the fact that $\mathcal{P}(I)/\mathcal{D}$ has little freedom (as measured by the fact that any maximal antichain is countable) does not prevent extending $\mathcal{D}$ to an ultrafilter $\mathcal{D}_1$ on I such that $M^I/\mathcal{D}_1$ is λ^+-saturated whenever M is a model of the theory of the random graph. This result has catalyzed our subsequent work on ultrapowers of simple unstable theories, and we briefly discuss some future directions of this work.

Keywords: unstable model theory, saturation of ultrapowers, Keisler's order

MSC: Primary: 03C20, 03C45, Secondary: 03E05

***Corresponding Author: Maryanthe Malliaris:** Department of Mathematics, University of Chicago, 5734 S. University Avenue, Chicago, IL 60637, USA and Einstein Institute of Mathematics, Edmond J. Safra Campus, Givat Ram, The Hebrew University of Jerusalem, Jerusalem, 91904, Israel, e-mail: mem@math.uchicago.edu
Saharon Shelah: Einstein Institute of Mathematics, Edmond J. Safra Campus, Givat Ram, The Hebrew University of Jerusalem, Jerusalem, 91904, Israel, and Department of Mathematics, Hill Center - Busch Campus, Rutgers, The State University of New Jersey, 110 Frelinghuysen Road, Piscataway, NJ 08854-8019 USA, e-mail: shelah@math.huji.ac.il

1 Introduction

Keisler's order is a long-standing (and far-reaching) program for comparing the complexity of unstable theories, proposed in Keisler 1967 [6]. The measure of complexity, roughly speaking, is the relative difficulty of producing saturated regular ultrapowers. The order $\trianglelefteq$ already has significant connections to classification theory, and we believe the present investigations will shed further light on the structure of simple unstable theories.

The goal of the present paper is first, to give an expository introduction to recent progress on Keisler's order, and second, to prove a surprising theorem which has catalyzed our subsequent work on Keisler's order for simple theories. The importance of this theorem can be described as follows. Previously, in [17], we had shown that the difference between saturation for random graphs and for linear order was visible in an interval of the form $(\mu, \lambda = 2^\mu)$. Theorem 3.2 shows it is visible in an interval of the form $(\aleph_0, \lambda)$. Thus Theorem 3.2 improves [17] Theorem 11.1, although the current paper does not supercede [17]: we rely here on the substantial background technology built there, and moreover the property Qr_1 developed there appears useful for more general arguments. Still, this new opening leads to a (long and hard) program of analyzing the fundamental structure of simple theories which we sketch in §1.2.1 and which is currently work in progress.

The present section is primarily expository. In §1.1 below, we define Keisler's order and survey what was known. In §1.2 we present this paper's main result, Theorem 3.2, and explain the relevance for simple theories. In §1.3, we describe the known points of contact between regular ultrafilters and theories. In §1.4, we discuss several theorems of set theory which have come from this program. §2 contains definitions and other preliminaries. §3 contains the proof of the paper's main theorem.

1.1 Keisler's order and model theory

This subsection, written primarily for model theorists, aims to explain Keisler's order and our recent work.

For transparency, all languages are countable and all theories are complete. We say that the ultrapower $M^\lambda/\mathcal{D}$ is *regular* when $\mathcal{D}$ is a regular ultrafilter on λ, Definition 2.1 below. A key property of regularity is that:

Fact 1.1. (Keisler [6] Corollary 2.1 p. 30; see also Shelah [20].VI.1) *Let $M \equiv N$ in a countable signature, $\lambda \geq \aleph_0$, $\mathcal{D}$ a regular ultrafilter on λ. Then $M^\lambda/\mathcal{D}$ is λ^+-saturated if and only if $N^\lambda/\mathcal{D}$ is λ^+-saturated.*

1.1.1 What is Keisler's order and why is it model-theoretically interesting?

Questions of saturation have long been central to model theory. Morley's theorem is a fundamental example. Suppose we anachronistically define the *Łos order* on complete countable theories:

- $T_1 \trianglelefteq^L_\lambda T_2$ if
  - (for all $M_2 \models T_2$, $|M_2| = \lambda$, M_2 is saturated) implies
  - (for all $M_1 \models T_1$, $|M_1| = \lambda$, M_1 is saturated)
- $T_1 \trianglelefteq^L T_2$ if $T_1 \trianglelefteq^L_\lambda T_2$ for all $\lambda > \aleph_0$

Then Morley's theorem shows that $\trianglelefteq^L$ has a minimum class, a maximum class, and no other classes, i.e. either all uncountable models of some given countable theory are saturated, or else the theory has some unsaturated model of every uncountable size.

Keisler's order may be thought of as generalizing this hypothetical "Łos order" in the following powerful way: rather than considering all uncountable models, we consider only regular ultrapowers (so "saturated" becomes "λ^+-saturated"). This reveals a richer field of comparison: we compare not by cardinality, but by provenance. Each $M_1^\lambda/\mathcal{D} \models T_1$ is naturally compared to models $M_2^\lambda/\mathcal{D}$ of T_2, built using the same ultrafilter $\mathcal{D}$ on λ. More precisely:

Definition 1.2. *(Keisler [6])* Let T_1, T_2 be complete countable first-order theories.
1. Let $\mathcal{D}$ be a regular ultrafilter on λ. Write $T_1 \trianglelefteq_{\mathcal{D}} T_2$ when (for all $M_2 \models T_2$, $M_2^\lambda/\mathcal{D}$ is λ^+-saturated) implies (for all $M_1 \models T_1$, $M_1^\lambda/\mathcal{D}$ is λ^+-saturated).
2. Write $T_1 \trianglelefteq_\lambda T_2$ if for any regular ultrafilter $\mathcal{D}$ on λ, $T_1 \trianglelefteq_{\mathcal{D}} T_2$.
3. *(Keisler's order)* Write $T_1 \trianglelefteq T_2$ if for all infinite λ, $T_1 \trianglelefteq_\lambda T_2$.

By Fact 1.1, $\trianglelefteq$ is a pre-order on theories, usually thought of as a partial order on the equivalence classes. [Note that by 1.1, $T_1 \trianglelefteq_{\mathcal{D}} T_2$ is equivalent to "for all $M_1 \models T_1$, $M_2 \models T_2$, ($M_2^\lambda/\mathcal{D}$ is λ^+-saturated) implies ($M_1^\lambda/\mathcal{D}$ is λ^+-saturated)."]

Keisler proved that $\trianglelefteq$ had at least a minimum and maximum class, and asked:

Question 1.3. *(Keisler 1967)* Determine the structure of Keisler's order.

Surprising early work of Shelah established the model theoretic significance of $\trianglelefteq$ (note the independent appearance of dividing lines from classification theory).

Theorem A. (Shelah 1978 [19] Chapter VI)

$$\mathcal{T}_1 \triangleleft \mathcal{T}_2 \triangleleft \cdots\cdots ? \cdots\cdots \trianglelefteq \mathcal{T}_{max}$$

where $\mathcal{T}_1 \cup \mathcal{T}_2$ is precisely the class of countable stable theories, and:
1. *T_1, the minimum class, is the set of all T without the finite cover property.*

2. T_2, *the next largest class, is the set of all stable* T *with the f.c.p.*
3. *There is a maximum class* $\mathcal{T}_{max}$, *containing all linear orders (i.e. SOP, in fact* SOP_3 *[21]), however, its model-theoretic identity is not known.*

For many years there was little progress, and the unstable case appeared relatively intractable despite the flourishing of unstable model theory.

Very recently, work of Malliaris and Shelah has led to considerable advances in our understandings of how ultrafilters and theories interact (Malliaris [9]-[12], Malliaris and Shelah [14]-[17]). For an account of some main developments, see the introductory sections of [14] as well as [17]. Most of the seismic shifts are underground and not yet visible as divisions in the order. Still, one can update the diagram:

$$\mathcal{T}_1 \lhd \mathcal{T}_2 \lhd \mathcal{T}_x \cdots ? \cdots \lhd \cdots ? \cdots \mathcal{T}_{nl} \trianglelefteq \cdots ? \cdots \trianglelefteq \mathcal{T}_{max}$$

$\mathcal{T}_x$ denotes the minimum unstable class (not yet characterized, but contains the random graph). $\mathcal{T}_{nl}$ denotes the simple non low theories, which are not known to be an equivalence class but are all strictly above (=more complex than) $\mathcal{T}_x$. A more finely drawn diagram would also show that there is a minimum TP_2 theory, and that at limit λ SOP_2 suffices for maximality, see §1.4. Note that it is not yet known whether the order is finite, or linear.

1.2 Prolegomena to our main theorem

Earlier this year, building on our prior work described above, we had shown that the question of saturation of ultrapowers could be substantially recast in terms of a two-stage approach, involving a more set-theoretic stage [constructing a so-called *excellent* filter $\mathcal{D}$ on I admitting a surjective homomorphism $\mathbf{j} : \mathcal{P}(I) \rightarrow \mathfrak{B}$ onto a specified Boolean algebra, with $\mathbf{j}^{-1}(\{1_{\mathfrak{B}}\}) = \mathcal{D}$] followed by a more model-theoretic stage [constructing a so-called *moral* ultrafilter on the Boolean algebra $\mathfrak{B}$, a step defined in terms of "possibility patterns" of formula incidence represented in the theory].

This advance allowed us to give the first ZFC $\trianglelefteq$-dividing line among the unstable theories, and the first dividing line since 1978: separating the Keisler-minimum unstable theory, the random graph T_{rg}, from all non-low or non-simple theories.

Theorem B. (Malliaris and Shelah [17]) *Suppose* λ, μ *are given with* $\mu < \lambda \leq 2^{\mu}$. *Then there is a regular ultrafilter* $\mathcal{D}$ *on* λ *which saturates ultrapowers of all countable stable theories and of the random graph, but fails to saturate ultrapowers of any non-low or non-simple theory.*

Convention 1.4. We say that a regular ultrafilter $\mathcal{D}_1$ on λ "saturates ultrapowers of T" to mean that whenever $M \models T$, $M^\lambda/\mathcal{D}$ is λ^+-saturated.

Naturally this result raised many further questions. Chief among them was the role of $\mu < \lambda \leq 2^\mu$, which appeared to have a strong model-theoretic motivation by analogy to [18], and in terms of what might be called the "Engelking-Karłowicz property of the random graph:" in the monster model of the random graph, if $|B| = \lambda$ and $\mu < \lambda \leq 2^\mu$ then there is some set A such that every nonalgebraic $p \in S(B)$ can be finitely realized in A. The role of μ in the proof of Theorem B was as the size of a maximal antichain of a quotient Boolean algebra at the key transfer point in the inductive ultrafilter construction, i.e., the point where the excellent filter $\mathcal{D}$ has been built. When $\mu < \lambda$, this "lack of freedom" at the transfer point was sufficient to guarantee the non-saturation result for any ultrafilter $\mathcal{D}_1 \supseteq \mathcal{D}$, and as indicated, when $\lambda \leq 2^\mu$, model-theoretic considerations made it possible to guarantee saturation of the random graph for *some* $\mathcal{D}_1 \supseteq \mathcal{D}$.

In this paper we show that, contrary to initial expectations, the situation here is fundamentally different and the second restriction on μ is unnecessary. Rather it is, within the regime of so-called excellent filters on some arbitrary but fixed λ, *always* possible to extend to an ultrafilter which will saturate the random graph, even when $\mu = \aleph_0$ at the transfer point:

Main Theorem. *Let $\aleph_0 \leq \mu \leq \lambda = |I|$. Suppose that $\mathcal{G}_*$ and $\mathcal{D}$ satisfy:*
1. *$\mathcal{D}$ is a regular, excellent filter on I*
2. *$\mathcal{G}_* \subseteq {}^\lambda\mu$, $|\mathcal{G}_*| = 2^\lambda$*
3. *$\mathcal{G}_*$ is a $\mathcal{D}$-independent family of functions*
4. *$\mathcal{D}$ is maximal subject to (3)*

Then for some ultrafilter $\mathcal{D}_1 \supseteq \mathcal{D}$, and all $M \models T_{\mathbf{rg}}$, $M^I/\mathcal{D}_1$ is λ^+-saturated.

Moreover, the proof itself modifies the usual inductive construction of an ultrafilter by means of independent functions by introducing so-called approximations; this has been significant for our subsequent work.

Note that when $(I, \mathcal{D}, \mathcal{G})$ is $(\lambda, \aleph_0)$-good it is always possible to extend $\mathcal{D}$ to an ultrafilter $\mathcal{D}_1$ which saturates all stable theories. [It suffices to ensure that the coinitiality of ω in $(\omega, <)^I/\mathcal{D}_1$ is $\geq \lambda^+$.] Our theorem here shows that *if* $\mathcal{D}$ is excellent this is additionally always possible for the random graph. It complements our construction in [15] of a filter $\mathcal{D}$, necessarily not excellent, *no* extension of which is able to saturate the random graph.

1.2.1 What avenues of investigation does this suggest for simple theories?

The parameter μ measuring the freedom available in the underlying Boolean algebra during ultrafilter construction, i.e. μ in Definition 2.9 below, appears significant for "outside definitions" of simple theories. Given Theorem 3.2, the natural question is whether membership in the Keisler minimum unstable class is characterized by: whenever $\mathcal{D}$ is regular and excellent and $(I, \mathcal{D}, \mathcal{G})$ a $(\lambda, \aleph_0)$-good triple, some ultrafilter $\mathcal{D}_1 \supseteq \mathcal{D}$ saturates T. By our prior results, $\mu < \lambda$ blocks saturation of non-simple or non-low theories (this can be circumvented by introducing a complete filter on the quotient Boolean algebra). Thus, the current focus is simple low theories.

By increasing the range of possible μ, Theorem 3.2 opens the door to a stratification of simple low theories. That is, we will want to try to distinguish between classes of theories based on various cardinal invariants of the Boolean algebras $\mathcal{P}(I)/\mathcal{D}$. Our previous result, Theorem B above, distinguished between the random graph (suffices to have antichains of size μ) and non-low theories (necessary to have antichains of size λ) when $\mu < \lambda \leq 2^{\mu}$. So to see finer divisions in the unstable low theories, one might need to assume failures of GCH. In light of Theorem 3.2 this is no longer necessary.

The property $\mu = \aleph_0$ appears connected with a model-theoretic coloring property introduced in work in progress of the authors.

1.3 Translations between set theory and model theory

A fundamental part of investigating Keisler's order involves construction of ultrafilters, thus combinatorial set theory. Isolating "model-theoretically meaningful properties" and determining implications and nonimplications between them gives a useful perspective on ultrafilters. We now include two "translation" theorems from [14]. (Some definitions are given in the Appendix.)

Theorem C. (Malliaris and Shelah [14] Theorem F) *In the following table, for each of the rows* (1),(3),(5),(6) *the regular ultrafilter* $\mathcal{D}$ *on* λ *fails to have the property in the left column if and only if it omits a type in every formula with the property in the right column. For rows* (2) *and* (4), $\mathcal{D}$ *fails to have the property on the left then it omits a type in every formula with the property on the right.*

Set theory: properties of filters		***Model theory: properties of formulas***
(1) $\mu(\mathcal{D}) \geq \lambda^+$		*A. finite cover property*
(2) $\mathrm{lcf}(\aleph_0, \mathcal{D}) \geq \lambda^+$	**	*B. order property*
(3) *saturates* T_{rg}		*C. independence property*
(4) *flexible, i.e. λ-flexible*	**	*D. non-low*
(5) *good for equality*		*E.* TP_2
(6) *good, i.e. λ^+-good*		*F. strict order property*

Proof. The characterization of the maximum class via good ultrafilters and the definition of the f.c.p. are due to Keisler 1967 [6], see [14] for details. (1)-(2) Shelah 1978 [19] VI.5. (3) Straightforward by quantifier elimination. (4) Malliaris 2009 [9]. (5) Malliaris 2010 [11]. (6) Shelah 1978 [19] VI.2.6. □

Theorem D. (updated version of Malliaris and Shelah [14] Theorem 4.2) *Assume that $\mathcal{D}$ is a regular ultrafilter on λ (note that not all of these properties imply regularity). Then:*

(1) $\leftarrow$ (2) $\leftarrow$ (3) $\leftarrow$ (5) $\leftarrow$ (6), *with* (1) $\not\rightarrow$ (2), (2) $\not\rightarrow$ (3), (3) $\not\rightarrow$ (5), *and whether* (5) *implies* (6) *is open. Moreover* (1) $\leftarrow$ (4) $\leftarrow$ (5) $\leftarrow$ (6), *where* (3) $\not\rightarrow$ (4) *thus* (2) $\not\rightarrow$ (4), (4) $\not\rightarrow$ (3), *consistently* (4) $\not\rightarrow$ (5), *consistently* (4) $\not\rightarrow$ (6); *and* (4) *implies* (2) *is open.*

"Consistently" throughout Theorem D means assuming a measurable cardinal. One of the surprises of [14]-[15] was the relevance of measurable cardinals in constructing regular ultrafilters. This bridges a certain cultural gap between regular filters, typically used in model theory, and complete ultrafilters, used primarily in set theory. §1.4 gives some consequences.

1.4 Set-theoretic theorems and aspects of this program

We now discuss some set-theoretic results of our program, from [14] and [17].

First, "consistently (4) $\not\rightarrow$ (6)" in Theorem D above addressed a question raised in Dow 1985 [2] about whether OK filters (introduced by Kunen, Keisler) are necessarily good. We proved that consistently the gap may be arbitrarily large:

Theorem E. (Malliaris and Shelah [14] Theorem 6.4) *Assume $\kappa > \aleph_0$ is measurable and $2^\kappa \leq \lambda = \lambda^\kappa$. Then there exists a regular uniform ultrafilter $\mathcal{D}$ on λ such that $\mathcal{D}$ is λ-flexible, thus λ-OK, but not $(2^\kappa)^+$-good.*

Notably, the failure of goodness is "as strong as possible" given the construction: $\mathcal{D}$ will fail to $(2^\kappa)^+$ saturate the random graph, thus any unstable theory. See [14]

§1.2 for an account of this result. As explained there, this is a natural context in which to study further weakenings of goodness. The two papers [14]-[15] contain several ultrafilter existence theorems.

We now describe a program from our paper [16]. Because of the Keisler-maximality of linear order, but also for set theoretic reasons (e.g. cardinal invariants of the continuum), it is natural to study cuts by defining:

Definition 1.5. *([16])* Let $\mathcal{D}$ be a regular ultrafilter on I, $|I| = \lambda$. Let $\mathcal{C}(\mathcal{D})$ be the set of all (κ_1, κ_2) such that $\kappa_1, \kappa_2 \leq \lambda$ are regular and $(\mathbb{N}, <)^I/\mathcal{D}$ has a (κ_1, κ_2)-cut.

Definition 1.6. *([16])* Say that a regular ultrafilter $\mathcal{D}$ on λ has κ-treetops if for any tree $(\mathcal{T}, \trianglelefteq)$ and any infinite regular cardinal $\gamma < \kappa$, in $N = (\mathcal{T}, \trianglelefteq)^\lambda/\mathcal{D}$ any strictly $\trianglelefteq^N$-increasing γ-indexed sequence has a $\trianglelefteq^N$-upper bound.

Both model-theoretic and set-theoretic considerations pointed to the question of whether λ^+-treetops implies λ^+-good.

We had shown that an ultrafilter $\mathcal{D}$ on λ has λ^+-treetops if and only if $\mathcal{C}(\mathcal{D})$ has no symmetric cuts, so the question was:

Question 1.7. Given $\kappa \leq \lambda \implies (\kappa, \kappa) \notin \mathcal{C}(\mathcal{D})$, what are the possible $\mathcal{C}(\mathcal{D})$?

Theorem F. (Malliaris and Shelah [16]) *Suppose that for all $\kappa \leq \lambda$, $(\kappa, \kappa) \notin \mathcal{C}(\mathcal{D})$. Then $\mathcal{C}(\mathcal{D}) = \emptyset$, and $\mathcal{D}$ is good.*

In fact, the setup in [16] is somewhat more general: we consider what we call "cofinality spectrum problems," which cover a variety of problems including cofinalities of cuts in ultrapowers of linear order. In [16] we also prove two corollaries of this theorem, one model theoretic, one about cardinal invariants of the continuum. First, SOP_2 is maximal in Keisler's order. Second, $\mathfrak{p} = \mathfrak{t}$, where $\mathfrak{p}, \mathfrak{t}$ are the pseudointersection number and the tower number respectively.

This concludes the introduction. We now work towards Theorem 3.2.

2 Preliminaries

Here we define: regular filters, good filters, excellent filters, and good triples.

Definition 2.1. *(Regular filters)* Let $\mathcal{D}$ be a filter on an index set I of cardinality λ. A μ-regularizing family $\{X_i : i < \mu\}$ is a set such that:
- for each $i < \mu$, $X_i \in \mathcal{D}$, and
- for any infinite $\sigma \subset \mu$, we have $\bigcap_{i \in \sigma} X_i = \emptyset$

Equivalently, for any element $t \in I$, t belongs to only finitely many of the sets X_i. A filter $\mathcal{D}$ on an index set I of cardinality λ is said to be *μ-regular* if it contains a μ-regularizing family. $\mathcal{D}$ is called *regular* if it is λ-regular, i.e. $|I|$-regular.

Regular filters on λ always exist, see [1] or [17] top of p. 7.

Definition 2.2. *(Good filters, Keisler [5])* The filter $\mathcal{D}$ on I is said to be *μ^+-good* if every $f : \mathcal{P}_{\aleph_0}(\mu) \to \mathcal{D}$ has a multiplicative refinement, where this means that for some $f' : \mathcal{P}_{\aleph_0}(\mu) \to \mathcal{D}$, $u \in \mathcal{P}_{\aleph_0}(\mu) \implies f'(u) \subseteq f(u)$, and $u, v \in \mathcal{P}_{\aleph_0}(\mu) \implies f'(u) \cap f'(v) = f'(u \cup v)$.

Note that we may assume the functions f are monotonic.

$\mathcal{D}$ is said to be *good* if it is $|I|^+$-good.

Remark 2.3. The importance of good filters here arises from Keisler's observation that when $\mathcal{D}$ is regular, $A \subseteq N := M^\lambda/\mathcal{D}$, $p \in S(A)$, $|A| \leq \lambda$, $p = \{\varphi_i(x; \overline{a}_i) : i < \lambda\}$ then p is realized in N iff there is some multiplicative $f : \mathcal{P}_{\aleph_0}(\lambda) \to \mathcal{D}$ which refines the "existential" map $\sigma \mapsto \{t \in \lambda : M \models \exists x \bigwedge_{i \in \sigma} \varphi_i(x; \overline{a}_i[t])\}$.

The proof of Theorem 3.2 builds on a main innovation of Malliaris and Shelah [17], so-called excellent filters: 2.6 below. For the purposes of the current proof, the reader may take "excellent" to be a new characterization of "good" which is most interesting in the case where $\mathcal{D}$ is a filter and not an ultrafilter, as is proved in Theorem 12.3 of the appendix to [17]. (As noted there, however, it is not yet clear whether various useful restrictions of these two definitions will generally coincide.)

Suppose we are given a filter $\mathcal{D}$, a Boolean algebra $\mathfrak{B}$ and a surjective homomorphism $\mathbf{j} : \mathcal{P}(I) \to \mathfrak{B}$ such that $\mathbf{j}^{-1}(\{1_{\mathfrak{B}}\}) = \mathcal{D}$. Roughly speaking, if $\mathcal{D}$ is excellent we may assume that $\mathbf{j}$ is "accurate" in the sense that if certain distinguished Boolean terms hold on elements $\overline{\mathbf{b}}$ of $\mathfrak{B}$, the same terms will hold on some $\mathbf{j}$-preimage of $\overline{\mathbf{b}}$ *without* the a priori necessary qualifier "$\mod \mathcal{D}$." [For details and history, see [17] §2.] The main example for us here is:

Fact 2.4. *Let $\lambda \geq \aleph_0$ and let $\mathcal{D}$ be an excellent, i.e. λ^+-excellent filter on I, $|I| = \lambda$. Let*

$$\overline{A} = \langle A_u : u \in [\lambda]^{<\aleph_0} \rangle$$

be a sequence of elements of $\mathcal{P}(I)$ which is multiplicative $\mod \mathcal{D}$, i.e. for each $u, v \in [\lambda]^{<\aleph_0}$, $A_u \cap A_v = A_{u \cup v} \mod \mathcal{D}$.

Then there exists $\overline{B} = \langle B_u : u \in [\lambda]^{<\aleph_0} \rangle$ such that:

1. $u \in [\lambda]^{<\aleph_0} \implies B_u \subseteq A_u$
2. $u \in [\lambda]^{<\aleph_0} \implies B_u = A_u \mod \mathcal{D}$
3. $u, v \in [\lambda]^{<\aleph_0} \implies B_u \cap B_v = B_{u \cup v}$

i.e. $\overline{B}$ refines $\overline{A}$ and is truly multiplicative.

Proof. [17] Claim 4.9. □

The effect of excellence is to allow a so-called "separation of variables," [17] Theorem 5.11. That is, as sketched in 1.2, ultrafilter construction can now be done in two stages. First, one builds an excellent filter with a specified quotient $\mathfrak{B}$. One can then build ultrafilters directly on $\mathfrak{B}$ which ensure that so-called possibility patterns of a given theory (a measure of the complexity of incidence in φ-types) have multiplicative refinements. An immediate advantage of this separation is that it allows us to prove results like Theorem B above. Namely, a "bottleneck" is built in to the construction by arranging for the c.c. of $\mathcal{P}(I)/\mathcal{D}$ to be small by the time $\mathcal{D}$ is excellent. This prevents future saturation of some theories, but not all, and so gives a dividing line. This analysis also pushes one to understand how these "possibility patterns" reflect model-theoretic complexity.

Though we include the full definition of excellence from [17], Theorem 3.2 will only use Fact 2.4 above, which the reader may prefer to take as axiomatic.

Definition 2.5. Let $\mathfrak{B}$ be a Boolean algebra and $\bar{\mathbf{a}} = \langle \mathbf{a}_u : u \in [\lambda]^{<\aleph_0} \rangle$ be a sequence of elements of $\mathfrak{B}$. When u is a finite set, write $\overline{x}_{\mathcal{P}(u)} = \langle x_v : v \subseteq u \rangle$ for a sequence of variables indexed by subsets of u.

1. Define

$$N(\overline{a}\restriction_{\mathcal{P}(u)}) = \{ \langle a'_v : v \subseteq u \rangle : \text{ for some } w \subseteq u \text{ we have } a'_v = a_v \text{ if } v \subseteq w \text{ and } a'_v = 0_{\mathfrak{B}} \text{ otherwise} \}$$

2. Define $\Lambda_{\mathfrak{B},\bar{a}}$ to be the set

$$\{ \sigma(\overline{x}_{\mathcal{P}(u)}) : \sigma(\overline{x}_{\mathcal{P}(u)}) \text{ is a Boolean term such that } \mathfrak{B} \models \text{“}\sigma(\overline{a}') = 0\text{” whenever } \overline{a}' \in N(\overline{a}) \}$$

3. If $\mathcal{D}$ is a filter on $\mathfrak{B}$ then $\Lambda_{\mathfrak{B},\mathcal{D},\bar{a}} = \Lambda_{\mathfrak{B}_1,\bar{a}_1}$ where $\mathfrak{B}_1 = \mathfrak{B}/\mathcal{D}$ and $\bar{a}_1 = \langle a_v/\mathcal{D} : v \subseteq u \rangle$.
4. If $\mathcal{D}$ is a filter on a set I, then $\mathcal{D}$ determines I, so we write $\Lambda_{\mathcal{D},\bar{a}}$ for $\Lambda_{\mathcal{P}(I),\mathcal{D},\bar{a}}$.

Definition 2.6. *(Excellent filters, Malliaris and Shelah [17] Definition 4.6)* Let $\mathcal{D}$ be a filter on the index set I. We say that $\mathcal{D}$ is λ^+-*excellent* when: if $\overline{A} = \langle A_u : u \in [\lambda]^{<\aleph_0} \rangle$ with $u \in [\lambda]^{<\aleph_0} \implies A_u \subseteq I$, then we can find $\overline{B} = \langle B_u : u \in [\lambda]^{<\aleph_0} \rangle$ such that:
1. for each $u \in [\lambda]^{<\aleph_0}$, $B_u \subseteq A_u$
2. for each $u \in [\lambda]^{<\aleph_0}$, $B_u = A_u \mod \mathcal{D}$
3. *if* $u \in [\lambda]^{<\aleph_0}$ and $\sigma \in \Lambda_{\mathcal{D},\overline{A}|_u}$, so $\sigma(\overline{A}|_{\mathcal{P}(u)}) = \emptyset \mod \mathcal{D}$, *then* $\sigma(\overline{B}|_{\mathcal{P}(u)}) = \emptyset$

We say that $\mathcal{D}$ is ξ-excellent when it is λ^+-excellent for every $\lambda < \xi$.

Definition 2.7. Given a filter $\mathcal{D}$ on λ, we say that a family $\mathcal{F}$ of functions from λ into λ is *independent* $\mod \mathcal{D}$ if for every $n < \omega$, distinct $f_0, \dots f_{n-1}$ from $\mathcal{F}$ and choice of $j_\ell \in \text{Range}(f_\ell)$,

$$\{\eta < \lambda : \text{for every } i < n, f_i(\eta) = j_i\} \neq \emptyset \mod \mathcal{D}$$

Theorem G. (Engelking-Karłowicz [3] Theorem 3, see also Shelah [20] Theorem A1.5 p. 656) *For every $\lambda \geq \aleph_0$ there exists a family $\mathcal{F}$ of size 2^λ with each $f \in \mathcal{F}$ from λ onto λ such that $\mathcal{F}$ is independent modulo the empty filter (alternately, by the filter generated by $\{\lambda\}$).*

In particular, such families can be naturally thought of as Boolean algebras, so we introduce some notation:

Definition 2.8. Denote by $\mathfrak{B}^1_{\chi,\mu}$ the completion of the Boolean algebra generated by $\{x_{\alpha,\epsilon} : \alpha < \chi, \epsilon < \mu\}$ freely except for the conditions $\alpha < \chi \wedge \epsilon < \zeta < \mu \implies x_{\alpha,\epsilon} \cap x_{\alpha,\zeta} = 0$.

We follow the literature in using the term "good triple" for the following object, despite the name's ambiguity.

Definition 2.9. *Good triples (cf. [20] Chapter VI)* Let $\lambda \geq \kappa \geq \aleph_0$, $|I| = \lambda$, $\mathcal{D}$ a regular filter on I, and $\mathcal{G}$ a family of functions from I to κ.

1. Let $\text{FI}(\mathcal{G}) =$

$$\{h : h : [\mathcal{G}]^{<\aleph_0} \to \kappa \text{ and } g \in \text{Dom}(g) \implies h(g) \in \text{Range}(g)\}$$

2. Let $\text{FI}_s(\mathcal{G}) = \{A_h : h \in \text{FI}(\mathcal{G})\}$ where

$$A_h = \{t \in I : g \in \text{Dom}(g) \implies g(t) = h(g)\}$$

3. We say that triple $(I, \mathcal{D}, \mathcal{G})$ is (λ, κ)-pre-good when $I, \mathcal{D}, \mathcal{G}$ are as given, and for every $h \in \text{FI}(\mathcal{G})$ we have that $A_h \neq \emptyset \mod \mathcal{D}$.
4. We say that $(I, \mathcal{D}, \mathcal{G})$ is (λ, κ)-good when $\mathcal{D}$ is maximal subject to this condition.

Fact 2.10. *If $(I, \mathcal{D}, \mathcal{G})$ is a good triple, then $\text{FI}_s(\mathcal{G})$ is dense in $\mathcal{P}(I) \mod \mathcal{D}$.*

Observation 2.11. Let $\mathcal{D}$ be a filter on I, $\mathfrak{B} = \mathfrak{B}^1_{\xi,\mu}$ for some $\xi \leq 2^\lambda, \mu \leq \lambda$ and $\mathbf{j} : \mathcal{P}(I) \to \mathfrak{B}$ a surjective homomorphism such that $\mathbf{j}^{-1}(\{1_\mathfrak{B}\}) = \mathcal{D}$. Let $\mathcal{G} = \mathcal{G}_\mathfrak{B} = \{g_\alpha : \alpha < 2^\lambda\} \subseteq {}^\lambda\mu$ be given by $g_\alpha(\epsilon) = \mathbf{j}^{-1}(x_{\alpha,\epsilon})$. Then $(I, \mathcal{D}, \mathcal{G})$ is a (λ, μ)-good triple.

We will use the following existence theorem from [17].

Theorem H. (Existence theorem for excellent filters, [17]) *Let $\mu \leq \lambda$, $|I| = \lambda$ and let $\mathfrak{B}$ be a μ^+-c.c. complete Boolean algebra of cardinality $\leq 2^\lambda$. Then there exists a regular excellent filter $\mathcal{D}$ on I and a surjective homomorphism $\mathbf{j} : \mathcal{P}(I) \to \mathfrak{B}$ such that $\mathbf{j}^{-1}(\{1_\mathfrak{B}\}) = \mathcal{D}$.*

Corollary 2.12. *Let I, $|I| = \lambda \geq \aleph_0$, and μ with $\aleph_0 \leq \mu \leq \lambda$ be given. Let $\mathcal{D}$ be an excellent filter on I given by Theorem H in the case where $\mathfrak{B} = \mathfrak{B}^1_{2^\lambda,\mu}$. Let $\mathcal{G}_\mathfrak{B} \subseteq {}^\lambda\mu$ be given by Observation 2.11. Then $(I, \mathcal{D}, \mathcal{G}_\mathfrak{B})$ is a (λ, μ)-good triple.*

3 Main Theorem

We now prove that it is possible to saturate the theory T_{rg} of the random graph using only functions with range $\aleph_0$. While the construction is a natural evolution of our argument for (an ultrapower version of) the Engelking-Karłowicz property in [17] Lemma 9.9, the result was a surprise. Here, notably, we modify the usual "inductive construction via independent families" to allow a much finer degree of control. The calibrations are noted throughout the proof, beginning with 3.3.

Remark 3.1. Note that in Theorem 3.2, possibly $2^\mu << \lambda$; indeed, possibly $\mu = \aleph_0$ while $\lambda = |I|$ is arbitrary.

Theorem 3.2. *Suppose that we are given:*
1. $(I, \mathcal{D}, \mathcal{G}_*)$ *is a* (λ, μ)*-good triple*
2. $\aleph_0 \leq \mu \leq \lambda = |I|$
3. $\mathcal{D}$ *is* λ*-regular*
4. $\mathcal{D}$ *is* λ^+*-excellent*

Then there is an ultrafilter $\mathcal{D}_1 \supseteq \mathcal{D}$ such that for any $M \models T_{rg}$, $M^I/\mathcal{D}_1$ is λ^+-saturated.

We shall fix I, $\mathcal{D}$, $\mathcal{G}_*$, μ, λ as in the statement of Theorem 3.2 for the remainder of this section. Hypothesis (4), excellence, is used only at one point, in Step 13.

The infrastructure for the proof will be built via several intermediate claims and definitions. *Note*: When φ is a formula, we write φ^0 for $\neg\varphi$ and φ^1 for φ.

Remark 3.3. In contrast to the usual method, we will not complete the filter at each inductive step to a "good" triple. This is a crucial difference. Rather, we build a series of approximations to the final ultrafilter. Note that condition (c) on the size of the approximation is natural, though not needed.

Background Note. An account of why realizing types in ultrapowers amounts to finding a multiplicative refinement is given in [14] §1.2, or see 2.3 below.

1. Approximations For $\alpha \leq 2^\lambda$, let AP_α be the set of pairs $\mathbf{a} = (\mathcal{A}, \mathcal{G}) = (\mathcal{A}_\mathbf{a}, \mathcal{G}_\mathbf{a})$ such that:

(a) $\mathcal{G}_\mathbf{a} \subseteq \mathcal{G}_*$

(b) $|\mathcal{G}_\mathbf{a}| \leq |\alpha| + \lambda$

(c) $|\mathcal{A}_{\mathbf{a}}| \leq |\alpha| + \lambda$
(d) $\mathcal{A}_{\mathbf{a}} \subseteq \mathcal{P}(I)$
(e) $(\forall A \in \mathcal{A}_{\mathbf{a}})(A$ is supported by $\mathcal{G}_{\mathbf{a}}$ modulo $\mathcal{D})$
(f) $\emptyset \notin \text{fil}\langle \mathcal{D} \cup \mathcal{A}_{\mathbf{a}} \rangle$

The elements of each AP_α can be naturally partially ordered (by inclusion in both coordinates).

Convention 3.4. For $\mathbf{a} \in \text{AP}_\alpha$, denote by $\text{fil}\langle \mathcal{D} \cup \mathbf{a} \rangle$ the filter on I generated by $\mathcal{D} \cup \mathcal{A}_{\mathbf{a}}$.

2. Aim of the inductive step. Our aim in the key inductive step will be to prove the following. Note that from a certain point of view, this is about the Boolean algebra $\mathcal{P}(I)/\mathcal{D}$; however, it clarifies our presentation here to refer explicitly to I and to the given types as they are presented in the reduced product. By quantifier elimination, a type here is a choice of λ parameters and a function from λ to 2.

Claim 3.5. *If (A) then (B):*
(A) (a) $\mathbf{a} \in \text{AP}_\alpha$
  (b) $M \models T_{\mathbf{rg}}$
  (c) $h_i \in {}^I M$ *for* $i < \lambda$
  (d) $\eta \in {}^\lambda 2$
  (e) *For* $i < j < \lambda$, *if* $\eta(i) = \eta(j)$ *then* $A_{i,j} = I$, *and if* $\eta(i) \neq \eta(j)$ *let*

$$A_{i,j} = \{t \in I : h_i(t) \neq h_j(t)\} = I \mod \text{fil}\langle \mathcal{D} \cup \mathbf{a} \rangle$$

  (f) $p = \{(xR(h_i/\mathcal{D}_1))^{[\eta(i)]} : i < \lambda\}$ *is a type in* $M^I/\mathcal{D}_1$ *for every ultrafilter* $\mathcal{D}_1 \supseteq \text{fil}\langle \mathcal{D} \cup \mathbf{a} \rangle$
(B) *There are* $\mathbf{b}, \overline{B}$ *such that:*
  (a) $\mathbf{a} \leq \mathbf{b} \in \text{AP}_{\alpha+1}$
  (b) $\overline{B} = \langle B_i : i < \lambda \rangle$
  (c) $B_i \in \text{fil}\langle \mathcal{D} \cup \mathbf{b} \rangle$
  (d) $B_i \cap B_j \subseteq A_{i,j} \mod \mathcal{D}$ *for* $i < j < \lambda$ *such that* $\eta(i) \neq \eta(j)$

The proof will follow from Steps 11-13 below, following some intermediate definitions. In Step 13, we verify that Claim 3.5(B) is sufficient to realize the type.

Definition 3.6. Given any sequence $\overline{B} = \langle B_i : i < \lambda \rangle$ from Claim 3.5(B), let the sequence $\overline{B}_* = \langle B_u : u \in [\lambda]^{<\aleph_0} \rangle$ be given by: $B_{\{i\}} = B_i$ for $i < \lambda$, and $B_u = \bigcap\{B_{\{i\}} : i \in u\}$ for $u \in [\lambda]^{<\aleph_0}$, $|u| \geq 2$.

Discussion. Condition (A) corresponds to the data of a random graph type over the parameters $\{h_i : i < \lambda\}$. Define a sequence $\overline{A} = \langle A_u : u \in [\lambda]^{<\aleph_0} \rangle$ by $|u| = 1$ implies

$A_u = I$, $u = \{i, j\}$, $i \neq j$ implies $A_u = A_{i,j}$ as defined above, and $|u| \geq 3$ implies $A_u = \bigcap\{A_v : v \subseteq u, |v| = 2\}$. Then $\overline{A}$ gives a distribution for this type.

Note that $t \in A_{i,j}$ implies $\{(xR(h_i/\mathcal{D}_1))^{[\eta(i)]}, (xR(h_j/\mathcal{D}_1))^{[\eta(j)]}\}$ is consistent. Since random graph types have 2-compactness, and $(\exists x)xR(h_i/\mathcal{D}_1)$ is sent to I by the Łos map, item $(B)(d)$ implies that the corresponding sequence $\overline{B}_* = \langle B_u : u \in [\lambda]^{<\aleph_0}\rangle$ from Definition 3.6 refines $\overline{A}$ modulo $\mathcal{D}$ and is multiplicative $\mod \mathcal{D}$. This will be sufficient to realize the type given the background assumption of excellence, see Step 13 below.

Data for the key inductive step. Steps 3-10 below are in the context of Claim 3.5, meaning that we fix the data of Claim 3.5(A) and define the following objects based on it. Thus all objects defined here are implicitly subscripted by α and $\mathbf{a}$ and depend on the choice of sequence $\langle A_i : i < \lambda\rangle$ at this inductive step.

3. The set $\mathcal{F}^1_i = \mathcal{F}^1_{\alpha,\mathbf{a},i}$. For each $i < \lambda$, define $\mathcal{F}^1_i$ to be the set of all $f \in \mathrm{FI}(\mathcal{G}_\mathbf{a})$ such that for some $j \leq i$:

1. $h_j|_{A_f} = h_i|_{A_f} \mod \mathcal{D}$
2. $\gamma < j \implies h_\gamma|_{A_f} \neq h_i|_{A_f} \mod \mathcal{D}$
   i.e., $\gamma < j \implies$

$$\{t \in A_f : h_j(t) \neq h_i(t) \text{ or } h_\gamma(t) = h_i(t)\} = \emptyset \mod \mathcal{D}$$

Note that fixing A_f there is a minimal j such that $h_j|_{A_f} = h_i|_{A_f} \mod \mathcal{D}$. Since the ordinals are well ordered, we can choose a least j for which there is such a witness A_f.

4. The set $\mathcal{F}^2_i = \mathcal{F}^2_{\alpha,\mathbf{a},i}$. For each $i < \lambda$, choose $\mathcal{F}^2_i$ so that:

1. $\mathcal{F}^2_i \subseteq \mathcal{F}^1_i$
2. $f' \neq f'' \in \mathcal{F}^2_i \implies f', f''$ are incompatible functions
3. $\mathcal{F}^2_i$ is maximal under these restrictions

5. Density. Notice that for $\ell = 1, 2$ $\mathcal{F}^\ell_i$ is pre-dense, i.e. for every $f' \in \mathrm{FI}(\mathcal{G}_\mathbf{a})$ for some $f'' \in \mathcal{F}^1_i$ the functions f', f'' are compatible.

The set $\mathcal{F}^1_i$ is dense, meaning that for each $f' \in \mathrm{FI}(\mathcal{F}_\mathbf{a})$ there is $f'' \in \mathcal{F}^1_i$ such that $f' \subseteq f''$. Moreover, it is open, meaning that if $f'' \in \mathcal{F}^1_i$ and $f'' \subseteq f'''$ then $f''' \in \mathcal{F}^1_i$. Here by "$\subseteq$" we mean that the domain of the smaller function is contained in the domain of the larger function, and the two functions agree on their common domain. [Note that $f \subseteq f' \implies A_{f'} \subseteq A_f$.]

6. The collision function. For each $i < \lambda$ define the function $\rho_i : \mathcal{F}^2_i \to i$ by:

$$\rho_i(f) = \min\{j \leq i : h_i|_{A_f} = h_j|_{A_f} \mod \mathcal{D}\}$$

Note that by the definition of $\mathcal{F}_i^2$, this is the whole story in the sense that if $\rho_i(f) = j$ then for no $f' \supseteq f$ does there exist $j' < j$ such that $h_i|_{A_{f'}} = h_{j'}|_{A_{f'}}$ mod $\mathcal{D}$.

7. The new support. By induction on $i < \lambda$ choose $g_i \in \mathcal{G}_* \setminus \mathcal{G}_{\mathbf{a}} \setminus \{g_j : j < i\}$ (we make no further requirements on the sequence, but note the functions will be distinct).

8. The partition. For each $i < \lambda$, the sequence $\langle A_f : f \in \mathcal{F}_i^2 \rangle$ is (by definition) a sequence of pairwise disjoint sets.

9. The refinement. Let $\overline{B} = \langle B_i : i < \lambda \rangle$ where for $i < \lambda$

$$B_i = \bigcup \{ A_f \cap \left(g^{-1}_{\rho_i(f)}(\{0\}) \right)^{[\eta(i)]} : f \in \mathcal{F}_i^2 \}$$

where recall that for $B \subseteq I$, $B^{[1]} = B$, $B^{[0]} = I \setminus B$.

10. Definition of **b**. Finally, we define

$$\mathbf{b} = (\mathcal{G}_{\mathbf{a}} \cup \{g_i : i < \lambda\}, \mathcal{A}_{\mathbf{a}} \cup \{B_i : i < \lambda\})$$

Note that in contrast to the "usual" construction, here we have addressed the problem of realizing a type by using λ functions of range $\aleph_0$, rather than a single function of range λ. Moreover, we do not complete to a good triple at the end of the inductive step.

11. Proof of Claim 3.5(B)(a). We need to check that **b** as defined in Step 10 satisfies clauses (a)-(e) of the definition of approximation from Step 1. The only non-trivial part is proving that $\emptyset \notin \mathrm{fil}\langle \mathcal{D} \cup \mathcal{A}_{\mathbf{b}} \rangle$, recalling Convention 3.4.

Suppose we are given $C_0, \dots C_{k-1} \in \mathcal{A}_{\mathbf{a}}$ and $i_0, \dots i_{n-1} < \lambda$. It will suffice to prove that

$$\bigcap_{j<n} B_{i_j} \cap \bigcap_{\ell<k} C_\ell \neq \emptyset \mod \mathcal{D}$$

Informally speaking, we first try to find a $\mathcal{D}$-nonzero set on which the corresponding parameters h_{i_j} are distinct. On such a set, the instructions for each B_{i_j} are clearly compatible, so we can then find some $A_{f_*} \in \mathrm{FI}_s(\mathcal{G}_*)$, Definition 2.9, which is contained in their intersection mod $\mathcal{D}$. We now give the details.

First, by $(A)(e)$ of the inductive step and the fact that $\mathrm{fil}\langle \mathcal{D} \cup \mathcal{A}_{\mathbf{a}} \rangle$ is a filter, there is $A \in \mathrm{fil}\langle \mathcal{D} \cup \mathbf{a} \rangle$, $A \subseteq \bigcap \{C_\ell : \ell < k\}$ such that $j < j' < n \wedge \eta(j) \neq \eta(j') \implies A_{j,j'} \supseteq A$. Moreover, by the definition of approximation, any $A \in \mathrm{fil}\langle \mathcal{D} \cup \mathbf{a} \rangle$ contains a set which is supported by $\mathcal{G}_{\mathbf{a}}$. Let $f \in \mathrm{FI}(\mathcal{G}_{\mathbf{a}})$ be such that $A_f \subseteq A \mod \mathcal{D}$.

Second, recall that each $\mathcal{F}_i^2$ is pre-dense. So for each $j < n$, we may choose $f_{i_j} \in \mathcal{F}_{i_j}^2$ which is compatible with f. As we can increase f, without loss of generality, for $j < n$ there is $f_{i_j} \in \mathcal{F}_{i_j}^2$ such that $f_{i_j} \subseteq f$ (choose these by induction on $j < n$).

By choice of f, for no $j < j' < n$ is it the case that $h_{i_j} = h_{i_{j'}}$ on A_f. In other words,

$$j < j' < n \wedge \eta(j) \neq \eta(j') \implies \rho_{i_j}(f_{i_j}) \neq \rho_{i_{j'}}(f_{i_{j'}})$$

Thus (identifying functions with their graphs) the function f_* defined by

$$f_* = f \cup \bigcup\{f_{i_j} : j < n\} \cup \{(g_{k_j}, t_j) : k_j = \rho_{i_j}(fi_j), t_j = \eta(i_j)\}$$

is indeed a function, thus an element of $\mathrm{FI}(\mathcal{G}_*)$. Clearly $A_{f_*} \subseteq B_{i_j}$ for each $j < n$, and $A_{f_*} \neq \emptyset \mod \mathcal{D}$ by the hypothesis that $(I, \mathcal{D}, \mathcal{G}_*)$ is a good triple. This completes the proof of Step 11.

Now Claim 3.5(B)(b)-(c) obviously hold, so we are left with:

12. Proof of Claim 3.5(B)(d). We now show that if $i \neq j < \lambda$, $\eta(i) \neq \eta(j)$ then $B_i \cap B_j \subseteq A_{i,j} \mod \mathcal{D}$. Note that if $\eta(i) = \eta(j)$ the inclusion holds trivially, which is why we assume $\eta(i) \neq \eta(j)$ (so $i \neq j$).

Assume for a contradiction that $A_* := (B_i \cap B_j) \setminus A_{i,j} \neq \emptyset \mod \mathcal{D}$. As $B_i, B_j, A_{i,j}$ are supported by $\mathcal{G}_\mathbf{b}$, there is $f_* \in \mathrm{FI}(\mathcal{G}_\mathbf{b})$ such that $A_{f_*} \subseteq A_* \mod \mathcal{D}$.

As there is no problem increasing f_*, we may choose $f_i \in \mathcal{F}^2_i$, $f_j \in \mathcal{F}^2_j$ such that $f_i \subseteq f_* \wedge f_j \subseteq f_*$. In other words, $A_{f_*} \subseteq A_{f_i} \cap A_{f_j}$. Recall that

$$A_{i,j} = \{t \in I : h_i(t) \neq h_j(t)\}$$

By the definition of $A_{i,j}$, since $A_{f_*} \cap A_{i,j} = \emptyset \mod \mathcal{D}$ it must be that $h_i = h_j$ on $A_{f_*} \mod \mathcal{D}$. Because we chose f_i and f_j from $\mathcal{F}^2_i$ and $\mathcal{F}^2_j$, respectively, there must be $k_i \leq i$ and $k_j \leq j$ so that $h_i = h_{k_i}$ on A_{f_i}, and $h_j = h_{k_j}$ on A_{f_j}. Since equality is transitive, there is $k \leq \min\{i, j\}$ such that $h_i = h_j = h_k$ on A_{f_*}. In the notation of Step 6, $\rho_i(f_i) = \rho_j(f_j) = k$.

Now we look at the definition of B_i, B_j. Since $\eta(i) = \mathbf{t}_i \neq \eta(j) = \mathbf{t}_j$, we have that $B_i \cap A_{f_*} \subseteq (g_k^{-1}(0))^{\mathbf{t}_i}$ whereas $B_j \cap A_{f_*} \subseteq (g_k^{-1}(0))^{\mathbf{t}_j}$. Thus $B_i \cap B_j \cap A_{f_*} = \emptyset$, which is the desired contradiction.

13. Finishing the key inductive step. Why is Claim 3.5 sufficient to realize the type? Let us rephrase the problem as follows.

Corollary 3.7. *Let $\mathcal{D}$ be the excellent background filter defined at the beginning of the proof. Let $\mathbf{a}$, $\langle h_i : i < \lambda \rangle$, $\langle A_{i,j} : i < j < \lambda \rangle$, η, p be as given in Claim* 3.5(*B*) *at the inductive stage α. Let $\langle B_i : i < \lambda \rangle$, $\mathbf{b}$ be as given by Claim* 3.5(*A*).

Then in any ultrafilter $\mathcal{D}_1$ extending $\mathrm{fil}\langle \mathcal{D} \cup \mathbf{b} \rangle$, the type p is realized.

Proof. Let $\langle A_u : u \in [\lambda]^{<\aleph_0} \rangle$ be the sequence defined in the Discussion in Step 2, which corresponds to a distribution of the type p. Let $\overline{B}_* = \langle B_u : u \in [\lambda]^{<\aleph_0} \rangle$ be constructed from $\{B_i : i < \lambda\}$ as in Definition 3.6 above. By definition $\overline{B}_*$ is multiplicative. In step 12, it was shown that $\overline{B}_*$ refines $\overline{A} \mod \mathcal{D}$.

Define a third sequence $\overline{B}' = \langle B'_u : u \in [\lambda]^{<\aleph_0}\rangle$ by $B'_u = B_u \cap A_u$ for $u \in [\lambda]^{<\aleph_0}$. Then the sequence $\overline{B}'$ truly refines $\overline{A}$, but is only multiplicative modulo $\mathcal{D}$.

By the excellence of $\mathcal{D}$ [i.e. Fact 2.4 above] we may replace $\overline{B}'$ by a sequence $\langle B^{**}_u : u \in [\lambda]^{<\aleph_0}\rangle$ such that:

- $u \in [\lambda]^{<\aleph_0}$ implies $B^{**}_u \subseteq B'_u \subseteq A_u$
- $u \in [\lambda]^{<\aleph_0}$ implies $B^{**}_u = B'_u \mod \mathcal{D}$, thus $B^{**}_u \in \mathrm{fil}\langle \mathcal{D} \cup \mathbf{b}\rangle$
- $\langle B^{**}_u : u \in [\lambda]^{<\aleph_0}\rangle$ is multiplicative

Thus we may realize the type. □

Step 14: Adding subsets of the index set.

Claim 3.8. *Let $\alpha < 2^\lambda$, $\mathbf{a} \in \mathrm{AP}_\alpha$, $A \subseteq I$ be given. Then there is $\mathbf{b} \in \mathrm{AP}_{\alpha+1}$ such that either $A \in \mathrm{fil}\langle \mathcal{D} \cup \mathbf{b}\rangle$ or $I \setminus A \in \mathrm{fil}\langle \mathcal{D} \cup \mathbf{b}\rangle$.*

Proof. Let $X = A$ if $\emptyset \notin \mathrm{fil}\langle \mathcal{D} \cup \mathbf{a} \cup \{A\}\rangle$, otherwise let $X = I \setminus A$. By the hypothesis that $(I, \mathcal{D}, \mathcal{G}_*)$ is (λ, μ)-good, we may choose a partition $\langle A_{f_i} : i < \mu\rangle$ of $\mathcal{P}(I)/\mathcal{D}_1$ such that each $f_i \in \mathrm{FI}(\mathcal{G}_*)$. Let $\mathcal{G}_\mathbf{b} = \mathcal{G}_\mathbf{a} \cup \bigcup\{\mathrm{dom}(f_i) : i < \mu\}$, and let $\mathcal{A}_\mathbf{b} = \mathcal{A}_\mathbf{a} \cup \{X\}$. This suffices. □

Proof of Theorem 3.2. We now prove the theorem.

Proof. (of Theorem 3.2) Without loss of generality, $|M| \le 2^\lambda$.

Let $\langle C_\alpha : \alpha < 2^\lambda\rangle$ enumerate $\mathcal{P}(I)$. Let $\langle \overline{h}^\alpha, \eta^\alpha : \alpha < 2^\lambda\rangle$ enumerate all $\eta \in {}^\lambda 2$ and all $\overline{h} = \langle h_i : i < \lambda\rangle$ with $h_i \in {}^I M$, with each such pair appearing 2^λ times in the enumeration.

We build the ultrafilter by induction on $\alpha \le 2^\lambda$. That is, we choose $\mathbf{a}_\alpha \in \mathrm{AP}_\alpha$ by induction on $\alpha < 2^\lambda$ such that:

1. $\beta < \alpha \implies \mathbf{a}_\beta \le \mathbf{a}_\alpha$
2. if $\alpha = 2\beta + 1$ then either $C_\beta \in \mathrm{fil}\langle \mathcal{D} \cup \mathbf{a}_\alpha\rangle$ or $I \setminus C_\beta \in \mathrm{fil}\langle \mathcal{D} \cup \mathbf{a}_\alpha\rangle$
3. if $\alpha = 2\beta + 2$ and $(\mathbf{a}_{2\beta+1}, \overline{h}^\beta, \eta^\beta)$ satisfies Claim 3.5(A), then there is $\overline{B}$ which, along with $\mathbf{a}_\alpha$, satisfies Claim 3.5(B).
4. if α is a limit ordinal then $\mathbf{a}_\alpha$ is the least upper bound of $\{\mathbf{a}_\beta : \beta < \alpha\}$ in the natural partial order, i.e. given by taking the union in both coordinates.

The odd inductive steps are given by Claim 3.8 above, and clearly ensure that $\mathrm{fil}\langle \mathcal{D} \cup \mathbf{a}_{2^\lambda}\rangle$ is an ultrafilter. The even inductive steps are given by Claim 3.5 above, and Corollary 3.7 ensures saturation.

Thus letting $\mathcal{D}_1 = \mathrm{fil}\langle \mathcal{D} \cup \mathbf{a}_{2^\lambda}\rangle$, we complete the proof. □

Appendix

We include some definitions used mainly in §1.3. For details, see [14] §1.

Definition 3.9. *(Shelah [20] Definition III.3.5)* Let $\mathcal{D}$ be an ultrafilter on λ.

$$\mu(\mathcal{D}) := \min\left\{\prod_{t<\lambda} n_t/\mathcal{D} \,:\, n_t < \aleph_0, \prod_{t<\lambda} n_t/\mathcal{D} \geq \aleph_0\right\}$$

be the minimum value of the product of an unbounded sequence of cardinals mod $\mathcal{D}$.

Definition 3.10. *(Good for equality, Malliaris [12])* Let $\mathcal{D}$ be a regular ultrafilter. Say that $\mathcal{D}$ is *good for equality* if for any set $X \subseteq N = M^I/\mathcal{D}$, $|X| \leq |I|$, there is a distribution $d : X \to \mathcal{D}$ such that $t \in \lambda, t \in d(a) \cap d(b)$ implies that $(M \models a[t] = b[t]) \iff (N \models a = b)$.

Definition 3.11. *(Lower cofinality,* $\operatorname{lcf}(\kappa, D)$*)* Let D be an ultrafilter on I and κ a cardinal. Let $N = (\kappa, <)^I/\mathcal{D}$. Let $X \subset N$ be the set of elements above the diagonal embedding of κ. We define $\operatorname{lcf}(\kappa, D)$ to be the cofinality of X considered with the reverse order.

Acknowledgement: Malliaris was partially supported by NSF grant DMS-1001666 and by a Gödel fellowship. Shelah was partially supported by Israel Science Foundation grant 1053/11. This is paper number 1009 in Shelah's list of publications.

Thanks to the anonymous referee for useful comments which improved the article's presentation.

Bibliography

[1] C. C. Chang and H. J. Keisler, *Model Theory*, North-Holland, 1973.
[2] A. Dow, "Good and OK ultrafilters." Trans. AMS, Vol. 290, No. 1 (July 1985), pp. 145-160.
[3] R. Engelking and M. Karłowicz, "Some theorems of set theory and their topological consequences." Fund. Math. 57 (1965) 275–285.
[4] R. Grossberg, J. Iovino, O. Lessmann, "A primer of simple theories." Arch. Math. Logic 41 (2002), no. 6, 541–580.
[5] H. J. Keisler, "Good ideals in fields of sets." Annals of Math. (2) 79 (1964), 338–359.
[6] H. J. Keisler, "Ultraproducts which are not saturated." J. Symbolic Logic 32 (1967) 23–46.
[7] S. Kochen, "Ultraproducts in the theory of models." Annals of Math. (2) 74, No. 2 (1961), pp. 221–261.
[8] K. Kunen, "Ultrafilters and independent sets." Trans. Amer. Math. Soc. 172 (1972), 299–306.

[9] M. Malliaris, Ph. D. thesis, University of California, Berkeley (2009).
[10] M. Malliaris, "Realization of φ-types and Keisler's order." Ann. Pure Appl. Logic 157 (2009), no. 2-3, 220–224.
[11] M. Malliaris, "Hypergraph sequences as a tool for saturation of ultrapowers." Journal of Symbolic Logic, 77, 1 (2012) 195–223.
[12] M. Malliaris, "Independence, order and the interaction of ultrafilters and theories." Ann Pure Appl Logic, 163, 11 (2012) 1580–1595.
[13] M. Morley, "Categoricity in power." Trans. Amer. Math. Soc. 114, No. 2, (1965), 514–538.
[14] M. Malliaris and S. Shelah, "Constructing regular ultrafilters from a model-theoretic point of view." To appear, Trans. Amer. Math. Soc.
[15] M. Malliaris and S. Shelah, "Model-theoretic properties of ultrafilters built by independent families of functions." To appear, J. Symbolic Logic.
[16] M. Malliaris and S. Shelah, "Cofinality spectrum theorems in model theory, set theory, and general topology." Submitted. http://arxiv.org/abs/1208.5424.
[17] M. Malliaris and S. Shelah, "A dividing line within simple unstable theories." Advances in Math 249 (2013) 250–288.
[18] S. Shelah, "Simple unstable theories." Ann. Math. Logic, 19, 177–203 (1980).
[19] S. Shelah, *Classification Theory*, North-Holland, 1978.
[20] S. Shelah, *Classification Theory and the number of non-isomorphic models*, rev. ed. North-Holland, 1990.
[21] S. Shelah, "Toward classifying unstable theories." Annals of Pure and Applied Logic 80 (1996) 229–255.

Ilkka Niiniluoto

Constructive Realism in Mathematics

Ilkka Niiniluoto: Department of Philosophy, History, Culture and Art Studies, University of Helsinki, Finland, e-mail: ilkka.niiniluoto@helsinki.fi

1 Introduction

Traditional classifications of the main schools in the philosophy of mathematics are based upon two questionable presuppositions. First, it is assumed that a realist, who wishes to defend objective truth values of mathematical statements, has to be either a Platonist or a physicalist. Secondly, a constructivist, who regards mathematical entities as human constructs rather than pre-existing objects, has to be either a subjective mentalist or an objective idealist. In contrast to these alternatives and their many variants, this paper argues that it is possible to be a constructivist and a realist at the same time. Mathematical entities are public creations in the human-made domain of abstract artifacts; this domain Karl Popper called World 3. They are not completely transparent to their makers, so that mathematical constructs are full of open problems for us to study. The applicability of mathematics to physical, mental, and social reality can be explained by the Theory of Measurement, which also helps to show what is wrong and what is right with the empiricist approaches to mathematical knowledge.

2 Philosophy of Mathematics: Three Main Problems

The philosophy of mathematics is a rich specialty with its own experts and publications - comparable to fields like the philosophy of physics, the philosophy of biology, the philosophy of medicine, the philosophy of technology, and the philosophy of the social sciences. It draws significant inspiration from the developments of mathematics and the often surprising results of mathematical logic (see Hintikka, 1969). While metamathematics, or the application of logical and mathematical tools in the study of the foundations of mathematics, can be regarded as a branch of mathematics (see Rasiowa and Sikorski, 1968), the philosophy of mathematics is a subdiscipline of philosophy. By using general concepts and approaches of philosophy, it investigates problems of existence, truth, and knowledge in relation to mathematics. Thereby the philosophy of mathematics has traditionally concentrated on three main problems.

First, the *ontological* question concerns the existence of mathematical entities. Here the main issue is the pre-existence vs. construction of mathematical entities. The former view claims that mathematical objects and structures exist objectively, independent of human activities. Already Plato in his theory of forms argued that mathematical objects are abstract, eternal, indestructible, and not part of space and time. Stewart Shapiro (2005), p. 6, calls this *realism in ontology*, but we shall see that the realists have proposed other kinds of objective existence besides Platonism. The most popular alternative has been the naturalist reduction of the objects of mathematics to physical phenomena (see Maddy, 1997). The *constructivists* suggest that mathematical entities exist only as creations of mathematical activities. *Ontological anti-realists* deny the existence of specific mathematical objects altogether. One form of such anti-realism claims that mathematical entities are fictions (see Leng, 2010).

Secondly, the *semantical* question concerns the nature of mathematical truth. Here the main issue is the contrast between proof and truth. What Shapiro calls *realism in truth-value* claims that mathematical statements have truth-values independently of the minds, languages, and conventions of the mathematician. This view is typical to semantical and model-theoretical approaches to mathematical theories. Various kinds of *semantical anti-realists* reduce the assessment of mathematical statements to syntactical considerations of their provability in some formal system or their assertability conditions (see Dummett, 1977).

Thirdly, the *epistemological* question concerns the nature of mathematical knowledge. Here the main issue is between a priori vs. empirical knowledge. The former alternative was defended by Plato's rationalist account of non-sensuous intuition, but later it has been supported by doctrines of conceptual or analytic truth, i.e. truths based on definitions and meanings. J. S. Mill's 1843 thesis that the truths of arithmetic and geometry are inductive generalizations from empirical experience (see Mill, 1906) has recently been revived by some naturalist philosophers (see Kitcher, 1983).

In this paper, I argue that the most popular positions in the philosophy of mathematics ignore an important alternative that I call "constructive realism" (cf. Niiniluoto, 1984, 1992). My inspiration comes from Karl Popper's (1972) ontological doctrine of World 3. Popper is not mentioned at all in the recently published *Oxford Handbook of Philosophy of Mathematics and Logic* (2005), whose editor Stewart Shapiro promises "a clear picture of the state of the art".

Before outlining my view in the next sections, let me give you a preliminary hint of its direction by quoting Jouko Väänänen, the brilliant mathematical logician, model-theorist and set-theorist. In two papers in Finnish, he has expressed a highly interesting idea of mathematical reality. In a paper on Georg Cantor's concept of infinity, Väänänen notes that Cantor is usually interpreted as a Platonist who assumes fixed truth about an eternal and unchanging domain of mathematical objects, but instead Cantor in fact promotes mathematical practice as free intellectual activity which creates first an intrasubjective reality and then a developing transsubjective reality (Väänänen, 1982). In a paper on mathematical objects, Väänänen suggests that classical realism can be defended by assuming that mathematical entities are constructed by a "superbeing" or a "superman", capable of infinite generalizations of finite human constructions (Väänänen, 1984). As an example of an infinite construction, sufficient for the purposes of classical mathematics, he mentions the cumulative hierarchy of sets, obtained by successively applying to the empty set $\emptyset$ iterations of the power set operation and unions (cf. Väänänen, 2012).

3 Abstract Entities: Three Main Schools

When the philosophy of mathematics emerged as a specialty, largely inspired by the crisis in the foundations of mathematics in the first years of the Twentieth century, the field was divided by three main schools: Logicism, Formalism, and Intuitionism. This division, as the "received view", was well formulated in 1931 in an issue of *Erkenntnis*, the new journal of the logical empiricists, with papers by Rudolf Carnap, Johann von Neumann, and Arend Heyting. It was reinforced in 1964 in the first edition of the authoritative anthology *Philosophy of Mathematics*, edited by Paul Benacerraf and Hilary Putnam, with classical papers by the logicists Gottlob Frege and Bertrand Russell, the formalists David Hilbert and Haskell B. Curry, and the intuitionist L. E. J. Brouwer.

Logicism reduces mathematics to logic and set theory. Logical truths are tautologies without any factual content. They can be known a priori, as they are analytic truths, consequences of axioms and definitions. In contrast to axiomatic pure mathematics, which is analytic and a priori, interpreted mathematical theories (like applied geometry) are branches of natural science with synthetic a posteriori truths (see Nagel, 1961). Logicism as such can be neutral with respect to ontological questions, but some of its leading proponents (like Frege and Kurt Gödel) were strongly inclined to Platonism.

Formalism treats logic and mathematics as games of manipulating syntactical signs. From signs one can form expressions, and from sequences of expressions one can form proofs. Rigorous study of mathematical theories as formal deductive systems, including theorems about their consistency, can be given in proof theory.

Intuitionism contends that mathematics studies mental or introspective constructions in the mind of the mathematician. Mathematical objects are mind-dependent, and their existence presupposes a successful construction. Brouwer's philosophy had Kantian and phenomenological roots, but later intuitionists have formulated non-classical systems of logic and mathematical reasoning. In intuitionistic mathematics, all infinity is potential, and infinite classes are replaced with "laws" or processes generating "choice sequences".

It is interesting to observe that these three schools are closely parallel to three approaches in the great medieval controversy about universals (cf. Armstrong, 1978). The *realists* accepted universals (e.g., whiteness) as independently existing entities. Some of them followed Plato in placing such universals in a transcendent realm (*ante rem*), and others agreed with Aristotle in thinking that universals are immanent forms of objects, always instantiated in things (*in rebus*). The *nominalists*, who accepted only the existence of particular things, treated universals as mere words in a written or spoken language (*flatus voci*). The *conceptualists* claimed that universals have a mental existence as concepts (*in intellectu*).

4 Popper's Three Worlds

A new way of classifying ontological or metaphysical positions can be based on Karl Popper's terminology of three worlds (see Popper, 1972, 1974). *World 1* consists of physical things, events, and processes in space and time, including lawlike relations between such entities. This is the domain of inorganic and organic nature, studied by physics and biology. *World 2* includes subjective mental states and events (e.g., beliefs, emotions, and volitions) in individual human minds. This is the domain of human psyche, studied by psychology and cognitive science. *World 3* contains the products of human social action, such as cultural objects, social institutions, and abstract entities like concepts, propositions, arguments, theories, problems, and numbers. This domain is studied by cultural and social sciences, logic and mathematics.

Three monistic metaphysical doctrines can now be identified (cf. Niiniluoto, 1999). *Materialism* in its radical eliminative form claims that everything real belongs to World 1. Reductive materialism states that reality is reducible to World 1 entities and their complexes. Eliminative and reductive materialism are forms of *physicalism*. Emergent materialism takes World 1 as primary, but admits that sufficiently complex material systems may have "emergent" non-physical properties. *Subjective idealism* makes parallel claims about World 2. Its eliminative and reductive forms constitute the doctrine of *spiritualism*, but emergent idealism is also a possible view. *Objective idealism* in its classical versions has taken some non-material and non-subjective ideas (such as Plato's forms, thoughts of supernatural gods, and Hegel's objective spirit) as the ultimate source of all being, but more mundane variations could replace them by some abstract World 3 entities. Idealist views (e.g., phenomenalism, social constructivism) are ontologically antirealist, as they treat the material reality in World 1 as mind-dependent or human-made.

Besides such monistic views, *dualist* ontologies may accept World 1 and World 2 as two independent (but possibly causally interacting) domains of reality. This is the Cartesian tradition initiated by Descartes. Another kind of dualism could accept World 1 and World 3. Trialist doctrines accept the reality of all three worlds.

When Popper introduced his theory of the third world in a paper with a teasing title "On the Theory of the Objective Mind", Mario Bunge was shocked that in 1967 Popper had a sudden "conversion to objective idealism" (Bunge, 1981, p. 138). However, Popper's position can be interpreted as a form of emergent materialism (see Niiniluoto, 1999): in his evolutionary account he sees Worlds 2 and 3 as historical products of World 1. Neither of them could exist without the material World 1, but they have achieved a relatively independent status by being able to influence material entities by a causal "feedback mechanism".

Popper mentions Bernard Bolzano's "sentences in themselves" (*Sätze an sich*) and Frege's "thoughts" (*Gedanke*) as anticipations of his World 3. Frege stated in 1918 that "thoughts are neither things in the external world nor ideas", so that "a third realm has to be recognized" (Frege, 1984). So for Popper propositions as the "objective contents of thought" are paradigmatic examples of World 3 entities. Frege's position about propositions is Platonist, however, and Popper wants to dissociate his "man-made and changing" World 3 from Plato's "divine and unchanging third world". World 3 entities emerge as by-products of human social action, especially the creation of symbolic languages. Popper's theory of World 3 is thus a kind of "poor man's Platonism": as no higher being or god has created abstract objects for us, man must himself make them (see Niiniluoto, 1992).

Popper does not refer to Carnap's *Aufbau*, which in 1928 gave a three-fold classification of objects into physical, psychological, and cultural (see Carnap, 1967, cf. Niiniluoto, 2006). Cultural objects include concrete artifacts (e.g., books, paintings, coins) which have a material kernel in World 1, but also non-physical relational properties (e.g., meaning, value, price) through their relations to human practices. Cultural objects include also abstract artifacts (e.g., poems, symphonies) which can be copied, recorded, and reproduced by their embodiments that belong to Worlds 1 and 2. Using Carnap's terms, a symphony can be *documented* in World 1 by notes on a score, acoustic waves, recording tapes, or compact discs, and it can be *manifested* in World 2 as the composer's idea or the listener's experience. A virtue of the theory of World 3 is that it allows us to say that there is one and only one Beethoven's Eroica Symphony: it would be quite arbitrary to identify this masterpiece with any of its multiple documentations and interpretations in Worlds 1 and 2. We can also see that abstract cultural objects in World 3 exist in time but not in space: they are created at a certain moment of time, and they could be destroyed if all of their documentations and memories disappeared from Worlds 1 and 2.

Popper defends the reality of World 2 and 3 entities by their causal powers. For example, mental states influence our bodily behavior by "downward causation", and coins function as money via existing social institutions. It is less evident that purely abstract entities could have a causal impact on us. For such entities one can apply Charles Peirce's "medieval" definition of reality: real things have characters that "are independent of what anybody may think them to be" (see Peirce 1931-35, § 5.405). Popper's own example is prime numbers: the largest prime number known in 2013 is $2^{57885161} - 1$. Probably no one had ever thought about this number or written down its 17,425,170 digits, but still it had (even before it was found by The Great Internet Mersenne Prime Search on January 25, 2013) "objectively" the property of being prime. In the same spirit, Haim Gaifman (2012) characterizes the realist ontology of mathematics in terms of "factually meaningful questions" which have determinate answers, "independent of our knowing it or our abilities to find it".

Peirce's criterion is not unproblematic in sophisticated branches of mathematics like set theory. The definition of measurable cardinals was given Stanislaw Ulam in 1930, and a great number of theorems about them have been proved. For example, Ulam and Tarski showed that measurable cardinals are strongly inaccessible. Further, if κ is measurable, then κ is the κth weakly inaccessible cardinal. Solovay proved in 1967 that ZF together with the axiom of determinacy AD implies the measurability of the cardinal $\aleph_1$. On the other hand, it is known that the existence of measurable cardinals is independent of the axioms of ZFC. Scott proved in 1961 that the axiom of constructibility **V**=**L** implies the non-existence of measurable cardinals, but it is still unknown whether their existence is compatible with the axioms of ZFC. So in a sense, after 70 years of intensive studies, we still don't know whether these entities exist or not (see Kanamori, 2003). One possible conclusion for a realist, compatible with Peirce's criterion, is that mathematicians are able to rigorously investigate alternative universes of set theory – Hamkins (2012) defends a Platonist version of this "set-theoretic multiverse".

Peirce himself used his criterion to distinguish reality and fiction, and by the same principle we can distinguish real public entities in World 3 and merely fictional objects (e.g., Santa Claus and Anna Karenina). Thus, World 3 differs from "Meinong's jungle" of all possible conceivable objects (see Routley, 1980; cf. Niiniluoto, 2011).

5 Mathematics and World 3

The main schools in the philosophy of mathematics have their typical locations for mathematical entities in different categories of reality. This concerns not only the old schools of Logicism, Intuitionism, and Formalism, but also the new approaches well described in Shapiro's *Handbook* (2005).

The appeal of Platonism is still strong. In an influential article, Paul Benacerraf (1965) argued that arithmetic is not a science concerned with particular objects, the numbers, but instead studies the abstract structure that all progressions of a certain kind have in common. This view is called "eliminative structuralism" (see Parsons, 1990). Its weaker modal form, which asserts only the possibility of such structures, is defended by Geoffrey Hellman (1989). But instead of getting rid of mathematical entities, some realist structuralists have suggested that arithmetic studies numbers as "positions" and "patterns" (Resnik, 1997) or the natural number structure as an *ante rem* universal (Shapiro, 2005, p. 22). Neither of them has considered World 3 realism as an alternative to Platonism.

Intuitionists typically locate mathematical objects as mental constructions in the subjective World 2. But Brouwer's own account of mathematics as the temporally cumulating work of "the creative subject" (see Dummett, 1977) has a whiff of objective idealism. A limiting case of constructivism without ontological commitments is given by Charles Chihara (1990), whose "constructibility quantifiers" tell only what kinds of entities are possible to construct.

Formalism places numerals and other linguistic expressions as material signs in World 1. New naturalist and empiricist approaches, often influenced by W. V. O. Quine, have tried to analyze mathematics without abstract entities. The wish to defend realism without Platonism has led to physicalist accounts which in some sense reduce mathematical truth to scientific claims about World 1 (see Irvine, 1990; Maddy, 1997). Armstrong's (1978) physicalism accepts Aristotelian universals, and nominalism is also criticized by Burgess and Rosen (1998).

Some physicalists have tried to get rid of mathematical ontology altogether. Hartry Field's (1980) fictionalism argues that mathematical terms are dispensable in scientific theories, so that '2 + 2 = 4' can be taken to be true pretty much in the same sense as 'Oliver Twist lived in London'. But Field (1989) advocates a "substantival" view of space-time regions and points in his physical ontology (even though regions need not be conceived as sets of points), which makes it easy to dispense with any additional special mathematical ontology of real numbers.

Another version of fictionalism is defended by Mary Leng (2010). She starts from Quine's and Putnam's argument that naturalism, confirmational holism, and the indispensability of mathematical entities in successful physical theories leads to the conclusion that we ought to believe in the existence of mathematical objects. Instead of rejecting indispensability like Field, Leng attacks the holistic principle that "the confirmation our theories receive extends to all their statement equally". She concludes that mathematical statements are less confirmed than physical claims, so that it is sufficient merely to "pretend" or "make-believe" that there are mathematical objects satisfying the axioms of an empirically successful physical theory. But Leng's rejection of the egalitarian formulation of holism does not capture Quine's idea of a "web of belief" where logical and mathematical truths belong to its "interior" which is less susceptible to rejection by empirical evidence than observational statements in "the periphery" (see Quine and Ullian, 1970).

On the whole contemporary philosophy of mathematics seems to be based on two tacit assumptions. First, it is assumed that a realist, who wishes to defend objective truth values of mathematical statements, has to be either a Platonist or a physicalist. Secondly, a constructivist, who regards mathematical entities as human constructs rather than pre-existing objects, has to be either a subjective mentalist or an objective idealist. These assumptions ignore the possibility that one can be a realist and constructivist at the same time. This alternative is provided by the interpretation of mathematical entities as public creations in the human-made domain of abstract artifacts or the Popperian World 3 (see Niiniluoto, 1984, 1992).

Popper himself expressed explicitly his constructivism: "Pace Kronecker, I agree with Brouwer that the sequence of natural numbers is a human construction" (Popper, 1972, p, 118). Still, the third world is real or "largely autonomous, though created by us". However, Popper does not assume with Brouwer that our constructions are completely transparent to us. This is in sharp contrast to the Cartesian assumption that the mind has privileged access to its own contents. Kant expressed this tradition of "maker's knowledge" by the principle that we can "cognize in things *a priori* that which we ourselves place there" (cf. Niiniluoto, 1999, p. 88), For Brouwerian intuitionists and anti-realists the idea of unknown truths is an anathema, but for Popperian realists World 3 "transcends its makers", since "it creates its own autonomous problems" (e.g, the existence and distribution of prime numbers), and these open problems are left for us as a challenge to discover and study.

Popper does not explain in detail how such constructions are made, but we know that infinite mathematical structures (like **N** and **R**, Zermelo's cumulative hierarchy of sets **V**, Gödel's constructible hierachy of sets **L**, and Barwise's admissible sets with urelements) as World 3 entities can be defined by finite descriptions (e.g., recursive definitions and set-theoretical constructions with transfinite iterations). (See Barwise, 1975.) (This fact is not appreciated by the "finitist" arguments of some followers of Ludwig Wittgenstein, as pointed out in Niiniluoto (1999), p. 266.) Such constructions can often be given in several alternative ways - for example, the natural number 2 can be represented set-theoretically by $\{\emptyset, \{\emptyset\}\}$ (Zermelo) or $\{\{\emptyset\}\}$ (von Neumann) - but still in World 3 we have number 2 as a unique entity with its relational properties, in the same way as there is only one Eroica symphony in spite of its numerous documentations and interpretations. Similarly, the same structure of real numbers can be introduced either by Dedekind cuts or converging Cauchy sequences of rational numbers. These constructions and abstractions guarantee realism in truth value: mathematical statements about such World 3 structures have objective but possibly unknown truth values in the sense of Tarskian model theory, and thus they satisfy the principles of classical logic.

Donald Gillies (1990, 2010) has given a sympathetic account of Popper's philosophy of mathematics. He calls it "constructive Platonism", while I prefer to call it "constructive realism". The terminology of Gillies is based upon his own "constructive Aristotelianism". This variant of realism claims that numbers are embodied or instantiated in the material World 1 (and as well in Worlds 2 and 3). For example, the set of moons of Mars is an embodiment of the number 2. Thus, numbers are *in rebus* rather than *ante rem*. Popper instead insists that there are "unembodied" numbers in World 3, "more than will ever be pronounced by men, or used by computers", "materialized in libraries" or "grasped or understood by some people" (see Popper, 1974).

In my view, it is important to be cautious in talking about "unembodied" entities in World 3, in order to avoid collapsing this account back to Platonism (see Niiniluoto, 2006). Gillies may be right in saying against Popper that mathematical problems come into existence only when they are formulated. Similarly, World 3 should contain only actually composed symphonies, not all possible and conceivable musical compositions. On the other hand, it makes sense to claim that the class of natural numbers **N** as a well-defined totality belongs to World 3, and thereby the same is true about all of its known and still unknown members. Against Gillies' version of Aristotelianism, it seems clear that all members of the infinite set **N** will never be actually realized in World 1 (e.g., by paper and pencil, or represented by a computer) or in World 2 (by calculations or thoughts in the human mind). In fact, if Gillies wished to be a true Aristotelian, he should agree, as Aristotle himself in his *Physics* argued that infinity exists only potentially and never actually.

Note, however, that some physicalists, like Field (1989), would accept the space-time continuum in their physical ontology, so that for them World 1 could include counterparts of all natural and real numbers as well. Recall Galileo's statement that "the book of nature" is written in the language of mathematics. Against this view, which is a kind of naturalized Platonism, I think it is more appropriate to regard real numbers and functions as mathematical tools in World 3 that can be used in idealized descriptions of physical phenomena (see the remarks on measurement below). This agrees with Mill's remark that geometrical objects like "points without magnitude", "lines without breadth" and "perfect squares" exist "neither in nature nor in the human mind" (see Mill, 1906, p. 147-148).

Gillies (2010) also proposes to reduce the truth of mathematical statements to the material relations of their embodiments in World 1. He notes the implication that there are no truths about transfinite cardinals, even though some such correct claims are theorems of set theory. However, it follows as well that there are many simple arithmetical equations that fail to be true, since they concern numbers that are larger than those embodied in World 1. Thus, the position of Gillies concerning arithmetic is not sufficient to defend realism in truth value. The doctrine of World 3 instead allows us say that all statements in arithmetic have a truth value in the standard model **N** - and, by Gödel's incompleteness theorem, any recursive axiomatization of arithmetic fails to have all of these truths as its theorems.

6 Mathematical Knowledge

Our version of constructive realism helps to explain why mathematical knowledge is a priori and analytic, independent of sense experience. While scientific knowledge in most cases is based upon causal interaction between the subject and object of cognition (see Niiniluoto, 1999, p. 94), this causal theory of knowledge need not be applied in the case of abstract entities in mathematics (*pace* Benacerraf, 1965). Popper's suggestion that our grasping of a World 3 object means its "making" or "re-creation" is not at all plausible with respect to most concrete and abstract artifacts (e.g., painting, symphonies, social facts), but for mathematical entities it fits quite well. When a mathematician wants to solve some problem about a structure like **N**, this mathematical domain has to reconstructed to extent that it can be studied. The principles of such constructions can be included among the definitions and axioms which serve as premises of mathematical proofs. Such proofs may concern singular statements about particular structures (e.g., $2 + 3 = 5$), general statements about particular structures (e.g., $m + n = n + m$ for all m, n in **N**), and generalizations over types of structures (e.g., every metric space is a Hausdorff space).

The ultimate basis of mathematical knowledge in deductive proofs, as emphasized both by the logicist and formalist traditions, is compatible with the fact that sometimes non-demonstrative inferences (abduction and induction) may have a role in the discovery and justification of mathematical claims, just as George Polya and Imre Lakatos have pointed out (see Niiniluoto, 1992). In such cases, we may be able to prove some instances of a claim (e.g. Goldbach's Conjecture, Riemann's Hypothesis) without a proof of the generalization. It is also possible that, under certain conditions, empirical observations give support to purely mathematical claims. To see why, we need an account of the problem of why mathematics is applicable to reality.

In his otherwise illuminating paper on the applicability problem, Mark Steiner (2005) does not mention at all the concept of measurement, which in my view is the key to the relations between empirical relations in World 1 and mathematical facts in World 3 (see Niiniluoto, 1992, 2012). In fact, mathematics can be applied to all the three worlds, but here I restrict my remarks to quantitative descriptions of physical reality. The modern Theory of Measurement (or metrization) gives representation theorems which justify a link from World 1 to World 3 , and thereby justifies the assignment of *quantitative attributes* to a class E of physical objects (see Krantz et al., 1971). However, representation theorems assume that the mathematical structures (such as **N** or **R**) are already available, so that they don't give arguments for the existence of mathematical entities.

For example, the axioms of *extensive measurement* give sufficient and necessary conditions for a comparative relation ▶ to be represented by a quantitative real-valued function $m : E \to \mathbf{R}$ such that, for any two objects a and b in E, a ▶ b iff $m(a) \geq m(b)$ and, for the combined object $a \circ b$, we have $m(a \circ b) = m(a) + m(b)$. If a and b are physical objects, and a ▶b means that a is at least as low as b in an equal arm balance, then m is the quantitative mass function. This representation theorem justifies quantitative assertions like

$$2\,\text{kg} + 3\,\text{kg} = 5\,\text{kg}. \quad (1)$$

The quantitative concept of length can be justified in the same way. The scale of measurement determines what kinds of mathematical operations (such as addition and multiplication) with quantities are allowed. Such applications of the quantitative method in natural science thus presuppose that certain factual assumptions are correct.

A similar justification can be given to factual statements like

2 pebbles and 3 pebbles together are 5 pebbles. (2)

This presupposes that the objects satisfy factual regularity assumptions: they do not lose their identity, split or merge with other objects, when collected as an aggregate. The class of such aggregates, together with the operation $\circ$ of heaping up, and the comparative relation “at least as many members as” (defined independently of counting with numbers) constitutes an extensive system. Then the function m, whose existence is guaranteed by the representation theorem, is the cardinality of the aggregate. The principle (2) is valid for pebbles, making empirical “pebble arithmetic” a possibility (see Mill, 1906; Bigelow, 1988), but not correct for water drops. Factual statements like (1) and (2) differ from the genuine arithmetical equation

$$2 + 3 = 5, \tag{3}$$

which is a provable claim about numbers in World 3 without any factual assumptions. Frege was thus right in arguing that Mill’s empiricist interpretation of mathematics confuses applications of arithmetic with the pure mathematical proposition itself (Frege, 1959, p. 13).

Empirical arithmetic may concern also the operations of physical devices, such as pocket calculators and computers. Factual statements like

According to my calculator, $741 \times 852 = 631\,332$ (4)

may give empirical confirmation to a purely mathematical statement, if we have good reasons for believing that the calculator functions without mistakes and disturbances. Computers have also succeeded to verify important steps in mathematical proofs, as the case of the four-color theorem 4CT shows (see Tymoczko, 1986).

Similar remarks are valid on the relations of empirical and axiomatic geometry. Measurement of lengths may give approximate confirmation to Pythagoras’ Theorem, but in its exact form it can be proved by Euclid’s axioms.

7 Conclusion

Mathematics is not waiting us in the Platonist heaven or in physical reality, but constructed in the human-made World 3. Genuine mathematical truths are a priori, but some of them can be interpreted and supported by factual and empirical results. This constructive realism allows a pluralist picture of mathematics, as set theory may have alternative models or universes - besides the cumulative hierarchy as its standard model. But the construction and reconstruction of public mathematical entities, and the inquiry into the still unknown depths of such structures, is a joint enterprise which does not happen in the mind of an individual scholar. So Jouko Väänänen's vision of a superman can be understood as the scientific community of mathematicians as a collective.

Bibliography

[1] Armstrong, D. (1978), *Universals and Scientific Realism*, vols. 1-2, Cambridge University Press, Cambridge.

[2] Barwise, J. (1975), *Admissible Sets and Structures: An Approach to Definability Theory*, Springer-Verlag, Berlin.

[3] Benacerraf, P. (1965), "What Numbers Could not Be?", *Philosophical Review* 74, 47-73. Reprinted in Benacerraf & Putnam (1983), pp. 272-294.

[4] Benacerraf, P. & Putnam, H. (eds.) (1964), *Philosophy of Mathematics*, Prentice-Hall, Englewood Cliffs, NJ. (2nd ed., Cambridge University Press, Cambridge, 1983.)

[5] Bigelow, J. (1988), *The Reality of Numbers: A Physicalist's Philosophy Mathematics*, Oxford University Press, Oxford.

[6] Bunge, M. (1981), *Scientific Materialism*, D. Reidel, Dordrecht.

[7] Burgess, J. & Rosen, G. (1998), *A Subject with No Object*, Oxford University Press, New York.

[8] Carnap, R. (1967), *The Logical Structure of the World & Pseudoproblems in Philosophy*, University of California Press, Berkeley.

[9] Chihara, Ch. (1990), *Constructibility and Mathematical Existence*, Oxford University Press, Oxford.

[10] Dummett, M. (1977), *Elements of Intuitionism*, Clarendon Press, Oxford.

[11] Field, H. (1980), *Science without Numbers*, Princeton University Press, Princeton.

[12] Field, H. (1989), *Realism, Mathematics and Modality*, Blackwell, Oxford.

[13] Frege, G. (1959), *The Foundations of Arithmetic*, 2nd ed., Northwestern University Press, Evanston.

[14] Frege, G. (1984), "Logical Investigations. Part I: Thoughts", in *Collected Papers on Mathematics, Logic, and Philosophy*, ed. by B. McGuinness, Blackwell, Oxford.

[15] Gaifman, H. (2012), "On Ontology and Realism in Mathematics", *The Review of Symbolic Logic* 5, 480-512.

[16] Gillies, D. (1990), "Intutionism vs. Platonism: A 20th Century Controversy Concerning the Nature of Numbers", in F. Gil (ed.), *Scientific and Philosophical Controversies*, Fragmentos, Lisbon, pp. 299-314.

[17] Gillies, D. (2010), "Informational Realism and World 3", *Knowledge, Technology & Policy* 23: 1-2, 7-24.

[18] Hamkins, J. D. (2012), "The Set-Theoretic Multiverse", *The Review of Symbolic Logic* 5, 416-449.

[19] Hellman, G. (1989), *Mathematics without Numbers: Towards a Modal-Structural Interpretation*, Oxford University Press, Oxford.

[20] Hintikka, J. (1969 (ed.), *The Philosophy of Mathematics*, Oxford University Press, Oxford.

[21] Irvine, A. D. (ed.) (1990), *Physicalism in Mathematics*, Kluwer, Dordrecht.

[22] Kanamori, A. (2003), *The Higher Infinite: Large Cardinals in Set Theory from Their Beginnings*, Springer, Berlin.

[23] Kitcher, P. (1983), *The Nature of Mathematical Knowledge*, Oxford University Press, Oxford.

[24] Krantz, D., Luce, R., Suppes, P. & Tversky, A. (1971), *Foundations of Measurement*, vol. 1, Academic Press, New York.

[25] Leng, M. (2010), *Mathematics and Reality*, Oxford University Press, Oxford.

[26] Maddy, P. (1997), *Naturalism in Mathematics*, Clarendon Press, Oxford.

[27] Mill, J. S. (1906), *A System of Logic*, 8th edition, Longmans, Green, and Co., London.

[28] Nagel, E. (1961), *The Structure of Science: Problems in the Logic of Scientific Explanation*, Routledge & Kegan Paul, London.

[29] Niiniluoto, I. (1984), "Realism, Worldmaking, and the Social Sciences", in *Is Science Progressive?*, D. Reidel, Dordrecht, pp. 211-225.

[30] Niiniluoto, I. (1992), "Reality, Truth, and Confirmation in Mathematics - Reflections on the Quasi-Empiricist Programme", in J. Echeverria, A. Ibarra & T. Mormann (eds.), *The Space of Mathematics: Philosophical, Epistemological and Historical Explorations*, Walter de Gruyter, Berlin, pp. 60-78.

[31] Niiniluoto, I. (1999), *Critical Scientific Realism*, Oxford University Press, Oxford.

[32] Niiniluoto, I. (2006), "World 3: A Critical Defence", in I. Jarvie, K. Milford & D. Miller (eds.), *Karl Popper: A Centenary Assessment, vol. II: Metaphysics and Epistemology*, Ashgate, Aldershot, pp. 59-69.

[33] Niiniluoto, I. (2011), "Virtual Worlds, Fiction, and Reality", *Discusiones Filosoficas* 12, 13-28.

[34] Niiniluoto, I. (2012), "Quantities as Realistic Idealizations", in D. Dieks, W. Gonzalez, S. Hartmann, M. Stöltzner & M. Weber (eds.), *Probabilities, Laws, and Structures*, Springer, Dordrecht, pp. 297-306.

[35] Parsons, Ch. (1990), "The Structuralist View of Mathematical Objects", *Synthese* 84, 303-346.

[36] Peirce, C. S. (1931-35), *Collected Papers 1-6*, ed. by C. Hartshorne & P. Weiss, Harvard University Press, Cambridge, Mass.

[37] Popper, K. R. (1972), *Objective Knowledge*, Oxford University Press, Oxford.

[38] Popper, K. R. (1974), "Replies to my Critics", in P. A. Schilpp (ed.), *The Philosophy of Karl Popper*, Open Court, La Salle, Ill., pp. 961-1197.

[39] Quine, W. V and Ullian, J. S. (1970), *The Web of Belief*, Random House, New York.

[40] Rasiowa, H. & Sikorski, R. (1968), *The Mathematics of Metamathematics*, 2nd edition, Polish Scientific Publishers, Warsaw.

[41] Resnik, M.(1997), *Mathematics as a Science of Patterns*, Oxford University Press, Oxford.

[42] Routley, R. (1980), *Exploring Meinong's Jungle and Beyond*, Australian National University, Camberra.

[43] Shapiro, S. (ed.) (2005), *The Oxford Handbook in the Philosophy of Mathematics and Logic*, Oxford University Press, Oxford.

[44] Steiner, M. (2005), "Mathematics - Application and Applicability", in Shapiro (2005), pp. 625-650.

[45] Tymoczko, T. (ed.) (1986), *New Directions in the Philosophy of Mathematics*, Birkhäuser, Boston.

[46] Väänänen, J. (1982), "Äärettömyyden käsite Georg Cantorilla", in J. Oikkonen (ed.), *Katsauksia matematiikan historiaan*, Gaudeamus, Helsinki, pp. 88-101.
[47] Väänänen, J. (1984), "Matemaattiset oliot", in L. Haaparanta (ed.), *Olio*, Reports of the Department of Philosophy, University of Helsinki, Helsinki, pp. 142-146.
[48] Väänänen, J. (2012), "Second Order Logic or Set Theory?", *The Bulletin of Symbolic Logic* 18, 91-121.

Jeff B. Paris* and Alena Vencovská

The Twin Continua of Inductive Methods

Dedicated to our good friend Jouko

Abstract: After dominating the subject of Inductive Logic for over 50 years Carnap's Continuum of Inductive Methods has in the past decade had its monopoly challenged by a second continuum of inductive methods, the NP-Continuum, which is also based on arguably rational principles. Does this mean there are (at least) two distinct notions of rational or logical probability? We describe the bases and key properties of both continua.

Keywords: uncertain reasoning, inductive logic, logical probability, Carnap's continuum, NP-continuum

MSC: 03B42, 03B48, 03A10

***Corresponding Author: Jeff B. Paris, Alena Vencovská:** School of Mathematics, University of Manchester, e-mail: jeff.paris@manchester.ac.uk

1 Introduction

A good way to understand the underlying problem that Inductive Logic addresses is as follows: Imagine an agent who inhabits a first order structure M for a language L with relation symbols $R_1, R_2, \ldots, R_q$ and constants $a_1, a_2, a_3, \ldots$ but no function symbols nor equality. Everything in the universe of M is the interpretation of at least one of the a_n so we may assume that the set of a_n actually is the universe of M with each of the constants interpreted as itself. The agent knows s/he is in such a structure for L but that is the sum total of the agent's knowledge.

The question now is: Given $\theta \in SL$, where SL is the set of first order sentences of L (and for future reference $QFSL$ is the set of quantifier free sentences of L), what degree of belief, as subjective probability, should the agent give to θ being true in M? Or to take this a step further, what probability function $w : SL \to [0,1]$ should the agent adopt to reflect his/her beliefs, where by a *probability function on L* we mean here that w satisfies for $\theta, \phi, \exists x\, \psi(x) \in SL$,

(P1) $\models \theta \;\Rightarrow\; w(\theta) = 1$

(P2) $\theta \models \neg\phi \;\Rightarrow\; w(\theta \vee \phi) = w(\theta) + w(\phi)$

(P3) $w(\exists x\, \psi(x)) = \lim_{n\to\infty} w(\bigvee_{i=1}^{n} \psi(a_i))$

Although we assume that the agent has no further knowledge of M we should allow that the agent might further consider 'What probability should I give to θ were I to learn that ϕ was true in M?' The standard convention in the subject is that the agent should in that case take the resulting *conditional probability* of θ given ϕ, to be

$$w(\theta \mid \phi) = \frac{w(\theta \wedge \phi)}{w(\phi)}.$$

(A problem seems to arise here when $w(\phi) = 0$ but we can satisfactorily side step it by agreeing to the convention that an expression such as $w(\theta \mid \phi) = b$ is actually short for $w(\theta \wedge \phi) = bw(\phi)$, so since $w(\phi) = 0$ implies $w(\theta \wedge \phi) = 0$, we can in the case $w(\phi) = 0$ assign $w(\theta \mid \phi)$ an arbitrary value, for the sake of argument say 1.)

Of course the agent is free to choose *any* probability function to quantify his/her beliefs. However the name of the game here is that the agent should make a choice which is *logical* or *rational*. What that means is that w should be required to satisfy certain principles, principles which we feel it is logical or rational to obey, or more pointedly perhaps which it would be *irrational* to flout.

For example suppose that $\theta(a_{k_1}, a_{k_2}, \ldots, a_{k_n}) \in SL$, where the notation is intended to indicate that all the constant symbols appearing in θ are amongst the (distinct) $a_{k_1}, a_{k_2}, \ldots, a_{k_n}$, and we form $\theta(a_{j_1}, a_{j_2}, \ldots, a_{j_n})$ by the obvious simultaneous substitution. Then since the agent has no material reason for thinking that $\theta(a_{k_1}, a_{k_2}, \ldots, a_{k_n})$ is any more likely to be true in M than $\theta(a_{j_1}, a_{j_2}, \ldots, a_{j_n})$ s/he should give them each equal probability.

This principle is, unsurprisingly, referred to as Constant Exchangeability, Ex, and it is so widely assumed in this subject that from now on we will assume that all the probability functions we consider satisfy it without further mention. One simplifying consequence of this is that in stating principles we can take the 'arbitrary' $a_{k_1}, a_{k_2}, \ldots, a_{k_n}$ to be $a_1, a_2, \ldots, a_n$ without any loss of generality.

A second principle that we shall assume from the start, both because of its evident rationality and because its presence simplifies the results we wish to state later is:

Regularity, Reg

If $\theta \in QFSL$ is consistent then $w(\theta) > 0$.

The justification here is that given our agent's supposed lack of any specific knowledge about the ambient structure M it would be irrational of him/her to assign probability zero to any consistent quantifier free sentence, in other words to assert that as far as s/he was concerned it was impossible for that sentence to hold.[1]

Ex and Reg were principles that W.E.Johnson, [12], and R. Carnap, [1], [2], [3], [4], [5], founders of the subject as a branch of mathematical logic, both employed, Johnson in a period in the late 1920's early 1930's and Carnap, independently, a decade or so later.

The principles Ex and Reg of course give no particular significance to the arity of the relation symbols $R_1, R_2, \ldots, R_q$. However, both Johnson and Carnap concentrated almost exclusively on *unary* L, that is when the R_i were all predicates or unary relations. (Indeed it has only been in this millenium that significant advances have been made in *Polyadic Inductive Logic*, see for example [13].) Like these founding fathers we shall henceforth assume that the R_i are all unary.

2 Carnap's Continuum of Inductive Methods

A third principle which both Johnson and Carnap proposed became in time known as *Johnson's Sufficientness Postulate, JSP* after a suggestion by I.J.Good (it started life as Johnson's Sufficiency Postulate but that threatened being confused with the notion of a 'sufficient statistic'). In order to state this principle, and for future use, we first need to introduce a little notation.

Since all our relation symbols are now being taken as unary it follows that everything there is to know about a single constant a_i is given by the atom it satisfies, where the *atoms* of L are those 2^q formulae $\alpha_1(x), \alpha_2(x), \ldots, \alpha_{2^q}(x)$ of the form

$$\pm R_1(x) \wedge \pm R_2(x) \wedge \ldots \wedge \pm R_q(x)$$

($\pm R(x)$ stands for $R(x)$ or $\neg R(x)$). In turn everything there is to know about $a_{k_1}, a_{k_2}, \ldots, a_{k_n}$ is determined by the *state description*

$$\bigwedge_{i=1}^{n} \alpha_{h_i}(a_{k_i})$$

that they satisfy.

1 The question of whether Regularity should also hold for *quantified* sentences, in particular consistent sentences of the form $\forall x\, \theta(x)$ with θ quantifier free, has sparked a considerable literature, as a lead in see [6, page 87], [10].

Johnson's Sufficientness Postulate, JSP

$$w\left(\alpha_j(a_{n+1}) \mid \bigwedge_{i=1}^{n} \alpha_{h_i}(a_i)\right)$$

should only depend on n and $m_j = |\{i \mid h_i = j\}|$, i.e. the number of times that $\alpha_j(x)$ is instantiated in the state description $\bigwedge_{i=1}^{n} \alpha_{h_i}(a_i)$.

One might find an intuitive justification for this postulate by imagining that M was formed by repeated picking (with replacement) copies of atoms $\alpha_{h_1}(x), \alpha_{h_2}(x), \alpha_{h_3}(x), \ldots$ from an urn and requiring that

$$M \vDash \alpha_{h_1}(a_1), \alpha_{h_2}(a_2), \alpha_{h_3}(a_3), \ldots$$

With that picture it would seem reasonable that the probability that atom $\alpha_j(x)$ would be picked at the $n+1$'st stage would only depend on the number of previous picks which had been made and the number of those which had yielded a copy of $\alpha_j(x)$.

A reasonably straightforward consequence[2] of JSP, which we mention because it will figure in what follows, is:

Atom Exchangeability, Ax

$$w\left(\bigwedge_{i=1}^{n} \alpha_{h_i}(a_i)\right)$$

depends only on the multiset $\{m_1, m_2, \ldots, m_{2^q}\}$ where $m_j = |\{i \mid h_i = j\}|$ (as usual).

Equivalently, if σ is a permutation of $\{1, 2, \ldots, 2^q\}$ then

$$w\left(\bigwedge_{i=1}^{n} \alpha_{h_i}(a_i)\right) = w\left(\bigwedge_{i=1}^{n} \alpha_{\sigma(h_i)}(a_i)\right).$$

Interestingly Ax would normally be seen as a principle based on, and indeed rationally justified by, considerations of *symmetry* whereas JSP is based on *irrelevance*.

What really makes JSP a major principle in the subject however is the following result which was first proved[3] by Johnson in [12] and later by Kemeny [11].

Theorem 2.1. *Provided L has at least 2 relation symbols (i.e. $q \geq 2$) the only probability functions on L satisfying JSP + Reg are the c_λ^L for $0 < \lambda \leq \infty$, the so called Carnap's Continuum of Inductive Methods, given by:*

$$c_\lambda^L\left(\alpha_j(a_{n+1}) \,\middle|\, \bigwedge_{i=1}^{n} \alpha_{h_i}(a_i)\right) = \frac{m_j + 2^{-q}\lambda}{n + \lambda} \tag{1}$$

2 For details of the proofs of the results stated in this paper collected together in one place see [16] or the forthcoming [18].

3 See [23] for the history of this result and its more general versions.

As implied already by the statement of this theorem just the identities afforded by (1) are enough to determine c_λ^L on the whole of SL (though the final extension from $QFSL$ to SL requires a later result due to Gaifman, see [7], or [16, Theorem 7]). We should mention here that if we drop the assumption of Regularity in this theorem then a further solution[4] appears, c_0^L, defined by

$$c_0^L\left(\bigwedge_{i=1}^{n} \alpha_{h_i}(a_i)\right) = \begin{cases} 2^{-q} & \text{if } h_1 = h_2 = \ldots = h_n, \\ 0 & \text{otherwise.} \end{cases} \tag{2}$$

Going to the other extreme, when $\lambda = \infty$, we have that

$$c_\infty^L\left(\bigwedge_{i=1}^{n} \alpha_{h_i}(a_i)\right) = 2^{-nq},$$

so c_∞^L treats all the $R_j(a_i)$ as stochastically independent, each with probability $1/2$. For $0 < \lambda < \infty$ the right hand side of (1) looks like the proportion of α_j amongst the α_{h_i} moderated by fixed terms in λ whose influence tends to zero as $n \to \infty$. (For a fuller discussion on this and related points see the account by Zabell in [23].)

Theorem 2.1 must be considered one of the most important results in the subject. For it seems to go very far along the road of justifying the idea that considerations of rationality will determine a very tightly restricted choice of probability functions for our agent, indeed a single choice if only λ could be fixed unequivocally. Although no generally persuasive argument for any particular value for λ has ever emerged, nevertheless we might rightly claim that the continuum of c_λ^L are *logical* on the grounds that any other rational being will be led inexorably as Johnson and Carnap were to these same choices.

The influence of this position on the later developments of Inductive Logic has been enormous. Most subsequent proposals for logical probability functions to satisfy additional rationality principles not immediately satisfied by the c_λ^L have been based on taking combinations, or adaptations, of the c_λ^L. And in justification of this position the c_λ^L do have some attractive features in the context of rational belief.

For example a property that the c_λ^L for $0 < \lambda < \infty$ satisfy (directly from (1)) is[5]:

Reichenbach's Axiom, RA

If $\alpha_{h_1}(x), \alpha_{h_2}(x), \alpha_{h_3}(x), \ldots$ is an infinite sequence of atoms of L and $m_j(n)$ is the number of times that $\alpha_j(x)$ appears amongst $\alpha_{h_1}(x), \alpha_{h_2}(x), \ldots, \alpha_{h_n}(x)$ then

$$\lim_{n\to\infty}\left(w\left(\alpha_j(a_{n+1}) \,\middle|\, \bigwedge_{i=1}^{n} \alpha_{h_i}(a_i)\right) - \frac{m_j(n)}{n}\right) = 0$$

4 Sometimes also included in Carnap's Continuum, though not here in this note.

5 Attributed to H.Reichenbach after a suggestion by H.Putnam, see [4, p120].

In spirit then, as a recommendation, RA asserts that in the limit the subjective probabilities should converge to frequency. In other words if 34% of the many people I have ever met have had blue eyes then I should give probability close to 34/100 to the next new person I meet having blue eyes.

A second attractive property is that for fixed λ the c_λ^L form a **Language Invariant family**[6], meaning that if $L \subseteq L'$ (so $SL \subseteq SL'$) then the restriction of $c_\lambda^{L'}$ to SL is c_λ^L.

This seems clearly justified on the grounds that whilst at any one time our agent might be inhabiting a structure for L with relation symbols $R_1, R_2, \ldots, R_q$ there is no reason to say that these are all the relation symbols there could ever be, and if at a later time the language L was expanded to L' then this should not require the agent to revise his/her beliefs about the truth or otherwise of the sentences of the original language L.

A final attractive property of the c_λ^L which we mention[7] is:

The Principle of Instantial Relevance, PIR

For an atom $\alpha_j(x)$ and state description $\Theta(a_1, a_2 \ldots, a_n)$,

$$w(\alpha_j(a_{n+2}) \mid \alpha_j(a_{n+1}) \wedge \Theta(a_1, a_2, \ldots, a_n)) \geq w(\alpha_j(a_{n+1}) \mid \Theta(a_1, a_2, \ldots, a_n))$$

This principle reflects the very heart of what we understand in the real world by 'induction': That the more times I have observed something in the past the more times I should expect to observe it in the future. For example suppose I arrive in a foreign country late at night and waking at 6.00 the next morning I find that it is raining. Shouldn't this increase my belief, or at least not diminish it, that it will also be raining at 6.00 tomorrow morning?

In fact as it turns out the c_λ^L can hardly be said to be particularly distinguished by satisfying PIR since by a result of Gaifman, [8]:

Theorem 2.2. *For $\theta(a_{n+1}), \psi(a_1, a_2, \ldots, a_n) \in QFSL$ and w satisfying Ex,*

$$w(\theta(a_{n+2}) \mid \theta(a_{n+1}) \wedge \psi(a_1, a_2, \ldots, a_n)) \geq w(\theta(a_{n+1}) \mid \psi(a_1, a_2, \ldots, a_n))$$

Whilst this does not particularly promote the primacy of the c_λ^L as such it is an important result because its discovery suggested that rational principles are not only consistent with each other but even to some extent interderivable (and so obliquely that the c_λ^L are their embodiment).

---

6 This notion is usually employed in the polyadic context when we do not restrict our attention to unary predicates so strictly speaking we should be talking here about *Unary* Language Invariance.

7 This is a slight simplification of a principle of the same name proposed by Carnap, [3].

Viewing the above results we might well be led to conclude that in Carnap's Continuum we had uncovered some universal truth about rationality. It all fits together simply and elegantly, isn't that just what we expect of a true theory? Furthermore another characteristic we associate with 'fundamental truths' is that they reoccur as the result of seemingly different approaches (take for example the explication of 'effective process'). That is the case here, there are several principles other than JSP which characterize the c_λ^L, see for example [18].

3 Another Continuum

Given so many positive features it is scarcely surprising then that for many decades Inductive Logic largely saw Carnap's Continuum as *the* base camp from which to explore further. In the early 2000's however C. J. Nix and the first author, [14], made a somewhat unexpected discovery.

To explain the background recall the example we gave of observing rain at 6.00 to motivate PIR but now suppose instead that I oversleep and waking at 7.00 note that the streets are wet, though it is not actually raining. Shouldn't that also increase my belief, or at least not diminish it, that it will be raining at 6.00 tomorrow morning? In other words observing a consequence of rain, if not actual rain itself, should still be viewed as supporting evidence for rain tomorrow.

This sentiment might be captured in a principle generalizing Theorem 2.2:

The Generalized Principle of Instantial Relevance, GPIR

For $\theta(a_{n+2}), \phi(a_{n+1}), \psi(a_1, a_2, \ldots, a_n) \in QFSL$, if $\theta(x) \models \phi(x)$ then

$$w(\theta(a_{n+2}) \mid \phi(a_{n+1}) \wedge \psi(a_1, a_2, \ldots, a_n)) \geq w(\theta(a_{n+1}) \mid \psi(a_1, a_2, \ldots, a_n))$$

(As an aside here we mention that this is the same principle if instead we require $\theta(x)$ to be a *consequence* of $\phi(x)$ rather than the other way round.)

The following theorem, which will appear in [18], is simply a reformulation of results in [14]:

Theorem 3.1. *Provided $q \geq 2$ the only probability functions on L satisfying GPIR + Ax + Reg are the w_L^δ for $-(2^q - 1)^{-1} < \delta < 1$ given by:*

$$w_L^\delta\left(\bigwedge_{i=1}^{n} \alpha_{h_i}(a_{k_i})\right) = 2^{-q} \sum_{j=1}^{2^q} \gamma^{n-m_j}(\gamma + \delta)^{m_j} \tag{1}$$

where $\gamma = 2^{-q}(1 - \delta)$ and (as usual) m_j is the number of times that $\alpha_j(x)$ is instantiated in the state description $\bigwedge_{i=1}^{n} \alpha_{h_i}(a_{k_i})$.

Interestingly the key power of GPIR here is in allowing additional information to be present in the form of $\psi(a_1, a_2, \ldots, a_n)$. Without this GPIR holds for all probability functions satisfying Ax, see [18].

So here we have another 'continuum of inductive methods' sometimes referred to as the Nix-Paris (see [22]) or NP-Continuum. As with Carnap's Continuum it is based on rational principles, though the only member that both continua have in common is c^L_∞, w^0_L.[8]

For a fixed δ the w^δ_L do not form a Language Invariant family for $-(2^q - 1)^{-1} < \delta < 0$, this is shown in [15], but they do for $0 \leq \delta < 1$ and for this reason it is now common practice (which we henceforth follow) to take the NP-continuum to be the w^δ_L with δ restricted to this range.

In addition to Language Invariance the w^δ_L satisfy two other arguably attractive properties. Firstly they satisfy:

The Weak Irrelevance Principle, WIP

If $\theta, \phi \in QFSL$ have no relation or constant symbols in common then

$$w(\theta \mid \phi) = w(\theta).$$

The intended justification for this principle seems clear: If θ and ϕ come from different languages whyever should knowing the truth of ϕ influence one's belief in θ?

This property does not provide a characterization of the w^δ_L, there are other probability functions satisfying Ax + Li which satisfy WIP though they are in a sense all built from the w^δ_L, see [17]. A property which does however characterize the w^δ_L is:

Recovery

A probability function w on L satisfies Recovery, or is Recoverable, if whenever $\Psi(a_1, a_2, \ldots, a_n)$ is a state description then there is another state description $\Phi(a_{n+1}, a_{n+2}, \ldots, a_h)$ such that $w(\Phi \wedge \Psi) \neq 0$ and for any quantifier free sentence $\theta(a_{h+1}, a_{h+2}, \ldots, a_{h+g})$,

$$w(\theta(a_{h+1}, a_{h+2}, \ldots, a_{h+g}) \mid \Phi \wedge \Psi) = w(\theta(a_{h+1}, a_{h+2}, \ldots, a_{h+g})). \tag{2}$$

In other words w is Recoverable if given any 'past history' as a state description Ψ there is a possible 'future' state description Φ which will take us right back to where we started, at least as far as the quantifier free properties of the currently unobserved constants $a_{h+1}, a_{h+2}, \ldots$ are concerned.

8 In some accounts (1) is extended to the case $\delta = 1$, at which point $w^1_L = c^L_0$, so meeting again the 'extended Carnap's Continuum' referred to earlier.

It is debateable whether Recoverability is an obviously desirable property of a rational probability function, though if we think of it as a manifestation of simplicity or at least 'simplifiability', of being able to wipe out the influence of the past, then that could perhaps be argued for. The following result from [20] shows that even a weak form of Recovery (in suitable company at least) actually characterizes the NP-continuum:

Theorem 3.2. *Suppose that the probability functions w_L satisfy Reg + Ax and form a Language Invariant family. Then w_L has the property that for some state description $\Phi(a_1, a_2, \ldots, a_n)$ with $n > 0$, and for all $\theta(a_{n+1}, a_{n+2}, \ldots, a_{n+g}) \in QFSL$,*

$$w(\theta(a_{n+1}, a_{n+2}, \ldots, a_{n+g}) \mid \Phi(a_1, a_2, \ldots, a_n)) = w(\theta(a_{n+1}, a_{n+2}, \ldots, a_{n+g}))$$

just if $w_L = w_L^\delta$ for some $0 \leq \delta < 1$.

4 Two Rationalities?

We have seen that we now have two continua, the very well established Carnap's Continuum and the much more recent NP-continuum. Both are 'justified' by satisfying several apparently desirable properties. Indeed it currently seems that most rational properties that we might propose are satisfied by at least one of them.

Comparing these two continua[9], they have in common that they both satisfy Ax, Reg and Language Invariance. As for their differences the c_λ^L satisfy Johnson's Sufficientness Postulate, which of course the w_L^δ do not, and similarly the c_λ^L do not satisfy the Generalized Principle of Instantial Relevance except at $\lambda = \infty$.

A further difference is that apart from the end point c_∞^L where the two continua meet the c_λ^L do not satisfy the Weak Irrelevance Property, WIP. As a particular instance of this latter failure one can check that for $0 < \lambda < \infty$,

$$c_\lambda^L(R_2(a_3) \wedge R_2(a_4) \mid R_1(a_1) \wedge R_1(a_2)) > c_\lambda^L(R_2(a_3) \wedge R_2(a_4)). \tag{1}$$

So according to c_λ^L, $R_1(a_1) \wedge R_1(a_2)$ adds support to $R_2(a_3) \wedge R_2(a_4)$ despite these two sentences having no relation nor constant symbols in common. However one might argue that this is actually a manifestation of *induction* which c_λ^L recognizes. Namely the information $R_1(a_1) \wedge R_1(a_2)$ *suggests* that the constants look alike, since both a_1 and a_2 have been found to satisfy $R_1(x)$, and this supports belief in a_3, a_4 also being alike and so both satisfying $R_2(x)$. [Actually it turns out that we will

9 Further comparisons are given in [21].

always have ≥ in (1) for any probability function that is a member of a Language Invariant family of functions satisfying Ax, see [9], so from this point of view the w_L^δ might be criticized for *not* picking up this version of relevance.]

Unlike the c_λ^L (for $\lambda < \infty$) the w_L^δ do not satisfy Reichenbach's Axiom, RA. To give a specific example of this failure consider the sequence of atoms $\alpha_{h_i}(x)$ where $h_i = 1$ if $i = 0 \bmod 3$ and $h_i = 2$ otherwise. Then with $j = 1$ and the notation of the statement of RA, $m_j(n)/n \to 1/3$ as $n \to \infty$. However

$$w_L^\delta\left(\alpha_1(a_{3n}) \mid \bigwedge_{i=1}^{3n-1} \alpha_{h_i}(a_i)\right) = \frac{\gamma(\gamma+\delta)^{2n} + \gamma^{n+1}(\gamma+\delta)^n + (2^q-2)\gamma^{2n+1}}{(\gamma+\delta)^{2n} + \gamma^{n+1}(\gamma+\delta)^{n-1} + (2^q-2)\gamma^{2n}}$$

which tends to γ as $n \to \infty$, so RA fails. (Of course if $\gamma = 1/3$ we can just change the frequency of the $\alpha_1(x)$.) In defense of the w_L^δ, there would seem to be no reason per se for our agent to suppose that the $m_j(n)/n$ would be a convergent sequence. Such a supposition would seem to be based on the rather unwarranted assumption that there was some fixed process, such as picking from an urn with replacement, which was deciding whether or not a_n satisfied $\alpha_1(x)$. Without this assumption there would seem to be little, or at least less, reason for the agent to slavishly attempt to close in on the 'limit' of the $m_j(n)/n$.

Overall though the c_λ^L seem to us to edge it in a straight contest with the w_L^δ. However that is not the main issue. The main issue is in explaining the alternate concepts of 'rationality' which underpin them.

One might object here that if we accept as our rational probability function the only member that the two continua have in common, i.e. $c_\infty^L = w_L^0$, then we get the best of both worlds and there needs be no dilemma. Unfortunately it is rather that we seem get the *worst* of both worlds. In particular the failure of RA for the w_L^0 is especially dramatic: w_L^0 does not respond at all to the evidence $\bigwedge_{i=1}^n \alpha_{h_i}(a_i)$ and just gives $\alpha_j(a_{n+1})$ probability 2^{-q} regardless. In other words w_L^0 is entirely devoid of induction and largely in consequence of this few in the area would nowadays wish to promote it as 'the rational choice'.[10,11]

Turning back then to the question of their underlying versions of rationality one might hazard that the difference is this: The principles underlying the c_λ^L require the agent to picture his/her ambient structure M as the result of some regular statistical process. To repeat ourselves, the example of picking balls from an urn

10 But see [22] for a spirited counter argument.

11 Another possibility is to extend the continua to c_0^L, w_L^1 where they again agree and make this choice. This has certainly been proposed in its time (and is even inevitable if one pushes symmetry to the limit with purely unary languages, see [19]) but suffers the opposite defect from c_∞^L of being too much influenced by the evidence, as shown already in (2).

in the case of JSP. So the notion of rationality being taken here is of regularity, predictability, pattern. On the other hand the version of rationality being exploited, and subsequently manifest, in the case of the w_L^δ might be seen as simplicity or economy. For example the consideration that information should be ignored, or have the potential to be ignored, where possible leads to GPIR, WIP and Recoverability.

5 Conclusion

In this paper a number of rationality principles have been proposed each of which is satisfied by, or even characterizes, the probability functions in one of two 'Continua of Inductive Methods', Carnap's classical continuum and the recently explicated NP-continuum. We would suggest that what is significant here is that it implies that there are at least two separate notions of rationality at work and that they are essentially incompatible, or at least their common ground does not provide acceptable rational or logical probability functions to most people's way of thinking.

Bibliography

[1] Carnap, R., *Logical Foundations of Probability*, University of Chicago Press, Chicago, Routledge & Kegan Paul Ltd., 1950.

[2] Carnap, R., *The Continuum of Inductive Methods,* University of Chicago Press, 1952.

[3] Carnap, R. & Jeffrey, R.C. eds., *Studies in Inductive Logic and Probability,* Volume I, University of California Press, 1971.

[4] Carnap, R., A Basic System of Inductive Logic, in *Studies in Inductive Logic and Probability*, Volume II, ed. R. C. Jeffrey, University of California Press, 1980, pp7-155.

[5] Carnap, R. & Stegmüller, W., *Induktive Logik und Wahrscheinlichkeit,* Springer Verlag, Wien, 1959.

[6] Earman, J., *Bayes or Bust?*, MIT Press, 1992.

[7] Gaifman, H., Concerning Measures on First Order Calculi, *Israel Journal of Mathematics*, 1964, **2**:1-18.

[8] Gaifman, H., Applications of de Finetti's Theorem to Inductive Logic, in *Studies in Inductive Logic and Probability*, Volume I, eds. R.Carnap & R.C.Jeffrey, University of California Press, Berkeley and Los Angeles, 1971, pp235-251.

[9] Hill, A. & Paris, J.B., The Counterpart Principle of Analogical Support by Structural Similarity, to appear in *Erkenntnis*.

[10] Hintikka, J. & Niiniluoto, I., An Axiomatic Foundation of the Logic of Inductive Generalization, in *Formal Methods in the Methodology of the Empirical Sciences,* eds. M.Przelecki, K.Szaniawski & R.Wójcicki, Synthese Library 103, Kluwer, Dordrecht, 1976, pp57-81.

[11] Kemeny, J.G., Carnap's theory of probability and induction, in The Philosophy of Rudolf Carnap, ed. P.A.Schilpp, La Salle, Illinois, Open Court (1963), pp 711-738.

[12] Johnson, W.E., Probability: The deductive and inductive problems, *Mind*, 1932, **41**(164):409-423.

[13] Landes, J., Paris, J.B. & Vencovská, A survey of some recent results on Spectrum Exchangeability in Polyadic Inductive Logic, *Synthese*, 2011, **181**(Supplement 1), pp19-47.

[14] Nix, C.J. & Paris, J.B., A Continuum of Inductive Methods arising from a Generalized Principle of Instantial Relevance, *Journal of Philosophical Logic*, 2006, **35**(1): 83-115.

[15] Nix, C.J., Nix, C.J., *Probabilistic Induction in the Predicate Calculus* Doctoral Thesis, Manchester University, Manchester, UK, 2005.
See http://www.maths.manchester.ac.uk/~jeff/

[16] Paris, J.B., *Guangzhou Winter School course on Pure Inductive Logic, 2010*, available at http://www.maths.manchester.ac.uk/~jeff/lecture-notes/Guangzhou.pdf

[17] Paris, J.B. & Vencovská, A., A Note on Irrelevance in Inductive Logic, *Journal of Philosophical Logic*, 2011, **40**(3): 357-370

[18] Paris, J.B. & Vencovská, A., *Pure Inductive Logic*, to appear in the ASL series *Perspectives in Mathematical Logic*, Cambridge University Press, 2014.

[19] Paris, J.B. & Vencovská, Symmetry's End?, *Erkenntnis*, 2011, **74**(1): 53-67.

[20] Paris, J.B. & Waterhouse, P., Atom Exchangeability and Instantial Relevance, *Journal of Philosophical Logic*, 2009, **38**(3):313-332.

[21] Williamson, J., Inductive Influence, *British Journal for the Philosophy of Science*, 2007, **58**:689-708.

[22] Williamson, J., An Objective Baysian Account of Confirmation, in *Explanation, Prediction, and Confirmation. New Trends and Old Ones Reconsidered*, Eds. D.Dieks, W.J.Gonzalez, S.Hartmann, T.Uebel & M.Weber. Volume 2 in the series *The Philosophy of Science in a European Perspective*, Springer, 2011, pp53-81.

[23] Zabell, S. L., Carnap and the Logic of Induction, in: *Handbook of the History and Philosophy of Logic*, Volume 10, eds. D.Gabbay, S.Hartmann & J.Woods, Elsevier, 2011.

Saharon Shelah

A.E.C. with Not Too Many Models

Dedicated to Jouko Väänänen honouring his 60th birthday

Abstract: Consider an a.e.c. (abstract elementary class), that is, a class K of models with a partial order refining $\subseteq$ (submodel) which satisfy the most basic properties of an elementary class. Our test question is trying to show that the function $\dot{I}(\lambda, K)$, counting the number of models in K of cardinality λ up to isomorphism, is "nice", not chaotic, even without assuming it is sometimes 1, i.e. categorical in some λ's. We prove here that for some closed unbounded class $\mathbf{C}$ of cardinals we have (a),(b) or (c) where

(a) for every $\lambda \in \mathbf{C}$ of cofinality $\aleph_0$, $\dot{I}(\lambda, K) \geq \lambda$

(b) for every $\lambda \in \mathbf{C}$ of cofinality $\aleph_0$ and $M \in K_\lambda$, for every cardinal $\kappa \geq \lambda$ there is N_κ of cardinality κ extending M (in the sense of our a.e.c.)

(c) $\mathfrak{k}$ is bounded; that is, $\dot{I}(\lambda, K) = 0$ for every λ large enough (equivalently $\lambda \geq \beth_{\delta_*}$ where $\delta_* = (2^{\mathrm{LST}(\mathfrak{k})})^+$).

Recall that an important difference of non-elementary classes from the elementary case is the possibility of having models in K, even of large cardinality, which are maximal, or just failing clause (b).

Keywords: model theory, classification theory, a.e.c. (abstract elementary classes), categoricity, number of non-isomorphic models

MSC: Primary: 03C45, 03C48; Secondary: 03C55, 03C75

Saharon Shelah: Einstein Institute of Mathematics, Edmond J. Safra Campus, Givat Ram, The Hebrew University of Jerusalem, Jerusalem, 91904, Israel, and , Department of Mathematics, Hill Center - Busch Campus , Rutgers, The State University of New Jersey , 110 Frelinghuysen Road , Piscataway, NJ 08854-8019 USA, e-mail: shelah@math.huji.ac.il

Annotated Content

1 Introduction to the subject

We would like to have classification theory for non-elementary classes K and more specifically to generalize stability. Naturally we use the function $\dot{I}(\lambda, K)$ = number of models up to isomorphism, as a major test problem. Now "non-elementary" has more than one interpretation, we shall start with the infinitary logics $\mathbb{L}_{\lambda,\kappa}$.

There are other directions; mostly where compactness in some form holds (e.g. a.e.c. with amalgmation, see about those in [Sh:E53], and on a try to blend with descriptive set theory see [Sh:849]). We had held that for $\kappa > \aleph_0$ the above cannot be developed as, e.g. if $\mathbf{V} = \mathbf{L}$ or just $\mathbf{V} \models$ "$0^\#$ does not exist", then there is $\psi \in \mathbb{L}_{\aleph_1,\aleph_1}$ such that if $\mathrm{cf}(\mu) = \aleph_0 \wedge (\forall \alpha < \mu)(|\alpha|^{\aleph_0} < \mu)$ <u>then</u> $M \models \psi, \|M\| = \mu$ <u>iff</u> $M \cong (\mathbf{L}_\mu, \in)$. However, lately [Sh:1019] gives evidence that for θ a compact cardinal, we can generalize to $\mathbb{L}_{\theta,\theta}$ some theorems of [Sh:c, Ch.VI] on saturation of ultra-powers and Keisler's order. This shows that stability theory for $T \subseteq \mathbb{L}_{\theta,\theta}$ exists, but it is still not clear how far we can go including $A = |N|, N \prec M$ and $\cup\{M_u : u \subset n\}$ when $\langle M_u : u \subset n\rangle$ is a so called stable $\mathscr{P}^-(n)$-system.

Anyhow (for the purposes of this history, and the present paper) we now concentrate on $\mathrm{Mod}_\psi, \psi \in \mathbb{L}_{\lambda^+,\aleph_0}$ so $\kappa = \aleph_0$. Here we have both downward LST theorems, even using $\le \lambda$ finitary Skolem functions. Also we have the upward LST theorem, using EM models.

Naturally all works started with assuming categoricity in some cardinal, except some dealing with the $\aleph_n$'s for $\psi \in \mathbb{L}_{\aleph_1,\aleph_0}$. In this case we may many times deal with $\psi \in \mathbb{L}_{\aleph_1,\aleph_0}(Q)$. Some works apeared in the eighties (see the books [Bal09], and [Sh:h], [Sh:i]).

Definition 1.1. Let $\dot{I}(\lambda, K)$ be the cardinality of $\{M/\cong: M \in K$ of cardinality $\lambda\}$ where K is a class of $\tau(K)$-models (e.g. $K = K_{\mathfrak{k}}$ where $\mathfrak{k} = (K_{\mathfrak{k}}, \leq_{\mathfrak{k}})$).

First, in ZFC, answering a question of Baldwin, it was proved that ψ cannot be categorical, moreover if $\dot{I}(\aleph_1, \psi) = 1$ then $\dot{I}(\aleph_2, \psi) \geq 1$. Also if $\dot{I}(\aleph_1, \psi) < 2^{\aleph_1}$, then for some countable first order T with an atomic model $K_T = \{M : M$ an atomic model of $T\}$ is $\subseteq \mathrm{Mod}_\psi$, but $1 \leq \dot{I}(\aleph_1, K_T)$. Fix T for awhile, now if $2^{\aleph_n} < \aleph_{n+1}, \dot{I}(\aleph_n, T) < \mu_{\mathrm{wd}}(\aleph_{n+1}, 2^{\aleph_n})$ for[1] every n then K_T is excellent which means it is quite similar to the class of models of an $\aleph_0$-stable countable complete first order theory. For this we consider $\mathbf{S}^m(A, M)$ for $A \subseteq M \in K_T$, only for some "nice" A. On the other hand for any n for some such T_n, K_{T_n} is categorical in every $\lambda \leq \aleph_n$ but $\dot{I}(\lambda, T) = 2^\lambda$ for λ large enough. However, we do not know:

Conjecture 1.2. (Baldwin) If K_T is categorical in $\aleph_1$, then K_T is $\aleph_0$-stable, equivalently is absolutely categorical.

Related is the:

Conjecture 1.3. If K_T is categorical in $\aleph_1$ but not $\aleph_0$-stable then $\dot{I}(2^{\aleph_0}, K_T) = \beth_2$.

See work in preparation Baldwin-Laskowski-Shelah ([BaLaSh:1037]) on such K_T's; it certainly says there is a positive theory for such classes (e.g. pseudo minimal types exist). We recently have changed our mind and now think:

Conjecture 1.4. If K_T is categorical in every $\aleph_n$ then K_T is excellent.

This means that the present counter-examples are best possible. As this seems very far we may consider a weaker conjecture.

Conjecture 1.5. Assume $\mathbb{P}$ is a c.c.c. forcing notion of cardinality λ such that $\Vdash_{\mathbb{P}}$ "MA + $2^{\aleph_0} = \lambda$" and $\lambda = \lambda^{<\lambda} > \beth_{\omega_1}$. If K_T is categorical in every $\lambda < 2^{\aleph_0}$ then K_T is excellent.

There is more to be said, see [Sh:F1273].

* * *

In another direction, the investigation of models of cardinality $\aleph_1$ does not point to a canonical choice of logic for which the theorems on $\dot{I}(\psi, \aleph_1) = 1$ holds. This had motivated the definition of a.e.c. $\mathfrak{k} = (K_{\mathfrak{k}}, \leq_{\mathfrak{k}})$ which has the "bottom" property of elementary class $K = (\mathrm{Mod}_T, \prec)$, T a complete first order theory (i.e. $K_{\mathfrak{k}}$, a class of $\tau_{\mathfrak{k}}$-models, $\leq_{\mathfrak{k}}$ a partial order on it, both closed under isomorphism, union under $\leq_{\mathfrak{k}}$-directed systems of member of $K_{\mathfrak{k}}$ belong to $K_{\mathfrak{k}}$, moreover is a $\leq_{\mathfrak{k}}$-lub (= union

1 note that $\mu_{\mathrm{wd}}(\lambda^+, 2^\lambda)$ is essentially 2^{λ^+}.

of a directed system of $\leq_{\mathfrak{k}}$-submodels of N is a $\leq_{\mathfrak{k}}$-submodel of N), existence of a LST number and $M_1 \subseteq M_2 \wedge M_1 \leq_{\mathfrak{k}} N \wedge M_2 \leq_{\mathfrak{k}} N \Rightarrow M_1 \leq_{\mathfrak{k}} M_2$).

Thesis 1.6. 1. The framework of a.e.c. $\mathfrak{k}$ is wider and not too far and better than the family of $(\mathrm{Mod}_\psi, \prec_{\mathrm{sub}(\psi)})$ where $\psi \in \mathbb{L}_{\lambda^+,\aleph_0}$.
2. The right generalization of types in this context is orbital types.

Why? The "wider" in 1.6(1) is obvious. The "not too far" is by the representation theorem which says that for some vocabulary $\tau_1 \supseteq \tau(\mathfrak{k})$ of cardinality $\leq \lambda$, λ the LST-number $+|\tau(\mathfrak{k})|$ and set Γ of quantifier free 1-types, $K_{\mathfrak{k}} = \mathrm{PC}(\emptyset, \Gamma) = \{M \restriction \tau_{\mathfrak{k}} : M$ a τ_1-model omitting every $p(x) \in \Gamma\}$; similarly $\leq_{\mathfrak{k}}$. We can deduce the upward LST, and so existence of suitable $\Phi \in \Upsilon^{\mathrm{lin}}[\mathfrak{k}]$ so we have EM-models. For $\mathfrak{k}$ with $\mathrm{LST}_{\mathfrak{k}} = \aleph_0$ it is natural to restrict ourselves to the case "Γ is countable" above for both $K_{\mathfrak{k}}$ and $\leq_{\mathfrak{k}}$, then we say $\mathfrak{k}$ is $\aleph_0$-presentable. So we may wonder for such $\mathfrak{k}$ <u>if</u> $n < \omega \Rightarrow 2^{\aleph_n} + \dot{I}(\aleph_{n+1}, K_{\mathfrak{k}}) < \mu_{\mathrm{wd}}(\aleph_{n+1}, 2^{\aleph_n})$ implies $\mathfrak{k}$ satisfies the parallel of being excellent? The answer is yes by [Sh:h], [Sh:i], but the way is long. Also, we may replace $\aleph_0$ by any λ provided that $I(\lambda, K_{\mathfrak{k}}) = 1 = I(\lambda^+, K_{\mathfrak{k}})$ and $1 \leq \dot{I}(\lambda^{++}, K_{\mathfrak{k}}) < \mu_{\mathrm{wd}}(\lambda^{++}, 2^{\lambda^+})$, see more in [Sh:E53].

A central notion there is "$\mathfrak{s}$ is a good λ-frame", $\mathfrak{k}_{\mathfrak{s}} = \mathfrak{k}$, $\mathrm{LST}_{\mathfrak{k}} \leq \lambda$, this is "bare bones superstable".

This is enough for proving

(∗) if ($\mathfrak{k}$ is an a.e.c.), $\mathrm{LST}_{\mathfrak{k}} \leq \lambda, 2^{\lambda^{+n}} < 2^{\lambda^{+n^+}}$ and $\dot{I}(\lambda^{+n}, K_{\mathfrak{k}}) = 1$ for every n and $K_{\mathfrak{k}}$ has models of cardinality $\geq \beth_{(2^{\mathrm{LST}(\mathfrak{k})})^+}$, <u>then</u> $K_{\mathfrak{k}}$ is categorical in every $\mu \geq \lambda$.

However

Conjecture 1.7. If $\mathfrak{k}$ is an a.e.c., $K_{\mathfrak{k}}$ is categorical in some λ large enough than $\mathrm{LST}_{\mathfrak{k}}$, <u>then</u> $K_{\mathfrak{k}}$ is categorical in every $\mu \geq \lambda$.

Note that [Sh:734] is a step ahead: in the context of 1.7, for many $\mu = \beth_\mu \in [\mathrm{LST}_{\mathfrak{k}}, \lambda)$, there is a good μ-frame $\mathfrak{s}_\mu$ such that $\mathfrak{k}_{\mathfrak{s}} = K_\mu^{\mathfrak{k}}$. If we have this for ω successive μ's we shall be done by [Sh:600], but in [Sh:734] the family of such μ's is scattered; a beginning is [Sh:842].

A much harder conjecture is:

Conjecture 1.8. 1. The main gap theorem holds for a.e.c. $K_{\mathfrak{k}}$ for λ large enough.
2. The class $\sup-\lim_{\mathfrak{k}} = \{\lambda$: there is a super-limit $M \in K_\lambda^{\mathfrak{k}}\}$ is "nice", e.g. contains every large enough λ <u>or</u> contains no large enough λ.

We are continuing this work in [Sh:F1302].

* * *

We may wonder

Question 1.9. 1. Maybe there is a natural logic which is the natural framework for categoricity spectrum.
2. Also for the super-limit spectrum.

We expect such logic to be stronger than $\mathbb{L}_{\lambda^+,\aleph_0}$ but weaker than $\mathbb{L}_{\lambda,\lambda}$. This may remind us of [Sh:797]. The logic discovered there is $\mathbb{L}^1_{<\lambda}$ for $\lambda = \beth_\lambda$, it is between $\mathbb{L}^{-1}_{<\lambda} = \cup\{\mathbb{L}_{\mu^+,\aleph_0} : \mu < \lambda\}$ and $L^0_{<\lambda,\mu} = \cup\{\mathbb{L}_{\mu^+,\mu^+} : \mu < \lambda\}$, in a strong way well ordering is not well defined and it can be characterized (as Lindström theorem characterize first order logic) and has interpolation. In addition, for λ a compact cardinal $\mathbb{L}^1_{<\lambda}$-equivalence of M_1, M_2 is equivalent to having isomorphism ω-limit ultra-powers by λ-complete ultrafilters, see [Sh:1019].

However, probably the characterization in [Sh:797] was by "the maximal logic such that ...". So maybe we should restrict the logic further such that "EM model can be constructed".

We conjecture there is a logic characterized by being maximal under this stronger demand, and in it we can say at least something on the function $\dot{I}(\lambda, \psi)$, and maybe much. This is interesting also from the point of view of soft model theory: we conjecture that there are many such intermediate logics with characterization (and the related interpolation theorem).

2 Introduction to the paper

In this section, we begin by motivating our line of investigation. See notation in §(1D) below (and more self contained introduction in §(1B), §(1C)).

2.1 Motivation/Content

We knew of old (see: [Sh:c, Ch.XIII,4.15]):

Theorem 2.1. *For a countable complete first order theory T, one of the following holds:*
(a) T is categorical in every $\lambda > \aleph_0$
(b) $\dot{I}(\lambda, T) = \beth_2$ for every cardinal $\lambda \geq 2^{\aleph_0}$
(c) $\dot{I}(\aleph_\alpha, T) \geq 1 + |\alpha|$ for every ordinal α.

For a.e.c. we have something when $\mathfrak{k}$ is categorical in some λ's ([Sh:734], [Sh:600]) and something about $\dot{I}(\aleph_1, \mathfrak{k})$, ([Sh:88r], about when $1 \leq \dot{I}(\aleph_1, \mathfrak{k}) < 2^{\aleph_1}$, particularly when $2^{\aleph_0} < 2^{\aleph_1}$ and then on higher cardinals) but nothing for general a.e.c. $\mathfrak{k}$. The current paper is motivated by hopes of finding something like 2.1 for a.e.c.'s. Recall the history.

Our approach here assumes/relies on:

Thesis 2.2. Reasonable to concentrate on cardinals from $\mathbf{C}_{\mathrm{fp}} = \{\lambda : \lambda = \beth_\lambda\}$, where fp stands for "fixed points".

Why? If $\lambda \in \mathbf{C}_{\mathrm{fp}}, \lambda > \mathrm{LST}(\mathfrak{k})$ and $M \in K^{\mathfrak{k}}_\lambda$ <u>then</u> for every $\theta \in [\mathrm{LST}(\mathfrak{k}), \lambda)$ and $N \leq_{\mathfrak{k}} M, \|N\| = \theta$ there is $\Phi \in \Upsilon_{\mathfrak{k},\theta}$ so $|\tau(\Phi)| = \theta$ such that for any linear order I, e.g. $I = \lambda$ we have $N \leq_{\mathfrak{k}} \mathrm{EM}_{\tau(\mathfrak{k})}(I, \Phi)$. So in $K^{\mathfrak{k}}_\lambda$ we have many models of the form $\mathrm{EM}_{\tau(\mathfrak{k})}(I, \Phi), \Phi \in \Upsilon_{\mathfrak{k},<\lambda}$. If $\dot{I}(\lambda, \mathfrak{k}) < \lambda$, many of them will be isomorphic. Hence for many $\theta_1 < \theta_2 < \lambda, \theta_1 \geq \mathrm{LST}(\mathfrak{k})$, every $N \leq_{\mathfrak{k}} M$ of cardinality θ_2 can be $\leq_{\mathfrak{k}}$-embedded into some $\mathrm{EM}_{\tau(\mathfrak{k})}(\lambda, \Phi), \Phi \in \Upsilon^{\mathrm{or}}_\kappa[\mathfrak{k}]$.

Informally, the point is it allows us to use EM models. The key point is finding a suitable template, set Φ of quantifier free types, which requires finding enough indiscernible sequences. When $K_{\mathfrak{k}}$ is an a.e.c. (as opposed to an elementary or pseudo elementary class) we must go through the Presentation Theorem to find an indiscernible sequence, i.e. we require sufficiently large models omitting the types in Γ.

To further motivate our approach, consider a not so strong conjecture, still enough to exemplify "the function $\lambda \mapsto \dot{I}(\lambda, \mathfrak{k})$ cannot be too wild".

Conjecture 2.3. 1. Letting $\mathbf{C}^{\mathrm{fp}}_{\aleph_0} = \{\lambda : \lambda = \beth_\lambda$ and $\mathrm{cf}(\lambda) = \aleph_0\}$ and fixing an a.e.c. $\mathfrak{k}$, not both of the following classes are stationary (or restrict yourself to some strongly inaccessible μ and "stationary" means below it):
   (a) $\mathbf{S}_1 = \{\lambda \in \mathbf{C}^{\mathrm{fp}}_{\aleph_0} : \dot{I}(\lambda, \mathfrak{k}) < \lambda\}$
   (b) $\mathbf{S}_2 = \{\lambda \in \mathbf{C}^{\mathrm{fp}}_{\aleph_0} : \dot{I}(\lambda, \mathfrak{k}) \geq \lambda\}$.
2. A weaker conjecture (presented in the abstract) is replacing clause (b) by
   (b)′ $\mathbf{S}_3 = \{\lambda \in \mathbf{C}^{\mathrm{fp}}_{\aleph_0}$: for every $M \in K^{\mathfrak{k}}_\lambda$ has $\leq_{\mathfrak{k}}$-extensions N of any cardinality $> \lambda\}$.

Why "$\mathrm{cf}(\lambda) = \aleph_0$"? First, trying to prove $\lambda \in \mathbf{S}_3$, we can approximate N by $\Phi \in \Upsilon^{\mathrm{or}}_{\lambda_n}[\mathfrak{k}], \lambda_n < \lambda$ as we can approximate M by $N' \leq_{\mathfrak{k}} M, \|N'\| = \lambda_n$ where $\lambda_n < \lambda_{n+1} < \lambda = \Sigma\{\lambda_m : m\}$. Second, for $\lambda \in \mathbf{C}^{\mathrm{fp}}_{\aleph_0}$ it is enough to show that $\{M/\equiv_{\mathbb{L}_{\infty,\lambda}} : M \in K^{\mathfrak{k}}_\lambda\}$ is small because it is well known that if $\mathrm{cf}(\lambda) = \aleph_0$ and M_1, M_2 are of cardinality λ and $\mathbb{L}_{\infty,\lambda}$-equivalent then they are isomorphic; on such logics see, e.g. [Dic85].

Thesis 2.4. There are, for a.e.c. $\mathfrak{k}$, meaningful dichotomy theorems for $\dot{I}(\lambda, K_{\mathfrak{k}})$ when K is a class of $\tau(\mathfrak{k})$-models, $K = K_{\mathfrak{k}}$ and $\mathfrak{k} = (K_{\mathfrak{k}}, \leq_{\mathfrak{k}})$.

This is a more concrete thesis than "considering a.e.c.'s is a good frame for model theory"; even more concrete is the "main gap conjecture". It had been proved that if $K_{\mathfrak{k}}$ is the class of models of a complete countable first order theory <u>then</u> it satisfies the "main gap", i.e. either $\dot{I}(\lambda, K)$ is large, even $= 2^\lambda$ for all uncountable λ or $\dot{I}(\aleph_\alpha, K)$ is small, even $< \beth_{\omega_1}(|\alpha|)$ for all $\alpha > 0$; see [Sh:c, Ch.XII], "The book's main theorem". In general for a class K of τ-models the "main gap" will say that either $\dot{I}(\lambda, K)$ is large (i.e. 2^λ or $\geq \lambda^+$) for every λ large enough or it is small for every λ large enough say $\dot{I}(\aleph_\alpha, K)$ is $\leq \beth_{1,n}(|\alpha|)$ for some $n = n(K) < \omega$.

We are far away from this, still, until now for the a.e.c. the categoricity case was almost alone, i.e. we start assuming $\dot{I}(\lambda, K) = 1$ in some λ, see below, but we try here to look "higher".

The contribution of the present paper is to show that in the much more general context of a.e.c.'s for some $\aleph_0$-closed unbounded class $\mathbf{C}$ of cardinals, we have $\lambda \in \mathbf{C} \Rightarrow \dot{I}(\lambda, K_\mathfrak{k}) \geq \lambda$, a non-structure result, <u>or</u> $\lambda \in \mathbf{C} \wedge M \in K^\mathfrak{k}_\lambda \Rightarrow M$ has arbitrary large $\leq_\mathfrak{k}$-extensions. Note that the latter property is now taken for granted for elementary classes but is a real gain for a.e.c.

As noted in §1, in [Sh:734] and [Sh:600] we obtained results on $\dot{I}(\lambda, K)$ for a.e.c.'s assuming categoricity in some λ's. However, nothing was known for general a.e.c.'s under weaker few models assumption.

On abstract elementary classes, see [Sh:88r], [Bal09] and [Sh:E53]. We will make essential use of the Presentation Theorem, which says that every a.e.c. can be represented as a PC class, say $\mathrm{PC}(T, \Gamma)$, see [Sh:88r, §1].

We thank the audience in the lecture in the Hebrew University seminar 2/2005 for their comments on an earlier version of this paper and Maryanthe Malliaris for helping much in improving §2 and some corrections in fall 2011 - winter 2012 and Will Boney for some further corrections (fall 2013).

2.2 Discussion

We give some further details regarding §(2.1).

In Thesis 2.2 the result on EM models needed is: [Sh:394, Claim 0.6], [Sh:394, Claim 8.6], the "a.e.c. omitting types theorem" and [Sh:394, Lemma 8.7,p.46].

Fact 2.5. *Let $\mathfrak{k}$ be an a.e.c. If $\lambda \in \mathbf{C}_{\mathrm{fp}}, \lambda > \mathrm{LST}_\mathfrak{k}$ and $M \in K^\mathfrak{k}_\lambda$ <u>then</u> for every $\theta \in [\mathrm{LST}_\mathfrak{k}, \lambda)$ and $N \leq_\mathfrak{k} M$ of cardinality θ there is $\Phi \in \Upsilon[\mathfrak{k}]$ such that:*

(a) $|\tau(\Phi)| = \theta$

(b) *for any linear order I, in particular $I = \lambda$,* without loss of generality $N \leq_\mathfrak{k} \mathrm{EM}_{\tau(\mathfrak{k})}(I, \Phi)$ *where this denotes the reduct of the EM model to the vocabulary of* $\mathfrak{k}$.

<u>*Comment:*</u>

Let us repeat, the two points when $\mathrm{cf}(\lambda) = \aleph_0$ *may be as required:*

(a) *downward large depth in §4,*

(b) *if we like to find large* $N \leq_{\mathfrak{k}}$*-extending* M *for a given* $M \in K^{\mathfrak{k}}_{\lambda}$*, if* $\mathrm{cf}(\lambda) = \aleph_0$ *we can get it as an* ω*-limit of* $M' <_{\mathfrak{k}} M, \|M'\| < \lambda$.

Such considerations further lead us to

Question 2.6. Let $\Phi \in \Upsilon_\theta[\mathfrak{k}]$ and κ be a cardinal.

Sort out the functions

(a) $\lambda \mapsto |\{\mathrm{EM}_{\tau(\mathfrak{k})}(I, \Phi)/\cong : I$ a linear order of cardinality $\lambda\}|$

(b) $\lambda \mapsto \dot{I}_{\tau(\mathfrak{k})}(\lambda, \kappa, \Phi) := |\{\mathrm{EM}_{\tau(\mathfrak{k})}(I, \Phi)/\equiv_{\mathbb{L}_{\infty,\kappa}} : I$ a linear order of cardinality $\lambda\}$.

Recall, by [Sh:11] restricting ourselves to cardinals $\lambda = \lambda^{<\kappa}$, that the function in clause (b) of 2.6 is "nice", more specifically: if $\theta \leq \lambda_1 = \lambda_1^{<\kappa} < \lambda_2$ then $\dot{I}_{\tau(\mathfrak{k})}(\lambda_1, \kappa, \mathfrak{k}) \geq \min\{\lambda_1^+, \dot{I}(\lambda_2, \kappa, \mathfrak{k})\}$.

What occurs if $\lambda_1 < \lambda_1^{<\kappa}$? The case $\lambda_1 = \beth_\delta, \mathrm{cf}(\delta) = \aleph_0$ is more approachable than the general case, see 5.2.

Our hope is to get "bare bones superstability", i.e. good λ-frames inside $\mathfrak{k}$, (as in [Sh:600],[Sh:734]).

Another point concerning the function $\dot{I}(\lambda, \kappa, \mathfrak{k})$ is: for a model M, cardinal θ and logic $\mathscr{L}$ we can define the depth of M for $(\mathscr{L}, \theta)$ as $\min\{\alpha$: if $\bar{a}, \bar{b} \in {}^{\varepsilon}M, \varepsilon < \theta$ and $\bar{a}, \bar{b}$ realizes the same formulas of $\mathbb{L}_{\infty,\theta}$ (or $\mathbb{L}_{\infty,\theta}[\mathfrak{k}]$) of depth $< \alpha$ then they realize the same $\mathbb{L}_{\infty,\theta}$-formulas$\}$; of course, only formulas in $L_{\|M\|^{<\theta},\theta}$ are relevant. This is a good way to "slice" the equivalence and it is easier for LST considerations.

2.3 What is Done

A phenomena making the investigation of general a.e.c. hard is having $\leq_{\mathfrak{k}}$-maximal models of large cardinality. As with amalgamation, we may consider the property

$(*)^1_\lambda$ if $M \in K^{\mathfrak{k}}_\lambda$ then M is not $\leq_{\mathfrak{k}}$-maximal.

In investigations like [Sh:E46] and [Sh:576], which look at $\cup\{K^{\mathfrak{k}}_{\lambda+\ell} : \ell < 4\}$ this is relevant. But in investigations as in [Sh:734], looking at $\cup\{K^{\mathfrak{k}}_\lambda : \lambda = \beth_\lambda\}$, it is more natural to consider

$(*)^2_\lambda$ if $M \in K^{\mathfrak{k}}_\lambda$ then for any $\mu > \lambda$ there is $N \in K^{\mathfrak{k}}_\mu$ which $\leq_{\mathfrak{k}}$-extends N.

In §4 we consider a $\lambda = \beth_\lambda$ of cofinality $\aleph_0$ which is more than strong limit and try to prove non-structure from $\neg(*)^2_\lambda$. Given $N \in K^{\mathfrak{k}}_\lambda$ we try to build an EM model (that is construct the Φ) $\leq_{\mathfrak{k}}$-extending N by an increasing chain of approximations:

given $\lambda_n \to \lambda, M_n \to N, M_n \in K^{\mathfrak{k}}_{\lambda_n}$. The n-th approximation Φ_n to Φ has to have "Φ_n in a suitable sense is represented in N say of size λ_{n+1}".

Being stuck should be a reason for non-structure. For simplicity we consider only cardinals $\mu = \beth_\mu$, the gain without this restriction seems minor.

Concerning the results of §4 it would be nicer to make one more step concerning 4.15, 4.14 and deal also with $\lambda = \beth_\lambda$ instead of $\lambda = \beth_{1,\lambda}$, but a more central question is to get the non-structure result for every $\lambda' > \lambda$. It is natural to try given $\Phi \in \Upsilon^{\text{sor}}_\kappa[\mathfrak{k}_M]$ and $M \leq_{\mathfrak{k}} N$, to define a "depth" for approximation of the existence of a $\leq_{\mathfrak{k}}$-embedding of standard $\text{EM}_{\tau(\mathfrak{k})}(I, \Phi)$ into N (see Definition 3.2(2)), so that depth infinity give existence. But this does not work for us, so Definition 4.2 is a substitute, moreover we need "indirect evidence", see Definition 4.7.

Our main theorem is

Theorem 2.7. *For any a.e.c. for some closed unbounded class of cardinals* C, *if* $(\exists\lambda \in \mathrm{C})[\mathrm{cf}(\lambda) = \aleph_0 \wedge \dot{I}(\lambda, K_{\mathfrak{k}}) < \lambda]$ *and* $M \in K_{\mathfrak{k}}$ *of cardinality* $\mu \in \mathrm{C}$ *of cofinality* $\aleph_0$, *then* M *has a proper* $<_{\mathfrak{k}}$*-extension, and even ones of arbitrarily large cardinality.*

The natural next steps are

Conjecture 2.8. 1. In Theorem 4.16, i.e. what is promised in the abstract we can choose C as an end segment of $\{\mu : \mu = \beth_{1,\mu}\}$ or just choose C as $\{\mu : \mu = \beth_{2,\mu} > \mathrm{LST}_{\mathfrak{k}}\}$.

2. For every a.e.c. $\mathfrak{k}$ for some closed unbounded class C of cardinals, we have $M \in K^{\mathfrak{k}}_\lambda \wedge \lambda \in \mathrm{C} \wedge \mathrm{cf}(\lambda) = \aleph_0 \Rightarrow \Upsilon^{\text{or}}_\lambda[\mathfrak{k}_M] \neq \emptyset$ or $\lambda \in \mathrm{C} \wedge \mathrm{cf}(\lambda) = \aleph_0 \Rightarrow \dot{I}(\lambda, K_{\mathfrak{k}}) \geq 2^\lambda$ or at least $\geq \lambda^+$.

We intend to deal with part (1) in a continuation.

2.4 Recalling Definitions and Notation

Notation 2.9. Let Card be the class of infinite cardinals.

Definition 2.10. 1. Let $\beth_{0,\alpha}(\lambda) = \beth_\alpha(\lambda) := \lambda + \Sigma\{2^{\beth_\beta(\lambda)} : \beta < \alpha\}$. Let $\beth_{\varepsilon,\alpha}(\lambda)$ be defined by induction on $\varepsilon > 0$ and for each ε by induction on $\alpha : \beth_{\varepsilon,0}(\lambda) = \lambda$, for limit β we let $\beth_{\varepsilon,\beta} = \sum_{\gamma<\beta} \beth_{\varepsilon,\gamma}$ and for $\varepsilon = \zeta + 1$ let $\beth_{\zeta+1,\beta+1}(\lambda) = \beth_{\zeta,\mu}(\lambda)$ where[2]

2 why not, e.g. $\mu = \beth_{1,\beta}(\lambda)^+$? Not a serious difference as for limit α we shall get the same value and in 2.14(1) this simplifies the notation.

$\mu = (2^{\beth_{\zeta,\beta}(\lambda)})^+$, lastly for limit ε let $\langle \beth_{\varepsilon,\alpha} : \alpha \in \text{Ord}\rangle$ list in increasing order the closed unbounded class $\bigcap_{\zeta<\varepsilon}\{\beth_{\zeta,\alpha} : \alpha \in \text{Ord}\}$.

2. Let $\lambda \gg \kappa$ mean $(\forall \alpha < \lambda)(|\alpha|^\kappa < \lambda)$.

Convention 2.11. 1. $\mathfrak{k} = (K_{\mathfrak{k}}, \le_{\mathfrak{k}})$ is an a.e.c., with vocabulary $\tau_{\mathfrak{k}} = \tau(\mathfrak{k})$ and $\text{LST}(\mathfrak{k}) = \text{LST}_{\mathfrak{k}}$ its Löwenheim-Skolem-Tarski number, see [Sh:88r, §1]. If not said otherwise, we assume $|\tau_{\mathfrak{k}}| \le \text{LST}_{\mathfrak{k}}$.

2. $K^{\mathfrak{k}}_\lambda = K_{\mathfrak{k},\lambda} = \{M \in K_{\mathfrak{k}} : \|M\| = \lambda\}$.
3. If $K = K_{\mathfrak{k}}$ we may write $\mathfrak{k}$ instead of K; also we may write K or K_λ omitting $\mathfrak{k}$ when (as usually here) $\mathfrak{k}$ is clear from the context.

Definition 2.12. For a class K of τ-models:

(*a*) for a cardinal λ, let $\dot{I}(\lambda, K)$ be the cardinality of $\{M/\cong : M \in K$ has cardinality $\lambda\}$

(*b*) for a cardinal λ and a logic $\mathscr{L}$, let $\dot{I}(\lambda, \mathscr{L}, K) = \{M/\equiv_{\mathscr{L}(\tau)} : M \in K$ has cardinality $\lambda\}$.

Definition 2.13. 1. Φ is a template proper for linear orders <u>when</u>:

(*a*) for some vocabulary $\tau = \tau_\Phi = \tau(\Phi)$, Φ is an ω-sequence, with the n-th element a complete quantifier free n-type in the vocabulary τ,

(*b*) for every linear order I there is a τ-model M denoted by $\text{EM}(I, \Phi)$, generated by $\{a_t : t \in I\}$ such that $s \ne t \Rightarrow a_s \ne a_t$ for $s, t \in I$ and $\langle a_{t_0}, \dots, a_{t_{n-1}}\rangle$ realizes the quantifier free n-type from clause (a) whenever $n < \omega$ and $t_0 <_I \dots <_I t_{n-1}$. We call $(M, \langle a_t : t \in I\rangle)$ a Φ – EM-pair or EM-pair for Φ; so really M and even $(M, \langle a_t : t \in I\rangle)$ are determined only up to isomorphism but abusing notation we may ignore this and use $I_1 \subseteq J_1 \Rightarrow \text{EM}(I_1, \Phi) \subseteq \text{EM}(I_2, \Phi)$. We call $\langle a_t : t \in I\rangle$ "the" skeleton of M; of course again "the" is an abuse of notation as it is not necessarily unique.

1A. If $\tau \subseteq \tau(\Phi)$ then we let $\text{EM}_\tau(I, \Phi)$ be the τ-reduct of $\text{EM}(I, \Phi)$.

2. $\Upsilon^{\text{or}}_\kappa[\mathfrak{k}]$ is the class of templates Φ proper for linear orders satisfying clauses $(a)(\alpha), (b), (c)$ of Claim 2.14(1) below and $|\tau(\Phi)\backslash\tau_{\mathfrak{k}}| \le \kappa$; normally we assume $\kappa \ge |\tau_{\mathfrak{k}}| + \text{LST}_{\mathfrak{k}}$ but using $\mathfrak{k}_M$ we do not assume $\kappa \ge \|M\|$, see 3.1. The default value of κ is $\text{LST}_{\mathfrak{k}}$ and then we may write $\Upsilon^{\text{or}}_{\mathfrak{k}}$ or $\Upsilon^{\text{or}}[\mathfrak{k}]$ and for simplicity if not said otherwise $\kappa \ge \text{LST}_{\mathfrak{k}}$ (and so $\kappa \ge |\tau_{\mathfrak{k}}|$). We may omit $\mathfrak{k}$ and or omit "or" when clear from the context.

3. For a class K of so called index models, we define "Φ proper for K" similarly when in clause (b) of part (1) we demand $I \in K$, so K is a class of τ_K-models, i.e.

(*a*) Φ is a function, giving for any complete quantifier free n-type in τ_K realized in some $M \in K$, a quantifier free n-type in τ_Φ

$(b)'$ in clause (b) of part (1), the quantifier free type which $\langle a_{t_0}, \dots, a_{t_{n-1}} \rangle$ realizes in M is $\Phi(\text{tp}_{\text{qf}}(\langle t_0, \dots, t_{n-1} \rangle, \emptyset, I))$ for $n < \omega, t_0, \dots, t_{n-1} \in I$.

Fact 2.14. *1. Let $\mathfrak{k}$ be an a.e.c. and $M \in K_{\mathfrak{k}}$ be of cardinality $\geq \lambda = \beth_{1,1}(\text{LST}_{\mathfrak{k}})$ recalling we may assume $|\tau_{\mathfrak{k}}| \leq \text{LST}_{\mathfrak{k}}$ as usual.*
<u>Then</u> there is a Φ such that Φ is proper for linear orders and:
- *(a) (α) $\tau_{\mathfrak{k}} \subseteq \tau_{\Phi}$,*
  (β) $|\tau_{\Phi}| = \text{LST}_{\mathfrak{k}} + |\tau_{\mathfrak{k}}|$
- *(b) for any linear order I the model $\text{EM}(I, \Phi)$ has cardinality $|\tau(\Phi)| + |I|$ and we have $\text{EM}_{\tau(\mathfrak{k})}(I, \Phi) \in K_{\mathfrak{k}}$*
- *(c) for any linear orders $I \subseteq J$ we have $\text{EM}_{\tau(\mathfrak{k})}(I, \Phi) \leq_{\mathfrak{k}} \text{EM}_{\tau(\mathfrak{k})}(J, \Phi)$; moreover, if $M \subseteq \text{EM}(J, \Phi)$ then $(M \restriction \tau_{\mathfrak{k}}) \leq_{\mathfrak{k}} \text{EM}_{\tau(\mathfrak{k})}(J, \Phi)$*
- *(d) for every finite linear order I, the model $\text{EM}_{\tau(\mathfrak{k})}(I, \Phi)$ can be $\leq_{\mathfrak{k}}$-embedded into M.*

1A. Moreover, assume in (1) also $\lambda = \beth_{1,1}(\kappa), \kappa \geq \text{LST}_{\mathfrak{k}} + |\tau_{\mathfrak{k}}|$ so not necessarily assuming $\text{LST}_{\mathfrak{k}} \geq |\tau_{\mathfrak{k}}|$, M^+ is an expansion of M with $\tau(M^+)$ of cardinality $\leq \kappa$ and $b_\alpha \in M$ for $\alpha < \lambda$ are pairwise distinct. <u>Then</u> there is Φ proper for linear orders such that:
- *(a) (α) $\tau(M^+) \subseteq \tau_{\Phi}$ hence $\tau(\mathfrak{k}) \subseteq \tau_{\Phi}$*
  (β) τ_{Φ} has cardinality κ
- *(b), (c)has in part (1)*
- *(d) if I is a finite linear order and $t_0 <_I \dots <_I t_{n-1}$ list its elements and $M_I = \text{EM}(I, \Phi)$ with skeleton $\langle a_{t_i} : t \in I \rangle$, then for some ordinals $\alpha_0 < \dots < \alpha_{n-1} < \lambda$ there is an embedding of M_I into M^+ mapping a_{t_ℓ} to b_{α_ℓ} for $\ell < n$.*

2. If $\text{LST}_{\mathfrak{k}} < |\tau_{\mathfrak{k}}|$ and there is $M \in K_{\mathfrak{k}}$ of cardinality $\geq \beth_{1,1}(2^{\text{LST}_{\mathfrak{k}}})$, <u>then</u> there is $\Phi \in \Upsilon^{\text{or}}_{\text{LST}(\mathfrak{k})+|\tau(\Phi)|}[\mathfrak{k}]$ such that $\text{EM}(I, \Phi)$ has cardinality $\leq \text{LST}_{\mathfrak{k}}$ for I finite and $\tau_{\Phi} \backslash \tau(M)$ has cardinality $\text{LST}_{\mathfrak{k}}$. Note that $\mathscr{E}$ has $\leq 2^{\text{LST}_{\mathfrak{k}}}$ equivalence classes where $\mathscr{E} = \{(P_1, P_2) : P_1, P_2 \in \tau_{\Phi}$ and $P_1^{\text{EM}(I,\Phi)} = P_2^{\text{EM}(I,\Phi)}$ for every linear order $I\}$ hence above "$\geq \beth_{1,1}(2^{\text{LST}(\mathfrak{k})})$" suffice.

3. We can combine parts (1A) and (2). Also in both cases having a model of cardinality $\geq \beth_\alpha$ for every $\alpha < (2^{\text{LST}(\mathfrak{k})+|\tau(\mathfrak{k})|})^+$ suffice in parts (1),(1A) and for every $\alpha < \beth_2(\text{LST}_{\mathfrak{k}})^+$ suffice in part (2).

We add

Claim 2.15. *For every cardinal μ and strong limit $\chi \leq \mu$ there is a dense κ-saturated linear order $I = I_\mu$ of cardinality μ such that:*
$(*)$ if $\theta < \text{cf}(\partial) = \partial < \mu, 2^\theta \leq \chi$ <u>then</u>
$(*)_{I,\chi,\partial,\theta}$ we have $2^\theta \leq \chi$ and $\theta < \partial = \text{cf}(\partial)$ and $(A) \Rightarrow (B)$ where:
- *(A) (a) $I_0 \subseteq I$*
  (b) I_0 has cardinality $\leq \theta$

(c) I_1 *is a linear order extending* I_0

(d) $u_n \subseteq u_{n+1} \subseteq \theta = \bigcup_n u_n$

(e) $\bar{t}^1_\alpha \in {}^\theta(I_1)$ *for* $\alpha < \partial$ *and* $\langle \bar{t}^1_\alpha : \alpha < \partial \rangle$ *is an indiscernible sequence in* I_1 *over* I_0 *(for quantifier free formulas)*

(f) *for every* n, $I_{1,n} = I_1 \restriction (\{t^1_{\alpha,i} : i \in u_n, \alpha < \partial\} \cup I_0)$ *is embeddable into* I *over* I_0

(B) *there is* $\langle \bar{t}_\alpha : \alpha < \mu \rangle$ *such that*

(a) $\bar{t}_\alpha \in {}^\theta I$

(b) $\langle \bar{t}_\alpha : \alpha < \mu \rangle$ *is an indiscernible sequence over* I_0 *into* I *(for quantifier free formulas)*

(c) *the quantifier free type of* $\bar{t}_0 {}^\wedge \ldots {}^\wedge \bar{t}_n$ *over* I_0 *in* I *is equal to the quantifier free type of* $\bar{t}^1_0 {}^\wedge \ldots {}^\wedge \bar{t}^1_n$ *over* I_0 *in* I_1 *for every* n

$(B)^+$ *moreover we can replace* $\langle \bar{t}_\alpha : \alpha < \mu \rangle$ *by* $\langle \bar{t}_s : s \in I \rangle$.

Remark 2.16. 1. We may consider replacing (A)(e) by

$(e)'$ $\alpha = \beth_2(\theta)^+, u_n \subseteq u_{n+1} \subseteq \theta = \bigcup_n u_n$ and $I_{1,n} = \{t^1_{\alpha,\varepsilon} : \alpha < \partial, \varepsilon \in u_n\}$ and there is $\bar{f} = \langle f_\eta : \eta \in \Lambda \rangle$ such that f_η embeds $I_{1,\ell g(\eta)}$ into I_1 over I_0 and $\nu \triangleleft \eta \Rightarrow f_\nu \subseteq f_\eta$ where $\Lambda = \{\eta : \eta$ is a decreasing sequence of ordinals $< \alpha\}$.

2. Clauses (A)(d),(e) can be weakened to:

$\oplus$ if $i, j < \theta$ <u>then</u> $I_1 \restriction (\{t^1_{\alpha,i} \alpha = 0, 1$ and $i < \theta\} \cup I_0)$ can be embedded into I over I_0.

But the present form fits our application.

Proof.. First we give a sufficient condition for $(*)_{I,\chi,\partial,\theta}$

⊞ the linear order I satisfies $(*)_{I,\chi,\partial,\theta}$ when: $\chi > \partial = \mathrm{cf}(\partial) > \theta$ and

(a) I is a linear order of cardinality μ

(b) if $I_0 \subseteq I, |I_0| \le \theta$ <u>then</u> the set $I_0^+ = \{t \in I : t \notin I_0$ and there is no $t' \in I \backslash I_0 \backslash \{t\}$ realizing the same cut of I_0 in $I\}$ has cardinality $< \partial$, so if $\partial = (2^\theta)^+$ this holds

(c) if $a <_I b$ then I is embeddable into $(a, b)_I$

(d) every linear order of cardinality $\le \theta$ is embeddable into I

(e) in I there is a decreasing sequence of length μ and an increasing sequence of length μ

(f) to get $(B)^+$ we need: if J is a linear order of cardinality $\le \theta$ then we can embed $I \times J$ (ordered lexicographically into I).

It is obvious that there is such linear order. It is also easy to see that if I satisfies (a)-(d) then $(*)_{I,\partial,\theta}$. $\square_{2.15}$

3 More on Templates

Why do we need $\Upsilon_\kappa^{\rm sor}[M, \mathfrak{k}]$? Remember that such Φ's are witnesses to M having $\leq_{\mathfrak{k}}$-extensions in every $\mu > \mathrm{LST}_{\mathfrak{k}} + \|M\|$ so proving existence is a major theme here. First, why do we need below $\Upsilon_\kappa^{\rm sor}$? Because "$\Upsilon_\kappa^{\rm sor}[M, \mathfrak{k}] \neq \emptyset$" is equivalent to M being not $\leq_{\mathfrak{k}}$-maximal; moreover has $\leq_{\mathfrak{k}}$-extensions of arbitrarily large cardinality so proving this for every $M \in K_\lambda^{\mathfrak{k}}$ indicates "$\mathfrak{k}$ is nice, at least in λ". Second, why do we need various partial orders on $\Upsilon_\kappa^{\rm sor}[M, \mathfrak{k}]$'s?

In a major proof here to build $\Phi \in \Upsilon_\kappa^{\rm sor}[M, \mathfrak{k}]$ we use $\leq_{\mathfrak{k}}$-increasing M_n with union M and try to choose $\Phi_n \in \Upsilon_\kappa^{\rm sor}[M_n, \mathfrak{k}]$ increasing with n. For this we assume $\|M_n\| = \lambda_n, \lambda_n \ll \lambda_{n+1}$ and we use an induction hypothesis that Φ_n has a say λ_{n+5}-witness in M.

Of course, it is nice if $\mathrm{EM}_{\tau(\mathfrak{k})}(\lambda_{n+5}, \Phi_n)$ is $\leq_{\mathfrak{k}}$-embeddable into M over M_n but for this we do not have strong enough existence theorem. To fine tune this and having a limit ($\Phi \in \Upsilon_\kappa^{\rm sor}[M, \mathfrak{k}]$) we need some orders.

Definition 3.1. For $\mathfrak{k}$ an a.e.c. and $M \in K_{\mathfrak{k}}$ let $\mathfrak{k}_M = \mathfrak{k}[M]$ be the following a.e.c.:
(a) vocabulary $\tau_{\mathfrak{k}} \cup \{c_a : a \in M\}$ where the c_a's are pairwise distinct new individual constants
(b) $N \in K_{\mathfrak{k}_M}$ <u>iff</u> $N \restriction \tau_{\mathfrak{k}} \in K_{\mathfrak{k}}$ and $a \mapsto c_a^N$ is a $\leq_{\mathfrak{k}}$-embedding of M into $N \restriction \tau_{\mathfrak{k}}$;
(c) $N_1 \leq_{\mathfrak{k}_M} N_2$ <u>iff</u>
  (α) N_1, N_2 are $\tau_{\mathfrak{k}_M}$-models from $K_{\mathfrak{k}_\mu}$
  (β) $N_1 \subseteq N_2$
  (γ) $(N_1 \restriction \tau_{\mathfrak{k}}) \leq_{\mathfrak{k}} (N_2 \restriction \tau_{\mathfrak{k}})$.

Definition 3.2. 1. We call $N \in K_{\mathfrak{k}_M}$ standard <u>when</u> $M \leq_{\mathfrak{k}} N \restriction \tau_{\mathfrak{k}}$ and $a \in M \Rightarrow c_a^N = a$.
2. If $N^1 \in K_{\mathfrak{k}_M}$ is standard and $N^0 = N^1 \restriction \tau_{\mathfrak{k}}$ <u>then</u> we write $N^1 = N^0_{[M]}$.
3. We call $\Phi \in \Upsilon_{\mathfrak{k}}^{\rm or}$ standard <u>when</u> $M = \mathrm{EM}_{\tau(\mathfrak{k})}(\emptyset, \Phi)$ implies $N \leq_{\mathfrak{k}} M \restriction \tau_{\mathfrak{k}}$ when N is the submodel[3] of $M \restriction \tau_{\mathfrak{k}}$ with universe $\{c^M : c \in \tau(\Phi)$ an individual constant$\}$. We call Φ fully standard when above $N = M \restriction \tau_{\mathfrak{k}}$.
4. Let $\Upsilon_\kappa^{\rm sor}[\mathfrak{k}]$ be the class of standard $\Phi \in \Upsilon_\kappa^{\rm or}[\mathfrak{k}]$.
5. For $M \in K_{\mathfrak{k}}$ let $\Upsilon_\kappa^{\rm sor}[\mathfrak{k}_M]$ be the class of κ-standard $\Phi \in \Upsilon_\kappa^{\rm or}[\mathfrak{k}_M]$ which[4] means:
  (a) letting $\kappa_1 = \kappa + \|M\|$, we have $\Phi \in \Upsilon_{\kappa_1}^{\rm sor}[\mathfrak{k}]$

3 Note that we have not said "$\Phi \in \Upsilon_{\mathfrak{k}[N]}^{\rm or}$" but by renaming this follows.

4 So though such Φ belongs to $\Upsilon_\kappa^{\rm or}[\mathfrak{k}]$, being standard for $\Upsilon_\kappa^{\rm sor}[\mathfrak{k}_M]$ is a different demand than being standard for $\Upsilon_\kappa^{\rm or}[\mathfrak{k}]$ as for the latter possibly $\{c_a : a \in M\} \subsetneqq \{c \in \tau_\Phi : c$ an individual constant$\}$.

(*b*) $\{c_a : a \in M\} = \{c \in \tau(\Phi) : c \text{ an individual constant}\}$.
(*c*) $N = \mathrm{EM}(\emptyset, \Phi) \Rightarrow |N| = \{c^N : c \in \tau_\Phi\}$
(*d*) $\tau'_\Phi := \tau_\mathfrak{k} \backslash \{c \in \tau_\Phi$ is an individual constant$\}$ has cardinality $\leq \kappa$
(*e*) if $N = \mathrm{EM}(I, \Phi)$ and N_1 is a submodel of $N \restriction \tau'_\Phi$ then $N_1 \restriction \tau_\mathfrak{k} \leq_\mathfrak{k} N \restriction \tau_\mathfrak{k}$.

5A. We may omit κ in part (5) when $\kappa = \mathrm{LST}_\mathfrak{k} + |\tau_\mathfrak{k}|$. We may write $\Upsilon^{\mathrm{sor}}_\kappa[M, \mathfrak{k}]$ instead of $\Upsilon^{\mathrm{sor}}_\kappa[\mathfrak{k}_M]$, useful when $\mathfrak{k}$ is not clear from the context.

Observation 3.3. 1. If $\Phi \in \Upsilon^{\mathrm{sor}}_\kappa[\mathfrak{k}, M]$ then $\Phi \in \Upsilon^{\mathrm{or}}_{\kappa+\|M\|}[\mathfrak{k}]$ but not necessarily the inverse.

2. If $\Phi \in \Upsilon^{\mathrm{sor}}_\kappa[\mathfrak{k}, M]$ then Φ is a fully standard member of $\Upsilon^{\mathrm{or}}_\kappa[\mathfrak{k}_M]$.

Claim 3.4. *Assume $\mathfrak{k}$ is an a.e.c. and $M \in K_\mathfrak{k}$ and $\mathfrak{k}_1 = \mathfrak{k}_M$ then:*
(*a*) *$\mathfrak{k}_1$ is an a.e.c.*
(*b*) $\mathrm{LST}_{\mathfrak{k}_1} = \mathrm{LST}_\mathfrak{k} + \|M\|$
(*c*) *applying 2.14 to $\mathfrak{k}_1$, we can add "$\Phi \in \Upsilon^{\mathrm{sor}}_\kappa[\mathfrak{k}_M]$".*

Proof.. Straightforward. $\square_{3.4}$

Definition 3.5. Assume J is a linear order of cardinality λ and $\lambda \to (\mu)^n_\theta$. We define the ideal $\mathscr{I} = \mathrm{ER}^n_{J,\mu,\theta}$ on the set $[J]^\mu$ by:

- $\mathscr{S} \subseteq [J]^\mu$ belongs to $\mathscr{I}$ iff for some $\mathbf{c} : [J]^{\leq n} \to \theta$ there is no $s \in \mathscr{S}$ such that $\mathbf{c} \restriction [s]^n$ is constant.

Observation 3.6. 1. If $|J| = \lambda$ and $\lambda \to (\mu)^n_\theta$ then $\mathrm{ER}^n_{J,\mu,\theta}$ is indeed an ideal, i.e. $J \notin \mathrm{ER}^n_{J,\mu,\theta}$.

2. If $\theta = \theta^{<\kappa}$ then this ideal is κ-complete.

Definition 3.7. 1. For vocabularies τ_1, τ_2 we say that $\mathbf{h}$ is an isomorphism from τ_1 onto τ_2 when $\mathbf{h}$ is a one-to-one function from the non-logical symbols of τ_1 (= the predicates and function symbols) onto those of τ_2 such that:
(*a*) if $P \in \tau_1$ is a predicate then $\mathbf{h}(P)$ is a predicate of τ_2 and $\mathrm{arity}_{\tau_1}(P) = \mathrm{arity}_{\tau_2}(\mathbf{h}(P))$
(*b*) if $F \in \tau_1$ is a function symbol[5] then $\mathbf{h}(F)$ is a function symbol of τ_2 and $\mathrm{arity}_{\tau_1}(F) = \mathrm{arity}_{\tau_2}(\mathbf{h}(F))$.

2. If $\mathbf{h}$ is an isomorphism from the vocabulary τ_1 onto the vocabulary τ_1 and M_1 is a τ_1-model then $M_1^{[\mathbf{h}]}$ is the unique M_2 such that:
(*a*) M_2 is a τ_2-model
(*b*) $|M_2| = |M_1|$
(*c*) $P_2^{M_2} = P_1^{M_1}$ when $P_1 \in \tau_1$ is a predicate and $P_2 = \mathbf{h}(P_1)$
(*d*) $F_2^{M_2} = F_1^{M_1}$ when $F_1 \in \tau_1$ is a function symbol and $F_2 = \mathbf{h}(F_1)$.

5 this includes individual constants

3. We say $\mathbf{h}$ is an isomorphism from τ_1 onto τ_2 over τ <u>when</u> $\tau \subseteq \tau_1 \cap \tau_2$, $\mathbf{h}$ is an isomorphism from τ_1 onto τ_2 and $\mathbf{h} \restriction \tau$ is the identity.
4. If $\Phi_1 \in \Upsilon_\kappa^{\rm or}$ and $\mathbf{h}$ is an isomorphism from the vocabulary $\tau_1 := \tau(\Phi)$ onto the vocabulary τ_2 <u>then</u> $\Phi^{[\mathbf{h}]}$ is the unique $\Phi_2 \in \Upsilon_\kappa^{\rm or}$ such that: if I is a linear order, $M_1 = {\rm EM}(I, \Phi_1)$ with skeleton $\langle a_t : t \in I\rangle$ then $M_1^{[\mathbf{h}]}$ is the model $({\rm EM}(I, \Phi_2))^{[\mathbf{h}]}$ with the same skeleton.

Observation 3.8. 1. In 3.7(2), $M_2 = M_1^{[\mathbf{h}]}$ is indeed a τ_2-model. If in addition $\mathbf{h}$ is over τ (i.e. $\tau \subseteq \tau_1 \cap \tau_2$ and $\mathbf{h} \restriction \tau = {\rm id}_\tau$) <u>then</u> $M_1 \restriction \tau = M_2 \restriction \tau$.
2. In 3.7(4), indeed $\Phi_2 \in \Upsilon_\kappa^{\rm or}$.
3. If $\mathbf{h}$ is an isomorphism from τ_1 onto τ_2 over $\tau_\mathfrak{k}$ so $\tau_\mathfrak{k} \subseteq \tau_1 \cap \tau_2$ and $\Phi_1 \in \Upsilon_\kappa^{\rm or}[\mathfrak{k}]$, $\tau_1 = \tau(\Phi_1)$ <u>then</u> $\Phi_2 = \Phi_1^{[\mathbf{h}]}$ belongs to $\Upsilon_\kappa^{\rm or}[\mathfrak{k}]$.
4. In part (3) if in addition $M \in K_\mathfrak{k}$ and $\Phi_1 \in \Upsilon_\kappa^{\rm sor}[M, \mathfrak{k}]$ and $a \in M \Rightarrow \mathbf{h}(c_a) = c_a$ <u>then</u> $\Phi_2 = \Phi_1^{[\mathbf{h}]}$ belongs to $\Upsilon_\kappa^{\rm sor}[M, \mathfrak{k}]$.

Proof.. Straightforward. $\square_{3.8}$

Next we recall the partial orders $\leq_\kappa^1, \leq_\kappa^2$ and define an equivalence relation and some quasi-orders on $\Upsilon_\kappa^{\rm or}[\mathfrak{k}]$.

Definition 3.9. Fixing $\mathfrak{k}$, we define partial orders $\leq_\kappa^\oplus = \leq_\kappa^1 = \leq_{\mathfrak{k},\kappa}^1$ and $\leq_\kappa^\otimes = \leq_\kappa^2 = \leq_{\mathfrak{k},\kappa}^2$ on $\Upsilon_\kappa^{\rm or}[\mathfrak{k}]$ (for $\kappa \geq {\rm LST}_\mathfrak{k}$):
1. $\Psi_1 \leq_\kappa^\oplus \Psi_2$ <u>if</u> $\tau(\Psi_1) \subseteq \tau(\Psi_2)$ and ${\rm EM}_{\tau(\mathfrak{k})}(I, \Psi_1) \leq_\mathfrak{k} {\rm EM}_{\tau(\mathfrak{k})}(I, \Psi_2)$ and ${\rm EM}(I, \Psi_1) = {\rm EM}_{\tau(\Psi_1)}(I, \Psi_1) \subseteq {\rm EM}_{\tau(\Psi_1)}(I, \Psi_2)$ for any linear order I (so, of course, same a_t's, etc.).
   Again for $\kappa = {\rm LST}_\mathfrak{k} + |\tau_\mathfrak{k}|$ we may drop the κ.
2. For $\Phi_1, \Phi_2 \in \Upsilon_\kappa^{\rm or}$, we say Φ_2 is an inessential extension of Φ_1 and write $\Phi_1 \leq_\kappa^{\rm ie} \Phi_2$ <u>if</u> $\Phi_1 \leq_\kappa^\oplus \Phi_2$ and for every linear order I, we have

$$ {\rm EM}_{\tau(\mathfrak{k})}(I, \Phi_1) = {\rm EM}_{\tau(\mathfrak{k})}(I, \Phi_2). $$

(note: there may be more function symbols in $\tau(\Phi_2)$!)

2A. We define the two-place relation $\mathrm{E}^{\text{æ}}$ on $\Upsilon_\mathfrak{k}^{\rm or}$ as follows $\Phi_1 \mathrm{E}^{\text{æ}} \Phi_2$ <u>iff</u> $\tau(\Phi_1) = \tau(\Phi_2)$ and for some unary function symbol $F \in \tau(\Phi_1)$ or F is just a (finite) composition[6] of such function symbols, if $M = {\rm EM}(I, \Phi_1)$ with skeleton $\langle a_t^1 : t \in I\rangle$ and we let $a_t^2 = F^M(a_2^1)$ for $t \in I$ then:
   - $F^M(a_t^2) = a_t^1$
   - M is ${\rm EM}(I, \Phi_2)$ with skeleton $\langle a_t^2 : t \in I\rangle$;

6 but abusing our notation we may still write $F \in \tau_\Phi$

"æ" stands for almost equal.

2B. Above we say $\Phi_2 \mathrm{E}^{æ} \Phi_2$ is witnessed by F.

2C. We define the two-place relation $\mathrm{E}^{\mathrm{ie}}_\kappa$ on $\Upsilon^{\mathrm{or}}_{\mathfrak{k}}$ by: $\Phi_1 \mathrm{E}^{\mathrm{ie}}_\kappa \Phi_2$ iff for some $\Phi_3, \Phi_1 \leq^{\mathrm{ie}}_\kappa \Phi_3$ and $\Phi_2 \leq^{\mathrm{ie}}_K \Phi_3$.

2D. We define a two-place relation $\mathrm{E}^{\mathrm{ai}}_\kappa$ on $\Upsilon^{\mathrm{or}}_\kappa[\mathfrak{k}]$ by $\Phi_1 \mathrm{E}^{\mathrm{ai}}_\kappa \Phi_3$ iff for some $\Phi_2 \in \Upsilon^{\mathrm{ai}}_\kappa[\mathfrak{k}]$ we have $\Phi_1 \mathrm{E}^{æ}_\kappa \Phi_2$ and $\Phi_2 \mathrm{E}^{\mathrm{ie}}_K \Phi_3$.

3. Let $\Upsilon^{\mathrm{lin}}_\kappa$ be the class of Ψ proper for linear order and producing linear orders, that is, such that:
   (a) $\tau(\Psi)$ has cardinality $\leq \kappa$,
   (b) $\mathrm{EM}_{\{<\}}(I, \Psi)$ is a linear order which is an extension of I which means $s <_I t \Rightarrow \mathrm{EM}(I, \Psi) \models$ "$a_s < a_t$"; in fact we can have $[t \in I \Rightarrow a_t = t]$.

4. $\Phi_1 \leq^{\otimes}_\kappa \Phi_2$ <u>iff</u> there is Ψ such that:
   (a) $\Psi \in \Upsilon^{\mathrm{lin}}_\kappa$
   (b) $\Phi_\ell \in \Upsilon^{\mathrm{or}}_\kappa$ for $\ell = 1, 2$
   (c) $\Phi'_2 \leq^{\mathrm{ie}}_\kappa \Phi_2$ where $\Phi'_2 = \Psi \circ \Phi_1$, i.e.
   $$\mathrm{EM}_{\tau(\Phi_1)}(I, \Phi'_2) = \mathrm{EM}(\mathrm{EM}_{\{<\}}(I, \Psi), \Phi_1).$$

(So we allow further expansion by functions definable from earlier ones (composition or even definition by cases), as long as the number is $\leq \kappa$).

It is not a real loss to restrict ourselves to standard Φ because

Claim 3.10. 1. *For every $\Phi_1 \in \Upsilon^{\mathrm{or}}_\kappa[\mathfrak{k}]$ there is a standard $\Phi_2 \in \Upsilon^{\mathrm{or}}_\kappa[\mathfrak{k}]$ such that $\Phi_1 \leq^{\mathrm{ie}}_\kappa \Phi_2$; moreover $M = \mathrm{EM}(\emptyset, \Phi_2) \Rightarrow |M| = \{c^M : c \in \tau(\Phi_2)$ an individual constant$\}$, that is Φ_2 is fully standard.*

2. *Assume $\Phi_1 \in \Upsilon^{\mathrm{or}}_\kappa[\mathfrak{k}], F \in \tau(\Phi)$ is a unary function symbol such that $M = \mathrm{EM}(I, \Phi_1) \wedge t \in I \Rightarrow F^M(F^M(a_t)) = a_t$. <u>Then</u> for a unique $\Phi_2, \Phi_1 \mathrm{E}^{æ} \Phi_2$ as witnessed by F and $\Phi_1 \in \Upsilon^{\mathrm{sor}}_\kappa[\mathfrak{k}_M] \Leftrightarrow \Phi_2 \in \Upsilon^{\mathrm{sor}}_\kappa[\mathfrak{k}_M]$.*

3. *$\mathrm{E}^{\mathrm{x}}_\kappa$ is an equivalence relation on $\Upsilon^{\mathrm{or}}_K[\mathfrak{k}]$ for $\mathrm{x} \in \{æ, \mathrm{ie}, \mathrm{ai}\}$ all refining $\mathrm{E}^{\mathrm{ai}}_\kappa$.*

Proof.. Obvious. $\square_{3.10}$

Observation 3.11. Let $\ell = 1, 2$.

1. The relation $\leq^\ell_\kappa$ is a partial order on $\Upsilon^{\mathrm{or}}_\kappa[\mathfrak{k}]$.
2. If $\langle \Phi_\alpha : \alpha < \delta \rangle$ is $\leq^\ell_\kappa$-increasing with δ a limit ordinal $< \kappa^+$ <u>then</u> $\bigcup_{\alpha<\delta} \Phi_\alpha$ naturally defined is a $\leq^\ell_\kappa$-lub.
3. $\mathrm{E}^{æ}$ is an equivalence relation on Υ^{or}.
4. If $\Upsilon^{\mathrm{or}}_{\kappa_1}[\mathfrak{k}] \subseteq \Upsilon^{\mathrm{or}}_{\kappa_2}[\mathfrak{k}]$ <u>then</u> $\kappa_1 \leq \kappa_2$. If $\kappa_1 \leq \kappa_2$ and $\iota \in \{1, 2\}$ and $\Phi, \Psi \in \Upsilon^{\mathrm{or}}_{\kappa_1}$ <u>then</u> $[\Phi \leq^\iota_{\kappa_1} \Psi \Leftrightarrow \Phi \leq^\iota_{\kappa_2} \Psi)$.
5. Similarly for $\Upsilon^{\mathrm{sor}}_\kappa[\mathfrak{k}_M]$ defined in 3.2(5).

Definition 3.12. 1. For $\kappa \geq \mathrm{LST}_{\mathfrak{k}} + |\tau_{\mathfrak{k}}|$, we define $\leq_\kappa^{\odot} = \leq_\kappa^3$, in full $\leq_{\mathfrak{k},\kappa}^3$, a two-place relation on $\Upsilon_\kappa^{\mathrm{sor}}[\mathfrak{k}]$, recalling Definition 3.2(5) as follows:
Let $\Phi_1 \leq_\kappa^3 \Phi_2$ mean that: for every linear order I_1 there are a linear order I_2 and $\leq_{\mathfrak{k}}$-embedding h of $\mathrm{EM}_{\tau(\mathfrak{k})}(I_1, \Phi_1)$ into $\mathrm{EM}_{\tau(\mathfrak{k})}(I_2, \Phi_2)$, moreover every individual constant c of $\tau(\Phi_1)$ is an individual constant of $\tau(\Phi_2)$ and $h(c^{\mathrm{EM}(I_1,\Phi_1)}) = c^{\mathrm{EM}(I_2,\Phi_2)}$.

2. We define $\leq_\kappa^4 = \leq_{\mathfrak{k},\kappa}^4$; a two-place relation on $\Upsilon_\kappa^{\mathrm{sor}}[\mathfrak{k}]$ as follows.
Let $\Phi_1 \leq_\kappa^4 \Phi_2$ mean that: for some F we have:
   (a) $\Phi_1, \Phi_2 \in \Upsilon_\kappa^{\mathrm{sor}}[\mathfrak{k}]$
   (b) – $\tau(\Phi_1) \subseteq \tau(\Phi_2)$
   – $F \in \tau(\Phi_2)$ is a unary function symbol or as in 3.9(2A)
   (c) if I is a linear order and $M_2 = \mathrm{EM}(I, \Phi_2)$ with skeleton $\langle a_s^2 : s \in I\rangle$ then there is $M_1 = \mathrm{EM}(I, \Phi_1)$ with skeleton $\langle a_s^1 : s \in I\rangle$ such that
   – $a_s^1 = F^{M_2}(a_s^2)$ for $s \in I$
   – $a_s^2 = F^{M_2}(a_s^1)$ for $s \in I$
   – $M_1 \subseteq M_2 \restriction \tau_{\Phi_1}$ so $\tau(\Phi_1) \subseteq \tau(\Phi_2)$
   – $(M_1 \restriction \tau_{\mathfrak{k}}) \leq_{\mathfrak{k}} (M_2 \restriction \tau_{\mathfrak{k}})$
   – $c^{M_1} = c^{M_2}$ when $c \in \tau(\Phi_1)$ is an individual constant.

Remark 3.13. So $\leq_\kappa^4$ is like $\leq_\kappa^1$ but we demand less as $a_s^1 = a_s^2$ is weakened by using the function symbol F.

Claim 3.14. 1. *$\leq_\kappa^3$ is a partial order on $\Upsilon_\kappa^{\mathrm{sor}}[\mathfrak{k}]$ as well as $\leq_\kappa^4$; also for $\Phi_1, \Phi_2 \in \Upsilon_\kappa^{\mathrm{sor}}[\mathfrak{k}]$ and $\ell = 1, 2, 4$ we have $\Phi_1 \leq^2 \Phi_2 \Rightarrow \Phi_1 \leq^1 \Phi_2 \Rightarrow \Phi_1 \leq^4 \Phi_2 \Rightarrow \Phi_1 \leq^3 \Phi_3$.*
2. *Assume $\Phi_1, \Phi_2 \in \Upsilon_\kappa^{\mathrm{sor}}[\mathfrak{k}]$ have the same individual constants. Then $\Phi_1 \leq_\kappa^3 \Phi_2$ iff as in 3.12(1) restricting ourselves to $I = \beth_{1,1}(\kappa)$ iff $\Phi_1, \Phi_2 \in \Upsilon_\kappa^{\mathrm{sor}}[\mathfrak{k}]$ and for some F and $\Phi_1', \Phi_2' \in \Upsilon_\kappa^{\mathrm{sor}}[\mathfrak{k}]$ and we have $\Phi_1 \leq_\kappa^4 \Phi_1'$ witnessed by F and $\Phi_1' \mathrm{E}^{\mathrm{æ}} \Phi_2'$ witnessed by F and for some $\tau_*, \mathbf{h}$ we have $\tau(\mathfrak{k}) \subseteq \tau_* \subseteq \tau(\Phi_1')$, $\mathbf{h}$ is an isomorphism from $\tau(\Phi_2)$ onto τ_* over $\tau(\mathfrak{k}) \cup \{c : c \in \tau(\Phi_1)\}$ and $\Phi_2^{[\mathbf{h}]} \leq_\kappa^{\mathrm{ie}} \Phi_2'$ iff for some $\Phi' \in \Upsilon_\kappa^{\mathrm{sor}}[\mathfrak{k}]$ we have $\Phi_1 \leq^3 \Phi'$ and $\Phi' \mathrm{E}_\kappa^{\mathrm{ai}} \Phi$, see 3.9(2).*
3. *If $\Phi_n \in \Upsilon_\kappa^{\mathrm{sor}}[\mathfrak{k}]$ and $\Phi_n \leq_\kappa^3 \Phi_{n+1}$ then there is $\Phi_\omega \in \Upsilon_\kappa[\mathfrak{k}]$ such that $n < \omega \Rightarrow \Phi_n \leq_\kappa^3 \Phi$; moreover, $\mathrm{EM}_{\tau(\mathfrak{k})}(\emptyset, \Phi)$ is the union of the $\leq_{\mathfrak{k}}$-increasing sequence $\langle \mathrm{EM}_{\tau(\mathfrak{k})}(\emptyset, \Phi_n) : n < \omega\rangle$.*
4. *Similarly for $\leq_\kappa^4$.*

Proof.. 1. Obvious.
2. First clause implies second clause
Holds trivially.

Second clause implies the third clause
Let $I_1 = (\lambda, <), \lambda$ large enough, e.g. $\lambda = \beth_{1,1}(\kappa)$. Let $M_1 = \mathrm{EM}(I_1, \Phi_1)$ be with skeleton $\langle a_t^1 : t \in I_1\rangle$. As $\Phi_1 \leq_\kappa^3 \Phi_2$, there is a linear order I_2 and

$M_2 = \mathrm{EM}(I_2, \Phi_2)$ with skeleton $\langle a^2_t : t \in I_2 \rangle$ and $\le_{\mathfrak{k}}$-embedding f from $M_1 \restriction \tau(\mathfrak{k})$ into $M_2 \restriction \tau(\mathfrak{k})$ such that $c \in \tau(\Phi_1) \Rightarrow c \in \tau(\Phi_2) \wedge f(c^{M_1}) = c^{M_2}$; so by renaming without loss of generality $f \restriction \mathrm{Sk}(\emptyset, M_1)$ is the identity and $|I_2| > \lambda$. As $\|M_2\| > \|M_1\| \ge \lambda > \kappa \ge |\tau(M_2)|$, clearly we can find pairwise distinct $t_\alpha \in I_2$ for $\alpha < \lambda$ such that $\{a^2_{t_\alpha} : \alpha < \lambda\} \cap \{f(a^1_\alpha) : \alpha < \lambda\} = \emptyset$.
Let $\tau_1 = \tau(\Phi_1)$ and[7] let the pair $(\mathbf{h}, \tau_3)$ be such that: $\mathbf{h}$ is an isomorphism from the vocabulary $\tau_2 = \tau(\Phi_2)$ onto τ_3 over $\tau(\mathfrak{k}) \cup \{c : c \in \tau(\Phi_1)\}$ such that $\tau_1 \cap \tau_3 = \tau(\mathfrak{k}) \cup \{c : c \in \tau(\Phi_2)\}$ and let $M_3 = M_2^{[\mathbf{h}]}$, so $\tau(M_3) = \tau_3, \Phi_3 = \Phi_2^{[\mathbf{h}]}$ so $\tau(M_3) = \tau_3 = \tau(\Phi_3)$ and M_3 is an $\mathrm{EM}(I_2, \Phi_3)$ model with skeleton $\langle a^2_t : t \in I_2 \rangle$.
Let $\tau_4 = \tau_3 \cup \tau_1 \cup \{F, P_l : l = 1,2,3,4\}$ with F a one place function symbol and $P_\ell, F \notin \tau_3 \cup \tau_1$ and P_ℓ one place predicates for $\ell = 1,2,3,4$. We define a τ_4-model M_4:

- $\bullet_1$ it has universe $|M_3|$
- $\bullet_2$ $F^{M_4}(a^2_{t_\alpha}) = f(a^1_\alpha)$ and $F^{M_4}(f(a^1_\alpha)) = a^2_{t_\alpha}$
- $\bullet_3$ $P_1^{M_4} = \{a^1_t : t \in I_1\}, P_2^{M_4} = \{a^2_t : t \in I_2\}, P_3^{M_4} = \{f(a^1_t) : t \in I_1\}, P_4^{M_4} = \mathrm{Rang}(f)$
- $\bullet_4$ $M_4 \restriction \tau_3 = M_3$
- $\bullet_5$ f embeds M_1 into $M_4 \restriction \tau_1$.

Clearly there is no problem to do this and we apply 2.14(1A) with $M_4 \restriction \tau(\mathfrak{k}), M_4, \langle a^2_{t_\alpha} : \alpha < \lambda \rangle$, here standing for $M, M^+, \langle b_\alpha : \alpha < \lambda \rangle$ there and get Φ_4 standing for Φ there. Now by inspection (see Definition 3.12(2)):
$(*)_1$ $\Phi_1 \le^4_\kappa \Phi_4$
$(*)_2$ $\Phi_3 \le^{\otimes}_\kappa \Phi_4$; moreover $\Phi_3 \le^{\mathrm{ie}} \Phi_4$.
We derive Φ_5 from Φ_4 by 3.10(2) using our F so $\Phi_4 \mathrm{E}^{\mathrm{æ}} \Phi_5$. To show that the third clause of part (2) indeed holds, we just note that $\Phi_1', \Phi_2', \mathbf{h}, \tau_*$, there can stand for $\Phi_4, \Phi_5, \mathbf{h}, \tau_3$ here, so we are done.

The third clause implies the first clause:
So we are given F and $\Phi_1, \Phi_2 \in \Upsilon^{\mathrm{sor}}_\kappa[\mathfrak{k}], \Phi_1', \Phi_2' \in \Upsilon^{\mathrm{sor}}_\kappa[\mathfrak{k}], \tau_* \subseteq \tau(\Phi_2')$ including $\tau(\mathfrak{k})$ and an isomorphism $\mathbf{h}$ from $\tau(\Phi_2)$ onto τ_* over $\tau_{\mathfrak{k}} \cup \{c : c \in \tau(\Phi_1)\}$ such that $\Phi_1 \le^4_\kappa \Phi_2'$ witness by F, $\Phi_1' \mathrm{E}^{\mathrm{æ}} \Phi_2'$ witness by F and $\Phi_2^{[\mathbf{h}]} \le^{\otimes}_\kappa \Phi_2'$.
Let $\Psi \in \Upsilon^{\mathrm{lin}}_\kappa$ witness $\Phi_2^{[\mathbf{h}]} \le^{\otimes}_\kappa \Phi_2'$; and for uniformity of notation we let $\Phi_3 := \Phi_2'$.
We have to prove $\Phi_1 \le^3_\kappa \Phi_2$ so let I_1 be a linear order.
Let $M_1^* = \mathrm{EM}(I_1, \Phi_1)$ be with skeleton $\langle a^1_t : t \in I_1 \rangle$, let $I_2 = \mathrm{EM}_{\{<\}}(\Psi, I_1)$ so with skeleton $\langle t : t \in I_1 \rangle$. Let $M_1 \subseteq M_2$ be defined by $M_\ell = \mathrm{EM}(I_\ell, \Phi_2)$ with

7 The reason is that there may be a symbol in $\tau(\Phi_2) \cap \tau(\Phi_c)$ but not from $\tau(\mathfrak{k}_1) \cup \{c : c \in \tau(\Phi_1)\}$. We eliminate this "accidental equality". Only now $\tau_3 \cup \tau_1$ "makes sense".

skeleton $\bar{a}^\ell = \{a_t^2 : t \in I_\ell\}$ for $\ell = 1, 2$ and let $M_3 = \mathrm{EM}(I_1, \Phi_1')$ be with skeleton $\{a_t^3 : t \in I_1\}$.

By the choice of Ψ and of I_2 without loss of generality $M_2^{[\mathrm{h}]} = M_3 \restriction \tau_*$.

Lastly, there is a unique embedding f of M_1^* into $M_3 \restriction \tau(\Phi_1)$ mapping a_t^1 to $F^{M_3}(a_t^2)$ for $t \in I_1$. Easily f is a $\leq_{\mathfrak{k}}$-embedding of $M_1 \restriction \tau(\mathfrak{k})$ into $M_3 \restriction \tau(\mathfrak{k})$ mapping c^{M_1} to c^{M_2} for $c \in \tau(\Phi_1)$ and $M_3 \restriction \tau(\mathfrak{k}) = M_2 \restriction \tau(\mathfrak{k})$ and $c \in \tau(\Phi_1) \Rightarrow c \in \tau(\Phi_2) \wedge f(c^{M_1^*}) = c^{M_2}$.

We leave the fourth clause to the reader.

3. By parts (2) and (4) or directly using 2.14(1) and the definition of $\leq_\kappa^3$.
4. So assume that $n < \omega \Rightarrow \Phi_n \leq_\kappa^4 \Phi_{n+1}$ as witnessed by $F_n \in \tau(\Phi_{n+1})$. For any infinite linear order I we can choose $M_n = \mathrm{EM}(I_n, \Phi_n)$ with skeleton $\langle a_t^n : t \in I\rangle$. Let $\tau_\omega = \cup\{\tau(\Phi_n) : n < \omega\}$. Without loss of generality $M_n \subseteq M_{n+1} \restriction \tau(\Phi_n), F_n^{M_{n+1}}(a_t^{n+1}) = a_t^n$ and $F_n^{M_{n+1}}(a_t^n) = a_t^{n+1}$. For each n we define $M_{\omega,n} = \cup\{M_{n+k} \restriction \tau_n : k \in [n, \omega)\}$, so $n_1 < n_2 \Rightarrow M_{\omega,n_1} = M_{\omega,n_2} \restriction \tau(\Phi_{n_1})$. Hence letting $\tau_\omega = \cup\{\tau(\Phi_n) : n < \omega\}$ there is a τ_ω-model M_ω with universe $|M_{\omega,0}|$ such that $M_\omega \restriction \tau_n = M_{\omega,n}$ for $n < \omega$. Now define Φ by $\Phi(n) = \mathrm{tp}_{\mathrm{qf}}(\langle a_{t_0}^0, \ldots, a_{t_{n-1}}^0\rangle, \emptyset, M_\omega)$ whenever $t_0 <_I \ldots <_I t_n$.
   Clearly $M_\omega = \mathrm{EM}(I, \Phi)$ with skeleton $\langle a_t^0 : t \in I\rangle$ and $F_{n-1} \circ \ldots \circ F_1 \circ F_0$ witness $\Phi_n \leq_\kappa^4 \Phi_\omega$, here we need composition of unary functions.

$\square_{3.14}$

Claim 3.15. *For $M \in K_{\mathfrak{k}}$ of cardinality $\kappa \geq \mathrm{LST}_{\mathfrak{k}} + |\tau_{\mathfrak{k}}|$ the following conditions are equivalent:*

(a) $\Upsilon_\kappa^{\mathrm{or}}[\mathfrak{k}_M] \neq \emptyset$

(b) *for every $\lambda \geq \kappa$ there is N such that $M \leq_{\mathfrak{k}} N \in K_\lambda^{\mathfrak{k}}$*

(c) *for every $\alpha < (2^\kappa)^+$ there is $N \in K_{\geq\beth_\alpha}^{\mathfrak{k}}$ which $\leq_{\mathfrak{k}}$-extend M*

(d) *there is $\Phi \in \Upsilon_\kappa^{\mathrm{or}}[\mathfrak{k}_M]$ such that if $N = \mathrm{EM}(I, \Phi)$ and $N \restriction \tau_{\mathfrak{k}_M}$ is standard then $M = (N \restriction \tau_{\mathfrak{k}}) \restriction \{c^N : c \in \tau_\Phi$ an individual constant$\}$*

(e) *$\Upsilon_\kappa^{\mathrm{sor}}[\mathfrak{k}_M]$ is non-empty.*

Proof.. For (d) note that we can replace an individual constant by a unary function which is interpreted as being a constant function. More generally an n-place function F^N by functions F_1, F_2 where

- F_1 is a $(n+1)$-place function
- if $\bar{a} = \langle a_\ell : \ell \leq n\rangle \in {}^{n+1}N \backslash {}^{n+1}M$ then $F_2(\bar{a}) = F^N(\bar{a} \restriction n)$
- if $\bar{a} \in {}^{n+1}M$ then $F_1(\bar{a}) = a_0$

$\square_{3.15}$

Claim 3.16. *If (A) then (B) when:*

(A) (a) $M_1 \leq_{\mathfrak{k}} M_2$

(b) Φ_1, Ψ_1 *are from* $\Upsilon^{\rm sor}_\kappa[\mathfrak{k}_{M_1}]$ *so are* κ*-standard*

(c) $\Psi_2 \in \Upsilon^{\rm sor}_\kappa[\mathfrak{k}_{M_2}]$

(d) $\Phi_1 \le^4_\kappa \Psi_1$

(e) $\Psi_1 \le^1_\kappa \Psi_2$

(f) $\{c_a : a \in M_2\} \cap \tau(\Psi_1) = \{c_a : a \in M_1\}$

(B) *there is* Φ_2 *such that*

(a) $\Phi_2 \in \Upsilon^{\rm sor}_\kappa[\mathfrak{k}_{M_2}]$

(b) $\Phi_1 \le^1_\kappa \Phi_2$

(c) $\Phi_2 \le^4_\kappa \Psi_2$.

Proof.. Straightforward: let I be an infinite linear order, $M_2 = {\rm EM}(I, \Psi_2)$ be with skeleton $\langle a^2_t : t \in I\rangle$. Let the unary function symbol F witness $\Phi_1 \le^4_\kappa \Psi_1$ so $F \in \tau(\Psi_1) \subseteq \tau(\Psi_2)$ and let $a^1_t = F^{M_2}(a^2_t)$. Clearly $\langle a^1_t : t \in I\rangle$ is indiscernible for quantifier formulas in M_2 and generate it hence for some $\Phi_2 \in \Upsilon^{\rm or}_\kappa$ we have $M_2 = {\rm EM}(I, \Phi_2)$ with skeleton $\langle a^1_t : t \in I\rangle$. Clearly $\Phi_2 \in \Upsilon^{\rm sor}_\kappa[\mathfrak{k}]$. Also $\Phi_2 {\rm E}^{\rm æ} \Phi_2$ hence $\Phi_2 \le^4_\kappa \Psi_2$ and $\Phi_1 \le^{\oplus}_\kappa \Phi_2$ as required. $\square_{3.16}$

* * *

The following will be used when applied to a tree of approximations to embedding of EM-models to a model. In fact, we use only 3.18 for the case $\mathscr{S} = \mathscr{T} \setminus \max(\mathscr{T})$, see background in 3.19.

Definition 3.17. 1. We say $\mathbf{i} = (\mathscr{T}, \bar{\mathbf{I}}) = (\mathscr{T}_{\rm i}, \bar{\mathbf{I}}_{\rm i})$ is pit (partially idealized tree) <u>when</u>:

(a) $\mathscr{T}$ is a tree with $\le \omega$ levels and
- for transparency it is a set of finite sequences ordred by $\triangleleft$, closed under initial segments
- let ${\rm lev}(\eta, \mathscr{T}) = {\rm lev}_{\mathscr{T}}(\eta)$ be the level of $\eta \in \mathscr{T}$ in $\mathscr{T}$, that is $|\{\nu \in \mathscr{T} : \nu \triangleleft \eta\}|$
- let ${\rm rt}_{\mathscr{T}}$ be the root
- the n-level of $\mathscr{T}$ is the set $\{\eta : {\rm lev}_{\mathscr{T}}(n) = n\}$ so we have
- ${\rm lev}_{\mathscr{T}}(\eta) = \ell g(\eta)$ and ${\rm rt}_{\mathscr{T}} = \langle\rangle$

(b) $\mathbf{I} = \langle \mathbf{I}_\eta : \eta \in \mathscr{S}\rangle$ where $\mathscr{S} \subseteq \mathscr{T} \setminus \max(\mathscr{T})$, we may write $\mathscr{S}_{\rm i} = \mathscr{S}$

(c) $\mathbf{I}_\eta$ is an ideal on ${\rm suc}_{\mathscr{T}}(\eta) := \{\rho : \nu \in \mathscr{T}, \eta <_{\mathscr{T}} \rho$ and there is no $\nu \in \mathscr{T}$ satisfying $\eta <_{\mathscr{T}} \nu <_{\mathscr{T}} \rho\}$ or just an ideal on a set which $\supseteq {\rm suc}_{\mathscr{T}}(n)$ such that ${\rm suc}_{\mathscr{T}}(\eta) \notin \mathbf{I}_\eta$; we may write $\mathbf{I}_{{\rm i},\eta}$.

1A. If $\mathbf{I}_\eta = \{\{s : \eta^\wedge\langle s\rangle \in X\} : X \in \mathbf{I}'_\eta\}$ for some ideal $\mathbf{I}'_\eta$ on some set <u>then</u> abusing notation we may write $\mathbf{I}'_\eta$ instead of $\mathbf{I}_\eta$.

2. Let $(\mathscr{T}_1, \bar{\mathbf{I}}_1) \le (\mathscr{T}_2, \bar{\mathbf{I}}_2)$ <u>when</u> (each is a pit and):

(a) $\mathscr{T}_1 \subseteq_{tr} \mathscr{T}_2$ which means:

(α) $\eta \in \mathscr{T}_2 \Rightarrow \eta_1 \in \mathscr{T}_1 \wedge \text{lev}(\eta, \mathscr{T}_2) = \text{lev}(\eta, \mathscr{T}_1) \wedge \text{suc}(\eta, \mathscr{T}_2) \subseteq \text{suc}(\eta, \mathscr{T}_1)$

(β) $\leq_{\mathscr{T}_1} = <_{\mathscr{T}_2} \restriction \mathscr{T}_1$

(b) $\bar{\mathbf{I}}_2 = \bar{\mathbf{I}}_1 \restriction \mathscr{T}_2$, i.e. $\bar{\mathbf{I}}_1 \restriction \{\eta \in \mathscr{T}_1 : \eta \in \text{Dom}(\mathbf{I}_1)$ and $\eta \in \mathbf{I}_2\}$

(c) if $\eta \in \mathscr{T}_2 \backslash \mathscr{S}_{\mathbf{i}_2}$ then $\text{suc}(\eta, \mathscr{T}_2) = \text{suc}(\eta, \mathscr{T}_1)$.

2A. Let $(\mathscr{T}_1, \bar{\mathbf{I}}_1) \leq_{pr} (\mathscr{T}_2, \bar{\mathbf{I}}_2)$ <u>when</u> (each is a pit and)

(a), (b), (c) as above

(d) if $\eta \in \text{Dom}(\bar{\mathbf{I}}_2)$ then $\text{suc}_{\mathscr{T}_1}(\eta) \backslash \text{suc}_{\mathscr{T}_2}(\eta) \in \mathbf{I}_{1,\eta}$.

3. We say $(\mathscr{T}, \bar{\mathbf{I}})$ is κ-complete <u>when</u> every ideal $\mathbf{I}_\eta$ is.

4. For $\mathbf{i} = (\mathscr{T}, \bar{\mathbf{I}})$ we define $\text{Dp}_{\mathbf{i}} = \text{Dp}_{\mathscr{T}, \bar{\mathbf{I}}} : \mathscr{T} \to \text{Ord} \cup \{\infty\}$ by (stipulate $\infty + 1 = \infty$) defining when $\text{Dp}_{\mathscr{T}, \bar{\mathbf{I}}}(\eta) \geq \alpha$ by induction on α as follows:

(a) if $\eta \in \max(\mathscr{T})$ <u>then</u> $D_{\mathscr{T}, \mathbf{I}}(\eta) \geq \alpha$ iff $\alpha = 0$

(b) if $\eta \in \mathscr{T} \backslash \max(\mathscr{T})$ and $\eta \in \mathscr{S}_{\mathbf{i}} = \text{Dom}(\bar{\mathbf{I}})$ <u>then</u> $\text{Dp}_{\mathscr{T}, \mathbf{I}}(\eta) \geq \alpha$ iff $(\forall \beta < \alpha)(\exists X \subseteq \text{suc}_{\mathscr{T}}(\eta))[X \in \mathbf{I}^+_\eta \wedge (\forall \nu \in X)(\text{Dp}_{(\mathscr{T}, \bar{\mathbf{I}})}(\nu) \geq \beta)]$

(c) if $\eta \in \mathscr{T} \backslash \max(\mathscr{T}) \backslash \mathscr{S}_{\mathbf{i}}$ <u>then</u> $\text{Dp}_{\mathbf{i}}(\eta) \geq \alpha$ <u>iff</u> $(\forall \nu)(\nu \in \text{suc}_{\mathscr{T}}(\eta) \Rightarrow \text{Dp}_{\mathbf{i}}(\nu) \geq \alpha)$.

6. If $\mathbf{i} = (\mathscr{T}, \bar{\mathbf{I}})$ is a pit and $\eta \in \mathscr{T}$ let $\text{proj}(\eta, \mathbf{i}) = \text{proj}_{\mathbf{i}}(\eta)$ is the sequence ν of length $\ell g(\eta)$ such that:

- $\ell < \ell g(\eta) \wedge \eta \restriction \ell \in \text{Dom}(\bar{\mathbf{I}}) \Rightarrow \nu(\ell) = -1$
- $\ell < \ell g(\eta) \wedge \eta \restriction \ell \notin \text{Dom}(\bar{\mathbf{I}}) \Rightarrow \nu(\ell) = \eta(\ell)$.

7. For $\mathbf{i} = (\mathscr{T}, \bar{\mathbf{I}})$ a pit let $\text{proj}(n, \mathbf{i}) = \text{proj}_{\mathbf{i}}(n) = \{\text{proj}_{\mathbf{i}}(\eta) : \eta \in \mathscr{T}$ has length $n\}$ and $\text{proj}_{\mathbf{i}} = \text{proj}(\mathbf{i})$ is $\cup\{\text{proj}_{\mathbf{i}}(\eta) : \eta \in \mathscr{T}\}$.

8. If $\mathbf{i}_\ell$ is a pit for $\ell < n$ then

(a) $\prod\limits^*_{\ell < \mathbf{n}} \mathscr{T}_{\mathbf{i}_\ell}$ is $\{\bar{\eta} : \bar{\eta} = \langle \eta_\ell : \ell < \mathbf{n} \rangle$ is such that $\ell < \mathbf{n} \Rightarrow \eta_\ell \in \mathscr{T}_{\mathbf{i}_\ell}$ and moreover for some n called $\text{lev}(\bar{\eta})$ we have $(\forall \ell < \mathbf{n})(\text{lev}_{\mathscr{T}_{\mathbf{i}(\ell)}}(\eta_\ell) = n)\}$.

Theorem 3.18. *There are a* pit $\mathbf{i}_2$ *and* $\langle c_\eta : \eta \in \text{proj}(\mathbf{i}_1) \rangle$ *such that:* $\mathbf{i}_1 \leq \mathbf{i}_2$, $\text{Dp}_{\mathbf{i}_2}(\text{rt}_{\mathbf{i}_2}) \geq \gamma_2$ *and* $\eta \in \mathscr{T}_{\mathbf{i}_2} \Rightarrow \mathbf{c}(\eta) = \mathbf{c}_{\text{proj}(\eta, \mathbf{i}_1)}$ <u>*when*</u>*:*

(a) $\mathbf{i}_1 = (\mathscr{T}_1, \bar{\mathbf{I}}_1)$ *is a* pit

(b) $\mathbf{i}_1$ *is* λ*-complete* pit

(c) $2^{\kappa^\theta} < \lambda$ *where* $\theta = |\text{proj}_{\mathbf{i}_1}|$, $\kappa + \theta$ *is infinite for transparency*[8]

(d) $\mathbf{c}$ *is a colouring of* $\mathscr{T}_1$ *by* $\leq \kappa$ *colours*

(e) $\gamma_1 = \gamma_2 = (2^{(\kappa^\theta)})^+$ *or just*

(α) $\gamma_1 \leq \text{Dp}_{\mathbf{i}_1}(\text{rt}_{\mathbf{i}_1})$, γ_1 *is a regular cardinal,*

(β) γ_2 *has cofinality* $> \kappa^\theta$ *and* $\gamma < \gamma_2 \Rightarrow |\gamma|^{\kappa^\theta} < \gamma_1$.

8 If κ and θ are finite, the computations are somewhat different. Note that $\kappa = 0$ is impossible and if $\kappa = 1$ then $\mathbf{i}_2 = \mathbf{i}_1$ will do so, without loss of generality $\kappa \geq 2$.

Remark 3.19. 1. This relates on the one hand to the partition theorem of [Sh:f, Ch.XI] continuing Rubin-Shelah [RuSh:117], Shelah [Sh:f, Ch.XI] and on the other hand to Komjath-Shelah [KoSh:796]; the latter is continued in Gruenhut-Shelah [GhSh:909] but presently this is not used.

2. Now 3.18 is what we use but we can get a somewhat more general result - see 3.21.
3. In 3.18 the case $\gamma_1 = \gamma_2 > |\mathscr{T}_1|$ is equivalent to $\gamma_1 = \gamma_2 = \infty$.

Proof.. Let $\mathscr{C} = \{\bar{c} : \bar{c} = \langle c_\varrho : \varrho \in \text{proj}_{\mathbf{i}_1}\rangle, c_{<>} = \mathbf{c}(\text{rt}(\mathscr{T}_1))$ and where $c_\varrho \in \text{Rang}(\mathbf{c})$ or just $(\exists \eta \in \mathscr{T}_1)(\varrho = \text{proj}_{\mathbf{i}_2}(\eta) \wedge c_\varrho = \mathbf{c}(\eta))\}$. For transparency without loss of generality we assume $\text{Rang}(\mathbf{c} \restriction \max(\mathscr{T}_{\mathbf{i}_1})), \text{Rang}(\mathbf{c} \restriction (\mathscr{T}_{\mathbf{i}_1} \setminus \max(\mathscr{T}_{\mathbf{i}_1}))$ are disjoint. Clearly $|\mathscr{C}| \leq \kappa^{|\text{proj}(\mathbf{i}_1)|} = \kappa^\theta < \lambda$.

Fix for a while $\bar{c} \in \mathscr{C}$, first let $\mathscr{T}_{\bar{c}} = \{\eta \in \mathscr{T}_1$: if $\nu \trianglelefteq \eta$ then $\mathbf{c}(\nu) = \mathbf{c}_{\text{proj}(\nu, \mathbf{i}_1)}\}$ so a subtree of $\mathscr{T}_1$, i.e. a downward closed subset noting that $\text{rt}_{\mathscr{T}_1} \in \mathscr{T}_{\bar{c}}$.

Second, for $\eta \in \mathscr{T}_1$, let $X^1_{\bar{c},\eta}$ be $\text{suc}_{\mathscr{T}_{\bar{c}}}(\eta)$ if $\eta \in \mathscr{T}_{\bar{c}} \cap \text{Dom}(\bar{\mathbf{I}}_1)$ and this set is $\in \mathbf{I}_{1,\eta}$ and be $\emptyset$ otherwise. Let $\mathscr{T}'_{\bar{c}} = \{\eta \in \mathscr{T}_{\bar{c}}$: if $\ell < \ell g(\eta)$ and $\eta \restriction \ell \in \text{Dom}(\mathbf{I}_1)$ then $\eta \restriction (\ell + 1) \notin X^1_{\bar{c},\eta}$, i.e. $\text{suc}_{\mathscr{T}_{\bar{c}}}(\eta \restriction \ell) := \{\nu \in \text{suc}_{\mathscr{T}_1}(\eta) : \nu \in \mathscr{T}_{\bar{c}}\} \neq \emptyset \mod \mathbf{I}_{1,\eta}\}$, again $\mathscr{T}'_{\bar{c}}$ is a subtree of $\mathscr{T}_{\bar{c}}$, moreover $\mathbf{i}_{2,\bar{c}} = (\mathscr{T}'_{\bar{c}}, \bar{\mathbf{I}} \restriction \mathscr{T}'_{\bar{c}})$ is a pit.

Third, for $\eta \in \mathscr{T}'_{\bar{c}}, \text{Dp}_{\mathbf{i}_1}(\eta) \in \text{Ord} \cup \{\infty\}$ is well defined and, now for $\eta \in \mathscr{T}_1$, let $X^2_{\bar{c},\eta}$ be $\{\nu \in \text{suc}_{\mathscr{T}'_{\bar{c}}}(\eta) : \text{Dp}_{\mathbf{i}_{2,\bar{c}}}(\nu) \geq \text{Dp}_{\mathbf{i}_{2,\bar{c}}}(\eta)\} = \emptyset \mod \mathbf{I}_{1,\eta}\}$ if $\eta \in \mathscr{T}'_{\bar{c}} \cap \text{Dom}(\bar{\mathbf{I}}_1), \text{Dp}_{\mathbf{i}_{2,\bar{c}}}(\eta) < \infty$ and be $\emptyset$ otherwise.

If for some $\bar{c} \in \mathscr{C}, \text{Dp}_{\mathbf{i}_{2,\bar{c}}}(\text{rt}_{\mathscr{T}'_{\bar{c}}}) \geq \gamma_2$ easily we are done, so toward a contradiction assume this is not the case, so recalling $\text{cf}(\gamma_2) > |\mathscr{C}|$ clearly $\gamma_* = \sup\{\text{Dp}_{\mathbf{i}_{2,\bar{c}}}(\text{rt}_{\mathscr{T}'_{\bar{c}}}) + 1 : \bar{c} \in \mathscr{C}\} < \gamma_2$. Now for each $\eta \in \text{Dom}(\bar{\mathbf{I}}_1)$ clearly all $X^1_{\bar{c},\eta}, X^2_{\bar{c},\eta}$ are from $\mathbf{I}_{1,\eta}$ and their number is $\leq 2|\mathscr{C}| < \lambda$ hence $X_\eta := \cup\{X^1_{\bar{c},\eta} \cup X^2_{\bar{c},\eta} : \bar{c} \in \mathscr{C}\}$ belong to $\mathbf{I}_{1,\eta}$.

Hence $\mathbf{i}_3$ is an pit and $\mathbf{i}_1 \leq \mathbf{i}_3$ where $\mathbf{i}_3 = \mathbf{i}(3) := \mathbf{i}_1 \restriction \{\eta \in \mathscr{T}_1$: if $\ell < \ell g(\eta)$ and $\eta \restriction \ell \in \text{Dom}(\bar{\mathbf{I}}_1)$ then $\eta \restriction (\ell + 1) \notin X_\eta\}$; moreover by the definition of $\text{Dp}_{\mathbf{i}_3}$ and the choice of $\mathbf{i}_3$, clearly

$(*)_1$ (*a*) $\mathbf{i}_3$ is a pit; moreover $\mathbf{i}_1 \leq_{\text{pr}} \mathbf{i}_3$ hence

(*b*) $\eta \in \mathscr{T}_{\mathbf{i}_3} \Rightarrow \text{Dp}_{\mathbf{i}_3}(\eta) = \text{Dp}_{\mathbf{i}_1}(\eta)$.

Define *h* by

$(*)_2$ *h* is a function from $\mathscr{T}_{\mathbf{i}_1} \times \mathscr{C}$ defined by

- $h(\eta, \bar{c})$ is -1 if $\eta \in \mathscr{T}_{\mathbf{i}_1} \setminus \mathscr{T}'_{\bar{c}}$
- $h(\eta, \bar{c})$ is $\text{Dp}_{\mathbf{i}_{2,\bar{c}}}(\eta)$ if $\eta \in \mathscr{T}'_{\bar{c}}$ and $\text{Dp}_{\mathbf{i}_{2,\bar{c}}}(\eta) < \gamma_*$
- $\text{Dp}(\eta, \bar{c}) = \gamma_*$ if none of the above.

We now choose $(\mathbf{c}_n, h_n, \mathscr{X}_n, \bar{\mathscr{Y}}_n, \mathscr{S}_n)$ by induction on n such that:

⊞ (*a*) (*α*) $\mathscr{X}_n$ is a subset of $\cup\{\text{proj}_{\mathbf{i}_1}(m) : m \leq n\}$

(β) if $n = k+1$ then $\mathscr{X}_k = \mathscr{X}_n \cap (\cup\{\mathrm{proj}_{\mathrm{i}_1}(m) : m \le k\})$
(γ) $\mathscr{S}_n \subseteq \mathscr{X}_n$

(b) (α) h_n is a function with domain $\mathscr{X}_n \times \mathscr{C}$ to $\gamma_* + 1$
(β) $\mathbf{c}_n$ is a function from $\mathscr{X}_n$ to $\mathrm{Rang}(\mathbf{c})$

(c) $\bar{\mathscr{Y}}_n = \langle \mathscr{Y}_{n,\gamma} : \gamma < \gamma_1 \rangle$

(d) (α) $\mathscr{Y}_{n,\gamma}$ is a subset of $\mathscr{T}_{\mathrm{i}_3}$, downward closed of cardinality $\le \theta$
(β) if $\eta \in \mathscr{Y}_{n,\gamma}$ then $\ell g(\eta) \le n$
(γ) if $\eta \in \mathscr{Y}_{n,\gamma}$ then $\mathrm{Dp}_{\mathrm{i}_3}(\eta) = \mathrm{Dp}_{\mathrm{i}_1}(\eta) \ge \gamma$
(δ) if $\eta \in \mathscr{Y}_{n,\gamma}$ and $\ell g(\eta) < n$ and $\eta \notin \mathrm{Dom}(\bar{\mathbf{I}}_1)$ then $\mathrm{suc}_{\mathscr{T}_{\mathrm{i}_3}}(\eta) = \mathrm{suc}_{\mathscr{T}_{\mathrm{i}_1}}(\eta)$ is $\subseteq \mathscr{Y}_{n,\gamma}$
(ε) if $\eta \in \mathscr{Y}_{n,\gamma}$ and $\ell g(\eta) < n$ and $\eta \in \mathrm{Dom}(\bar{\mathbf{I}}_1)$ <u>then</u> $\mathrm{suc}_{\mathscr{T}_{\mathrm{i}_3}}(\eta)$ is a singleton
(ζ) if $\gamma < \gamma_2$ <u>then</u> $\mathscr{X}_n = \{\mathrm{proj}_{\mathrm{i}_1}(\eta) : \eta \in \mathscr{Y}_{n,\gamma}\}$
(η) if $\eta \in \mathscr{Y}_{n,\gamma}$ and $\nu = \mathrm{proj}_{\mathrm{i}_3}(\eta)$ <u>then</u>:
- $\bullet_1$ $\mathbf{c}(\eta) = \mathbf{c}_n(\nu)$
- $\bullet_2$ $h_n(\nu, \bar{c}) = h(\eta, \bar{c})$ for every $\bar{c} \in \mathscr{C}$
- $\bullet_3$ $\eta \in \mathrm{Dom}(\bar{\mathbf{I}}_1)$ <u>iff</u> $\nu \in \mathscr{S}_n$.

(θ) follows: the function $\eta \mapsto \mathrm{proj}_{\mathrm{i}_3}(\eta)$ on $\mathscr{Y}_{n,\gamma}$ is one to one.

<u>Why this is possible:</u>

For $n = 0$ this is trivial.

For $n = m+1$ for every $\gamma < \gamma_1$, choose $\bar{\varrho}_{n,\gamma} \in \Pi\{\mathrm{suc}_{\mathscr{T}_{\mathrm{i}_3}}(\eta) : \eta \in \mathscr{Y}_{m,\gamma+1}, \ell g(\eta) = m, \eta \in \mathrm{Dom}(\bar{\mathbf{I}}_1)\}$ such that if $\eta \in \mathrm{Dom}(\bar{\varrho}_{n,\gamma})$ then $\mathrm{dp}_{\mathrm{i}_1}(\eta) \ge \gamma$, possible as $\eta \in \mathrm{Dom}(\bar{\varrho}_{n,\gamma}) \Rightarrow \mathrm{dp}_{\mathrm{i}_1}(\eta) \ge \gamma+1$. Let $\mathscr{Y}'_{n,\gamma} = \mathscr{Y}_{m,\gamma+1} \cup \{\nu$: for some $\eta \in \mathscr{Y}_{n,\gamma+1}$ we have $\ell g(\eta) = m$ and we have $\eta \notin \mathrm{Dom}(\bar{\mathbf{I}}_1) \Rightarrow \nu = \varrho_{n,\gamma}(\eta)$ and $\eta \notin \mathrm{Dom}(\bar{\mathbf{I}}_1) \Rightarrow \nu \in \mathrm{suc}_{\mathscr{T}_{\mathrm{i}_3}}(\eta)\} \cup \mathrm{Rang}(\bar{\varrho}_{n,\gamma})$.

Let $\mathscr{X}'_{n,\gamma} = \{\mathrm{proj}_{\mathrm{i}_1}(\eta) : \eta \in \mathscr{Y}'_{n,\gamma}\}$ and let the function $\mathbf{c}'_{n,\gamma} : \mathscr{X}'_{n,\gamma} \to \mathrm{Rang}(\mathbf{c})$ be defined by $\eta \in \mathscr{Y}'_{n,\gamma} \Rightarrow \mathbf{c}'_{n,\gamma}(\mathrm{proj}_{\mathrm{i}_3}(\eta))) = \mathbf{c}(\eta)$, well defined as in $\boxplus(d)(\eta)$ and let $\mathscr{S}_{n,\gamma} = \{\mathrm{proj}_{\mathrm{i}_1}(\eta) : \eta \in \mathscr{Y}'_{n,\gamma}$ and $\eta \in \mathrm{Dom}(\bar{\mathbf{I}}_1)\}$. Let $h_{n,\gamma} : \mathscr{X}'_{n,\gamma} \to \gamma_* + 1$ be defined by : if $\bar{c} \in \mathscr{C}, \nu = \mathrm{proj}_{\mathrm{i}_{2,\bar{c}}}(\eta)$ and $\eta \in \mathscr{Y}_{n,\gamma}$ then $\eta \notin \mathscr{T}'_{\bar{c}} \Rightarrow h_{n,\gamma}(\nu) = \gamma, \eta \in \mathscr{T}'_{\bar{c}} \Rightarrow h_{n,\gamma}(v) = \mathrm{Dp}_{\mathrm{i}_{2,\bar{c}}}(\eta)$.

Now $\mathscr{X}'_{n,\gamma}$ is a subset of $\mathrm{proj}_{\mathrm{i}_1}$, a set of cardinality $\le \theta$ and $\mathbf{c}'_{n,\gamma}$ is a funtion from $\mathscr{X}'_{n,\gamma}$ into $\mathrm{Rang}(\mathbf{c})$, a set of cardinality $\le \kappa$ and $h_{n,\gamma}$ is a function from $\mathscr{X}'_{n,\gamma} \subseteq \mathrm{proj}_{\mathrm{i}_1}$ into γ_*. But $\gamma_* < \gamma_2, \gamma_* + \kappa < \gamma_1, \gamma_1$ is a regular cardinal (recalling clause (e) of the theorem) and $(|\gamma_*| + \kappa)^\theta < \mathrm{cf}(\gamma_1) = \gamma_1$ hence for every $\gamma < \gamma_1$ we have $|\{(X'_{n,\gamma}, \mathbf{c}_{n,\gamma}, h_{n,\gamma}) : \gamma < \gamma_1\}| \le 2^\theta \cdot \kappa^\theta \cdot |\gamma_*|^\theta < \mathrm{cf}(\gamma_1) = \gamma_1$ hence for some $\mathbf{c}_n, h_n, \mathscr{X}_n$ the set $S_n := \{\gamma < \gamma_1 : \mathbf{c}'_{n,\gamma} = \mathbf{c}_n$ and $h_{n,\gamma} = h_n, \mathscr{X}'_{n,\gamma} = \mathscr{X}_n$ and $\mathscr{S}_{n,\gamma} = \mathscr{S}_n\}$ is unbounded in γ_1.

Lastly, let $\mathscr{Y}_{n,\gamma} = \mathscr{Y}'_{n,\min(S_n \setminus \gamma)}$, clearly $\mathbf{c}_{n+1}, h_{n+1}, \langle \mathscr{Y}_{n,\gamma} : \gamma < \gamma_2 \rangle$ are as required; so we can carry the induction.

Why this is enough:

Let $\mathscr{X} = \cup\{\mathscr{X}_n : n < \omega\} \subseteq \text{proj}(\mathbf{i}_1)$ and $\mathscr{S} = \cup\{\mathscr{S}_n : n < \omega\}$ and $\mathbf{c} = \cup\{\mathbf{c}_n : n < \omega\}$ and $\mathbf{h} = \cup\{h_n : n < \omega\}$ so by $\boxplus(d)(\eta)$ clearly there is $\bar{c}^* \in \mathscr{C}$ such that $c^*_\varrho = \mathbf{c}(\varrho)$ when the latter is defined, so:

$\odot_1$ if $n < \omega, \gamma < \gamma_1, \eta \in \mathscr{Y}_{n,\gamma}$ and $\nu = \text{proj}(\mathbf{i}_1) \in \mathscr{X}$ then

(a) $\mathbf{c}(\eta) = \mathbf{c}_n(\text{proj}_{\mathbf{i}_1}(\eta))$

(b) $\text{Dp}_{\mathbf{i}_{2,\bar{c}}}(\eta) = h(\eta, \bar{c}) = h_n(\nu, \bar{c})$

(c) $\text{Dp}_{\mathbf{i}_1}(\eta) \geq \gamma$

Also

$\odot_2$ $\mathscr{X} \subseteq \text{proj}_{\mathbf{i}_1}$ is a set of finite sequences, closed under initial segments with no $\triangleleft$-maximal member.

[Why? Straight, e.g. if $\nu \in X$ choose $n = \ell g(\nu) + 2$ let $\gamma < \gamma_1$ and choose $\eta \in \mathscr{Y}_{n,\gamma+1}$ such that $\text{proj}_{\mathbf{i}_1}(\eta) = \nu$, now by clause (c) of $\odot_1$ we know that $\text{Dp}_{\mathbf{i}_1}(\eta) \geq \gamma + 1$, hence there is $\eta_1 \in \text{suc}_{\mathscr{T}_{\mathbf{i}_1}}(\eta)$ in $\mathscr{Y}_{n,\gamma+1}$ hence $\nu_1 = \text{proj}_{\mathbf{i}_1}(\eta_1)$ is in $\text{suc}_{\mathscr{X}}(\nu)$, i.e. successor of η in $\mathscr{X}_{n+1}$ hence in $\mathscr{X}$.]

$\odot_3$ if $\nu \in \mathscr{X}$ then $\mathbf{h}(\nu, \bar{c}) \neq -1$.

[Why? Let $n > \ell g(\nu)$, let $\gamma < \gamma_2$. Now by $\boxplus(d)(\zeta)$ there is $\eta \in \mathscr{Y}_{n,\gamma}$ such that $\text{proj}_{\mathbf{i}_1}(\eta) = \nu$.

Next by $(*)_2$ we have $h(\eta, \bar{c})$ is -1 iff $\eta \notin \mathscr{T}'_{\bar{c}}$. However, $\eta \in \mathscr{T}_{\bar{c}}$ by the definition of $\mathscr{T}_{\bar{c}}$ and the choice of $\bar{c}$ and $\boxplus(d)(\eta)$; moreover $\eta \in \mathscr{T}'_{\bar{c}}$ by the definition of $\mathscr{T}'_{\bar{c}}$ and of $\mathbf{i}_3$ and clause $\boxplus(d)(\alpha)$.

By the last two sentences $h(\eta, \bar{c}) \neq -1$ hence by the choice of η, i.e. as $\text{proj}_{\mathbf{i}_1}(\eta) = \nu$, clause $\boxplus(d)(\eta)$ tells us $\mathbf{h}(\nu, \bar{c}) = h(\eta, \bar{c})$ so together $\mathbf{h}(\nu, \bar{c}) \neq -1$ as promised.]

$\odot_4$ $0 \leq \text{Dp}_{\mathbf{i}_{2,\bar{c}}}(\langle\rangle) < \gamma_*$ hence $\mathbf{h}(<>, \bar{c}) < \gamma_*$.

[Why? Similarly using $\boxplus(d)(\eta)\bullet_3$.]

$\odot_5$ if $\nu \in \mathscr{X} \backslash \mathscr{S}$ and $0 \leq \mathbf{h}(\nu, \bar{c}) < \gamma_*$ <u>then</u> for some $\rho \in \text{suc}_{\mathscr{X}}(\nu)$ we have $0 \leq \mathbf{h}(\rho, \bar{c}) < \mathbf{h}(\nu, \bar{c}) < \gamma_*$.

[Why? Similarly using $\boxplus(d)(\delta)$.]

$\odot_6$ if $\nu \in \mathscr{S}(\subseteq \mathscr{X})$ and $0 \leq \mathbf{h}(\nu, \bar{c}) < \gamma_*$ <u>then</u> for the unique $\rho \in \text{suc}_{\mathscr{X}}(\nu)$ we have $0 \leq \mathbf{h}(\rho, \bar{c}) < \mathbf{h}(\nu, \bar{c}) < \gamma_*$.

[Why? Similarly using $\boxplus(d)(\varepsilon)$.]

By $\odot_4, \odot_5, \odot_6$ together we get a contradiction. $\square_{3.18}$

We may prefer the following variant of 3.18.

Definition 3.20. For a pit $\mathbf{i} = (\mathscr{T}, \bar{\mathbf{I}})$ and partition $\bar{\mathscr{S}} = (\mathscr{S}_0, \mathscr{S}_1)$ of $\mathscr{S}_{\mathbf{i}}$ (or just $\bar{\mathscr{S}} = (\mathscr{S}_0, \mathscr{S}_1)$ such that $\mathscr{S}_0 \cap \mathscr{S}_1 = \emptyset$ and $\mathscr{S}_{\mathbf{i}} \subseteq \mathscr{S}_0 \cup \mathscr{S}_1$) we define $\mathrm{Dp}_{\mathbf{i},\mathscr{S}} : \mathscr{T} \to \mathrm{Ord} \cup \{\infty\}$, stipulating $\infty + 1 = \infty$ by defining when $\mathrm{Dp}_{\mathbf{i},\lambda}(\eta) \geq \alpha$ by induction on the ordinal α (compare with 3.17(4)):

(a) if $\eta \in \max(\mathscr{T})$ <u>then</u> $\mathrm{Dp}_{\mathbf{i},\lambda}(\eta) \geq \alpha$ iff $\alpha = 0$

$(b)_0$ if $\eta \in \mathscr{S}_0$ hence $\eta \in \mathscr{S}, \eta \notin \max(\mathscr{T})$ <u>then</u> $\mathrm{Dp}_{\mathbf{i},\bar{\mathscr{S}}}(\eta) \geq \alpha$ iff for every $\beta < \alpha$ the set $\{\nu \in \mathrm{suc}_{\mathscr{T}}(\eta) : \mathrm{Dp}_{\mathbf{i},\bar{\mathscr{S}}}(\nu) \geq \beta\}$ belong to $\mathbf{I}^+_\eta$

$(b)_1$ if $\eta \in \mathscr{S}_1$ <u>then</u> $\mathrm{Dp}_{\mathbf{i},\bar{\mathscr{S}}}(\eta) \geq \alpha$ iff $\{\nu \in \mathrm{suc}_{\mathscr{T}}(\eta) : \mathrm{Dp}_{\mathbf{i},\bar{\mathscr{S}}}(\nu) \geq \alpha\}$ belongs to $\mathbf{I}^+_\eta$

(c) if $\eta \in \mathscr{T} \backslash \mathscr{S} \backslash \max(\mathscr{T})$ <u>then</u> $\mathrm{Dp}_{\mathbf{i},\bar{\mathscr{S}}}(\eta) \geq \alpha$ iff for every $\nu \in \mathrm{suc}_{\mathscr{T}}(\eta)$ we have $\mathrm{Dp}_{\mathbf{i},\bar{\mathscr{S}}}(\eta)$.

Theorem 3.21. *There are a pit* $\mathbf{i}_2$ *and* $\bar{c} = \langle c_\eta : \eta \in \mathrm{proj}(\mathbf{i}_1)\rangle$ *such that* $\mathbf{i}_1 \leq \mathbf{i}_2, \mathrm{Dp}_{\mathbf{i}_2,\bar{\mathscr{S}}}(\mathrm{rt}_{\mathbf{i}_2}) \geq \gamma_2$ *and* $\eta \in \mathscr{T}_{\mathbf{i}_2} \Rightarrow \mathbf{c}(\eta) = \mathbf{c}_{\mathrm{proj}(\eta,\mathbf{i}_1)}$ <u>*when*</u>*:*

(a) – (e) as in 3.18 replacing $\mathrm{Dp}_{\mathbf{i}_2}$ *by* $\mathrm{Dp}_{\mathbf{i}_2,\bar{\mathscr{S}}}$ *in (e)(α)*

(f) $\bar{\mathscr{S}} = (\mathscr{S}_0, \mathscr{S}_1)$ *is a partition of* $\mathscr{S}_{\mathbf{i}}$.

Proof.. Similarly. $\square_{3.21}$

4 Approximation to EM models

In the game below the protagonist tries to exemplify in a weak form that the standard $\mathrm{EM}_{\tau(\mathfrak{k})}(\lambda, \Phi)$ is $\leq_{\mathfrak{k}}$-embeddable into N over M. We may consider games in which the protagonist tries to exemplify a weak form of isomorphism, this is connected to logics which have EM models, continuing [Sh:797], but not for now.

Here we do not try to get the best cardinal bounds; just enough for the result promised in the abstract.

Definition 4.1. Assume $\lambda > \kappa \geq \mathrm{LST}_{\mathfrak{k}} + |\tau_{\mathfrak{k}}|$ and $M \in K^{\mathfrak{k}}_\kappa$ and $M \leq_{\mathfrak{k}} N$ and γ is an ordinal.

1. We say Φ is an $(M, \lambda, \kappa, \gamma)$-solution of N or is an $(N, M, \lambda, \kappa, \gamma)$-solution <u>when</u> $\Phi \in \Upsilon^{\mathrm{sor}}_\kappa(\mathfrak{k}_M)$ and in the game $\partial^1_{N,M,\lambda,\Phi,\gamma}$ the protagonist has a winning strategy.
2. Assume $\Phi \in \Upsilon_\kappa(\mathfrak{k}_M)$ recalling Definition 3.1 fixing $M_\lambda = \mathrm{EM}(\lambda, \Phi)$ and $M_I = \mathrm{EM}(I, \Phi)$ for $I \subseteq \lambda$ and without loss of generality every M_I (equivalently some M_I) is standard, hence in particular $M \leq_{\mathfrak{k}} M_I \restriction \tau(\mathfrak{k})$. We define the game $\partial^1_{N,M,\lambda,\Phi,\gamma}$, a play last $< \omega$ moves, in the n-th move $\lambda_n, J_n, \bar{h}_n, \gamma_n$ are chosen such that:

$\boxplus_n$ (a) $\lambda_0 = \lambda$

(b) if $n = m + 1$ then $\kappa < \lambda_n < \lambda_m$ moreover $\lambda_m \to (\lambda_n)^n_{2^\kappa}$

(c) $J_0 = \lambda$, and if $n = m + 1$ then $J_n \subseteq J_m$

(d) $|J_n| = \lambda_n$

(e) $\bar{h}_n = \langle h_u : u \in [J_n]^n \rangle$

(f) if $u \in [J_n]^n$ <u>then</u> h_u is a $\leq_{\mathfrak{k}}$-embedding of M_u into N extending h_v whenever $v \subseteq u$

[Explanation: note if $v \subset u, |v| = m$ then $v \in [J_n]^m \subseteq [J_m]^n$ hence h_v was defined; this says then for $u_1, u_2 \in [J_n]^n, h_{u_2}, h_{u_2}$ are compatible functions]

(g) $\gamma_0 = \gamma$ and γ_{n+1} is an ordinal $< \gamma_n$.

In the n-th move:

(A) if $n = 0$ the antagonist chooses $\lambda_0 = \lambda, J_0 = \lambda, \gamma_0 = \gamma$ and the protagonist chooses $\bar{h}_0$

(B) if $n = m + 1$ then

(a) the antagonist chooses an ordinal $\gamma_n < \gamma_m$ and $\lambda_n > \kappa$ such that $\lambda_m \to (\lambda_n)^n_{\beth_2(\kappa)}$

(b) the protagonist chooses $\bar{h}'_n = \langle h_u : u \in [J_m]^n \rangle$ and $\mathscr{S}_n \in (\mathrm{ER}^n_{J_m, \lambda_n, \beth_2(\kappa)})^+$, i.e. $\mathscr{S}_n \subseteq [\lambda_m]^{\lambda_n}$ and $\mathscr{S}_n$ is not from this ideal, see Definition 3.5

(c) the antagonist chooses $J_n \in \mathscr{S}_n \subseteq [J_m]^{\lambda_n}$ and we let $\bar{h}_n = \bar{h}'_n \restriction [J_n]^n$.

(C) the play ends when a player has no legal move and then this player loses.

Another presentation:

Definition 4.2. Assume $M \leq_{\mathfrak{k}} N$ and $\mathrm{LST}_{\mathfrak{k}} + |\tau_{\mathfrak{k}}| \leq \theta, \|M\| + \theta \leq \kappa < \lambda$ and $\Phi \in \Upsilon^{\mathrm{or}}_\theta[M, \mathfrak{k}]$.

1. Below we omit γ if (a) or (b), where:
   (a) $\gamma = \mathrm{cf}(\lambda)$, λ strong limit and $\alpha < \mathrm{cf}(\lambda) \Rightarrow |\alpha|^{2^{\kappa+\|M\|}} < \mathrm{cf}(\lambda)$
   (b) not (a) but γ is maximal such that $\gamma = \omega\gamma$ is infinite and $\beth_\gamma(\kappa + \|M\|) \leq \lambda$ and λ is strong limit of cofinality $> \beth_2(\kappa)$
   (similarly in all such definitions).
2. We say that $\mathbf{x}$ is a direct witness for $(N, M, \lambda, \kappa, \gamma, \Phi)$ <u>when</u> $\mathbf{x}$ consists of:
   (a) $N, M, \Phi, \lambda, \kappa$ and γ
   (b) $\mathscr{T}$ is a non-empty set of finite sequences closed under initial segments
   (c) if $\eta \in \mathscr{T}$ <u>then</u>:
      (α) $\eta(2n)$ is a cardinal when $2n < \ell g(\eta)$
      (β) $\eta(2n+1)$ is a subset of λ of cardinality $\eta(2n)$ when $2n+1 < \ell g(\eta)$
      (γ) $\eta(2n+1) \supseteq \eta(2n+3)$ when $2n+3 < \ell g(\eta)$
      (δ) $\eta(2n) \geq \eta(2n+2)$, moreover $\eta(2n) \to (\eta(2n+2))^{2n+1}_{\beth_2(\kappa)}$ when $2n+2 < \ell g(\eta)$
   (d) I_η, λ_η for $\eta \in \mathscr{T}$ are defined by:
      (α) if $\ell g(\eta) = 0$ then $I_\eta = \lambda, \lambda_\eta = \lambda$
      (β) if $\ell g(\eta) = 2n+1$ then $I_\eta = I_{\eta \restriction (2n)}$, see (α) or (γ) and $\lambda_\eta = \eta(2n)$
      (γ) if $\ell g(\eta) = 2n+2$ then $I_\eta = \eta(2n+1), \lambda_\eta = \eta(2n) = \lambda_{\eta \restriction (2n+1)}$, see (α) or (β)

(e) if $\eta \in \mathscr{T} \setminus \max(\mathscr{T})$ has length $2n+1$ then: the set $\mathscr{S}_\eta = \{I_\nu : \nu \in \mathrm{suc}_{\mathscr{T}}(\eta)\} \subseteq [I_\eta]^{\lambda_\eta}$ is not from the ideal $\mathrm{ER}^{\lfloor \ell g(\eta)/2 \rfloor}_{I_\eta, \lambda_\eta, \beth_2(\kappa)}$

(f) if $\eta \in \mathscr{T}$ then:

(α) $\bar{h}_\eta = \langle h_{\eta,u} : u \in [I_\eta]^{\leq \lfloor \ell g(\eta)/2 \rfloor} \rangle$

(β) $h_{\eta,u}$ is a $\leq_{\mathfrak{k}}$-embedding of $\mathrm{EM}_{\tau(\mathfrak{k})}(u, \Phi)$ into N for $u \in [I_\eta]^{\leq \lfloor \ell g(\eta)/2 \rfloor}$

(γ) $u_1 \subseteq u_2 \in [I_\eta]^{\leq \lfloor \ell g(\eta)/2 \rfloor} \Rightarrow h_{\eta,u_1} \subseteq h_{\eta,u_2}$

(δ) if $u \in [I_\eta]^{\leq \lfloor \ell g(\eta)/2 \rfloor}$ and $\nu \triangleleft \eta$ and $\ell g(\nu) \geq 2|u|$, then $h_{\eta,u} = h_{\nu,u}$

(ε) if $\ell g(\eta) = 2n+2$ and $u \in [I_\eta]^{\leq n}$ then $h_{\eta,u} = h_{\eta \restriction (2n+1),u}$

(ζ) there is $\bar{a} = \bar{a}_{\mathbf{x}} = \langle a_\alpha : \alpha < \lambda \rangle \in {}^\lambda N$ such that $\alpha \in u \in [I_\eta]^{\leq \lfloor \ell g(\eta) \rfloor /2} + h_{\eta,u}(\alpha) = a_\alpha$ and $\bar{a}$ is with no repetitions

(g) $\mathrm{Dp}_{\mathbf{x}}(<>) \geq \gamma$ where $\mathrm{Dp}_{\mathbf{x}}(\eta)$ is defined as $\mathrm{Dp}_{\mathbf{i}(\mathbf{x})}(\eta)$, see Definition 3.17, where $\mathbf{i} = \mathbf{i}(\mathbf{x}) = \mathbf{i}_{\mathbf{x}}$ is defined by:

- $\mathscr{T}_{\mathbf{i}} = \mathscr{T}$
- $\mathscr{S}_{\mathbf{i}} = \{\eta \in \mathscr{T} : \eta$ is not $\triangleleft$-maximal in $\mathscr{T}$ and $\ell g(\eta)$ is odd$\}$
- if $\eta \in \mathscr{S}_{\mathbf{i}}$ and $\ell g(\eta)$ is odd then $\mathbf{I}_{\mathbf{i},\eta} = \mathrm{ER}^{\lfloor \ell g(\eta) \rfloor}_{I_\eta, \lambda_\eta, \beth_2(\kappa)}$ recalling 3.17(1A)
- if $\eta \in \mathscr{S}_{\mathbf{i}}$ and $\ell g(\eta)$ is even then $\mathscr{I}_{\mathbf{i},\eta} = \{\emptyset\}$.

Definition 4.3. 1. We say $\mathbf{x}$ is a pre-$\mathfrak{k}$-witness of $(N, M, \lambda, \kappa, \delta)$ when it as in 4.2 omitting $\bar{h}$, i.e. clause (f), so N, M are irrelevant.

2. We say $\mathbf{x}$ is a semi-$\mathfrak{k}$-witness of $(N^+, M, \lambda, \kappa, \delta)$ when : it consists of:

(a) N^+ expands a model from $K_{\mathfrak{k}}$, $M \leq_{\mathfrak{k}} (N^+ \restriction \tau(\mathfrak{k}))$, $\lambda \geq \kappa \geq (\tau(N^+))$

(b) – (e) as in 4.2(2)

(f) $\bar{a} = \langle a_\alpha : \alpha < \lambda \rangle$

(g) as in 4.2(2).

Claim 4.4. 1. *The definitions 4.1, 4.2 are equivalent.*

2. *In Definition 4.2, $\mathbf{i}_{\mathbf{x}}$ is indeed a pit.*

3. *If $\Phi_1 \mathrm{E}^{\mathrm{ai}}_\kappa \Phi_2, \Phi_\ell \in \Upsilon^{\mathrm{sor}}_\kappa[M, \mathfrak{k}]$ for $\ell = 1, 2$ and Φ_1 has a (N, M, λ, κ)-witness then Φ_2 has a (N, M, λ, κ)-witness.*

Proof.. Straightforward. $\square_{4.2}$

Claim 4.5. *If $\Phi_\ell \in \Upsilon^{\mathrm{sor}}_\kappa[\mathfrak{k}_M], \kappa \geq \tau(\mathfrak{k}) + \|M\|$ and $M_\ell = \mathrm{EM}_{\tau(\mathfrak{k})}(\lambda, \Phi_\ell)$ for $\ell = 1, 2$ and λ is strong limit of cofinality μ where $\mu = (\beth_2(\kappa))^+$ or μ is regular such that $(\forall \alpha < \mu)(|\alpha|^{2^\kappa} < \mu)$ and the protagonist wins in the game $\Game^1_{M_2, M, \lambda, \Phi_1, \mu}$ (equivalently some $\mathbf{x}$ is a witness for $(M_2, M, \lambda, \kappa, \Phi_1)$) then $\Phi_1 \leq^3_\kappa \Phi_2$, see Definition 3.12.*

Proof.. Straightforward by 3.18 and the definitions of the ideal ER in 3.5. See details in a similar case in the proof of 4.6(1) below. $\square_{4.5}$

Claim 4.6. *Assume $M \leq_{\mathfrak{k}} N, \kappa \geq \|M\| + \theta, \theta \geq \mathrm{LST}_{\mathfrak{k}} + |\tau_{\mathfrak{k}}|$ and $\|N\| \geq \lambda$, λ strong limit of cofinality μ and $\mu = (\beth_2(\kappa))^+$ or μ is regular such that $(\forall \alpha < \mu)(|\alpha|^{2^\kappa} < \mu)$.*

1. *There are* $\mathbf{x}, \Phi$ *such that:*
   (a) $\Phi \in \Upsilon^{\rm sor}_\theta(\mathfrak{k}_M)$
   (b) $\mathbf{x}$ *is a direct witness of* $(N, M, \lambda, \kappa, \Phi)$.
2. *If* $M_1 = M, \Phi_1 \in \Upsilon^{\rm sor}_\theta[\mathfrak{k}_{M_1}]$ *and* $\mathbf{x}_1$ *a direct witness for* $(N, M_1, \lambda, \kappa, \Phi_1)$ *and* $M_1 \leq_{\mathfrak{k}} M_2 \leq_{\mathfrak{k}} N$ *and* $\|M_2\| \leq \kappa$ *then there are* $\Phi_2, \mathbf{x}_2$ *such that:*
   (a) $\Phi_2 \in \Upsilon^{\rm sor}_\theta[M_2]$
   (b) $\Phi_1 \leq^1_\kappa \Phi_2$ *and* $\Phi_1 \leq^4_\kappa \Phi_2$
   (c) $\mathbf{x}_2$ *is a direct witness* $(N, M_2, \lambda, \kappa, \Phi_2)$.
3. *If in part (1) we change the assumption on* λ *to* $\lambda = \beth_{\omega \cdot \gamma}(\kappa)$ *then there are* $\Phi, \mathbf{x}$ *such that:*
   (a) $\Phi \in \Upsilon^{\rm sor}_\kappa[M, \mathfrak{k}]$
   (b) $\mathbf{x}$ *is a direct witness of* $(N, M, \Phi, \lambda, \kappa, \gamma, \Phi)$.
4. *Also part (2) has a version with* (γ_1, γ_2) *as in 3.18.*

Proof.. 1) Let $\langle a_\alpha : \alpha < \lambda\rangle$ be a sequence of pairwise distinct members of N.

Now

$(*)_1$ let $\mathscr{T}$ be the set of finite sequences η satisfying clauses (b),(c) of Definition 4.2

$(*)_2$ let $\bar{\mathbf{I}} = \langle \mathbf{I}_\eta : \eta \in \mathscr{S}\rangle$ where
- $\mathscr{S} = \{\eta \in \mathscr{T} : \eta$ is not $\triangleleft$-maximal in $\mathscr{T}\}$
- if $\eta \in \mathscr{S}, \ell g(\eta) = 2n+1$ then $\mathbf{I}_\eta = {\rm ER}^n_{I_\eta, \lambda_\eta, \beth_2(\kappa)}$
- if $\eta \in \mathscr{S}$ and $\ell g(\eta) = 2n$ then $\mathbf{I}_\eta = \{\emptyset\}$, the trivial ideal

$(*)_3$ $\mathbf{i}_1 = \mathbf{i}(1) = (\mathscr{T}, \bar{\mathbf{I}})$ is a pit and is $(2^\kappa)^+$-complete and ${\rm Dp}_{\mathbf{i}_1}(<>) \geq (\beth_2(\kappa))^+$.

[Why? Just read Definition 3.17(3) and the ideal ER is from Definition 3.5 and it is $(2^\kappa)^+$-complete by 3.6 and as for the depth recall $\mu = (\beth_2(\kappa))^+$.]

$(*)_4$ Let M^+ be such that:
   (a) M^+ is an expansion of N
   (b) $|\tau(M^+)| \leq \kappa$ and $\tau' := \tau(M^+) \backslash \{c_a : a \in M\}$ has cardinality $\leq \theta$
   (c) if $M_1^+ \restriction \tau' \subseteq M^+ \restriction \tau'$ then $M_1^+ \restriction \tau(\mathfrak{k}) \leq M^+ \restriction \tau(\mathfrak{k})$
   (d) $|M| = \{c^{M^+} : c \in \tau(M^+)\}$.

[Why M^+ exists? By the representation theorem, [Sh:88r, §1] except clause (d) which as before is easy.]

We like to apply Theorem 3.18 but before this we need

$(*)_5$ there is a pit $\mathbf{i}_2 = \mathbf{i}(2)$ such that $\mathbf{i}(1) \leq_{\rm pr} \mathbf{i}(2)$ (see 3.17(2A)) so ${\rm Dp}_{\mathbf{i}(2)}(\eta) = {\rm Dp}_{\mathbf{i}(1)}(\eta)$ for $\eta \in \mathscr{T}_{\mathbf{i}(2)}$ and:
- if $\eta \in \mathscr{T}_{\mathbf{i}(2)}, \ell g(\eta) = 2n+1$ and $\nu \in {\rm suc}_{\mathscr{T}_{\mathbf{i}(2)}}(\eta)$ then $\langle a_\alpha : \alpha \in \nu(2n+1)\rangle$ is an n-indiscernible sequence in M^+ for quantifier free formulas, may add: and $N \restriction \{\sigma_\varepsilon(a_{\alpha_0}, \ldots, a_{\alpha_{n-1}}) : \varepsilon < \zeta\} \leq_{\mathfrak{k}} N$ where $\zeta < \kappa^+$ and σ_ε is a $\tau(M^+)$-term.

[Why such $\mathbf{i}(2)$ exists? By the definition of the ideal $\mathbf{I}_\eta$, see $(*)_2$ above and by Definition 2.14. That is, for $\eta \in \mathrm{Dom}(\mathbf{I}_{\mathbf{i}_1})$ of length $2n+1$ let $X_\eta = \{\nu : \nu \in \mathrm{suc}_{\mathscr{T}}(\eta), \langle a_\alpha : \alpha \in \nu(2n+1)\rangle$ is n-indiscernible in M^+ for quantifier free formulas$\}$, recalling $\mathrm{Dom}(\mathbf{I}_{\mathbf{i}_1,\eta}) = \{u \subseteq I_\eta : |u| = \eta(2n)\}$. By 3.5 clearly $X_\eta = [\lambda_\eta]^{\eta(2n)} \mod \mathrm{ER}^{\mathrm{n}}_{\lambda_\eta,\eta(2),\beth_2(\kappa)}$, see Definition 3.17(1A).

Now let $\mathscr{T}' = \{\eta \in \mathscr{T}$: if $2n+1 < \ell g(\eta)$ then $\eta \restriction (2n+2) \in X_\eta\}$ and $\mathbf{i}_2 = \mathbf{i}_1 \restriction \mathscr{T}'$, so clearly $\mathbf{i}_1 \leq_{\mathrm{pr}} \mathbf{i}_2$, see Definition 3.17(2A).]

Next

$(*)_6$ define a function $\mathbf{c}$ with domain $\mathscr{T}_{\mathbf{i}_2}$ as follows:
- if $\eta \in \mathscr{T}, \ell g(\eta) = 2n+2$, <u>then</u> $\mathbf{c}(\eta)$ is the quantifier type in M^+ of $\langle a_\ell : \ell < n\rangle$ for any $\alpha_0 < \alpha_1 < \ldots < \alpha_{n-1}$ from $\eta(2n+1)$
- if $\eta \in \mathscr{T}, \ell g(\eta) = 2n+1$ or $\ell g(\eta) = 0$, <u>then</u> $\mathbf{c}(\eta) = 0$.

Clearly

$(*)_7$ $\mathrm{Rang}(\mathbf{c})$ has cardinality $\leq 2^\kappa = 2^\kappa$.

So by 3.18 (with a degenerate projection; so κ, θ there stands for $2^\kappa, \aleph_0$ here):

$(*)_8$ there are $\mathbf{i}(3) = \mathbf{i}_3 \geq \mathbf{i}_2$ and $\langle c_n : n < \omega\rangle$ such that:

(a) $\eta \in \mathscr{T}_{\mathbf{i}_3} \Rightarrow \mathbf{c}(\eta) = c_{\ell g(\eta)}$

(b) $\mathrm{Dp}_{\mathbf{i}_3}(<>) \geq \beth_2(\kappa)$.

The rest should be clear.

2) Similar proof, this time in M^+ we have individual constants for every member of M_2 and we start with the witness $\mathbf{x}_1$ so X_η have fewer elements still positive modulo the ideal.

3),4) Similarly. $\square_{4.6}$

Definition 4.7. We say $\mathbf{x}$ is an indirect witness for $(N, M, \lambda, \kappa, \gamma, \Phi)$, recalling 4.2(1), <u>when</u> for some Ψ:

(a) $M, N, \lambda, \kappa, \gamma, \Phi$ are as in Definition 4.2

(b) $\Psi \in \Upsilon^{\mathrm{sor}}_\kappa[\mathfrak{k}_M]$ and $\Phi \leq^4_\kappa \Psi$, see Definition 3.12

(c) $\mathbf{x}$ is a direct witness of $(N, M, \lambda, \kappa, \gamma, \Psi)$.

Remark 4.8. Why do we need the indirect witnesses? As if we use direct witness only in the proof of 4.14 it is not clear how to get many non-isomorphic models.

Claim 4.9. *Assume $I = I_\chi$ is as in 2.15.*

If (A) then (B) where:

(A) (a) $\mathrm{LST}_{\mathfrak{k}} + |\tau_{\mathfrak{k}}| \leq \kappa < \chi_1 < \chi_2 < \chi_3 \leq \chi$ *and for* $\ell = 1, 2, \chi_{\ell+1}$ *is strong limit of cofinality* $> \beth_2(\chi_\ell)$

(b) $N = \mathrm{EM}_{\tau(\mathfrak{k})}(I, \Phi_1)$ *where* $\Phi_1 \in \Upsilon^{\mathrm{sor}}_\kappa[M_1, \mathfrak{k}], \|M_1\| \leq \chi_1$

(c) $M_2 \leq_{\mathfrak{k}} N$ *and* $\|M_2\| \leq \chi_1$
(d) $\Phi_2 \in \Upsilon^{\rm sor}_\kappa[M, \mathfrak{k}]$
(e) Φ_2 *has a witness for* (N, M_2, χ_2, κ)

(B) (a) Φ_2 *has a witness for* (N, M_2, χ_3, κ)
(b) *if in addition* $M_2 \leq_{\mathfrak{k}} M_1$ *then* $\Phi_2 \leq^3_\kappa \Phi_1$
(c) *we can* $\leq_{\mathfrak{k}}$*-embed* $\mathrm{EM}_{\tau(\mathfrak{k})}(I_\chi, \Phi_2)$ *into* N.

Proof.. As in the proof of 4.6 recalling the choice of I in 2.15; for (B)(c) we use Clause $(B)^+$ of 4.6. $\square_{4.9}$

Remark 4.10. In fact, in 4.9, $\chi_2 = \beth_{1,1}(\chi_1)$ and $\chi_3 = \beth_{\omega\gamma}(\chi_1)$ suffices so, of course, in (B)(a) we use $(N, M_1, \chi_3, \kappa, \gamma)$.

Claim 4.11. *If (A) then (B) where:*

(A) (a) $M_1 \leq_{\mathfrak{k}} M_2 \leq_{\mathfrak{k}} N$
(b) (α) M_ℓ *has cardinality* κ_ℓ
(β) $\|N\| \geq \lambda$
(γ) $\kappa_\ell \geq \kappa \geq \mathrm{LST}_{\mathfrak{k}} + |\tau_{\mathfrak{k}}|$
(c) $\Phi_1 \in \Upsilon^{\rm sor}_\kappa(M_1, \mathfrak{k})$
(d) λ *is strong limit and* $\mathrm{cf}(\lambda) = (\beth_2(\kappa_2))^+$ *or just* $(\forall \alpha < \mathrm{cf}(\lambda))(|\alpha|^{2^\kappa} < \mathrm{cf}(\lambda))$
(e) $\mathbf{x}_1$ *is an indirect witness for* $(N, M_1, \lambda, \kappa, \Phi_1)$

(B) *there are* $\Phi_2, \mathbf{x}_2$ *such that:*
(a) $\Phi_2 \in \Upsilon^{\rm sor}_\kappa(\mathfrak{k}_{M_2})$
(b) $\Phi_1 \leq^1_{\kappa_2} \Phi_2$
(c) $\mathbf{x}_2$ *is an indirect witness for* $(N, M_2, \lambda, \kappa_2, \Phi_2)$.

Proof.. By clause $(A)(e)$ of the assumption and the definition of indirect witness in 4.7 there is Ψ_1 such that:

$(*)_1$ (a) $\Psi_1 \in \Upsilon^{\rm or}_{\kappa_1}[\mathfrak{k}_{M_1}]$ which is standard
(b) $\mathbf{x}_1$ is a direct witness of $(N, M_1, \lambda, \kappa_1, \Psi_1)$
(c) $\Phi_1 \leq^4_{\kappa_1} \Psi_1$.

By claim 4.6(2) there are $\mathbf{x}_2, \Psi_2$ such that

$(*)_2$ (a) $\Psi_2 \in \Upsilon^{\rm sor}_{\kappa_2}[\mathfrak{k}_{M_2}]$
(b) $\Psi_1 \leq^1_{\kappa_2} \Psi_2$
(c) $\mathbf{x}_2$ is a direct witness of $(N, M_2, \lambda, \kappa_2, \Psi_2)$.

Lastly, by 3.16 applied to our Φ_1, Ψ_1, Ψ_2 and get Φ_2 such that

$(*)_3$ (a) $\Phi_1 \in \Upsilon^{\rm sor}_\kappa[\mathfrak{k}_{M_2}]$
(b) $\Phi_1 \leq^1_\kappa \Phi_2$
(c) $\Phi_2 \leq^4_\kappa \Psi_2$.

So we have gotten Clause (B) as promised. $\square_{4.11}$

Claim 4.12. *If (A) + (B) then (C) where:*

(A) (a) $\lambda_n \geq \mathrm{LST}_{\mathfrak{k}}$ *is strong limit,* $\mathrm{cf}(\lambda_n) = (\beth_2(\mathrm{LST}_{\mathfrak{k}} + \lambda_m))^+$ *if* $n = m + 1$

(b) $\lambda = \sum_n \lambda_n$ *and* $\lambda_n < \lambda_{n+1}$

(c) $N \in K^{\mathfrak{k}}_{\lambda}$

(d) $M_n \leq_{\mathfrak{k}} M_{n+1} <_{\mathfrak{k}} N$ *and* $\|M_n\| = \lambda_n$

(e) $N = \cup\{M_n : n < \omega\}$

(B) *there is no* $\Phi \in \Upsilon^{\mathrm{sor}}_{\lambda}[\mathfrak{k}_N]$, *see 3.15*

(C) *for some n and* Φ

(a) $\Phi \in \Upsilon^{\mathrm{sor}}_{\lambda_n}[\mathfrak{k}_{M_n}]$

(b) *there is an indirect witness*[9] *for* $(N, M_n, \lambda_{n+4}, \lambda_n, \Phi_n)$

(c) *there is no indirect witness for* $(N, M_n, \lambda_{n+5}, \lambda_n, \Phi_n)$.

Remark 4.13. 1. Later we shall weaken $(A)(a)$.

2. We may use $\Upsilon^{\mathrm{sor}}_{\kappa}[\mathfrak{k}_{M_n}]$ where $\lambda_0 \geq \kappa \geq \mathrm{LST}_{\mathfrak{k}} + |\tau_{\mathfrak{k}}|$ in 4.11 and in 4.12, also in 4.14.

Proof.. We assume $(A) + \neg(C)$ and shall prove $\neg(B)$, this suffices. We try to choose $(\Phi_n, \mathbf{x}_n)$ by induction on n such that:

$\otimes$ (a) $\Phi_n \in \Upsilon^{\mathrm{sor}}_{\lambda_n}[\mathfrak{k}_{M_n}]$

(b) $\{c_a : a \in N\} \cap \tau(\Phi_n) = \{c_a : a \in M_n\}$

(c) $\mathbf{x}_n$ is an indirect witness for $(N, M_n, \lambda_{n+4}, \lambda_n, \Phi_n)$

(d) if $n = m + 1$ then $\Phi_m \leq^1_{\lambda_n} \Phi_n$.

Now

$(*)_1$ if we succeed to carry the induction <u>then</u> there is $\Phi \in \Upsilon^{\mathrm{sor}}_{\lambda}[\mathfrak{k}_N]$.

[Why? Note that $\Phi_n \in \Upsilon^{\mathrm{sor}}_{\lambda_n}[\mathfrak{k}_{M_n}] \subseteq \Upsilon^{\mathrm{sor}}_{\lambda_n}[\mathfrak{k}]$ and as $\lambda_n \leq \lambda$ clearly $\Phi_n \in \Upsilon^{\mathrm{sor}}_{\lambda_n}[\mathfrak{k}] \subseteq \Upsilon^{\mathrm{sor}}_{\lambda}[\mathfrak{k}]$ and so by 3.11(2) there is $\Phi \in \Upsilon_{\lambda}[\mathfrak{k}]$ such that $n < \omega \Rightarrow \Phi_n \leq^1_{\lambda} \Phi$. Easily N is $\leq_{\mathfrak{k}}$-embeddable into every $\mathrm{EM}_{\tau(\mathfrak{k})}(I, \Phi)$, in fact, $\Phi \in \Upsilon_{\lambda}[\mathfrak{k}_N]$, contradiction to clause (B) of the assumption.]

$(*)_2$ we can choose $(\mathbf{x}_n, \Phi_n)$ for $n = 0$.

[Why? By 4.6(1).]

$(*)_3$ if $n = m + 1$ and we have chosen $(\mathbf{x}_m, \Phi_m)$ then we can choose $(\mathbf{x}_n, \Phi_n)$.

9 hence also a direct one; similarly in $\otimes(d)$ in the proof

[Why? If there is no indirect witness $\mathbf{y}_m$ for $(N, M_m, \lambda_{m+5}, \lambda_m, \Phi_m)$ we have gotten clause (C), so without loss of generality $\mathbf{y}_m$ exists. Now apply 4.11 with $(\mathbf{y}_n, M_m, M_n, \lambda_{n+5}, \lambda_n)$ here standing for $(\mathbf{x}_1, M_1, M_2, \lambda, \kappa, \Phi_1)$ there, so we get $\mathbf{x}_n, \Phi_n$ here stand for $\mathbf{x}_2, \Phi_2$ there.] $\square_{4.12}$

Claim 4.14. *We have $\dot{I}(\mu, K_{\mathfrak{k}}) \geq \chi$ when:*

⊕ (a) $\mathrm{LST}_{\mathfrak{k}} + |\tau_{\mathfrak{k}}| \leq \kappa \leq \chi_1 < \chi_2 < \chi_3 \leq \min\{\lambda, \mu\}$
 (b) $M \leq_{\mathfrak{k}} N$
 (c) $\|M\| \leq \kappa$ *and* $\|N\| \geq \lambda$
 (d) $\Phi \in \Upsilon^{\mathrm{sor}}_{\kappa}[\mathfrak{k}_M]$
 (e) $\mathbf{x}$ *is an indirect witness for* $(N, M, \chi_2, \chi_1, \Phi)$
 (f) *there is no indirect witness for* $(N, M, \chi_3, \chi_1, \Phi)$
 (g) χ_3 *is strong limit of cofinality* $(\beth_2(\chi_2))^+$
 (h) $\chi = |\{\theta : \theta = \beth_\theta$ *and* $\theta \in [\chi_1, \chi_2]\}|$

Proof.. Let γ_* be maximal such that $\beth_{\omega\cdot\gamma_*}(\chi_1) \leq \chi_2$. Let $\Psi \in \Upsilon^{\mathrm{sor}}_{\kappa}[\mathfrak{k}_M]$ be such that $\Phi \leq^4_\kappa \Psi$ and Ψ has a direct witness for $(N, M, \chi_2, \chi_1, \Psi)$ and choose such a witness $\mathbf{x}$.

Let M_2 be such that $M \leq_{\mathfrak{k}} M_2 \leq_{\mathfrak{k}} N$ and $\|M_2\| = \beth_{\omega\cdot\gamma_*}(\chi_1) \leq \chi_2$ and $\mathbf{x}$ is a direct witness for $(M_2, M, \beth_{\omega\cdot\gamma}(\chi_1), \chi_1, \gamma_*, \Psi)$.

As χ_3 is strong limit of cofinality $> \beth_2(\chi_2)$ there are $\Phi_3 \in \Upsilon^{\mathrm{sor}}_{\kappa}[\mathfrak{k}_{M_2}]$ and $\mathbf{y}$ which is a direct witness for $(N, M_2, \chi_3, \chi_2, \Phi_3)$ and so $\tau'_{\Phi_3} = \tau(\Phi_3)\backslash\{c_a : a \in M_2\}$ has cardinality κ. For each $\gamma < \gamma_*$ there are $M_{2,\gamma}, \mathbf{x}_\gamma$ such that:

$(*)_1$ (a) $M_{2,\gamma} \leq_{\mathfrak{k}} M_2$
 (b) $\|M_{2,\gamma}\|$ is $\geq \beth_{\omega\cdot\gamma}(\chi_1)$ but $< \beth_{\omega\cdot\gamma+\omega}(\chi_1)$; can get even $\|M_{2,\gamma}\| = \beth_{\omega\cdot\gamma}(\chi_1)$
 (c) $\mathbf{x}_\gamma$ is a direct witness for $(M_{2,\gamma}, M, \beth_{\omega\cdot\gamma}(\chi_1), \chi_1, \gamma, \Psi)$.

[Why? Try by induction on k to choose $\eta_k \in \mathscr{T}_{\mathbf{x}}$ such that $\ell g(\eta_k) = 2k+1, \eta_k(2k) \geq \beth_{\omega\cdot\gamma}(\chi_1)$ and $\ell < k \Rightarrow \eta_\ell \triangleleft \eta_k$. For $k = 0$, clearly $\eta_k = \langle\rangle$ is O.K., and as $\eta_\ell(2\ell) > \eta_{\ell+1}(2\ell+2)$, necessarily for some k we have η_k but cannot choose η_{k+1}; let $A_\gamma = \cup\{\mathrm{Rang}(h^{\mathbf{x}}_{\eta,u}) : \eta_k \trianglelefteq \eta \in \mathscr{T}_{\mathbf{x}}$ and $u \in [I^{\mathbf{x}}_\eta]^{\lfloor \ell g(\eta)/2 \rfloor}\}$ so $A_\gamma \subseteq M$ has cardinality $\eta_k(2k) \in [\beth_{\omega\cdot\gamma}(\chi_1), \beth_{\omega\cdot\gamma+\omega}(\chi_1))$. Without loss of generality if $N_* = \mathrm{EM}(\emptyset, \Phi_3)$ is standard (i.e. $M = N_* \restriction \tau_{M_2}$) then A_γ is closed under the functions of $N_* \restriction \tau'_{\Phi_3}$. Let $M_{2,\gamma} = M_2 \restriction A_\gamma$; it is $\leq_{\mathfrak{k}} M$ and it satisfies clauses (a),(b) and include A_γ. Then we can easily find $\mathbf{x}_\gamma$ as required in clause (c).]

Next we can find $\mathbf{y}_\gamma, \Phi_{3,\gamma}$ such that

$(*)_2$ (a) $\mathbf{y}_\gamma$ is a direct witness of $(N, M_{2,\gamma}, \chi_3, \|M_{2,\gamma}\|, \Phi_{3,\gamma})$
 (b) $\Phi_{3,\gamma} \in \Upsilon^{\mathrm{sor}}_{\kappa}[M_{2,\gamma}, \mathfrak{k}]$.

[Why? Recall $\tau(\Phi_3)\backslash\{c_a : a \in M_2\}$ has cardinality κ. Let $\tau_{2,\gamma} = \tau(\Phi_3)\backslash\{c_a : a \in M_2\backslash M_{2,\gamma}\}$ so has cardinality $\|M_{2,\gamma}\|$, let $\Phi_{3,\gamma} = \Phi_3 \restriction \tau_{2,\gamma}$, is as required in $(*)_2(k)$. As for $\mathbf{y}_\gamma$ we derived it form $\mathbf{y}$.]

Now let $I = I_\mu$ be a linear order of cardinality μ as required in 2.15.

Lastly, let $N_\gamma = \mathrm{EM}_{\tau(\mathfrak{k})}(\mu, \Phi_{3,\gamma})$ be standard hence $M_{2,\gamma} \leq_{\mathfrak{k}} N_\gamma \in K^{\mathfrak{k}}_\mu$.

We choose ∂_i by induction on i such that: if $i = 0$ then $\partial_i = \chi_1$, if i is limit then $\partial_i = \cup\{\partial_j : j < i\}$ and if $i = j + 1$ then $\partial_i = \beth_{\beth_2(\partial_j)^+}$ when it is $\leq \chi_2$ and undefined otherwise. Let ∂_i be defined iff $i < i(*)$ and let $\Theta = \{\partial_{i+1} : i + 1 < i(*)\}$. Now $|\Theta| \geq \chi$ so it suffices to prove that $\langle N_\theta : \theta \in \Theta\rangle$ are pairwise non-isomorphic.

So toward contradiction assume

$(*)_3$ $\theta_1 < \theta_2$ are from Θ and π is an isomorphism from N_{θ_2} onto N_{θ_1}.

We can find $M_* \leq_{\mathfrak{k}} N_{\theta_1}$ such that $\|M_*\| = \theta_2$ and $M \cup M_{2,\theta_1} \cup \pi(M_{2,\theta_2}) \subseteq M_*$ and without loss of generality we can find $I_* \subseteq \mu$ of cardinality θ_2 such that $M_* = \mathrm{EM}_{\tau(\mathfrak{k})}(I_*, \Phi_{3,\theta_1})$.

Let $I^*_1 \subseteq I_*$ be of cardinality θ_1 such that

$$M_{2,\theta_1} \cup \pi(M) \subseteq N'_{\theta_1} := \mathrm{EM}_{\tau(\mathfrak{k})}(I^*_1, \Phi_{3,\theta_1})$$

and let $N'_{\theta_2} = \pi^{-1}(N'_{\theta_1})$. By 4.6(2) we can find $\Psi' \in \Upsilon^{\mathrm{sor}}_\kappa(N'_{\theta_2}, \mathfrak{k})$ and $\mathbf{x}_{\theta_2}$ a witness for $(M_{2,\theta_2}, N'_{\theta_2}, \theta_2, \kappa, \Psi')$ such that $\Psi \leq^4_\kappa \Psi'$ and $\mathbf{x}_{\theta_2} \leq \mathbf{x}'_{\theta_2}$ where $\theta_2 = \beth_{\omega\cdot\gamma_2}(\chi_1)$.

Now clearly $N'_{\theta_1}, \Psi, \pi(\Psi'), \pi(\mathbf{x}'_{\theta_2})$ satisfies the parallel statements in N_{θ_1}. By 4.9(B)(a) and the choice of I_μ there is a witness for $(N_{\theta_1}, N'_{\theta_1}, \chi_3, \kappa, \pi(\Psi'))$, hence applying π^{-1} there is a witness $\mathbf{x}''_{\theta_2}$ for $(N_{\theta_1}, N'_{\theta_1}, \chi_3, \kappa, \Psi')$.

Hence by 4.9(B)(b), $\Psi' \leq^3_\kappa \Phi_{3,\theta_2}$ but together $\Phi \leq^4_\kappa \Psi \leq^4_\kappa \Psi' \leq^3_\kappa \Phi_{3,\theta_2}$ hence $\Phi \leq^3_{\theta_2} \Phi_{3,\theta_2}$ by 3.14(1) so by 3.14(2), the last clause, there is $\Phi'_{3,\theta_2} \in \Phi_{3,\theta_2}/\mathrm{E}^{\mathrm{ai}}_{\theta_2}$ such that $\Phi \leq^4_{\theta_2} \Phi'_{3,\theta_2}$. But as Φ_{3,θ_2} has a $(N, M_{2,\theta_2}, \chi_3, \theta_2)$ witness by 4.4(3) also Φ'_{3,θ_2} has hence Φ has an indirect witness for (N, M, χ_3, κ), contradiction. $\square_{4.14}$

Conclusion 4.15. Assume $\mathrm{cf}(\lambda) = \aleph_0$ and $\lambda = \beth_{1,\lambda}$.

1. If $\lambda > \dot{I}(\lambda, K_{\mathfrak{k}})$ then $M \in K^{\mathfrak{k}}_\lambda \Rightarrow \Upsilon^{\mathrm{sor}}_\lambda[\mathfrak{k}_M] \neq \emptyset$.
2. If $\mu \geq \lambda > \dot{I}(\mu, K_{\mathfrak{k}})$ <u>then</u> $M \in K^{\mathfrak{k}}_\lambda \Rightarrow \Upsilon^{\mathrm{sor}}_\lambda[\mathfrak{k}_M] \neq \emptyset$.

Moreover, at least one of the following holds:

(*a*) for some $\chi_1 < \lambda$ if $\chi_1 < \chi_2 = \beth_{2,\delta} \leq \min\{\lambda, \mu\}$ then $|\delta| \leq \dot{I}(\mu, K_{\mathfrak{k}})$

(*b*) $\Upsilon^{\mathrm{sor}}_\lambda[\mathfrak{k}_M] \neq \emptyset$ for every $M \in K^{\mathfrak{k}}_\lambda$.

Theorem 4.16. *The result from the abstract holds, that is, for every a.e.c. $\mathfrak{k}$ for some closed unbounded class* $\mathbf{C}$ *of cardinals we have (a) or (b) where*

(*a*) *for every $\lambda \in \mathbf{C}$ of cofinality $\aleph_0$, $\dot{I}(\lambda, K) \geq \lambda$*

(*b*) *for every $\lambda \in \mathbf{C}$ of cofinality $\aleph_0$ and $M \in K_\lambda$, for every cardinal $\kappa \geq \lambda$ there is N_κ of cardinality κ extending M (in the sense of our a.e.c.).*

Proof.. Let $\Theta = \{\mu : \mu = \beth_{2,\delta}$ and $|\delta| > \dot{I}(\mu, K_{\mathfrak{k}})$ for some limit ordinal $\delta\}$.

Case 1: Θ is an unbounded class of cardinals.

So $\mathbf{C} = \{\mu : \mu = \sup(\mu \cap \Theta)\}$ is a closed unbounded class of cardinals. Easily $\mu \in \mathbf{C} \Rightarrow \mu = \beth_{1,\mu}$ and by 4.15 + 3.15 for every $\mu \in \mathbf{C}$, clause (b) of 4.16 holds.

Case 2: Θ is a bounded class of cardinals.

So by the definition of Θ, $\mathbf{C} = \{\mu : \mu > \sup(\Theta), \mu = \beth_{2,\mu}\}$ is as required. $\square_{4.16}$

Also

Theorem 4.17. *For every* aec $\mathfrak{k}$ *one of the following holds:*

(*a*) *for some χ we have $\chi < \mu = \beth_{2,\mu} \Rightarrow \dot{I}(\mu, K_{\mathfrak{k}}) \geq \mu$ and $\chi < \mu = \beth_{1,\omega\cdot\gamma} \Rightarrow \dot{I}(\mu, K_{\mathfrak{k}}) \geq |\gamma|$*

(*b*) *for some closed unbounded class $\mathscr{C}$ of cardinals we have* $\mathrm{cf}(\lambda) = \aleph_0 \wedge \lambda \in \mathbf{C} \wedge M \in K^{\mathfrak{k}}_{\lambda} \Rightarrow \Upsilon^{\mathrm{sor}}[M, \mathfrak{k}] \neq \emptyset$.

Proof.. Similarly to 4.16, using Fodor lemma for classes of cardinals. $\square_{4.17}$

5 Concluding Remarks

Definition 5.1. For an ordinal γ, τ-models M_1, M_2 and cardinal λ we define a game $\eth = \eth_{\theta,\gamma}(M_1, M_2)$. A play lasts less than ω models is defined as in [Sh:797, 2.1].

Claim 5.2. 1. *Assume* $\mathrm{cf}(\lambda) = \aleph_0$ *and M_1, M_2 are τ-models of cardinality λ. If the isomorphism player wins in $\eth_{\lambda,\gamma}(M_1, M_2)$ for every γ or just $\gamma < (2^{<\lambda})^+$ then M_1, M_2 are isomorphic.*

1A. *If above λ is strong limit then* "$(2^{<\lambda})^+ = \lambda^+$".

2. *Assume λ is strong limit of cofinality $K = K_{\mathfrak{k}}$ and* $|\tau_{\mathfrak{k}}| + \mathrm{LST}_{\mathfrak{k}} \leq \lambda$ *and $K = \{M \restriction \tau : M \models \psi\}$ for some $\psi \in \mathbb{L}_{\lambda^+,\aleph_0}$.*
If $\dot{I}(\lambda, K) \leq \lambda$ then for every $M_1 \in K$ there is $M_2 \in K_{\leq\lambda}$ such that the isomorphism player wins in $\eth_{\lambda,\gamma}(M_1, M_2)$ for every γ.

Conjecture 5.3. For every a.e.c. $\mathfrak{k}$ letting $\kappa = \mathrm{LST}_{\mathfrak{k}} + |\tau_{\mathfrak{k}}|$, at least one of the following occurs:

(*a*) if $\lambda = \beth_{1,\lambda} > \kappa$ and $\mathrm{cf}(\lambda) = \aleph_0$, then $\Upsilon^{\mathrm{sor}}_{\kappa}[M, \mathfrak{k}] \neq \emptyset$

(*b*) if $\lambda = \beth_{1,\lambda} > \kappa$ and $\mathrm{cf}(\lambda) = \aleph_0$, then $\dot{I}(\lambda, K_{\mathfrak{k}}) = 2^{\lambda}$.

Acknowledgement: The author would like to thank the Israel Science Foundation for partial support of this research (Grant No. 242/03). Publication 893.
The author thanks Alice Leonhardt for the beautiful typing.

Bibliography

[Bal09] John Baldwin, *Categoricity*, University Lecture Series, vol. 50, American Mathematical Society, Providence, RI, 2009.

[Dic85] M. A. Dickman, *Larger infinitary languages*, Model Theoretic Logics (J. Barwise and S. Feferman, eds.), Perspectives in Mathematical Logic, Springer-Verlag, New York Berlin Heidelberg Tokyo, 1985, pp. 317–364.

[Sh:c] Saharon Shelah, *Classification theory and the number of nonisomorphic models*, Studies in Logic and the Foundations of Mathematics, vol. 92, North-Holland Publishing Co., Amsterdam, xxxiv+705 pp, 1990.

[Sh:f] Saharon Shelah, *Proper and improper forcing*, Perspectives in Mathematical Logic, Springer, 1998.

[Sh:h] Saharon Shelah, *Classification Theory for Abstract Elementary Classes*, Studies in Logic: Mathematical logic and foundations, vol. 18, College Publications, 2009.

[Sh:i] Saharon Shelah, *Classification Theory for Abstract Elementary Classes 2*, Studies in Logic: Mathematical logic and foundations, vol. 20, College Publications, 2009.

[Sh:11] Saharon Shelah, *On the number of non-almost isomorphic models of T in a power*, Pacific Journal of Mathematics **36** (1971), 811–818.

[Sh:E46] Saharon Shelah, *Categoricity of an abstract elementary class in two successive cardinals, revisited*, Chapter V of [Sh:i].

[Sh:E53] Saharon Shelah, *Introduction and Annotated Contents*, Chapter N of [Sh:h], 0903.3428.

[Sh:88r] Saharon Shelah, *Abstract elementary classes near $\aleph_1$*, Chapter I of [Sh:h]. 0705.4137. 0705.4137.

[RuSh:117] Matatyahu Rubin and Saharon Shelah, *Combinatorial problems on trees: partitions, Δ-systems and large free subtrees*, Annals of Pure and Applied Logic **33** (1987), 43–81.

[MMSh:318] Dugald Macpherson, Alan H. Mekler and Saharon Shelah, *The number of infinite substructures*, Math Proc Cambridge Philosophical Soc 109 (1991), 193–209.

[Sh:394] Saharon Shelah, *Categoricity for abstract classes with amalgamation*, Annals of Pure and Applied Logic **98** (1999), 261–294, math.LO/9809197.

[Sh:576] Saharon Shelah, *Categoricity of an abstract elementary class in two successive cardinals*, Israel Journal of Mathematics **126** (2001), 29–128, math.LO/9805146.

[Sh:600] Saharon Shelah, *Categoricity in abstract elementary classes: going up inductively*, Chapter II of [Sh:h], math.LO/0011215.

[Sh:734] Saharon Shelah, *Categoricity and solvability of A.E.C., quite highly*, Chapter IV of [Sh:h], 0808.3023.

[KoSh:796] Peter Komjath and Saharon Shelah, *A partition theorem for scattered order types*, Combinatorics Probability and Computing **12** (2003, no.5-6), 621–626, Special issue on Ramsey theory. math.LO/0212022.

[Sh:797] Saharon Shelah, *Nice infinitary logics*, Journal of the American Mathematical Society **25** (2012), 395–427, 1005.2806.

[Sh:842] Saharon Shelah, *Solvability and Categoricity spectrum of a.e.c. with amalgamation*, Preprint.

[Sh:849] Saharon Shelah, *Beginning of stability theory for Polish Spaces*, to be published in Israel Journal of Mathematics.

[GhSh:909] Esther Gruenhut and Saharon Shelah, *Uniforming n-place functions on well founded trees*, Set Theory and Its Applications, Contemporary Mathematics (CONM), vol. 533, Amer. Math. Soc., 2011, 0906.3055, pp. 267–280.

[BaLaSh:1037] John Baldwin, Chris Laskowski and Saharon Shelah.

[Sh:1019] Saharon Shelah, *Model theory for θ-complete ultra-powers*, Israel Journal of Mathematics, **submitted**.

[Sh:F1273] Saharon Shelah, *Categoricity spectrum for the atomic model of a countable T: thoughts*.

[Sh:F1302] Saharon Shelah, *AEC with many models*, in preparation.

Jouko Väänänen

Pursuing Logic without Borders

Jouko Väänänen: Department of Mathematics and Statistics, University of Helsinki, Finland, and Institute for Logic, Language and Computation, University of Amsterdam, The Netherlands

1 Introduction

I am too young to have followed Mostowski's influential lectures "Thirty Years of Foundational Studies" at the Vaasa Summer University in 1964. However, when I was a student I bought the book [22] from the Societas Philosophica Fennica office in the House of the Estates, an imposing Neo-Renaissance building in the centre of Helsinki. Like many in my generation, I was greatly impressed by the book. Mostowski covers classical and non-classical logic, model theory and set theory, recursion theory and proof theory, as if they really were all more or less equal areas of logic. Now, 40 years later, I worry about logic having become too fragmented, young researchers of each subfield becoming so immersed in the internal questions of their field that the connection to neighboring subfields is eroded. This may be necessary from a narrow perspective of solving ever more difficult and specific technical problems but it does not serve well the interdisciplinary goal of logic. Logic should not have borders between its subfields, but should rather promote active movement and interaction across its different parts.

In the early 1970s we had an improvised study group of logic in the Mathematics Department of the University of Helsinki. One of the papers we read was an early manuscript of [3], which we got because one of us, Seppo Miettinen, spent a fall semester in Oslo, where Barwise was visiting. Seppo, who was my first logic teacher, got us all interested in extensions of first order logic. We were persuaded by the idea that model theory, set theory and recursion theory are just different approaches to the same goal, understanding definability. Indeed, in the times of my student years there was a popular such approach, namely the theory of admissible sets, which unified (infinitary) model theory, (Kripke-Platek) set theory and (generalized) recursion theory. A whole group of us went to Oslo to hear Ronald Jensen's lectures on admissible sets and forcing.

I found generalized quantifiers, infinitary logics, and abstract model theory very intriguing, and most of my research ever since has been on these topics. The questions I studied led me more and more into set theory and I became identified as a set theorist, although I consider myself also a model theorist, perhaps a "set-theoretical model theorist".

2 Generalized quantifiers

Early on I started to investigate the overall structure of generalized quantifiers. One might say that quite arbitrary generalized quantifiers are not interesting because they do not occur "naturally". But both the quantifiers that arise in computer science, especially as database queries, and the quantifiers of natural language can be so complex that the requirement of occurring "naturally" becomes too restrictive. My first paper on the structure of generalized quantifiers [27] was an attempt to understand the difference between unary and binary generalized quantifiers. Examples of unary quantifiers (of type (1)) are $Q_1 x\phi(x)$, "there are uncountably many x such that $\phi(x)$", and the (type (1,1)) Härtig-quantifier $Ixy\phi(x)\psi(y)$, "There are as many x such that $\phi(x)$ as there are y such that $\psi(y)$". An example of a binary quantifier is $Q_1^E xy\phi(x, y)$, "$\phi(x, y)$ defines an equivalence relation with uncountably many classes". I showed that the binary quantifier Q_1^E could not be defined in terms of unary quantifiers. At the same time Xavier Caicedo [5] developed back-and-forth techniques to get the same and better results. Later Xavier became a close associate and brought me in touch with the Bogotá logic group with which my own logic group has developed a fruitful interaction.

I struggled for a long time to find a ternary quantifier that cannot be defined in terms of binary ones. This may sound like a scholarly question of little practical importance, but attempts to solve this question forced us to sharpen our tools and invent a method (the method of bijective games) which became much more important than the original problem. The problem of binary versus ternary quantifiers was eventually solved in co-operation with Michał Kynicki and Alistair Lachlan [18]. The ternary quantifier that we considered was able to compute the dimension of a vector space and we showed that it cannot be accomplished with binary quantifiers. Krynicki was my fellow student in Warsaw and became a good friend and colleague. His early passing in 2011 was a great loss. Further discussions with Lachlan helped me to develop a general method for proving non-expressibility results for n-ary generalized quantifiers for any fixed n. This approach was perfected by my then student Lauri Hella [6] with his concept of a bijective Ehrenfeucht-Fraïssé game. The fundamental idea is that of a bijection between structures with n-predicates, which—without being an isomorphism—preserves all definable predicates of fewer than n variables. Hella developed a version of the Ehrenfeucht-Fraïssé game where some moves were such bijections.

This became a trademark game of the Helsinki Logic Group with many applications, most notably the result of Hella that there is no logic for PTIME based on generalized quantifiers of bounded arity type [7].

The corresponding question of the existence of unary quantifiers of unary type $(1, \ldots, 1)$ not definable by unary quantifiers of smaller type needed completely different methods and was solved in [23] by means of combinatorics. A new Ramsey-type theorem was needed and was kindly provided by Noga Alon. Ramsey theory is useful in Ehrenfeucht-Fraïssé games because the different types of a player's moves can be interpreted as a colouring of a structure, and knowing that there is a large homogeneous set helps to decide what move to make. A player can move so that whatever the opponent moves, she has still a large enough set of good moves left. In finite models this is an intrinsically combinatorial question.

The completely general hierarchy result for generalized quantifiers of any type was proved with a counting argument in [8], following a suggestion of Per Lindström via Dag Westerståhl. Lindström's idea was to count how many different quantifiers of a given type there are in a fixed finite domain and then show that they cannot all, for cardinality reasons, be definable in terms of quantifiers of a given smaller type. There is a vague resemblance to the priority method and the hierarchy of degrees of unsolvability. It seems possible to take this analogy further and construct quantifiers with special relationships, such as minimality, maximality, comparability, incomparability, etc, to other quantifiers. We are used to thinking that generalized quantifiers should be familiar, such as "for all but five" or the quantifier in "most boys in my class and most girls in your class know each other", and quantifiers constructed by a priority-type argument fit poorly into this picture. But if we understand the general landscape of quantifiers, we can try to position the familiar ones in this picture and thereby reveal their true nature.

In 1981/1982 I was invited to two informal meetings for preparing the volume [4], one in Freiburg and another in Stanford, both of them incredibly important for me personally as I got to know and work with people I knew only by name, such as Jon Barwise, Dana Scott, Sol Feferman, John Baldwin, Phokion Kolaitis and many others. Around 1989 Kolaitis sent me an e-mail, probably one of the first e-mails I ever received, asking me about Ehrenfeucht-Fraïssé games for generalized quantifiers, were they known and would they work on finite models. I got very excited because the answer was very definitely positive and I had not come to think of that possibility at all. Phokion then visited me in Finland in 1990 and taught me and some other Finns, e.g. Hella, what finite model theory is really all about. As it happens, the back-and-forth method, i.e. Ehrenfeucht-Fraïssé game, extensively studied at that time in Helsinki, works perfectly well in finite models, unlike most other methods of infinite model theory. Moreover, the concept of generalized quantifiers and infinitary logics were equally, if not more, relevant in the finite context, than in the infinite context.

I ended up writing the paper [16] with Kolaitis on generalized quantifiers. We established the basic results on back-and-forth techniques, especially the pebble game, in finite models in the presence of generalized quantifiers. We were led to consider applications of Ramsey theory in finite model theory, a precursor to the later applications of Ramsey theory to generalized quantifiers. Finite model theory became one of my lasting interests. There is a curious similarity in the difficulties one faces when dealing with finite models on the one hand and uncountable models on the other hand. In both cases one constructs models, not step by step from smaller models, but by taking a vast collection of models of the desired size and throwing out the ones that are unfit, hoping that something is left.

In 1993 Dag Westerståhl came to Helsinki for a month and gave us inspiring lectures on generalized quantifiers in natural language. I had known Dag since our student days when he was a student of Lindström in Gothenburg. These lectures opened the world of mathematical linguistics to me. Our experience with Ehrenfeucht-Fraïssé games on models in Helsinki led to the papers [9] and [38] in which several definability and hierarchy questions concerning generalized quantifiers, motivated by natural language examples, were solved. I started to appreciate the complexity of natural language, and would still like to contribute to the mathematical modelling of it. Perhaps team semantics, the topic of Section 6, can lead to some new ideas in this direction.

3 Härtig quantifier

In my student days I stumbled on the Härtig quantifier "there are as many x with $\phi(x)$ as there are y with $\psi(y)$" and did not suspect at that time, if indeed anybody did, how complicated it is. It turned out to be even more complicated, although not more powerful, than full second order logic.

Second order logic is entangled with set theory in a precise way. For any ordinal that second order logic can characterise, and such ordinals can be quite large, e.g. bigger than measurable cardinals, second order logic is able to characterise the cumulative hierarchy up to that level.

The entanglement of the Härtig quantifier with set theory is much more subtle in that it depends on set theory itself: in some models of set theory the Härtig quantifier is able to capture the cumulative hierarchy, in some models not. My first observation in this direction, following a hint from my supervisor Peter Aczel, was that if $V = L$ then $L(I)$ is more or less the same as second order logic in that it can characterise the cumulative hierarchy V_α for very large ordinals α. Thus if $V = L$, and this is true of many other inner models as well, for example L^μ, the logic $L(I)$ is totally (up to the level Σ_2 in the Levy-hierarchy) entangled with set theory. The technique I used was not new, it was essentially the argument behind Gödel's proof of GCH in L, that is, Mostowski's Collapsing Lemma and Levy's Reflection Principle.

My second observation, which took some time to mature, was that it is consistent with ZFC that $L(I)$ is so far from second order logic that every consistent $L(I)$-sentence has a model of smaller size than the continuum. In this case the cumulative hierarchy V_α can be characterised in $L(I)$ only in the trivial case $\alpha \leq \omega$. Thus in my new models of set theory the logic $L(I)$ was free from symbiosis with set theory and could have such model theoretic properties as small Löwenheim number[1], small (relatively speaking) Hanf number[2], decision problem[3] of as low complexity as Δ^1_3, etc. Thus in this case $L(I)$ was just an "iota" more complex than the well-ordering quantifier

$$Wxy\phi(x, y) \iff \phi(\cdot, \cdot) \text{ well-orders the universe.}$$

My technique for obtaining these methods was to develop iterated forcing in order to make the set-theoretical universe in a precise sense *generically maximal* (see discussion below in Section 4). In such maximal universes, in a sense "fat" universes, there are so many sets that being able to distinguish between cardinals does not give much of a clue as to what the other sets are like. In Gödel's L subsets of any ordinal are definable from ordinals of the same cardinality in an absolute sense, so $L(I)$ knows in the end all sets.

My method of maximal universes left one problem unsolved: can the Löwenheim number of $L(I)$ be below the first weakly inaccessible cardinal? I had shown already in my doctoral thesis that if that was the case, then $0^\sharp$ (and much more) must exit. So one had to force over models with large cardinals. This however is not the problem. The problem is that one has to collapse cardinals and at the same time control the behaviour of $L(I)$. Thus one cannot use the preservation of $L(I)$-truth in the forcing in order to control the behaviour of $L(I)$. The difficulty of this problem bewildered me for 30 years until Menachem Magidor came to visit me in Helsinki and we attacked this problem. He came up with a method to collapse

1 The smallest cardinal κ such that every sentence which has a model has a model of cardinality $\leq \kappa$.

2 The smallest cardinal κ such that every sentence which has a model of cardinality $\geq \kappa$ has arbitrarily large models.

3 The set of Gödel numbers of valid sentences.

cardinals and at the same time leave gaps where cardinals are not collapsed. In those gaps we could use preservation of $L(I)$-truth. On the other hand, if we start with a supercompact cardinal, we can stretch the gaps to any lengths we needed. So the old problem was solved and we wrote the paper [19].

Upon learning little by little how entangled the Härtig quantifier was with set theory, I started to study a general theory of relationships between abstract logics and set theory [29, 30, 31]. I developed the concept of “symbiosis” between an abstract logic and a predicate of set theory. This enabled me to establish a perfect equivalence between definability in set theory and definability in abstract logics. The only difference that remains is that model theoretic definability is always up to isomorphism while set theoretic definability is up to identity. An abstract logic L^* is said to be “symbiotic” with a predicate P of set theory when (essentially) the truth predicate of L^* is $\Delta_1(P)$, and on the other hand a class of models canonically derived from P is implicitly definable in L^*.

4 Set theory

While I worked on my doctoral thesis in Manchester in the late 70s, I realised, as I have related above, that I needed forcing to show that certain questions about the Härtig quantifier are undecidable on the basis of ZFC alone. My supervisor Peter Aczel suggested to me to think in terms of Boolean valued extensions, rather than forcing extensions. It was an eye-opener, although there is very little difference between the forcing approach and the Boolean valued approach. Using Boolean valued models I developed a method to construct forcing extensions with maximality properties, which would nowadays be called generic absoluteness properties. I worked entirely in the context of generalized quantifiers, but I had already proved that an equivalent approach would be to work in set theory, as I wrote at the end of Chapter 3, since powerful generalized quantifiers are in perfect symbiosis with predicates of set theory.

I met Jonathan Stavi in the Wrocław Logic Colloquium in August 1977. I had given a talk on the result of my forth-coming doctoral dissertation, explaining my construction of a forcing extension which satisfies in a maximal way properties of the Härtig quantifier preserved under CCC-extensions. In particular this gave a model of set theory in which the Löwenheim number of the Härtig-quantifier is below the continuum, rendering the continuum very large. The idea is that continuum is so large that it is closed under any “arithmetic” operations on cardinals below it. Stavi had had similar ideas, not in model theory like myself, but directly in set theory proper. We had extensive discussions, first in Wrocław and then by correspondence during 1977/1978. This lead to the paper [24]. Although this paper was written in the early 1980s, and the results were announced in [31] and [10], also I lectured about the results during a tour of the US in the late 80s, the paper circulated only in unpublished form for over twenty years. The reason was that it was impossible to contact my co-author, who had meanwhile, as I later learnt, left academia, and I did not want to publish the paper without the permission of my co-author.

The generic maximality principle $MAX(CCC)$ we devised with Stavi says that every predicate of set theory, with parameters of hereditary cardinality $< 2^{\omega}$, provably persistent under CCC-forcing, which is true in some CCC-extension, is actually true. Restricted to Σ_1-predicates this means that for all CCC Boolean algebras $\mathbb{B}$ we have:

$$(H(2^{\omega}), \in)^{V} \prec (H(2^{\omega}), \in)^{V^{\mathbb{B}}},$$

which is equivalent to MA, a result that was subsequently but independently proved by Joan Bagaria [2]. My recent work with Magidor [19] can be seen as a continuation of [24], although we do not formulate our results in terms of generic absoluteness. A particularly interesting result in [24] is that $MAX(CCC)$ and $MAX(C)$, maximality of the universe with respect to Cohen algebras, are in contradiction with each other. Although generic maximality has since caught encouragingly the attention of other researchers, the field is far from exhausted.

5 Trees, games and models

In many ways $L_{\omega_1\omega}$ is the only infinitary language with a nice model theory. Both $L_{\omega_2\omega}$ and $L_{\omega_1\omega_1}$ lead to the necessity to understand models of size $\aleph_1$, which are obviously much more complicated than countable models. My senior colleagues and good friends Jaakko Hintikka and Veikko Rantala had suggested in [12] so called "infinitely deep languages" for the systematic study of uncountable models. I tried to understand how this would work and ended up using transfinite Ehrenfeucht-Fraïssé games rather than "infinitely deep languages". Obviously the Ehrenfeucht-Fraïssé game of length ω_1, let us denote it by EF_{ω_1}, characterises any model of size $\aleph_1$. But what is this game like? Very little was known about games of length ω_1. My idea was to use the class $\mathcal{T}_1$ of trees of height ω_1 and with no uncountable branches as analogues of ordinals. It was clear what the natural ordering of $\mathcal{T}_1$ should be: it is the existence of a strict order preserving mapping (which need not be one to one). But it seemed that nobody had studied $\mathcal{T}_1$ under this ordering. It later turned out that Kurepa had studied it and introduced the so called σ-operation $T \mapsto \sigma(T)$ which then turned out to be the right notion. As my then student Tapani Hyttinen realised, it is not enough, as in the case of countable models, to investigate trees of countable plays of EF_{ω_1}, but one should also focus on trees of countable sequences of *strategies* of the second player in EF_{ω_1}. Once this became clear a general infinitary Scott sentence[4]-type analysis of uncountable models started to emerge. The first results were published in a paper with Tapani Hyttinen [15].

I met Saharon Shelah for the first time in the Clermont-Ferrand Logic Colloquium in July 1975, where Shelah gave lectures on stable theories. I had proved some results about the compactness and Beth properties in abstract logics and told them to Shelah, who to my disappointment already knew them, or figured out the proofs on the spot. I have later over the years worked extensively with him and learnt that one does not tell Saharon new results but rather new open problems. I share with Saharon an interest in abstract logic, generalized quantifiers, infinitary languages, and also transfinite Ehrenfeucht-Fraïssé games, which Shelah had come to consider independently of me.

I had visited Lachlan in Vancouver in the mid 80s and met Alan Mekler, who became a close friend until his early passing in 1993. He saw immediately the power of EF_{ω_1} and the analogy between ordinals and $\mathcal{T}_1$. For several years we had wondered whether EF_{ω_1} could be non-determined in some models of cardinality ω_2 and Mekler solved this problem almost on the spot: If $V = L$, then such models exist among almost free abelian groups, a topic close to Alan's heart. Then Shelah got involved and proved that one can force a model over a measurable cardinal and get the consistency of the determinacy of EF_{ω_1} in all models of size $\aleph_2$ [20].

On the other hand, I came to think that there are some basic principles in uncountable models which have an almost topological flavour, and I considered the generalized Baire space $\omega_1^{\omega_1}$ as an analogue of the classical Baire space ω^ω. I proved a kind of Cantor-Bendixson theorem [32] and, with Alan Mekler, other basic results in this type of generalised descriptive set theory [21]. While the classical descriptive set theory arises from the foundations of classical analysis, the new generalized descriptive set theory of the non-separable generalized Baire spaces 2^{ω_1} and $\omega_1^{\omega_1}$ is motivated by model theory. It isolates purely set theoretical and topological properties of uncountable models arising from the analysis of games like EF_{ω_1}. Now, after 20 years, there is a delightful surge of interest in generalized Baire spaces.

The natural ordering of the class of trees with no uncountable branches turned out to be quite problematic and is still far from settled. I have written sev-

4 The Scott sentence of a countable model is an $L_{\omega_1\omega}$-sentence which characterises the model up to isomorphism.

eral papers, e.g. [26] with Todorcevic, on the topic and continue to think about it. A natural assumption in the study of $\mathcal{T}_1$ is CH, because then $\mathcal{T}_1$ is closed under the σ-operation. However, Todorcevic [25] used PFA to get strong structural results about "Lipschitz" trees is $\mathcal{T}_1$. There is a strong connection between properties of trees in $\mathcal{T}_1$, properties of models of cardinality $\aleph_1$, and Shelah's Main Gap theorem in stability theory [14].

I wrote a monograph [36] on the topic of games and models, emphasising what I call the *Strategic Balance of Logic*, by which I mean the fact that the three games of logic, the Semantic Game governing truth, the Ehrenfeucht-Fraïssé Game governing equivalence of models, and the Model Existence Game governing consistency, are really three versions of one and the same game. I hope the monograph raises interest in the game theoretic approach to logic. The book has of course a chapter on transfinite Ehrenfeucht-Fraïssé games and the related generalized descriptive set theory, based on $\mathcal{T}_1$.

6 Dependence logic

It took me some time to understand what Hintikka's Independence Friendly Logic (also called the IF logic) [11] is about. In the 1999 *Logic, Methodology and Philosophy of Science* Conference in Cracow I had heard a talk on IF logic by Krynicki. His approach, joint work with Caicedo, was based entirely on the seminal work of Wilfrid Hodges [13] on a compositional semantics for IF logic. I realised the importance of Hodges' semantics and wrote very quickly [34] in which I showed, that just as Hodges proved that the semantics of Independence Friendly Logic can be made compositional, the game theoretic semantics can be made determined by modifying the game.

During several hot summers after the turn of the century I was given, with the help of John Burgess, a quiet office in the Philosophy Department of Princeton University, where I laid down the foundations of a logic based on the semantics Hodges had introduced, and called the new logic Dependence Logic [35]. The reason for introducing my own logic was that Hintikka's logic, though equivalent to mine on the level of sentences, had aspects that I could not make sense of in a satisfactory way. I focused on "dependence" rather than on "independence" of variables, being convinced that it is then easier to keep the logic under control without losing anything essential. Moreover I separated "dependence" from quantifiers into a new kind of atomic formula

$$=(x_1, \ldots, x_n) : \quad \text{“}x_n \text{ depends only on } x_1, \ldots x_{n-1}\text{”}. \tag{1}$$

Such an atomic formula can be given meaning in sets X of assignments, following Hodges: A set X satisfies (1) if

$$\forall s, s' \in X((\bigwedge_{i=1}^{n-1} s(x_i) = s'(x_i)) \to s(x_n) = s'(x_n)). \tag{2}$$

Note that (2) would be always true if X consisted of a single assignment, as in classical first order logic, so it is indeed essential to base the semantics of (1) on *sets* of assignments. The fact that dependence logic is based on compositional semantics, which I started to call team semantics, means that *formulas* have a natural semantics, not only sentences. This is significant because the dependence and independence atoms, which I later introduced with Erich Grädel, actually occur in database theory, experimental science, social choice theory, etc. So it made sense to bring them to the forefront.

Dependence logic has become a promising research area and I think it has potential in a variety of application areas of logic. With Juha Kontinen I proved that dependence logic captures exactly the downward closed existential second order properties of teams [17]. My then student Pietro Galliani proved that all existential second order properties of teams, downward closed or not, are definable in independence logic. My paper [1] with Samson Abramsky brought new proof-theoretic ideas into dependence logic, in particular it meant the emergence of intuitionistic implication in dependence logic. In her doctoral thesis my student Fan Yang proves the surprising result that the extension of dependence logic with intuitionistic implication has full second order power [39].

Similarly as condition (2) gives meaning to (1) in teams, we can give meaning to all kinds of new atomic formulas of the type (1). Examples of what has been considered so far are independence atoms, exclusion atoms and inclusion atoms. In a recent very interesting result of Hella and Galliani inclusion logic is shown to be equivalent in finite models to the fixpoint logic that Kolaitis got me interested in over 20 years ago. Thus the $P = NP$ question translates in team semantics to the question whether inclusion logic is equivalent in finite ordered models to independence logic.

7 Second order logic

I have been interested in second order logic ever since my very early paper [28]. Before I got deeper into this powerful logic I had thought about the Härtig-quantifier, as I have related above, and shown that its properties depended on set theory in the sense that in some models of set theory the quantifier was quite weak and in others quite strong. It was clear that the properties of second order logic are more "absolute" in the sense that it is quite strong in all models of set theory. It is absurd to call second order logic "absolute" in any sense, as it contains a vast number of notoriously non-absolute logics. However, second order logic attains some absoluteness in a roundabout way, simply by way of being so incredibly strong. Namely, second order logic is able to fix the universe into a V_α with very large ordinals α, and then absoluteness follows from the uniqueness of such models.

Let us ponder this phenomenon more closely. Usually absoluteness of a concept means that the concept remains the same even if the universe of set theory is changed, within the room that the axioms allow. The first order axioms of set theory allow a lot of variation for the model, as testified by inner models and forcing extensions. So, for example, finiteness is absolute between transitive models of ZFC, as is well-foundedness and all Π^1_2-properties (by Shoenfield's Absoluteness Lemma). Numerous second order properties, like uncountability, equicardinality, Souslinity, etc, are not absolute between transitive models of ZFC. However, if we use a second order axiomatization ZFC^2 of ZFC, models are (up to isomorphism) of the form V_κ, κ strongly inaccessible. It goes without saying that all the concepts mentioned above are absolute between models of the form V_κ, κ strongly inaccessible.

I have developed a view that second order logic is for all practical purposes a fragment of set theory, and defended this view in [33]. As a clear indication of this circumstance I showed that the decision problem of second order logic is the complete Π_2-definable set of natural numbers. Similarly the Löwenheim number, the Hanf number and the decision problem of second order logic have a characterisation in terms of Π_2-formulas of set theory. In [37] I developed further my claim of the close relationship between second order logic and set theory. In a sense, set theory is a very high order logic and therefore contains second order logic as a proper part. First order set theory permits quantification over sets, which intuitively means quantification over subsets, sets of subsets, sets of sets of subsets, etc of smaller sets.

Much confusion exists in the literature of philosophy of mathematics about whether second order logic is "better" than first order set theory. In truth they are very close to each other. Both have axioms and a Completeness Theorem with respect to non-standard models. Both are able to characterise the common structures of classical mathematics: second order logic characterises up to isomorphism, set theory up to identity. Both have a clear concept of a standard model: in second order logic it is the full second order logic, which quantifies over all subsets and relations, in set theory it is the cumulative hierarchy, which builds on the assumption that in successor stages we take all subsets of the previous level. In both cases the word "all" is explained by giving comprehension-type axioms.

Second order logic cannot be better than set theory, for the reason that it is more expressive, as is often claimed, but perhaps rather because it is *less* expressive. Defenders of second order logic as the best foundation of mathematics may be "doing the right thing, but for a wrong reason".

8 Friends

I have described above a couple of scientific programs that I have tried to push forward. I would not have made much progress if I did not have such great friends. My logic friends have spent time all over the world thinking about questions that one of us would propose, bringing me into contact with areas of logic and methods of mathematics that I would otherwise perhaps never have come to think and learn. It has been a great privilege and pleasure to work with my friends and I hope to have this good life in the future, too. A special joy has been and continues to be talking about mathematics, philosophy, logic, and everything else with my wife Juliette.

Bibliography

[1] Samson Abramsky and Jouko Väänänen. From IF to BI: a tale of dependence and separation. *Synthese*, 167(2, Knowledge, Rationality & Action):207–230, 2009.
[2] Joan Bagaria. A characterization of Martin's axiom in terms of absoluteness. *J. Symbolic Logic*, 62(2):366–372, 1997.
[3] Jon Barwise. Axioms for abstract model theory. *Ann. Math. Logic*, 7:221–265, 1974.
[4] Jon Barwise and Solomon Feferman, editors. *Model-theoretic logics*. Perspectives in Mathematical Logic. Springer-Verlag, New York, 1985.
[5] Xavier Caicedo. Back-and-forth systems for arbitrary quantifiers. In *Mathematical logic in Latin America (Proc. IV Latin Amer. Sympos. Math. Logic, Santiago, 1978)*, volume 99 of *Stud. Logic Foundations Math.*, pages 83–102. North-Holland, Amsterdam, 1980.
[6] Lauri Hella. Definability hierarchies of generalized quantifiers. *Ann. Pure Appl. Logic*, 43(3):235–271, 1989.
[7] Lauri Hella. Logical hierarchies in PTIME. *Inform. and Comput.*, 129(1):1–19, 1996.
[8] Lauri Hella, Kerkko Luosto, and Jouko Väänänen. The hierarchy theorem for generalized quantifiers. *J. Symbolic Logic*, 61(3):802–817, 1996.
[9] Lauri Hella, Jouko Väänänen, and Dag Westerståhl. Definability of polyadic lifts of generalized quantifiers. *J. Logic Lang. Inform.*, 6(3):305–335, 1997.
[10] Heinrich Herre, Michał Krynicki, Alexandr Pinus, and Jouko Väänänen. The Härtig quantifier: a survey. *J. Symbolic Logic*, 56(4):1153–1183, 1991.
[11] Jaakko Hintikka. *The principles of mathematics revisited*. Cambridge University Press, Cambridge, 1996. Appendix by Gabriel Sandu.

[12] Jaakko Hintikka and Veikko Rantala. A new approach to infinitary languages. *Ann. Math. Logic*, 10(1):95–115, 1976.

[13] Wilfrid Hodges. Compositional semantics for a language of imperfect information. *Log. J. IGPL*, 5(4):539–563 (electronic), 1997.

[14] Tapani Hyttinen and Heikki Tuuri. Constructing strongly equivalent nonisomorphic models for unstable theories. *Ann. Pure Appl. Logic*, 52(3):203–248, 1991.

[15] Tapani Hyttinen and Jouko Väänänen. On Scott and Karp trees of uncountable models. *J. Symbolic Logic*, 55(3):897–908, 1990.

[16] Phokion G. Kolaitis and Jouko A. Väänänen. Generalized quantifiers and pebble games on finite structures. *Ann. Pure Appl. Logic*, 74(1):23–75, 1995.

[17] Juha Kontinen and Jouko Väänänen. On definability in dependence logic. *J. Log. Lang. Inf.*, 18(3):317–332, 2009.

[18] Michał Krynicki, Alistair Lachlan, and Jouko Väänänen. Vector spaces and binary quantifiers. *Notre Dame J. Formal Logic*, 25(1):72–78, 1984.

[19] Menachem Magidor and Jouko Väänänen. On Löwenheim-Skolem-Tarski numbers for extensions of first order logic. *J. Math. Log.*, 11(1):87–113, 2011.

[20] Alan Mekler, Saharon Shelah, and Jouko Väänänen. The Ehrenfeucht-Fraïssé-game of length ω_1. *Trans. Amer. Math. Soc.*, 339(2):567–580, 1993.

[21] Alan Mekler and Jouko Väänänen. Trees and Π^1_1-subsets of ${}^{\omega_1}\omega_1$. *J. Symbolic Logic*, 58(3):1052–1070, 1993.

[22] Andrzej Mostowski. Thirty years of foundational studies. Lectures on the development of mathematical logic and the study of the foundations of mathematics in 1930–1964. *Acta Philos. Fenn. Fasc.*, 17:1–180, 1965.

[23] Jaroslav Nešetřil and Jouko A. Väänänen. Combinatorics and quantifiers. *Comment. Math. Univ. Carolin.*, 37(3):433–443, 1996.

[24] Jonathan Stavi and Jouko Väänänen. Reflection principles for the continuum. In *Logic and algebra*, volume 302 of *Contemp. Math.*, pages 59–84. Amer. Math. Soc., Providence, RI, 2002.

[25] Stevo Todorcevic. Lipschitz maps on trees. *J. Inst. Math. Jussieu*, 6(3):527–556, 2007.

[26] Stevo Todorčević and Jouko Väänänen. Trees and Ehrenfeucht-Fraïssé games. *Ann. Pure Appl. Logic*, 100(1-3):69–97, 1999.

[27] Jouko Väänänen. Remarks on generalized quantifiers and second order logics. *Prace Nauk. Inst. Mat. Politech. Wrocław.*, (14 Ser. Konfer. No. 1):117–123, 1977.

[28] Jouko Väänänen. Two axioms of set theory with applications to logic. *Ann. Acad. Sci. Fenn. Ser. A I Math. Dissertationes*, (20):19, 1978.

[29] Jouko Väänänen. Abstract logic and set theory. I. Definability. In *Logic Colloquium '78 (Mons, 1978)*, volume 97 of *Stud. Logic Foundations Math.*, pages 391–421. North-Holland, Amsterdam, 1979.

[30] Jouko Väänänen. Abstract logic and set theory. II. Large cardinals. *J. Symbolic Logic*, 47(2):335–346, 1982.

[31] Jouko Väänänen. Set-theoretic definability of logics. In *Model-theoretic logics*, Perspect. Math. Logic, pages 599–643. Springer, New York, 1985.

[32] Jouko Väänänen. A Cantor-Bendixson theorem for the space $\omega_1^{\omega_1}$. *Fund. Math.*, 137(3):187–199, 1991.

[33] Jouko Väänänen. Second-order logic and foundations of mathematics. *Bull. Symbolic Logic*, 7(4):504–520, 2001.

[34] Jouko Väänänen. On the semantics of informational independence. *Log. J. IGPL*, 10(3):339–352, 2002.
[35] Jouko Väänänen. *Dependence logic*, volume 70 of *London Mathematical Society Student Texts*. Cambridge University Press, Cambridge, 2007. A new approach to independence friendly logic.
[36] Jouko Väänänen. *Models and games*, volume 132 of *Cambridge Studies in Advanced Mathematics*. Cambridge University Press, Cambridge, 2011.
[37] Jouko Väänänen. Second order logic or set theory? *Bull. Symbolic Logic*, 18(1):91–121, 2012.
[38] Jouko Väänänen and Dag Westerståhl. On the expressive power of monotone natural language quantifiers over finite models. *J. Philos. Logic*, 31(4):327–358, 2002.
[39] Fan Yang. Expressing second-order sentences in intuitionistic dependence logic. *Studia Logica*, 101(2):323–342, 2013.

A Radio Interview with Jouko Väänänen

Interview in Finnish Radio of Jouko Väänänen 24.8.2009 on the occasion of a new edition of a collection of pacifist texts edited by Väänänen. Translation by Juliette Kennedy and Jouko Väänänen.

Interviewer: Welcome, Jouko Väänänen!

JV: Thank you!

Interviewer: Today we talk about pacifism and the peace movement and about the arguments in favor of pacifism. Jouko Väänänen, for a start would you tell us a little bit about your own background. You graduated from high school in the legendary year 1968. What was your road to the peace movement?

Fig. A.1. Jouko Väänänen. Photo by Burton Randol.

JV: I read Bertrand Russell, the British philosopher, peace activist and writer, quite famous, who even got the Nobel Prize, and whose article is in this book. Undoubtedly he influenced me. Then I participated in the anti-Vietnam war demonstrations. I considered the Vietnam War really wrong, and still do. I was appalled by the bombing of a country only because it had a communist government. But my real activity in the peace movement started when I objected to the military service and was ordered to the so-called alternative service in the Järvenpää Social Hospital. From there I joined the Union of Conscientious Objectors and I was elected chair of the union. That was the beginning of my activities in peace organizations. On the intellectual level, my job being that of a logician, Bertrand Russell was a kind of role model for me, because he was both a logician and a peace activist, and he also put himself at risk, participated in demonstrations and did things that took him to prison. In this way he influenced me by his example.

Interviewer: So, the intellectual aspect of the peace movement was important to you in addition to having experienced the Vietnam War, how it spread...

JV: And I am sure I was influenced by the times, what you referred to as Peace, Love and Rock 'n Roll, kind of a time of liberation. I suppose every generation has such a moment and time of liberation, but in the sixties it was combined with rock music, which united people, the opposition to the Vietnam War, which united people in the peace movement, and then also the sexual revolution. Nothing brings people together as powerfully as an idea, a radical idea that people feel is right. Then it gets joined with everything, music, love and then also peace movement. Of course I am also a product of this.

Interviewer: Child of the sixties?

JV: Yes, child of the sixties.

Interviewer: You have edited a book called "Rauhaa - Peace". This new edition that I have here in my hand is completely fresh: Classics of Pacifism. You have in your hand a different looking book from the 1980s. Now a new edition has been made of this book. Let us talk a little about pacifism as an ideology. Jouko Väänänen, is pacifism understood correctly, what in fact is pacifism?

JV: Usually it is thought that pacifism is conscientious objection, objection to take part in a war. This is how it manifests itself and this is the correct view. In reality it makes, of course, more sense to fight the causes of war rather than object to the war at the last moment when a war has already broken out. In fact all notable thinkers in pacifism have tried to approach the question how to prevent wars. It is true, many pacifists have a religious background. Almost all religions are in principle pacifist. It is rather remarkable that as long as we have had books at all, they have expressed pacifist ideas. It does not seem at all that pacifism is the product of the sixties, it goes back 3000 years.

Interviewer: Yes, I was in fact going to ask you why the back cover of this book says "Pacifism for 2500 years"? Where does this 2500 years come from? Is there some particular writing which marks the beginning?

JV: One cannot say that some writing marked the beginning but in the beginning of this book we have Tao-te-king, Motse and Dhammapada and they are from 500 b.c., so about 2500 years ago. Tao-te-king, Motse and Dahmmapada all have the basic idea that one cannot overcome hatred by another hatred, that hatred only breeds new hatred, and some other solution has to be invented. In its most primitive form pacifism refuses to take part in this vicious cycle of hitting back when you are being hit. But the next step is what for example Gandhi strongly advocated, that when one refuses to take part, then one actually has a positive impact and influences the so-called enemy. The bible says that one should turn the other cheek but this does not mean that one should resign. This is very important. Often people think that pacifism is "passivism". It is not passivism although we have there a similarity of words. It is not passivity, accepting exploitation or violence. By turning the other cheek one turns the focus of interaction to a different level. If

I take a modern example, I would say that in the 9/11 incident in which terrorists hit buildings and thousands of people died, the US had a remarkable opportunity to take the step that instead of saying that we will strike back and you can be safe because we strike back, to say instead that those people who attacked the US were driven by a wrong picture; we are not enemies of Muslims; we do not hate Muslims; we are not going to attack; we are not going to punish; we are on a different level; we are on the level where these matters are solved by international agreements and police actions. At that moment there was powerful sympathy for the US all around the world, also in the Muslim world, because the US was the victim. They could have turned this sympathy into a historical moment in the history of mankind. Instead they hit back and are now up to their neck in trouble both here and there.

Interviewer: History took a different route.

JV: Yes.

Interviewer: How can you, having followed the world, if we count from the year 1968, it is quite a few decades, how do you have strength to believe in this pacifism, despite the fact that so many times hate has been answered with hate and war with war.

JV: Well, in fact it is not so difficult, for pacifism has strongly influenced the history of mankind throughout time. For example, think about the cold war. After the Second World War the cold war kept us at the brink of nuclear war. Total destruction was with us as a frightening possibility because these two nuclear powers were armed against each other. How did this end? A group of people decided to try to walk across a border. A border guard decided not to shoot them. And this started a collapse, and in the end people walked over the Berlin Wall, and these East German soldiers, who were thought to be shooting-robots, lowered their guns, and the cold war disappeared.

Interviewer: Yes, who would have believed?

JV: Yes, the cold war did not end in one finally attacking the other and destroying it. Instead people took control of the situation. Of course there are many factors involved, Gorbachev, Reagan, etc. But in the end it was the civil disobedience of people that overwhelmed the wall of the cold war. This civil disobedience is the method that pacifism offers as an alternative to what the armies try to do. If we think that armies are there to defend a country, then the alternative way to defend a country is to train people not to be provoked, as President Mauno Koivisto said, to train people to act independently, the way Solidarity acted, if occupied; to train to build a parallel nonviolent society within a society, one that is morally on a high level. Then this moral strength, what Gandhi emphasized in India, that when we do not attack the British, we are on a higher level, is something that millions and millions of people understand. And when millions of people suddenly understand what tremendous power lies in refusing to engage in violence, then nothing can stop them. Let us look at the civil rights movement in the US. Rosa Parks in the fifties refused to go to the back of the bus where African-American passengers were supposed to sit. This initiated an avalanche that now took Obama to the presidency. Martin Luther King emphasized that we have to proceed non-violently in order not to be on the same level as those who attack us.

Interviewer: Now, Jouko Väänänen, you have given some names, and this book indeed has writings from a period of 2500 years. Which ones would you pick out as such as have been classics of pacifism for you yourself, kind of classics of the classics. Which are the important ones?

JV: Here is a text by Russell. Although Russell is not in my opinion the deepest thinker in this book, he is so bitingly powerful a writer, and Tolstoi is another who is, although he was a Christian pacifist, what I am not, being myself more of a rational pacifist, but he is so good a writer, that these texts are undoubtedly my favorites. But I would also like to pick out William Penn's "An address to the Native Americans". It is only one page long. William Penn founded in the 17th century the State of Pennsylvania in the US. There is still a strong community of Quakers in the US. The address to the Native Americans is just one paragraph long and I shall read now the last lines: "I consider you the same flesh and blood as the Christians and the same as if the body of one person had been divided into two." So Penn brings up in this talk the notion of equality. Equality is very important in pacifism. Each person is equally precious, this is the starting point. Many people understand this idea; you think about it for a moment and realize that - indeed - a Russian is equally precious and the same kind of a human being as myself, he has a family, I have a family, he has sorrow, I have sorrow.

Interviewer: Same pain.

JV: Yes, same pain.

Interviewer: Same agony.

JV: Yes same agony. When one understands this, it is like, it just sweeps over you.

Interviewer: How would you answer the question which a pacifist surely is often asked, that would you defend yourself and your own country.

JV: Yes, that is the point, the goal of civil disobedience and non-violent action is that we should give up armies in the future. So then countries would defend themselves by civil disobedience. But as you point out, it is often asked whether I would remain passive if someone attacked me, what then. The answer is that I would probably defend myself, but it does not follow from this that I would support training for years to kill other people. Having said that, I emphasize that on a personal level it is part of pacifism that one tries to develop oneself so as not to be provoked by aggression. It is very important not to take every insult to the emotional level. This is what Vonnegut's text says in this book: sometimes one has to give up one's pride. Often one hears that the pride of a nation is wounded and it is necessary to retaliate or else the honor is not maintained or something like that. It is all nonsense to me. This talk that if I am attacked then my honor requires that I defend myself, this is a wrong idea.

Interviewer: Yes, Vonnegut...

JV: Of course I would like to say that if someone attacks me then I will not hit back. I would like to say that but I would probably be lying.

Interviewer: Yes, Vonnegut ends his colorful essay with the words: "Nothing lost but honor".

JV: Yes, yes.

Interviewer: We have a couple of minutes left to talk about how pacifism appeals to people today, Jouko Väänänen, does it have a market.

JV: It is not in fashion at the moment. I believe that movements like the peace movement come in waves, they obey a certain harmonic movement. It was more popular in the past, it is not so popular now, but it will come back. But if one just observes the world and contemplates the world from the pacifist point of view, and hopefully this book helps in that, one sees that pacifism is very relevant also today. But another question is what appeals to young people today. I am not the best person to talk about that, but undoubtedly the internet era changes all these questions. In fact, in the future it is possible that wars can be kind of internet wars, cyber wars, and I am not sure what will come out of it, but it will not change pacifism in any way. The fundamental idea remains: one does not overcome hatred by another hatred, whether cyber hatred or ordinary hatred.

Interviewer: Thank you, Jouko Väänänen. I end with the words: "Peace, man!"

JV: The same to you!

FSC
www.fsc.org
MIX
Papier | Fördert gute Waldnutzung
FSC® C083411

Kolibri 360 GmbH
Ferdinand-Jühlke-Straße 7
99095 Erfurt, Deutschland
produktsicherheit@kolibri360.de